Perspectives and Problems in Nolinear Science

A Celebratory Volume in Honor of Lawrence Sirovich

Springer
New York
Berlin
Heidelberg
Hong Kong
London
Milan
Paris
Tokyo

Ehud Kaplan
Jerrold E. Marsden
Katepalli R. Sreenivasan
Editors

Perspectives and Problems in Nolinear Science

A Celebratory Volume in Honor of Lawrence Sirovich

Springer

Ehud Kaplan
Department of Opthalmology
Mount Sinai School of Medicine
New York, NY 10029-6574
USA
kaplan@mail.rockefeller.edu

Jerrold E. Marsden
Control and Dynamical Systems 107-81
Caltech
Pasadena, CA 91125
USA
marsden@cds.caltech.edu

Katepalli R. Sreenivasan
Department of Mechanical Engineering
Mason Laboratory
Yale University
New Haven, CT 06520-8286
USA
katepalli.sreenivasan@yale.edu

Photograph of Lawrence Sirovich on page v: Sarah Merians Photography & Company

Mathematics Subject Classification (2001): 76-02, 92C20

Library of Congress Cataloging-in-Publication Data
Perspectives and problems in nonlinear science: a celebratory volume in honor of Lawrence Sirovich / editors, Ehud Kaplan, Jerrold E. Marsden, Katepalli R. Sreenivasan.
p. cm.
Includes bibliographical references and index.
ISBN 978-1-4684-9566-9
1. Nonlinear theories. I. Sirovich, L., 1933- II. Kaplan, Ehud. III. Marsden, Jerrold E. IV. Sreenivasan, Katepalli R.
QA427.P47 2003
511'.8—dc21 2002044509

Printed on acid-free paper.

ISBN 978-1-4684-9566-9 ISBN 978-0-387-21789-5 (eBook)
DOI 10.1007/978-0-387-21789-5

9 8 7 6 5 4 3 2 1 SPIN 10906200

Typesetting: Pages were created from author-prepared LaTeX manuscripts by the technical editors, Wendy McKay and Ross Moore, using modifications of a Springer LaTeX macro package and other packages for the integration of graphics and consistent stylistic features within articles from diverse sources.

www.springer-ny.com

Springer-Verlag New York Berlin Heidelberg
A member of BertelsmannSpringer Science+Business Media GmbH

To Larry Sirovich
On the occasion of his 70th birthday,
with much admiration and warmth
from his friends and colleagues worldwide.

Photo by Sarah Merians Photography & Co., New York.

Contents

Preface

Lawrence Sirovich will turn seventy on March 1, 2003. Larry's academic life of over 45 years at the Courant Institute, Brown University, Rockefeller University and the Mount Sinai School of Medicine has touched many people and several disciplines, from fluid dynamics to brain theory. His contributions to the kinetic theory of gases, methods of applied mathematics, theoretical fluid dynamics, hydrodynamic turbulence, the biophysics of vision and the dynamics of neuronal populations, represent the creative work of an outstanding scholar who was stimulated mostly by insatiable curiosity. As a scientist, Larry has consistently offered fresh outlooks on classical and difficult subjects, and moved into new fields effortlessly. He delights in what he knows and does, and sets no artificial boundaries to the range of his inquiry. Among the more than fifty or so Ph.D. students and postdocs that he has mentored, many continue to make first-rate contributions themselves and hold academic positions in the US and elsewhere. Larry's scientific collaborators are numerous and distinguished. Those of us who have known him well will agree that Larry's charm, above all, is his taste, wit, and grace under fire.

Larry has contributed immensely to mathematics publishing. He began his career with Springer by founding the *Applied Mathematical Sciences* series together with Fritz John and Joe LaSalle some 30 years ago. Later he co-founded the *Texts in Applied Mathematics* series and more recently the *Interdisciplinary Applied Mathematics* series. He has overseen with imagination the cross-fertilization of a broad range of ideas in applied mathematics—including his favorite subjects, fluid dynamics and problems in biology. His good taste and judgement as an editor has touched many scientists worldwide, and continues to do so.

This volume represents a token of the affection with which we and the contributors hold him, and marks Larry's influence as well as the continuing commitment to his work.

Ehud Kaplan, Mount Sinai School of Medicine
Jerry Marsden, California Institute of Technology
Katepalli Sreenivasan, University of Maryland

Preface

List of Contributors

Editors:

* EHUD KAPLAN
  Depts. of Ophthalmology, Physiology and Biophysics
  The Mount Sinai School of Medicine
  New York, NY, 10029
  ehud.kaplan@mssm.edu

* JERROLD E. MARSDEN
  Control and Dynamical Systems 107-81
  California Institute of Technology
  Pasadena, CA 91125-8100
  marsden@cds.caltech.edu

* KATEPALLI R. SREENIVASAN
  Institute for Physical Science and Technology
  University of Maryland
  College Park, MD 20742
  sreeni@ipst.umd.edu

Authors:

* HENRY D. I. ABARBANEL
  Department of Physics and Marine Physical Laboratory
  Scripps Institution of Oceanography
  9500 Gilman Drive
  La Jolla, CA 92093-0402
  hdia@jacobi.ucsd.edu

* MAXIMINO ALDANA
  James Franck Institute
  The University of Chicago
  5640 S. Ellis Avenue
  Chicago, Illinois 60637
  maximino@control.uchicago.edu

* ANATOLY P. ALDUSHIN
  Institute of Structural Macrokinetics & Materials Science
  Russian Academy of Sciences
  142432 Chernogolovka, Russia
  anatoly@aldushin.home.chg.ru

* CHARLES W. ANDERSON
  Department of Computer Science
  Colorado State University
  Fort Collins, CO 80523
  anderson@cs.colostate.edu

* ORIOL BATISTE
  Departament de Física Aplicada
  Universitat Politècnica de Catalunya
  c/Jordi Girona 1–3, Campus Nord
  oriol@fa.upc.es

* ALVIN BAYLISS
  Department of Engineering Sciences and Applied Mathematics
  Northwestern University
  2145 Sheridan Road
  Evanston, IL 60208-3125
  a-bayliss@northwestern.edu

* ERIC BROWN
  Program in Applied and Computational Mathematics
  Princeton University
  Princeton, New Jersey 08544
  ebrown@math.princeton.edu

* CARLO CERCIGNANI
  Dipartimento di Matematica
  Politecnico di Milano
  Piazza Leonardo da Vinci 32
  20133 Milano, Italy
  carcer@mate.polimi.it

* Susan N. Coppersmith
Department of Physics
University of Wisconsin-Madison
1150 University Avenue
Madison, WI 53706
snc@physics.wisc.edu

* Mark Edelman
Courant Institute of Mathematical Sciences
251 Mercer St.
New York NY 10012
edelman@cims.nyu.edu

* Razvan C. Fetecau
Applied and Computational Mathematics 217-50
California Institute of Technology
Pasadena, CA 91125
van@cds.caltech.edu

* Philip Holmes
Department of Mechanical and Aerospace Engineering
and Program in Applied and Computational Mathematics
Princeton University
Princeton, NJ 08544
pholmes@math.Princeton.edu

* Daniel D. Joseph
Department of Aerospace Engineering
University of Minnesota
Minnealpolis, MN 55455
joseph@aem.umn.edu

* Leo P. Kadanoff
The James Franck Institute
The University of Chicago
5640 S. Ellis Avenue
Chicago, IL 60637
l-kadanoff@uchicago.edu

* Ehud Kaplan
Depts. of Ophthalmology, Physiology and Biophysics
The Mount Sinai School of Medicine
New York, NY, 10029
ehud.kaplan@mssm.edu

* Michael J. Kirby
Department of Mathematics
Colorado State University
Fort Collins, CO 80523
kirby@math.colostate.edu

* Bruce Knight
Professor, Knight Laboratory of Biophysics
The Rockefeller University
1230 York Avenue
New York, NY 10021
knight@mail.rockefeller.edu

* Edgar Knobloch
Department of Physics
University of California, Berkeley
Berkeley, CA 94720
knobloch@physics.berkeley.edu
and
Department of Applied Mathematics
University of Leeds
Leeds LS2 9JT, UK
knobloch@amsta.leeds.ac.uk

* Jerrold E. Marsden
Control and Dynamical Systems 107-81
California Institute of Technology
Pasadena, CA 91125-8100
marsden@cds.caltech.edu

* Bernard J. Matkowsky
Department of Engineering Sciences and Applied Mathematics
Northwestern University
2145 Sheridan Road
Evanston, IL 60208-3125
b-matkowsky@northwestern.edu

* David W. McLaughlin
Director, Courant Institute of Mathematical Sciences
Professor of Mathematics and Neural Science
251 Mercer St.
New York, N.Y. 10012
dmac@cims.nyu.edu

* Jeff Moehlis
Program in Applied and Computational Mathematics
Princeton University
Princeton, NJ 08544
jmoehlis@math.princeton.edu

* Paul K. Newton
Department of Aerospace and Mechanical Engineering
University of Southern California
Los Angeles, CA 90089-1191
newton@spock.usc.edu

TEXnical Editors:

1

Reading Neural Encodings using Phase Space Methods

Henry D. I. Abarbanel
Evren Tumer

To Larry Sirovich, on the occasion of his 70th birthday.

ABSTRACT
Environmental signals sensed by nervous systems are often represented in spike trains carried from sensory neurons to higher neural functions where decisions and functional actions occur. Information about the environmental stimulus is contained (encoded) in the train of spikes. We show how to "read" the encoding using state space methods of nonlinear dynamics. We create a mapping from spike signals which are output from the neural processing system back to an estimate of the analog input signal. This mapping is realized locally in a reconstructed state space embodying both the dynamics of the source of the sensory signal and the dynamics of the neural circuit doing the processing. We explore this idea using a Hodgkin–Huxley conductance based neuron model and input from a low dimensional dynamical system, the Lorenz system. We show that one may accurately learn the dynamical input/output connection and estimate with high precision the details of the input signals from spike timing output alone. This form of "reading the neural code" has a focus on the neural circuitry as a dynamical system and emphasizes how one interprets the dynamical degrees of freedom in the neural circuit as they transform analog environmental information into spike trains.

Contents

1 Introduction

A primary task of nervous systems is the collection at its periphery of information from the environment and the distribution of that stimulus input to central nervous system functions. This is often accomplished through the production and transmission of action potentials or *spike* trains Rieke, Warland, de Ruyter van Steveninck, and Bialek [1997].

The book Rieke, Warland, de Ruyter van Steveninck, and Bialek [1997] and subsequent papers by its authors and their collaborators Brenner, Strong, Koberle, Bialek, and de Ruyter van Steveninck [2000] carefully lay out a program for interpreting the analog stimulus of a nervous system using ideas from probability theory and information theory, as well as a representation of the input/output or stimulus/response relation in terms of Volterra kernel functions. In Rieke, Warland, de Ruyter van Steveninck, and Bialek [1997] the authors note that when presenting a stimulus to a neuron, it is common "that the response spike train is not identical on each trial." Also they observe that "Since there is no unique response, the most we can say is that there is some probability of observing each of the different possible responses." This viewpoint then underlies the wide use of probabilistic ideas in describing how one can "read the neural code" through interpreting the response spike trains to infer the stimulus.

In this paper we take a different point of view and recognize that the neuron into which one sends a stimulus is itself a dynamical system with a time dependent state which will typically be different upon receipt of different realizations of identical stimulus inputs. Viewing the transformation of the stimulus waveform into the observed response sequence, as a result of deterministic dynamical action of the neuron one can attribute the variation in the response to identical stimuli to differing neuron states when the stimulus arrives. This allows us to view the entire transduction process of analog input (stimulus) to spike train output (response) as a deterministic process which can be addressed by methods developed in nonlinear dynamics for dealing with input/output systems Rhodes and Morari [1998].

Previous research on information encoding in spike trains has concentrated on nonlinear filters that convert analog input signals into spike trains. It has been shown that these models can be used to reconstruct the dynamical phase space of chaotic inputs to the filters using the spike timing information Sauer [1994]; Castro and Sauer [1997]; Pavlov, Sosnovtseva, Mosekilde, and Anishcenko [2000, 2001]. Using simple dynamical neuron models, Castro and Sauer Castro and Sauer [1999] have shown that aspects of a dynamical system can be reconstructed using interspike intervals (ISIs) properties. Experimental work has demonstrated the ability to discriminate between chaotic and stochastic inputs to a neuron Richardson, Imhoff, Grigg, and Collins [1998], as well as showing that decoding sensory information from a spike train through linear filtering Volterra series techniques can allow for large amounts of information to be carried by the precise timing of the spikes Rieke, Warland, de Ruyter van Steveninck, and Bialek [1997].

We discuss here the formulation of input/output systems from a dynamical system point of view, primarily summarizing earlier work Rhodes and Morari [1998]; Abarbanel [1996], but with a focus on recognizing that we may treat the response signals as trains of identical spikes. Since the modulation of the spike train must be carrying the information in the analog input presented to the neuron, if the spike pulse shapes are identical, all information must be encoded in the ISIs. We shall show that this is, indeed, the case.

What is the role of information theory in a deterministic chain of actions from stimulus to spiking response? The ideas of information theory, though often couched in terms of random variables, applies directly to distributed variation in dynamical variables such as the output from nonlinear systems. The use of concepts such as entropy and mutual information, at the basis of information theoretic descriptions of systems, applies easily and directly to deterministic systems. The understanding of this connection dates from the 1970s and 1980s where the work of Fraser Fraser [1989] makes this explicit, and the connection due to Pesin Pesin [1977] between positive Lyapunov exponents of a deterministic system and the Kolmogorov–Sinai entropy quantifies the correspondence.

In the body of this paper, we first summarize the methods used to determine a connection between analog input signals and spiking output, then we apply these methods to a Hodgkin–Huxley conductance based model of the R15 neuron of *Aplysia* Canavier, Clark, and Byrne [1990]; Plant and Kim [1976]. Future papers will investigate the use of these methods on biological signals from the H1 visual neuron of a fly and a stretch receptor in the tail of a crayfish Tumer, Wolfe, Wood, Abarbanel, Rabinovich, and Selverston [2002]

2 Input Estimation from State Space Reconstruction

The general problem we address is the response to stimuli of a neural circuit with N dynamical variables

$$\mathbf{x}(t) = \left[x_1(t), x_2(t), \ldots, x_N(t)\right].$$

When there is no time varying input, $\mathbf{x}(t)$ satisfies the ordinary differential equations

$$\frac{dx_a(t)}{dt} = F_a\big(\mathbf{x}(t)\big), \quad a = 1, 2, ..., N. \tag{2.1}$$

The $F_a(\mathbf{x})$ are a set of nonlinear functions which determine the dynamical time course of the neural circuit. The $F_a(\mathbf{x})$ could well represent a conductance based neural model of the Hodgkin–Huxley variety as in our example below.

When there is a time dependent external stimulus $s(t)$, these equations become

$$\frac{dx_a(t)}{dt} = F_a\big(\mathbf{x}(t), s(t)\big) , \tag{2.2}$$

and the time course of $\mathbf{x}(t)$ in this driven or non-autonomous setting can become rather more complicated than the case where $s(t) =$ constant.

If we knew the dynamical origin of the signal $s(t)$, then in the combined space of the stimuli and the neural state space $\mathbf{x}(t)$, we would again have an autonomous system, and many familiar Abarbanel [1996] methods for analyzing signals from nonlinear systems would apply. As we proceed to our "input signal from spike outputs" connection we imagine that the stimulus system is determined by some other set of state variables $\mathbf{z}(t)$ and that

$$\frac{d\mathbf{z}(t)}{dt} = \mathbf{G}\big(\mathbf{z}(t)\big) ; \quad s(t) = h\big(\mathbf{z}(t)\big) , \tag{2.3}$$

where $\mathbf{G}(\mathbf{z})$ are the nonlinear functions determining the time course of the state $\mathbf{z}(t)$ and $h\big(\mathbf{z}(t)\big)$ is the nonlinear function determining the input to the neuron $s(t)$.

With observations of just one component of the state vector $\mathbf{x}(t)$, the full dynamical structure of a system described by Equation 2.2 can be reconstructed in a proxy state space Mañé [1981]; Takens [1981]. Once the dynamics of the system is reconstructed, the mapping from state variable to input can be made in the reconstructed space. Assume the measured state variable, $r(t) = g\big(\mathbf{x}(t)\big)$, is sampled at times t_j, where j is an integer index. According to the embedding theorem Mañé [1981]; Takens [1981], the dynamics of the system can be reconstructed in an embedding space using time delayed vectors of the form

$$\begin{aligned} \mathbf{y}(j) &= \big[r(t_j), r(t_j + T\tau_s), \ldots, r\big(t_j + (d_E - 1)T\tau_s\big)\big] \\ &= \big[r(j), r(j + T), \ldots, r\big(j + (d_E - 1)T\big)\big] , \end{aligned} \tag{2.4}$$

where d_E is the dimension of the embedding, $t_j = t_0 + j\tau_s$, τ_s is the sampling time, t_0 is an initial time, and T is an integer time delay. If the dimension d_E is large enough these vectors can reconstruct the dynamical structure of the full system given in Equation 2.2. Each vector $\mathbf{y}(j)$ in the reconstructed phase space depends on the state of the input signal. Therefore a mapping should exist that associates locations in the reconstructed phase space $\mathbf{y}(j)$ to values of the input signal $s(t_j) \equiv s(j) : s(j) = H\big(\mathbf{y}(j)\big)$. The map $H(\mathbf{y})$ is the output-to-input relation we seek.

Without simultaneous measurements of the observable $r(t)$ and the input signal $s(t)$, this mapping could not be found without knowing the differential equations that make up Equation 2.2. But in a situation where a controlled stimulus is presented to a neuron while measuring the output, both $r(t)$ and $s(t)$ are available simultaneously. Such a data set with simultaneous measurements of spike time and input is split into two parts: the first part, called the training set, will be used to find the mapping $H(\mathbf{y}(j))$ between $\mathbf{y}(j)$ and $s(j)$. The second part, called the test set, will be used to test the accuracy of that mapping. State variable data from the training set $r(j)$ is used to construct time delayed vectors as given by

$$\mathbf{y}(j) = \left[r(j), r(j+T), \ldots, r(j+(d_E-1)T)\right]. \tag{2.5}$$

Each of these vectors is paired with the value of the stimulus at the midpoint time of the delay vector

$$s(j) = s\left(t_{j+\frac{1}{2}T(d_E-1)}\right). \tag{2.6}$$

We use state space values that occur before and after the input to improve the quality of the representation. The state variables and input values in the remainder of the data are organized in a similar way and used to test the mapping.

The phase space dynamics near a test data vector are reconstructed using vectors in the training set that are close to the test vector, where we use Euclidian distance between vectors. These vectors lie close in the reconstructed phase space, so they will define the dynamics of the system in that region and will define a **local** map from that region to a input signal value. In other words, we seek a form for $H(\mathbf{y}(j))$ which is local in reconstructed phase space to $\mathbf{y}(j)$. The **global** map over all of phase space is a collection of local maps.

The local map is made using the N_B nearest neighbors $\mathbf{y}^m(j)$, $m = 0 \ldots N_B$ of $\mathbf{y}^0(j) = \mathbf{y}(j)$. These nearest neighbor vectors and their corresponding input values $s^m(j)$ are used to find a local polynomial mapping between inputs $s^m(j)$ and vector versions of the outputs $r^m(j)$, namely $\mathbf{y}^m(j)$ of the form

$$s^m(j) = H(\mathbf{y}^m(j)) = M_0(j) + \mathbf{M}_1(j) \cdot \mathbf{y}^m(j) + \mathbf{M}_2(j) \cdot \mathbf{y}^m(j) \cdot \mathbf{y}^m(j) + \cdots, \tag{2.7}$$

which assume that the function $H(\mathbf{y})$ is locally smooth in phase space.

The scalar $M_0(j)$, the d_E-dimensional vector $\mathbf{M}_1(j)$, and the tensor $\mathbf{M}_2(j)$ in d_E-dimensions, etc are determined by minimizing the mean squared error

$$\sum_{m=0}^{N_B} \left| s^m(j) - M_0(j) + \mathbf{M}_1(j) \cdot \mathbf{y}^m(j) + \mathbf{M}_2(j) \cdot \mathbf{y}^m(j) \cdot \mathbf{y}^m(j) + \cdots \right|^2. \tag{2.8}$$

We determine $M_0(j), \mathbf{M}_1(j), \mathbf{M}_2(j), \ldots$ for all $j = 1, 2, \ldots$, and this provides a local representation of $H(\mathbf{y})$ in all parts of phase space sampled by the training set $\mathbf{y}(j)$, $j = 1, 2, \ldots, N_{\text{train}}$.

Once the least squares fit values of $M_0(j), \mathbf{M}_1(j), \mathbf{M}_2(j), \ldots$ are determined for our training set, we can use the resulting local map to determine estimates of the input associated with an observed output. This proceeds as follows: select a new output $r^{\text{new}}(l)$ and form the new output vector $\mathbf{y}^{\text{new}}(l)$ as above. Find the nearest neighbor in the training set to $\mathbf{y}^{\text{new}}(l)$. Suppose it is the vector $\mathbf{y}(q)$. Now evaluate an estimated input $s^{\text{est}}(l)$ as

$$s^{\text{est}}(l) = M_0(q) + \mathbf{M}_1(q) \cdot \mathbf{y}^{\text{new}}(l) + \mathbf{M}_2(q) \cdot \mathbf{y}^{\text{new}}(l) \cdot \mathbf{y}^{\text{new}}(l) + \cdots . \tag{2.9}$$

This procedure is applied for all new outputs to produce the corresponding estimated inputs.

3 R15 Neuron Model

To investigate our ability to reconstruct stimuli of analog form presented to a realistic neuron from the spike train output of that neuron, we examined a detailed model of the R15 neuron in *Aplysia* Canavier, Clark, and Byrne [1990]; Plant and Kim [1976], and presented this model neuron with nonperiodic input from a low dimensional dynamical system. This model has seven dynamical degrees of freedom. The differential equations for this model are

$$\begin{aligned} C\frac{dV_m(t)}{dt} &= \big(g_I y_2(t)^3 y_3(t) + g_T\big)\big(V_I - V(t)\big) + g_L\big(V_L - V(t)\big) \\ &\quad + \big(g_K y_4(t)^4 + g_A y_5(t) y_6(t) + g_P y_7(t)\big)\big(V_K - V(t)\big) \\ &\quad + I_0 + I_{\text{ext}} + I_{\text{input}}(t) , \end{aligned} \tag{3.1}$$

where the $y_n(t)$; $n = 2, 3, \ldots, 7$ satisfy kinetic equations of the form

$$\frac{dy_n(t)}{dt} = \frac{Y_n\big(V_m(t)\big) - y_n(t)}{\tau_n\big(V_m(t)\big)} , \tag{3.2}$$

which is the usual form of Hodgkin–Huxley models. The g_X, $X = I$, T, K, A, P, L are maximal conductances, the V_X, $X = I$, L, K are reversal potentials. $V_m(t)$ is the membrane potential, C is the membrane capacitance, I_0 is a fixed DC current, and I_{ext} is a DC current we vary to change the state of oscillation of the model. The functions $Y_n(V)$ and $\tau_n(V)$ and values for the various constants are given in Canavier, Clark, and Byrne [1990]; Plant and Kim [1976]. These are phenomenological forms of membrane voltage dependent gating variables, activation and inactivation of membrane ionic channels, and time constants for these gates. $I_{\text{input}}(t)$ is a time varying current input to the neural dynamics. Our goal will be to reconstruct $I_{\text{input}}(t)$ from observations of the spike timing in $V_m(t)$.

In Figure 3.1 we plot the bifurcation diagram of our R15 model. On the vertical axis we show the values of ISIs taken in the time series for $V_m(t)$ from the model; on the horizontal axis we plot I_{ext}. From this bifurcation plot we see that the output of the R15 model has regular windows

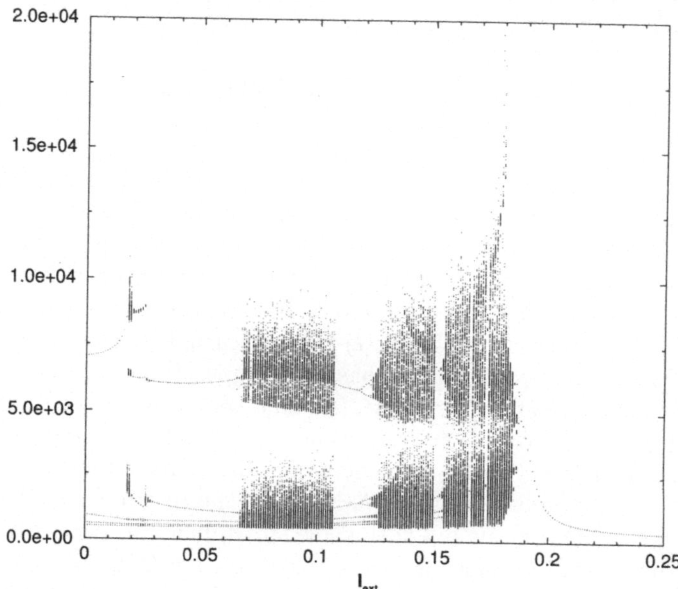

FIGURE 3.1. Bifurcation diagram for the R15 model with constant input current. This plot shows the values of ISIs which occur in the $V_m(t)$ time series for different values of I_{ext}.

for $I_{\text{ext}} < .07$ then chaotic regions interspersed with periodic orbits until $I_{\text{ext}} \approx 0.19$ after which nearly periodic behavior is seen. The last region represents significant depolarization of the neuron in which tonic periodic firing associated with a stable limit cycle in phase space is typical of neural activity. Periodic firing leads to a fixed value for ISIs, which is what we see. Careful inspection of the time series reveals very small fluctuations in the phase space orbit, but the resolution in Figure 3.1 does not expose this.

Other than the characteristic spikes, there are no significant features in the membrane voltage dynamics. In addition all the spikes are essentially the same, so we expect that all the information about the membrane voltage state is captured in the times between spikes, namely the interspike intervals: ISIs. The distribution of ISIs characterizes the output signal for information theoretic purposes.

We have chosen three values of I_{ext} at which to examine the response of this neuron model when presented with an input signal. At $I_{\text{ext}} = 0.1613$ we expect chaotic waveforms expressed as nonperiodic ISIs with a broad distribution. At $I_{\text{ext}} = 0.2031$ we expect nearly periodic spike trains. And at $I_{\text{ext}} = -0.15$ the neuron does not spike, the mebrane voltage remains at an equilibrium value.

For each $V_m(t)$ time series we evaluate the normalized distribution of ISIs which we call $P_{\mathrm{ISI}}(\Delta)$ and from this we compute the entropy associated with the oscillations of the neuron. Entropy is defined as

$$H(\Delta) = \sum_{\text{observed } \Delta} -P_{\mathrm{ISI}}(\Delta) \log\left(P_{\mathrm{ISI}}(\Delta)\right); \tag{3.3}$$

$H(\Delta) \geq 0$. The entropy is a quantitative measure Shannon [1948] of the information content of the output signal from the neural activity.

In Figure 3.2 we display a section of the $V_m(t)$ time series for $I_{\mathrm{ext}} = 0.1613$. The irregularity in the spiking times is clear from this figure and the distribution $P_{\mathrm{ISI}}(\Delta)$ shown in Figure 3.3. The $P_{\mathrm{ISI}}(\Delta)$ was evaluated from collecting 60,000 spikes from the $V_m(t)$ time series and creating a histogram with 15,000 bins. This distribution has an entropy $H(\Delta) = 12$. In contrast to this we have a section of the $V_m(t)$ time series for $I_{\mathrm{ext}} = 0.2031$ in Figure 3.4. Far more regular firing is observed with a firing frequency much higher than for $I_{\mathrm{ext}} = 0.1613$. This increase in firing frequency as a neuron is depolarized is familiar. With $I_{\mathrm{ext}} = 0.2031$ the distribution $P_{\mathrm{ISI}}(\Delta)$ is mainly concentrated in one bin with some small flucuations near that bin. Such a regular distribution leads to a very low entropy $H(\Delta) = 0.034$. If not for the slight variations in ISI, the entropy would be zero. If $P_{\mathrm{ISI}}(\Delta_0) = 1$ for some ISI value Δ_0, then $H(\Delta) = 0$.

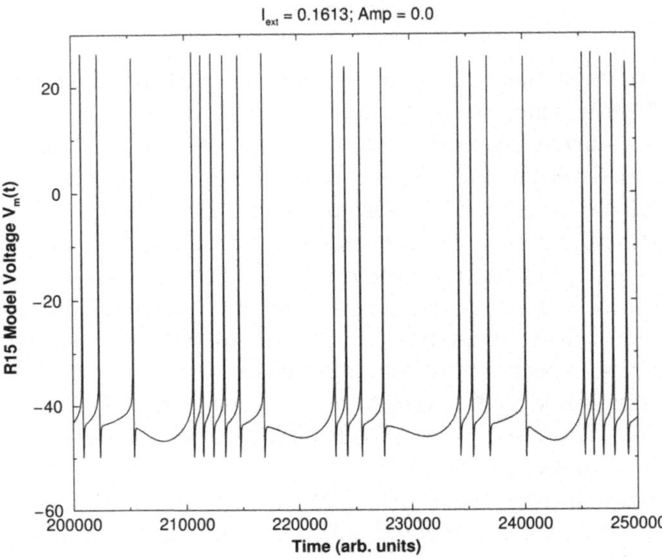

FIGURE 3.2. Membrane voltage of the R15 model with a constant input current $I_{\mathrm{ext}} = 0.1613$.

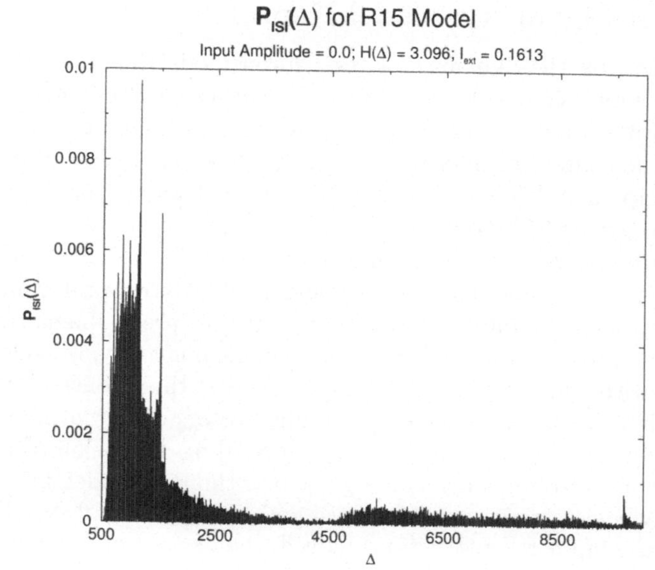

FIGURE 3.3. Normalized distribution $P_{\rm ISI}(\Delta)$ from the membrane voltage time series with $I_{\rm ext} = 0.1613$. The entropy for this distribution $H(\Delta) = 12$.

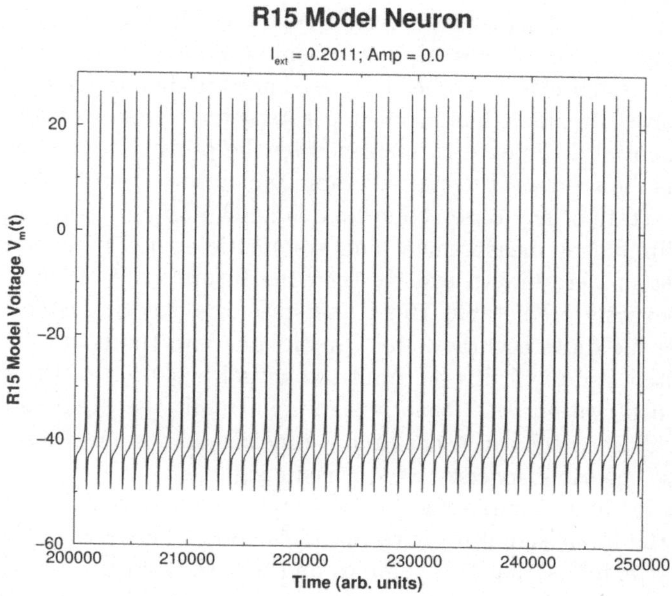

FIGURE 3.4. Membrane voltage of the R15 model with a constant input current $I_{\rm ext} = 0.2031$.

3.1 Input Signals to Model Neuron

In the last section the dynamics of the neuron model were examined using constant input signals. In studying how neurons encode information in their spike train, we must clarify what it means for a signal to carry information. In the context of information theory Shannon [1948], information lies in the unpredictability of a signal. If we do not know what a signal is going to do next, then by observing it we gain new information. Stochastic signals are commonly used as information carrying signals since their unpredictability is easily characterized and readily incorporated into the theoretical structure of information theory. But they are problematic when approaching a problem from a dynamical systems point of view, since they are systems with a high dimension. This means that the reconstruction of a stochastic signal using time delay embedding vectors of the form of Equation 2.4 would require an extremely large embedding dimension Abarbanel [1996]. If we are injecting stochastic signals into the R15 model, the dimension of the whole system would increase and cause practical problems in performing the input reconstruction. Indeed, the degrees of freedom in the stochastic input signal could well make the input/output relationship we seek to expose impossible to see.

An attractive input for testing the reconstruction method will have some unpredictability but have few degrees of freedom. If there are many degrees of freedom, the dimensionality of the vector of outputs $\mathbf{y}(j)$ above may be prohibitively large. This leads directly to the consideration of low dimensional chaotic systems. Chaos originates from local instabilities which cause two points initially close together in phase space to diverge rapidly as the system evolves in time, thus producing completely different trajectories. This exponential divergence is quantified by the positive Lyapunov exponents and is the source of the unpredictability in chaotic systems Abarbanel [1996]. The state of any observed system is known only to some degree of accuracy, limited by measurement and systematic errors. If the state of a chaotic system were known exactly then the future state of that system should be exactly predictable. But if the state of a chaotic system is only known to some finite accuracy, then predictions into the future based on the estimated state will diverge from the actual evolution of the system. Imperfect observations of a chaotic signal will limit the predictability of the signal. Since chaos can occur in low dimensional systems these signals do not raise the same concerns as stochastic signals.

We use a familiar example of a chaotic system, the Lorenz attractor Lorenz [1963], as the input signal to drive the R15 model. This is a well studied system that exhibits chaotic dynamics and will be used here as input to the R15 neuron model. The Lorenz attractor is defined by the differential equations

$$\begin{aligned}\kappa\frac{dx(t)}{dt} &= \sigma\big(y(t) - x(t)\big)\\ \kappa\frac{dy(t)}{dt} &= -x(t)z(t) + rx(t) - y(t)\\ \kappa\frac{dz(t)}{dt} &= x(t)y(t) - bz(t)\end{aligned} \tag{3.4}$$

For the simulations presented in this paper the parameters were chosen as $\sigma = 16$, $r = 45.92$ and $b = 4$. The parameter κ is used to change the time scale. An example times series of the $x(t)$ component of the Lorenz attractor is shown in Figure 3.5.

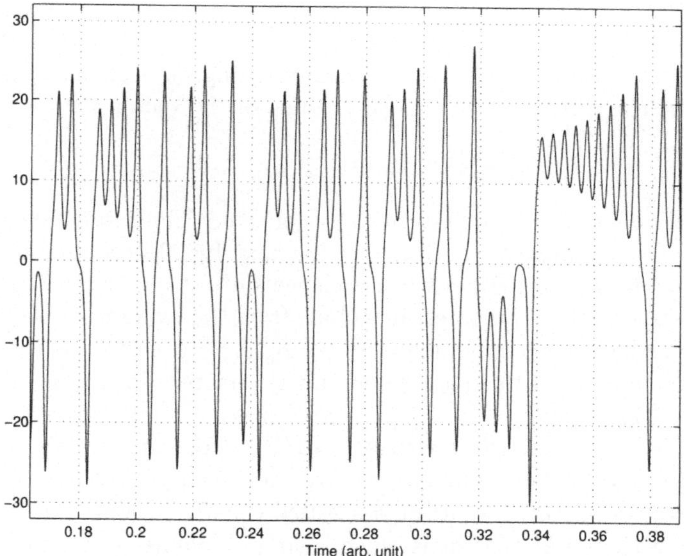

FIGURE 3.5. Small segment of the $x(t)$ component of the Lorenz attractor described in equations 3.4 with $\kappa = 10^4$.

3.2 Numerical Results

An input signal $s(t) = I_{\text{input}}(t)$ is now formed from the $x(t)$ component of the Lorenz system. Our goal is to use observations of the stimulus $I_{\text{input}}(t)$ and of the ISIs of the output signal $V_m(t)$ to learn the dynamics of the R15 neuron model in the form of a local map in phase space reconstructed from the observed ISIs. From this map we will estimate the input $I_{\text{input}}(t)$ from new observations of the output ISIs.

Our analog signal input is the $x(t)$ output of the Lorenz system, scaled and offset to a proper range, and then input to the neuron as an external current

$$I_{\text{input}}(t) = \text{Amp}\big(x(t) + x_0\big)\,, \tag{3.5}$$

where Amp is the scaling constant and x_0 is the offset. The R15 equations are integrated Press, Teukolsky, Vetterling, and Flannery [1992] with this input signal and the spike times t_j from the membrane voltage are recorded simultaneously with the value of the input current at that time $I_{\text{input}}(t_j)$. Reconstruction of the neuron plus input phase space is done by creating time delay vectors from the ISIs

$$\mathbf{y}(j) = \left[\text{isi}_j, \text{isi}_{j+1}, \ldots, \text{isi}_{j+(d_E-1)\tau}\right], \tag{3.6}$$

where

$$\text{isi}_j = t_j - t_{j-1}\,. \tag{3.7}$$

For each of these vectors there is a corresponding value of the input current which we chose to be at the midpoint time of the vector

$$s(j) = I_{\text{input}}\left(t_{j+\frac{1}{2}(d_E-1)\tau}\right). \tag{3.8}$$

In our work a total of 40000 spikes were collected. The first 30000 were used to create the training set vectors and the next 10000 were used to examine our input estimation methods. For each new output vector constructed from new observed ISIs, N_B nearest neighbors from the training set were used to generate a local polynomial map $\mathbf{y}(j) \to I_{\text{input}}^{\text{estimated}}(j)$. N_B was chosen to be twice the number of free parameters in the undetermined local coefficients $M_0, \mathbf{M}_1, \mathbf{M}_2, \ldots$.

We used the same three values, -0.15, 0.1613, and 0.2031, for I_{ext} as employed above in our simulations. We took Amp $= 0.001$, $\kappa = 10^4$, and $x_0 = 43.5$ for all simulations unless stated otherwise. This very small amplitude of the input current is much more of a challenge for the input reconstruction than large amplitudes. When Amp is large, the neural activity is entrained by the input signal and 'recovering' the input merely requires looking at the output and scaling it by a constant. Further, the intrinsic spiking of the neuron which is its important biological feature goes away when Amp is large. The large value of κ assures that the spikes sample the analog signal $I_{\text{input}}(t)$ very well.

For $I_{\text{ext}} = 0.1613$ we show a selection of both the input current I_{input} and the output membrane voltage $V_m(t)$ time series in Figure 3.6. The injected current substantially changes the pattern of firing seen for the autonomous neuron. Note that the size of the input current is numerically about 10^{-3} of $V_m(t)$, yet the modulation of the ISIs due to this small input is clearly visible in Figure 3.6.

Using the ISIs of this time series we evaluated $P_{\text{ISI}}(\Delta)$ as discussed above and from that the entropy $H(\Delta)$ associated with the driven neuron. The

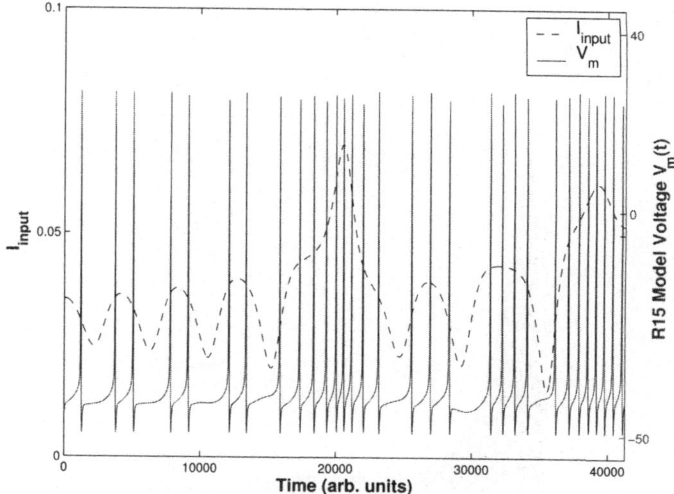

FIGURE 3.6. A segment of the R15 neuron model output $V_m(t)$ shown along with the scaled Lorenz system input current I_{input}. Here $I_{\text{ext}} = 0.1613$, Amp $= 0.001$, and $\kappa = 10^4$. Note the different scales for I_{input} (shown on the left axis) and $V_m(t)$ (shown on the right axis).

ISI distribution, $P_{\text{ISI}}(\Delta)$, shown in Figure 3.7, has an entropy $H(\Delta) = 8.16$. The effect of the input current has been to substantially narrow the range of ISIs seen in $V_m(t)$. This can be seen by comparison with Figure 3.3.

Figure 3.8 shows an example of input signal reconstruction which estimates I_{input} using ISI vectors of the described in Equation 3.6. We used a time delay $T = 1$, an embedding dimension $d_E = 7$, and a local linear map for $H(\mathbf{y}(j))$. The RMS error over the 10,000 reconstructed values of the input was $\sigma = 4.6 \cdot 10^{-4}$. The input signal is only reconstructed at times at which the neuron spikes. So each point is the reconstruction curve in Figure 3.8 corresponds to a spike in $V_m(t)$. Some features of the input are missed because no spikes occur during that time, but otherwise the reconstruction is very accurate. At places where the spike rate is high, interpolation seems to fill the gaps between spikes.

Different values of embedding dimension, time delay, and map order will lead to different reconstruction errors. For example, low embedding dimension may not unfold the dynamics and linear maps may not be able to fit some neighborhoods to the input. For the results shown here, there is little difference in the RMS reconstruction error if the embedding dimension is increased or quadratic maps are used instead of linear maps. This may not be true if lower embedding dimension is used.

The previous example probed the response of a chaotic neural oscillation to a chaotic signal. With $I_{\text{ext}} = 0.2031$ the neuron is in a periodic spiking regime and the input modulates the instantaneous firing rate of the neuron.

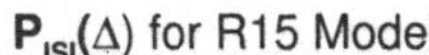

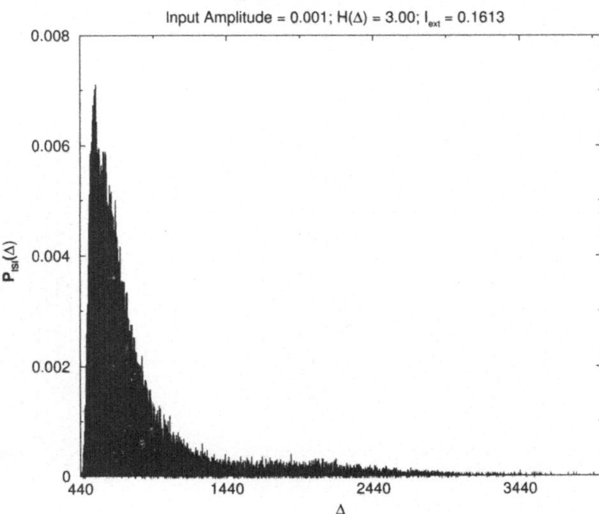

FIGURE 3.7. $P_{\mathrm{ISI}}(\Delta)$ for R15 model neuron output when a scaled $x(t)$ signal from the Lorenz system is presented with $I_{\text{ext}} = 0.1613$. The entropy of this distribution $H(\Delta) = 8.16$.

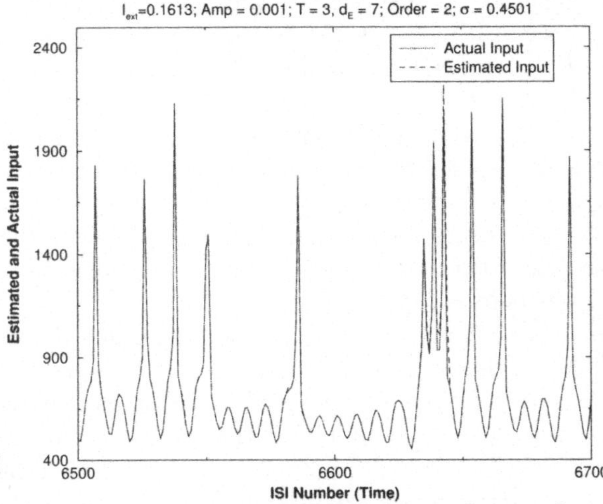

FIGURE 3.8. ISI ... neuron. The solid line is ... hed lines are the ISI reconstructions. The embedding dimension of the reconstruction d_E is 4, the time delay T is 1, $I_{\text{ext}} = 0.1613$, κ is 10^4, and a linear map was used. The RMS error of the estimates over 10,000 estimations is $\sigma = 4.6 \cdot 10^{-4}$ and the maximum error is about 0.01.

A sample of the input current and membrane voltage is shown in Figure 3.9. The distribution of ISIs, $P_{\text{ISI}}(\Delta)$, shown in Figure 3.10 and has an entropy $H(\Delta) = 9.5$. The effect of the input current is to substantially broaden the range of ISIs and increase its entropy as compared to the nearly periodic firing of the autonomous neuron with $I_{\text{ext}} = 0.2031$. The high spiking rate and close relationship between input current amplitude and ISI lead to very accurate reconstructions using low dimensional embeddings. A sample of the reconstruction using $d_E = 2$ and $T = 1$ is shown in Figure 3.11. The RMS reconstruction error of $\sigma = 6.1 \cdot 10^{-4}$ with a maximum error of 0.007.

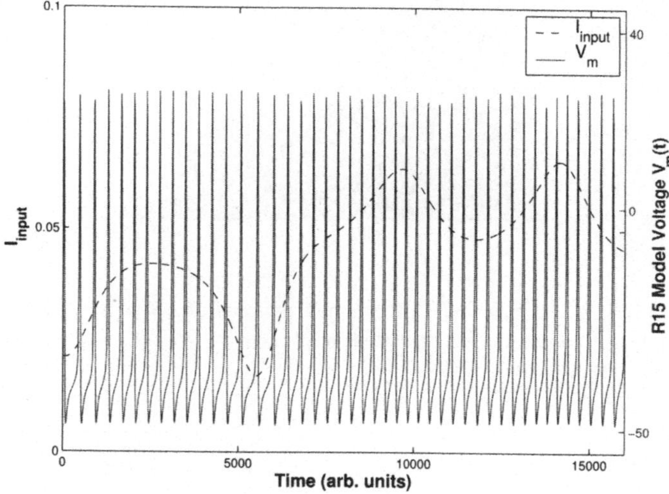

FIGURE 3.9. A segment of the R15 neuron model output $V_m(t)$ shown along with the scaled Lorenz system input current I_{input}. Here $I_{\text{ext}} = 0.2031$,Amp $= 0.001$, and $\kappa = 10^4$. Note the different scales for I_{input} and $V_m(t)$.

In a final example we show the reconstruction when the neuron is being driven with an input current below the threshold for spikes. With $I_{\text{ext}} = -0.15$, the autonomous R15 neuron will remain at an equilibrium level and not produce spikes. A Lorenz input injected into the neuron with Amp $= 0.002$ and $x_0 = 43.5$ is large enough to cause the neuron to spike. Figure 3.12 shows a sample of the membrane voltage time series along with the corresponding input current. Since the spiking rate of the neuron is much lower than before, κ is increased to $2 \cdot 10^5$. This slows down the dynamics of the Lorenz input relative to the neuron dynamics. Spikes occur during increasing portions of the input current and are absent for low values of input current. Figure 3.13 shows the distribution of ISIs which has an entropy $H(\Delta) = 5.3$. The low spiking rate shows up in the distribution in the form large numbers of long ISI. For the reconstruction of the input larger embedding dimensions were needed. An sample of the reconstruction

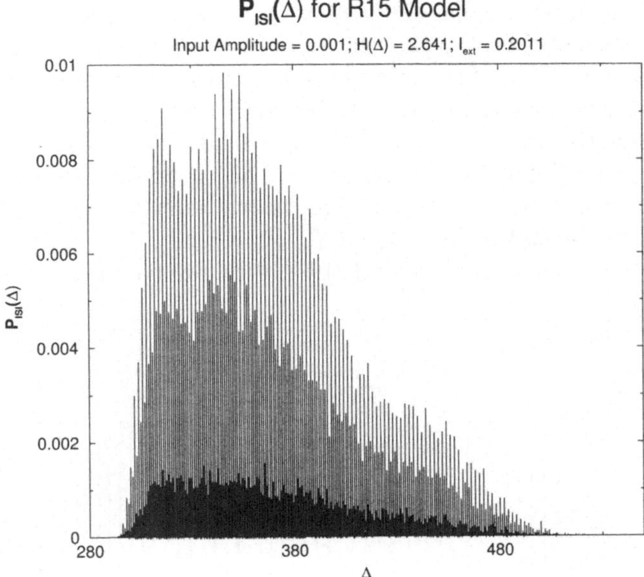

FIGURE 3.10. $P_{\mathrm{ISI}}(\Delta)$ for R15 model neuron output when a scaled $x(t)$ signal from the Lorenz system is presented with $I_{\mathrm{ext}} = 0.2031$. The entropy of this distribution $H(\Delta) = 9.5$.

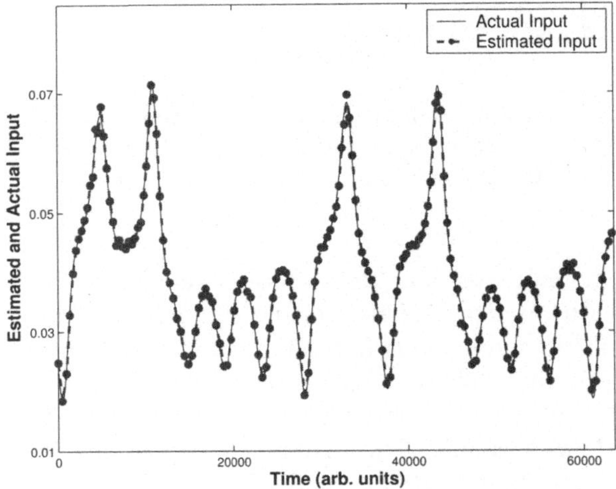

FIGURE 3.11. ISI Reconstruction of the input Lorenz signal to an R15 neuron. The solid line is the actual input to the neuron. The dots joined by dashed lines are the ISI reconstructions. The embedding dimension of the reconstruction d_E is 2, the time delay T is 1, $I_{\mathrm{ext}} = 0.2031$, κ is 10^4, and a linear map was used. The RMS error of the estimates over 10,000 estimations is $\sigma = 6.1 \cdot 10^{-4}$ and the maximum error is about 0.007.

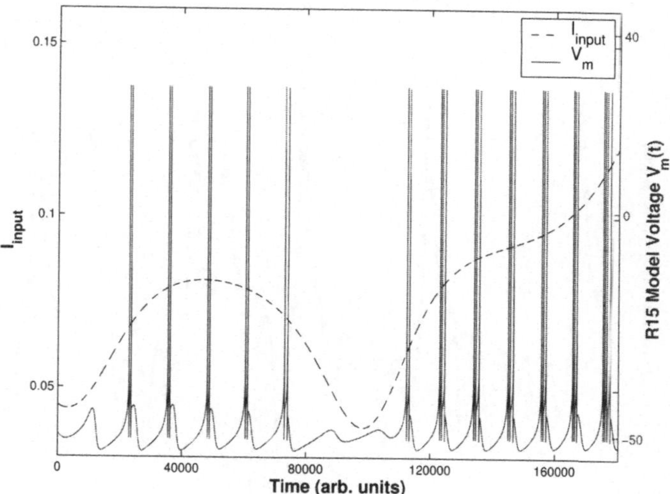

FIGURE 3.12. A segment of the R15 neuron model output $V_m(t)$ shown along with the scaled Lorenz system input current I_{input}. Here $I_{\text{ext}} = -0.15$.Amp $= 0.002$, and $\kappa = 2 \cdot 10^5$. Note the different scales for I_{input} and $V_m(t)$.

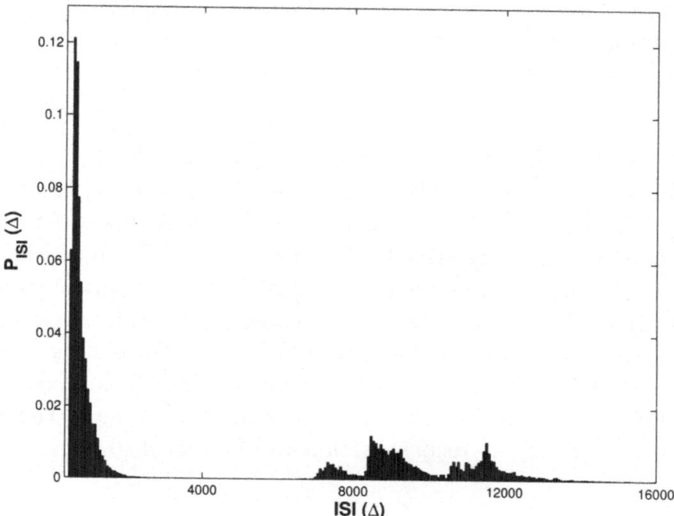

FIGURE 3.13. $P_{\text{ISI}}(\Delta)$ for R15 model neuron output when a scaled $x(t)$ signal from the Lorenz system is presented with $I_{\text{ext}} = -0.15$. The entropy of this distribution $H(\Delta) = 5.3$.

is shown in Figure 3.14 using $d_E = 7$ and $T = 1$. For this fit the RMS reconstruction error $\sigma = 0.0094$ with a maximum error of 0.03. These errors are noticeably higher than the previous two examples.

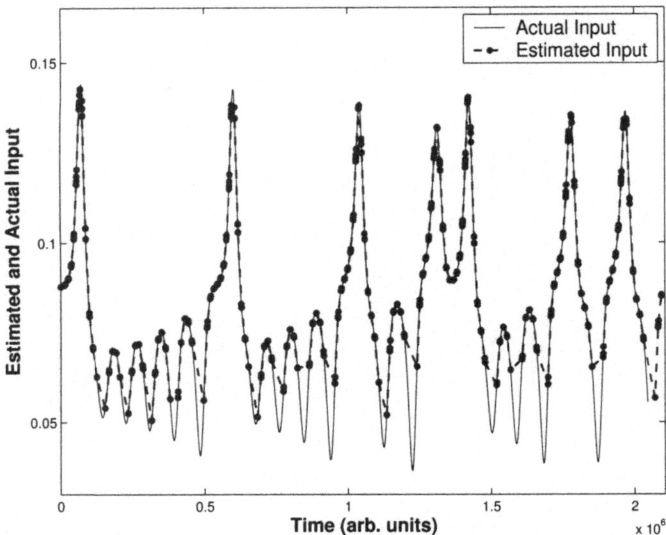

FIGURE 3.14. ISI Reconstruction of the input Lorenz signal to an R15 neuron. The solid line is the actual input to the neuron. The dots joined by dashed lines are the ISI reconstructions. The embedding dimension of the reconstruction d_E is 7, the time delay T is 1, $I_{\text{ext}} = -0.15$, κ is $2 \cdot 10^5$, and a linear map was used. The RMS error of the estimates over 10,000 estimations is $\sigma = 0.0094$ and the maximum error is about 0.03.

The accuracy of the reconstruction method depends on a high spiking rate in the neuron relative to the time scale of the input signal, since only one reconstructed input value is generated for each spike. If the spiking rate of the neuron is low relative to the time scales of the input signal, then the neuron will undersample the input signal and miss many of its features. This limitation can be demonstrated by decreasing the time scale parameter κ, thereby speeding up the dynamics of the input. During the longer ISIs the input current can change by large amounts. Though the reconstruction undersamples the input, but interpolation can fill in some of the gaps. As κ is increased further the reconstruction will further degrade.

4 Discussion

In previous research on the encoding of chaotic attractors in spikes trains, the spike trains were produced by nonlinear transformations of chaotic input signals. Threshold crossing neuron models have been used, which generate the spike times at upward crossings of a threshold. This is equivalent to a Poincare section of the input signal. Also integrate and fire neurons have been studied, which integrate the input signal and fire a spike when it crosses a threshold, after which the integral is reset to zero. Both of these models have no intrinsic complex dynamics; they can not produce entropy autonomously. All of the complex behavior is in the input signal. Even though the attractor of a chaotic input can be reconstructed from the ISIs, these models do not account for the complex behavior of real neurons. The input reconstruction method we have presented here allows for complex intrinsic dynamics of the neuron. We have shown that the local polynomial representations of input/output relations realized in reconstructed phase space can extract the chaotic input from the complex interaction between the input signal and neuron dynamics.

Other experimental works have used linear kernels to map the spike train into the input. They have shown that the precise timing of individual spikes can encode a lot of information about the input Rieke, Warland, de Ruyter van Steveninck, and Bialek [1997]. And the precise relative timing between two spikes can carry even more information than their individual timings combined Brenner, Strong, Koberle, Bialek, and de Ruyter van Steveninck [2000]. These results may be pointing toward a state space representation since the time delay embedding vectors used here take into account both the precise spike timing and the recent history of ISIs. From a dynamical systems perspective this is important because the state of the system at the time of the input will affect its response. This is a factor that linear kernels do not take into account.

The advantage of using local representations of input/output relations in reconstructed state space lies primarily in the insight it may provide about the underlying dynamics of the neural transformation process mapping analog environmental signals into spike trains. The goal of the work presented here is not primarily to show we can accurately recover analog input signals from the ISIs of spike output from neurons, though that is important to demonstrate. The main goal is to provide clues on how one can now model the neural circuitry which transforms these analog signals. The main piece of information in the work presented here lies in the size of the reconstructed space d_E which tells us something about the required dimension of the neural circuit. Here we see that a low dimension can give excellent results indicating that the complexity of the neural circuit is not fully utilized in the transformation to spikes. Another suggestion of this is in the entropy of the input and output signals. In the case where $I_{\text{ext}} = 0.1613$ the entropy of the analog input is 11.8 while the entropy of the ISI distribution of the output is 8.16. When $I_{\text{ext}} = 0.2031$ the output entropy is 9.5. This suggests, especially in the case of the larger current, that the signal into R15 neuron model acts primarily as a modulation on the ISI distribution. This modulation may be substantial, as in the case when $I_{\text{ext}} = 0.1613$ but reading the modulated signal does not require complex methods.

Our final example took $I_{\text{ext}} = -0.15$ at which value the undriven neuron has $V_m(t) =$ constant, so it is below threhold for production of action potentials. In this case the introduction of the stimulus drove the neuron above this threshold and produced a spike train which could be accurately reconstructed. This example is relevant to the behavior of biological neurons which act as sensors for various quantities: visiual stimuli, chemical stimuli (olfaction), etc. In the study of biological sensory systems Tumer, Wolfe, Wood, Abarbanel, Rabinovich, and Selverston [2002] the neural circuitry is quiet in the absence of input signals, yet as we now see the methods are equally valid and accurate.

Acknowledgments: This work was partially supported by the U.S. Department of Energy, Office of Basic Energy Sciences, Division of Engineering and Geosciences, under Grants No. DE-FG03-90ER14138 and No. DE-FG03-96ER14592, by a grant from the National Science Foundation, NSF PHY0097134, by a grant from the Army Research Office, DAAD19-01-1-0026, by a grant from the Office of Naval Research, N00014-00-1-0181, and by a grant from the National Institutes of Health, NIH R01 NS40110-01A2. ET acknowledges support from NSF Traineeship DGE 9987614.

References

Abarbanel, H. D. I. [1996], *The Analysis of Observed Chaotic Data.* Springer, New York.

Brenner, N., S. P. Strong, R. Koberle, W. Bialek, and R. de Ruyter van Steveninck [2000], Synergy in a Neural Code, *Neural Computation* **12**, 1531–52.

Canavier, C. C., J. W. Clark, and J. H. Byrne [1990], Routes to Chaos in a Model of a Bursting Neuron, *Biophys. J.* **57**, 1245–51.

Castro, R. and T. Sauer [1997], Correlation Dimension of Attractors Through Interspike Intervals, *Phys. Rev. E* **55**, 287–90.

Castro, R. and T. Sauer [1999], Reconstructing Chaotic Dynamics through Spike Filters, *Phys. Rev. E* **59**, 2911–17.

Fraser, A. M. [1989], *Information Theory and Strange Attractors*, PhD thesis, University of Texas, Austin.

Lorenz, E. N. [1963], Deterministic Nonperiodic Flow, *J. Atmos. Sci.* **20**, 130–41.

Mañé, R. [1981], On the Dimension of the Compact Invariant Sets of Certain Nonlinear Maps. In Rand, D. and L. S. Young, editors, *Dynamical Systems and Turbulence, Warwick, 1980*, volume 898, page 230, Berlin. Springer.

Pavlov, A. N., O. V. Sosnovtseva, E. Mosekilde, and V. S. Anishcenko [2000], Extracting Dynamics from Threshold-crossing Interspike Intervals: Possibilities and Limitations, *Phys. Rev. E* **61**, 5033–44.

Pavlov, A. N., O. V. Sosnovtseva, E. Mosekilde, and V. S. Anishcenko [2001], Chaotic Dynamics from Interspike Intervals, *Phys. Rev. E* **63**, 036205.

Pesin, Y. B. [1977], Lyapunov Characteristic Exponents and Smooth Ergodic Theory, *Usp. Mat. Nauk.* **32**, 55. English translation in *Russian Math. Survey,* Volume **72**, 55, (1977).

Plant, R. E. and M. Kim [1976], Mathematical Description of a Bursting Pacemaker Neuron by a Modification of the Hodgkin–Huxley Equations, *Biophys. J.* **16**, 227–44.

Press, W. H., S. A. Teukolsky, W. T. Vetterling, and B. P. Flannery [1992], *Numerical Recipes in FORTRAN.* Cambridge University Press, Cambridge.

Rhodes, C. and M. Morari [1998], Determining the Model Order of Nonlinear Input/Output Systems, *AIChE J.* **44**, 151–63.

Richardson, K. A., T. T. Imhoff, P. Grigg, and J. J. Collins [1998], Encoding Chaos in Neural Spike Trains, *Phys. Rev. Lett.* **80**, 2485–88.

Rieke, F., D. Warland, R. de Ruyter van Steveninck, and W. Bialek [1997], *Spikes: Exploring the Neural Code.* The MIT Press, Cambridge, MA.

Sauer, T. [1994], Reconstructing of Dynamical Systems from Interspike Intervals, *Phys. Rev. Lett.* **72**, 3811–14.

Shannon, C. E. [1948], A Mathematical Theory of Communication, *Bell Syst. Tech. J.* **27**, 379–423 and 623–656.

Takens, F. [1981], Detecting Strange Attractors in Turbulence. In Rand, D. and L. S. Young, editors, *Dynamical Systems and Turbulence, Warwick, 1980*, volume 898, page 366, Berlin. Springer.

Tumer, E. C., J. H. Wolfe, K. Wood, H. D. I. Abarbanel, M. I. Rabinovich, and A. I. Selverston [2002], *Reading Neural Codes : The Importance of Spike Patterns*, To be submitted to *Nature Neuroscience* September 2002, 2002.

2

Boolean Dynamics with Random Couplings

Maximino Aldana
Susan Coppersmith
Leo P. Kadanoff

To Larry Sirovich, on the occasion of his 70th birthday.

ABSTRACT This paper reviews a class of generic dissipative dynamical systems called N-K models. In these models, the dynamics of N elements, defined as Boolean variables, develop step by step, clocked by a discrete time variable. Each of the N Boolean elements at a given time is given a value which depends upon K elements in the previous time step. We review the work of many authors on the behavior of the models, looking particularly at the structure and lengths of their cycles, the sizes of their basins of attraction, and the flow of information through the systems. In the limit of infinite N, there is a phase transition between a chaotic and an ordered phase, with a critical phase in between. We argue that the behavior of this system depends significantly on the topology of the network connections. If the elements are placed upon a lattice with dimension d, the system shows correlations related to the standard percolation or directed percolation phase transition on such a lattice. On the other hand, a very different behavior is seen in the *Kauffman net* in which all spins are equally likely to be coupled to a given spin. In this situation, coupling loops are mostly suppressed, and the behavior of the system is much more like that of a mean field theory. We also describe possible applications of the models to, for example, genetic networks, cell differentiation, evolution, democracy in social systems and neural networks.

Contents

1 Introduction

In this review, we describe the dynamics of a set of N variables, or elements, which each have two possible values (say 0 and 1). These elements interact with each other according to some given interaction rules, specified through a set of Boolean coupling functions that determine the variables at the next time-step, and thereby give the dynamics of the system. Such a discrete stepping of a set of Boolean variables, also known in general terms as a *Boolean network*, is of potential interest in several different fields, ranging from gene regulation and control, to modeling democracy and social organization, to understanding the behavior of glassy materials.

The models were originally studied primarily for their biological interest, specifically by Stuart Kauffman who introduced the so-called *N-K model* in the context of gene expression and fitness landscapes in 1969 (Kauffman [1969, 1974, 1995, 1993, 1990, 1984]). Since Kauffman's original work, the scientific community has found a broad spectrum of applicability of these models. Specific biological problems studied include cell differentiation (Huang and Ingber [2000]), immune response (Kauffman and Weinberger [1989]), evolution (Bornholdt and Sneppen [1998]; Zawidzki [1998]; Bornholdt and Sneppen [2000]; Ito and Gunji [1994]), regulatory networks (Bornholdt and Rohlf [2000]) and neural networks (Wang, Pichler, and Ross [1990]; Derrida, Gardner, and Zippelius [1987]; Kürten [1988a]; Bornholdt and Rohlf [2000]). In the first two examples, the basic binary element might be a chemical compound, while in the last it might be the state of firing of a neuron. A computer scientist might study a similar set of models, calling the basic elements *gates*, and be thinking about the logic of computer design (Atlan, Fogelman-Soulie, Salomon, and Weisbuch [1981]; Lynch [1995]) or optimization (Lee and Han [1998]; Stauffer [1994]). Earlier work in the mathematical literature (Harris [1960]; Metropolis and Ulam [1953]) studied *random mapping models*, which are a subset of the models introduced by Kauffman. This same kind of problem has also drawn considerable attention from physicists interested in the development of chaos (Glass and Hill [1998]; Luque and Solé [1998, 1997a]; Kürten and Beer [1997]; Mestl, Bagley, and Glass [1997]; Bagley and Glass [1996]; Bhattacharjya and Liang [1996b]; Lynch [1995]) and also in problems associated with glassy and disordered materials (Derrida and Flyvbjerg [1986]; Derrida and Pomeau [1986]; Derrida [1987b]; Derrida and Flyvbjerg [1987a]). In these examples, the Boolean element might be an atomic spin or the state of excitation of a molecule. Kauffman models have even been applied to quantum gravity problems (Baillie and Johnston [1994]).

In some sense, the type of Boolean networks introduced by Kauffman can be considered as a prototype of generic dynamical system, as they present chaotic as well as regular behavior and many other typical structures of dynamical systems. In the thermodynamic limit $N \to \infty$, there can be "phase transitions" characterized by a critical line dividing chaotic from regular regions of state space. The study of the behavior of the system at and near the phase transitions, which are attained by changing the model-parameters, has been a very major concern.

As we shall describe in more detail below, these models are often studied in a version in which the couplings among the Boolean variables are picked randomly from some sort of ensemble. In fact, they are often called N-K models because each of the N elements composing the system, interact with exactly K others (randomly chosen). In addition, their coupling functions are usually picked at random from the space of all possible functions of K Boolean variables. Clearly this is a simplification of real systems as there is no particular problem which has such a generically chosen coupling. All real physical or biological problems have very specific couplings determined by the basic structure of the system in hand. However, in many cases the coupling structure of the system is very complex and completely unknown. In those cases the only option is to study the generic properties of generic couplings. One can then hope that the particular situation has as its most important properties ones which it shares with generic systems.

Another simplification is the binary nature of the variables under study. Nevertheless, many systems have important changes in behavior when "threshold" values of the dynamical variables are reached (e.g. the synapses firing potential of a neuron, or the activation potential of a given chemical reaction in a metabolic network). In those cases, even though the variables may vary continuously, the binary approach is very suitable, representing the above-below threshold state of the variables. The Boolean case is particularly favorable for the study of generic behavior. If one were to study a continuum, one would have to average the couplings over some rather complicated function space. For the Booleans, the function space is just a list of the different possible Boolean functions of Boolean variables. Since the space is enumerable, there is a very natural measure in the space. The averages needed for analytic work or simulations are direct and easy to define.

In addition to its application, the study of generic systems is of mathematical interest in and for itself.

1.1 Structure of Models

Any model of a Boolean net starts from N elements $\{\sigma_1, \sigma_2, \ldots, \sigma_N\}$, each of which is a binary variable $\sigma_i \in \{0,1\}$, $i = 1, 2, \ldots, N$. In the time stepping, each of these Boolean elements is given by a function of the other elements. More precisely, the value of σ_i at time $t+1$ is determined by the value of its K_i *controlling elements* $\sigma_{j_1(i)}, \sigma_{j_2(i)}, \ldots, \sigma_{j_{K_i}(i)}$ at time t. In symbols,

$$\sigma_i(t+1) = f_i\big(\sigma_{j_1(i)}(t), \sigma_{j_2(i)}(t), ..., \sigma_{j_{K_i}(i)}(t)\big)\,, \tag{1.1}$$

where f_i is a Boolean function associated with the ith element that depends on K_i arguments. To establish completely the model it is necessary to specify:

- the *connectivity* K_i of each element, namely, how many variables will influence the value of every σ_i;

- the *linkages* (or *couplings*) of each element, which is the particular set of variables $\sigma_{j_1(i)}, \sigma_{j_2(i)}, \ldots, \sigma_{j_{K_i}(i)}$ on which the element σ_i depends, and

- the *evolution rule* of each element, which is the Boolean function f_i determining the value of $\sigma_i(t+1)$ from the values of the linkages $\sigma_{j_1(i)}(t)$, $\sigma_{j_2(i)}(t)$, $\ldots$, $\sigma_{j_{K_i}(i)}(t)$.

Once these quantities have been specified, equation (1.1) fully determines the dynamics of the system. In the most general case, the connectivities K_i may vary from one element to another. However, throughout this work we will consider only the case in which the connectivity is the same for all the nodes: $K_i = K$, $i = 1, 2, \ldots, N$. In doing so, it is possible to talk about *the* connectivity K of the whole system, which is an integer parameter by definition. It is worth mentioning though that when K_i varies from one element to another, the important parameter is the *mean* connectivity of the system, $\langle K \rangle$, defined as

$$\langle K \rangle = \frac{1}{N} \sum_{i=1}^{N} K_i.$$

In this way, the mean connectivity might acquire non-integer values. Scale-free networks (Strogatz [2001]; Albert and Barabási [2002]), which have a very broad (power-law) distribution of K_i, can also be defined and characterized.

Of fundamental importance is the way the linkages are assigned to the elements, as the dynamics of the system both qualitatively and quantitatively depend strongly on this assignment. Throughout this paper, we distinguish between two different kinds of assignment: In a *lattice assignment* all the bonds are arranged on some regular lattice. For example, the K control elements $\sigma_{j_1(i)}, \sigma_{j_2(i)}, \ldots, \sigma_{j_K(i)}$ may be picked from among the $2d$ nearest neighbors on a d dimensional hyper-cubic lattice. Alternatively, in a *uniform assignment* each and every element has an equal chance of appearing in this list. We shall call a Boolean system with such a uniform assignment a *Kauffman net.* (See Figure 1.1.)

Of course, intermediate cases are possible, for example, one may consider systems with some linkages to far-away elements and others to neighboring elements. *Small-world networks* (Strogatz [2001]) are of this type.

For convenience, we will denote the whole set of Boolean elements $\{\sigma_1(t), \sigma_2(t), \ldots, \sigma_N(t)\}$ by the symbol Σ_t:

$$\Sigma_t = \{\sigma_1(t), \sigma_2(t), \ldots, \sigma_N(t)\}; \tag{1.2}$$

Σ_t represents then the state of the system at time t. We can think of Σ_t as an integer number which is the base-10 representation of the binary chain

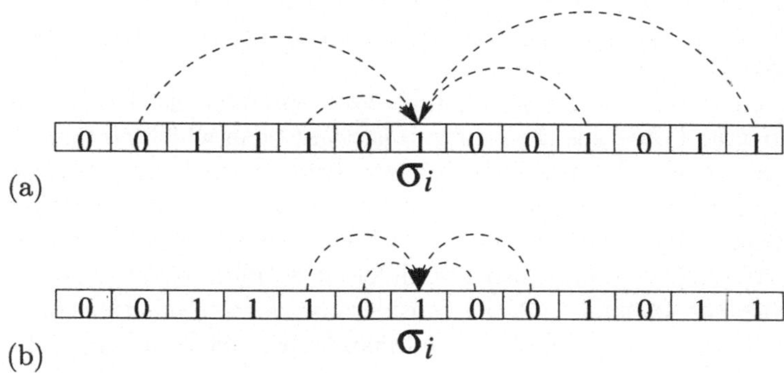

FIGURE 1.1. The different kinds of linkages in a one dimensional system. (a) In the Kauffman net the linkages of every element σ_i are chosen at random among all the other elements $\sigma_1 \ldots \sigma_N$. (b) In a completely ordered lattice, the linkages are chosen according to the geometry of the space. In the case illustrated in this figure, σ_i is linked to its first and second nearest neighbors.

$\{\sigma_1(t), \sigma_2(t), \ldots, \sigma_N(t)\}$. Since every variable σ_i has only two possible values, 0 and 1, the number of all the possible configurations is $\Omega = 2^N$, so that Σ_t can be thought of as an integer satisfying $0 \leq \Sigma_t < 2^N$. This collection of integers is the base-10 representation of the *state space* of the system. Although it is not essential for the understanding of the underlying dynamics of the network, this integer representation proves to be very useful in the implementation of computational algorithms used in numerical simulations (at least for small values of N).

A note of caution is relevant at this point. We should distinguish the purely Boolean model described in this work from Kauffman's N-K landscape model, which provides a description of fitness landscapes by including a fitness function to be optimized. We are not going to review on fitness landscapes since abundant literature already exists on this topic (Wilke, Ronnenwinkel, and Martinetz [2001]; Kauffman [1995, 1993]).

1.2 Coupling Functions

The arguments of the coupling functions $f_i(\sigma_{j_1(i)}, \ldots, \sigma_{j_K(i)})$ can take on 2^K different values. One specifies the functions by giving, for each of these values of the arguments, a value to the function. Therefore there are a total of

$$\aleph = 2^{2^K} \tag{1.3}$$

different possible functions. In Table 1.1 we give two examples of coupling functions for the case $K = 3$. There are $2^3 = 8$ configurations of the arguments $\sigma_{j_1(i)}$, $\sigma_{j_2(i)}$, $\sigma_{j_3(i)}$, and for each one of these configurations, the function f_i can acquire the values 1 or 0. For $K = 3$ there are $2^{2^3} = 256$ tables similar to the one shown in Table 1.1, one for each Boolean coupling function. Different tables differ in their assignments of 0's and 1's. If we assign a probability or weight to each of these functions, one gets an *ensemble* of possible couplings.

Possible ensemble choices abound. One ensemble used extensively by Kauffman and others is the *uniform distribution* in which all functions are weighted equally. Alternatively, a *magnetization bias*[1] may be applied by weighting the choice of functions with an outcome 0 with a probability p, and the outcome 1 with a probability $1-p$ (see, for example Bastolla and Parisi [1998b]). One may also give different weights to particular types of functions. For example, one can consider only *forcing functions* or *canalizing functions* (Stauffer [1987a]; Kauffman [1969, 1984]), in which the function's value is determined when just one of its arguments is given a specific value. The second function shown in Table 1.1 is a canalizing function. Another possibility is to specify the value of the function in order to simulate

[1] The word magnetization comes from the possibility of identifying each element with an atomic spin, which is a very small magnet.

σ_{j_1}	σ_{j_2}	σ_{j_3}	Random $f(\sigma_{j_1}, \sigma_{j_2}, \sigma_{j_3})$	Canalizing $f(\sigma_{j_1}, \sigma_{j_2}, \sigma_{j_3})$
0	0	0	0	1
0	0	1	1	1
0	1	0	1	1
0	1	1	0	1
1	0	0	1	0
1	0	1	0	1
1	1	0	1	0
1	1	1	1	0

TABLE 1.1. Illustration of two Boolean functions of three arguments. The first function is a particular random function, whereas the second one is a canalizing function of the first argument σ_1. When this argument is 0, the output of the function is always 1, while if $\sigma_1 = 1$, the output can be either 0 or 1.

the additive properties of neurons (Bornholdt and Rohlf [2000]; Genoud and Metraux [1999]; Cheng and Titterington [1994]; Wang, Pichler, and Ross [1990]; Kürten [1988a]; Derrida, Gardner, and Zippelius [1987]).

Here we enumerate some of the coupling functions occurring for different values of the connectivity K.

- For $K = 0$ there are but two functions, corresponding to the two possible values of a Boolean variables: tautology $f = 1$ and contradiction $f = 0$. Together these functions form a class which we might call $\mathcal{A}$.
- For $K = 1$, in addition to the class $\mathcal{A}$, there exists another class $\mathcal{B}$ in which $f(\sigma)$ can take on the value σ, called identity, and the value $\neg\sigma$, called negation. Thus there are a total of four functions, represented as columns in Table 1.2.

	Class $\mathcal{A}$		Class $\mathcal{B}$	
σ	$\mathcal{A}_0$	$\mathcal{A}_1$	$\mathcal{B}_I$	$\mathcal{B}_N$
0	0	1	0	1
1	0	1	1	0

TABLE 1.2. Boolean functions for $K = 1$. The first two functions form the class $\mathcal{A}$ of constant functions, $\mathcal{A}_0(\sigma) = 0$ and $\mathcal{A}_1(\sigma) = 1$. The other two functions form class $\mathcal{B}$ which consist in identity $\mathcal{B}_I(\sigma) = \sigma$ and negation $\mathcal{B}_N(\sigma) = \neg\sigma$.

- The situation for $K = 2$ has been particularly carefully studied. Here there are four classes of functions $f(\sigma_1, \sigma_2)$ (Lynch [1993b]; Coppersmith, Kadanoff, and Zhang [2001a]). Each class is invariant under making the interchange $0 \leftrightarrow 1$ in either arguments or value of f. The classes are $\mathcal{A}$ (two constant functions), $\mathcal{B}_1$ (four canalizing functions which depend on one argument), $\mathcal{B}_2$ (eight canalizing functions which depend on two arguments), $\mathcal{C}$, (two non-canalizing functions). These functions are explicitly shown in Table 1.3.

σ_{j_1}	σ_{j_2}	Class $\mathcal{A}$		Class $\mathcal{B}_1$				Class $\mathcal{B}_2$								Class $\mathcal{C}$	
0	0	1	0	0	1	0	1	1	0	0	0	0	1	1	1	1	0
0	1	1	0	0	1	1	0	0	1	0	0	1	0	1	1	0	1
1	0	1	0	1	0	0	1	0	0	1	0	1	1	0	1	0	1
1	1	1	0	1	0	1	0	0	0	0	1	1	1	1	0	1	0

TABLE 1.3. Boolean functions for the case $K = 2$. The 16 functions can be arranged in four different classes which differ in their symmetry properties (see text).

Several calculations have been done by giving different weights to the different classes (see for example Lynch [1995]; Stauffer [1987a]) .

1.3 The Updates

Once the linkages and the f_i's are given, one is said to have defined a *realization* of the model. Given the realization, one can define a dynamics by using equation (1.1) to update all the elements at the same time. This is called a *synchronous* update. In this paper, we assume a synchronous update unless stated otherwise. Alternatively, one may have a *serial* model in which one updates only one element at a time. This element may be picked at random or by some predefined ordering scheme.

Additional choices must be made. One can:

1. Keep the same realization through all time. We then have a *quenched* model.

2. Pick an entirely new realization after each time step. The model is then said to be *annealed.*[2]

3. Employ a *genetic algorithm* in which the system slowly modifies its realization so as to approach a predefined goal (Bornholdt and Sneppen [2000]; Stern [1999]; Stauffer [1994]).

4. Intermediate choices are also possible (Baillie and Johnston [1994]; Bastolla and Parisi [1996]).

Almost always, we shall regard the *real system* as one which updates synchronously and which is quenched so that the interactions are fixed for all time. The annealed and sequential models will be regarded as approximations which can provide clues to the behavior of this "real system". The

[2] These terms have been borrowed from the physics of alloys in which something which is cooled quickly so that it cannot change its configuration is said be be quenched, and something which is held at a high temperature for a long time so that it can respond to its environment is described as annealed. Hence these terms are applied to situations in which one wants to distinguish between problems with fixed versus changing interactions.

quenched model has time-independent dynamics describing motion within the state space of size $\Omega = 2^N$. One iterates the model through time by using equation (1.1) and thereby obtains a dynamics for the system. Each of the Ω different initial conditions will generate a motion, which will eventually fall into a cyclical behavior.

1.4 Symmetry Properties

Typically, each individual realization of these models shows little or no symmetry. However, the average over realizations has quite a large symmetry group, and the symmetry is important to model behavior. For example, the random mapping model (Harris [1960]), which is the $K \to \infty$ limit of the N-K model of the Kauffman net, has a full symmetry under the interchange of all states forming the state space. For finite values of K, the Kauffman net is symmetric under the interchange of any two basic elements. One can also have a symmetry under the interchange of the two values of each individual element if one chooses the couplings at random, or with the appropriate symmetry. One can use dynamics that have reversal symmetry (Harris [1960]; Coppersmith, Kadanoff, and Zhang [2001a,b]), and that choice will have a profound effect upon the structure of the cycles.

1.5 Outline of the Paper

To define fully the object of study, one must describe the dynamical process and the class of realizations that one wishes to study. For example, one can fix N and K, and study all of the properties of all realizations of frozen systems with those values. One might pick a realization at random among all possible linkages and functions, develop its properties. Then one would pick another realization, and study that. Many such steps would give us the properties of the frozen system averaged over realizations with a given N and K. What properties might we wish to study?

In the next section, we describe the gross information transfer through the system by describing how the system will respond to a change in initial data or couplings. There are three different *phases* that have qualitatively different information transfer properties. We argue that the Kauffman net, which can transfer information from any element to any other, is qualitatively different from lattice systems, in which the information transfer occurs through a d-dimensional space. We argue that the information transfer on the lattice is qualitatively, and even quantitatively, similar to the kind of information flow studied in percolation problems.

Section 3 is concerned with the temporal recurrence in the network as reflected in the statistical properties of its cycles. Here we base our arguments upon a discussion of two important limiting cases, $K = 1$, and very large K. The first case is dominated by situations in which there are a few short linkage loops. In the second, the Kauffman net shows a behavior which can be analyzed by comparison with a random walk through the state space. The distribution of cycle lengths is qualitatively different from any of the quantities that are commonly studied in percolation problems. So the results characterizing the cycles are substantially different from the behaviors usually studied in phase transition problems. We argue in addition that the cycles of Kauffman nets and of networks on d-dimensional lattices differ substantially.

Generic Kauffman models are dissipative in the sense that several different states may be mapped into one. Consequently, information is lost. Nonetheless, not all kind of problems are generic. In Section 4 we analyze the dynamics of *time-reversible* Kauffman models, in which every possible state is in exactly one cycle. Any cycle can be traversed equally well forward or backward. A phase transition also occurs in the time-reversible models, but as we will see, time-reversible orbits have quite different properties from dissipative ones. For example, in the chaotic phase, the number of different attractors in the reversible case grows exponentially with the system size, whereas in the dissipative case it grows linearly. Also, the interplay between the discrete symmetry and quenched randomness in the time-symmetric models can lead to enormous fluctuations of orbit lengths.

In Section 5 we discuss the robust properties of the N-K models. We argue that, due to the very large fluctuations present in such important quantities as the number of different cycles or the cycle lengths, a characterization of these models just via cycle properties does not reflect the robustness of the dynamics of the system. Therefore, another types of characterizations are needed. In particular, we focus our attention on a characterization via noise, initially proposed almost simultaneously by Golinelli et. al. and Miranda et. al. (Golinelli and Derrida [1989]; Miranda and Parga [1989]). By considering the time it takes for two trajectories in the state space to cross, it is apparent that the the stochastic dynamics of the Kauffman model evolving under the influence of noise is very robust in the sense that different Kauffman models exhibit the same kind of behavior.

As we already mentioned at the beginning of this section, the spectrum of applications of the Kauffman model comprises a very wide range of topics and fields. Therefore, in Section 6 we review only on a few of the applications (genetic networks and cell differentiation, evolution, organization in social systems and neural networks) which, from our point of view, stick out from this spectrum of varied and imaginative ideas.

2 Information Flow

2.1 Response to Changes

The first thing to study in an N-K model is its response to changes. This response is important because the actual values of the elements often do not matter at all. If, for example, we pick the functions f_i at random among the class of all Boolean functions of K variables, then the ensemble is invariant under flipping the value of the ith element. In that case, only changes matter, not values.

In computer studies, such changes can be followed quite simply. One follows a few different time developments of systems that are identical except for a small number of selected changes in the coupling functions or initial data, and sees how the differences between the configurations change in time. One can do this for two configurations or for many, studying pairwise differences between states, or things which remain identical across all the time-developments studied.

2.1.1 Hamming Distance and Divergence of Orbits

For simplicity, imagine starting out from two different possible initial states:

$$\Sigma_0 = \{\sigma_1(0), \sigma_2(0), \ldots, \sigma_N(0)\} \quad \tilde{\Sigma}_0 = \{\tilde{\sigma}_1(0), \tilde{\sigma}_2(0), \ldots, \tilde{\sigma}_N(0)\} \tag{2.1}$$

which differ in the values of a few elements. One can then observe the time-development of these configurations under the same dynamics, finding, for example, the distance $D(t)$ between the configurations as a function of time

$$D(t) = \sum_{i=1}^{N} \left(\sigma_i(t) - \tilde{\sigma}_i(t)\right)^2 . \tag{2.2}$$

If the transfer of information in the system is localized, this distance will never grow very large. If however, the system is sufficiently chaotic so that information may be transferred over the entire system, then in the limit of large N this Hamming distance can diverge for large times. Another interesting measure is the normalized overlap between configurations, $a(t)$, defined as

$$a(t) = 1 - N^{-1} D(t) . \tag{2.3}$$

One will wish to know whether a goes to unity or a lesser value as $t \to \infty$. If the overlap always goes to unity, independently of the starting states, then the system cannot retain a nonzero fraction of the information contained in its starting configuration. Alternatively, when $a(\infty)$ is less than unity, the system "remembers" a nonzero fraction of its input data.

2.1.2 Response to Damage

So far, we have considered the system's response to changes of the initial data. One can also attack the quenched problem by considering two systems, each with the same initial data, but with a few of the linkages varied. Then one can ask: given such "damage" to the system, how much do the subsequent states of the system vary? Do they become more and more alike or do they divergc? What is the likelihood of such a divergence?

These considerations of *robustness*—both to damage and to variation in initial data—are very important for the evaluation of the effectiveness of a network, either for computations or as part of a biological system. There have been fewer studies of the effect of damage than that of initial data. Usually the two types of robustness occur together (Luque and Solé [2000]; De Sales, Martins, and Stariolo [1997]). Damage has been studied for itself (Stauffer [1994]; Corsten and Poole [1988]).

2.2 Percolation and Phase Behavior

2.2.1 Percolation of Information

In the limiting case in which N approaches infinity, the different types of N-K models all show three different kinds of phases, depending upon the form of information transfer in the system. If the time development transfers information to a number of elements that grows exponentially in time, the system is said to be in a *chaotic phase.* Typically, this behavior occurs for larger values of K, up to and including $K = N$. If, on the other hand, a change in the initial data typically propagates to only a finite number of other elements, the system is said to be in a *frozen phase.* This behavior will arise for smaller values of K, most especially $K = 0$, and usually $K = 1$. There is an intermediate situation in which information typically flows to more and more elements as time goes on, but this number increases only algebraically. This situation is described as a *critical phase.*

When the linkages and the hopping among configurations are sufficiently random, one can easily perform a quite precise calculation of the boundary which separates these phases. Imagine starting with a state Σ_0, containing a very large number, N, of Boolean elements, picked at random. Imagine further another configuration $\tilde{\Sigma}_0$ in which the vast majority of the elements have the same value as in Σ_0, but nevertheless there are a large number of elements, picked at random, which are different. The Hamming distance at time zero, $D(0)$, is the number of changed elements. Now take the simplest N-K system in which all the linkages and the couplings are picked at random. On average, a change in a single element will change the argument of K functions, so there will be $KD(0)$ functions affected. Each of these will have a probability one half of changing their value. (The functions after all are quite random.) Thus the Hamming distance after the first time step will be

$$D(1) = 0.5KD(0)\,.$$

If the couplings and connections are sufficiently random, then at the start of the next step, the newly changed elements and their couplings will remain quite random. Then the same sort of equation will apply in the next time step, and the next. Just so long as the fraction of changed elements remains small, and the randomness continues, the Hamming distance will continue to change by a factor of $K/2$ so that

$$D(t+1) = 0.5KD(t),$$

which then has the solution

$$D(t) = D(0)\exp\left[t\ln(0.5K)\right]. \tag{2.4}$$

For $K > 2$ the number of changed elements will grow exponentially, for $K < 2$ it will decay exponentially, and for $K = 2$ there will be neither exponential growth nor decay, and the behavior will be substantially influenced by fluctuations. Thus, by varying the value of the connectivity, the system sets down into one of the three following phases:

- *Chaotic* ($K > 2$), the Hamming distance grows exponentially with time.
- *Frozen* ($K < 2$), the Hamming distance decays exponentially with time.
- *Critical* ($K_c = 2$), the temporal evolution of the Hamming distance is determined mainly by fluctuations.

In deriving equation (2.4) we have assumed that the coupling functions f_i of the system acquire the values 0 and 1 with the same probability $p = \frac{1}{2}$. Nonetheless, as we will see below, the chaotic, frozen and critical phases are also present in the more general case in which the coupling functions f_i evaluate to 0 and 1 with probabilities p and $1-p$ respectively. For a given value of p, there is a critical value $K_c(p)$ of the connectivity below which the system is in the frozen phase and above which the chaotic phase is attained. Conversely, for a given connectivity $K \geq 2$, a critical value $p_c(K)$ of the probability bias separates the chaotic and the frozen phases.

The behavior of the normalized Hamming distance, $D(t)/N$, can be seen in Figures 2.1 and 2.2, which respectively are for the Kauffman net and a two-dimensional lattice system. In both cases the system has $N = 10^4$ elements and the connectivity is $K = 4$. The linkages of every element of the Kauffman net are chosen randomly, whereas in the two-dimensional lattice each element receives inputs from its four nearest neighbors. Both figures contain three curves, with the parameter p picked to put the systems into the three different phases. For the Kauffman net $p_c = (1-\sqrt{1/2})/2$ (see equation (2.8) below). The value of p_c is not very well known for the two-dimensional lattice, but the best numerical estimations indicate that $p_c \simeq 0.29$ for an infinite lattice (Stauffer [1988]; Stölzle [1988]; Weisbuch and Stauffer [1987]; Stauffer [1987b]; Derrida and Stauffer [1986]). For finite lattices the value of p_c has been defined as the average over realizations of the value of p at which a first cluster spanning the whole net appears (Lam [1988]). For a 100×100 lattice this value is $p_c \simeq 0.27$.

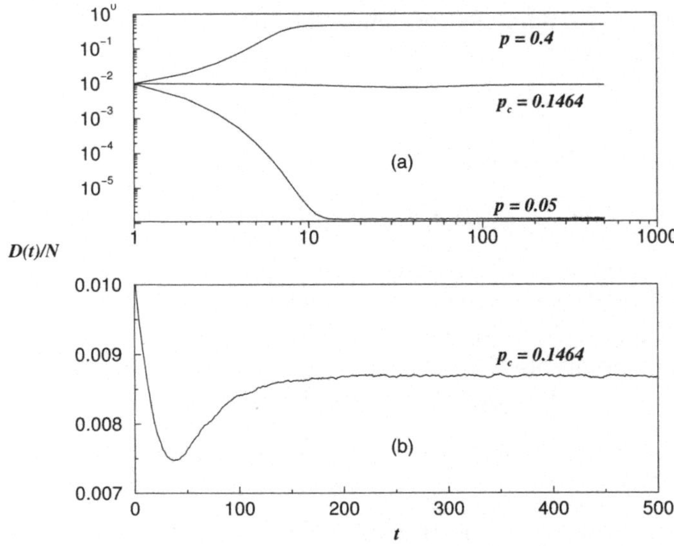

FIGURE 2.1. Hamming distance as a function of time for a Kauffman net composed of $N = 10000$ elements and connectivity $K = 4$. (a) Log-log graph showing the Hamming distance for the three different regimes of the system: frozen ($p = 0.05$), critical ($p_c = \frac{1}{2}(1 - \sqrt{1/2}) \simeq 0.1464$) and chaotic ($p = 0.4$). In all the cases the initial Hamming distance was $D(0) = 100$. (b) Hamming distance for the critical phase ($p = p_c$) but in a non-logarithmic graph. Note that the Hamming distance initially decreases, and then it rises again to saturate at a constant value that depends weakly on system size.

In the frozen phase the distance has an initial transient but then quite quickly approaches an asymptotic value. In the chaotic phase the distance shows an exponential rise followed by a similar saturation. These behaviors are almost the same for both the Kauffman net and the two-dimensional lattice. On the other hand, in the critical phase the behavior of the Hamming distance is very different in these two systems. In the Kauffman net the distance initially decreases and then increases again, asymptotically approaching a constant value that depends weakly on system size. In contrast, for the lattice the Hamming distance initially grows and then saturates. We will see later that, within the framework of the annealed approximation, the normalized Hamming distance for an infinite Kauffman net approaches zero monotonically in both the frozen and the critical phases (exponentially in the frozen phase, and algebraically in the critical phase). As far as we can tell, the non-monotonic behavior of the Hamming distance in finite systems at K_c shown in Figure 2.1b has not yet been explained.

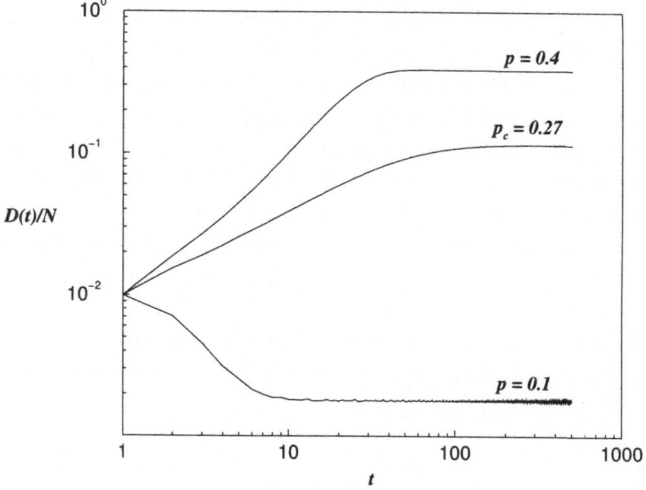

FIGURE 2.2. Hamming distance in a two-dimensional lattice composed of $N = 100 \times 100$ elements. Every node in the lattice receives inputs from its four first nearest neighbors ($K = 4$). The three curves (in log-log scale) show the behavior of the Hamming distance in the three regimes: frozen ($p = 0.1$), critical ($p_c = 0.27$) and chaotic ($p = 0.4$). Note that in the critical phase the Hamming distance initially increases algebraically and then saturates at a constant value that depends on system size.

2.2.2 Limitations on the mean field calculation

Let us look back at the argument which led to equation (2.4). Calculations like this, in which actual system properties are replaced by average system properties are in general called "mean field" calculations.

Naturally, the results derived as equation (2.4) depend crucially upon the assumptions made. The derivation follows from the assumption that the f_i's in each step are effectively random. (See also the derivations of equations (2.6) and (2.7) below, which also depend upon the randomness assumption.) The randomness will certainly be correct in the annealed situation, in which the couplings are reshuffled in each step. It will also be true in Kauffman net in the limiting situation in which $K = \infty$. In that case, information is spread out over the entire system and thus has a very small chance of correlation among the f_i's. The Kauffman net has a likely configuration that permits the replacement of the actual values of the f_i's by their statistical distribution (Hilhorst and Nijmeijer [1987]). However, the approximations used here will not always work. Specifically, they fail in all kinds of finite N situations, or in situations in which the linkages are arranged in a finite-dimensional lattice. In that case, the assumed randomness of the f_i does not hold, because their arguments are not random, and the derived equations will not work. To see this in the simplest example choose $K = N = 1$ with quenched couplings. A brief calculation shows that for any one of the four possible f_i's, after a one-step initial transient $a(t+2) = a(t)$. That does not agree with equation (2.6) derived below. In fact, for any finite dimension and linkages which involve short-range couplings, the overlap is not unity at long times even in the frozen phase.

More generally, if the system is put onto a finite dimensional lattice, or if the functions are not picked at random, or if the initial elements are not random, couplings initially used can be correlated with couplings used later on. Then the information transfer will be different and equation (2.4) will fail.

However, the principle that there can be a phase transition in the kind of information flow remains quite true for a d-dimensional lattice, and for other ensembles of coupling functions.

2.2.3 Connections to percolation problems

The transfer of information just described is quite similar to the transfer which occurs in a typical phase transition problem. Generically, these problems have three phases: ordered, critical, and disordered (Ma [1976]; Kadanoff [2000]). The bare bones of such an information transfer problem is described as a *percolation problem* (Stauffer [1985]).

In one sort of percolation problem one considers a lattice. Each bond or link of the lattice is picked to be either connected or disconnected. The choice is made at random and the probability of connecting a given link is picked to be q. Now one asks about the size of the structure obtained by considering sets of links all connected with one another. For small values of q, these sets tend to be small and isolated. As q is increased, the clusters tend to get larger. At sufficiently large values of q, one or more connected clusters may span the entire lattice. There is a critical value of the probability, denoted as q_c, at which such a spanning cluster just begins to form. Properties of the large clusters formed near that critical point have been studied extensively (Stauffer [1985]). The resulting behavior is "universal" in that for an isotropic situation, the critical properties depend only upon the dimensionality d of the lattice, at least when d is less than four. For $d > 4$, the percolating system obeys mean field equations. When the symmetry properties change, the critical behavior does change. For example, a system with directed percolation (Stauffer [1985]; Owezarek, Rechnitzer, and Guttmann [1997]) has the information flow move preferentially in a particular direction. The critical behavior of directed percolation is different from that of ordinary percolation. It is attractive to explore the connection between the phase transition for percolation, and the one for the N-K model.

Several authors have constructed this connection in detail. For example, Hansen [1988a] looked at the N-K model on lattices for $2, 3$ and 4 dimensions and demonstrated numerically that the phase transition occurred when the information transition probability reached the critical value for percolation on the corresponding lattice. Stölzle [1988] showed the connection to percolation for both sequential and parallel updating for a two dimensional lattice. However, Hansen [1988b] took very special forms of the connections, using only rules with connectivities of the form of a logical "or". This did not result in behavior like simple percolation but instead quite a different phase transition problem, related to the behavior of diodes and resistors. At roughly the same time, Stauffer [1988] indicated a close numerical correspondence to the two dimensional percolation problem both in p_c and also in the fractal dimension of the spanning cluster. (For p_c see also Lam [1988].) He also defined an order parameter, essentially a Hamming distance, that, when plotted as a function of $(p - p_c)$, looked just like a typical critical phenomena result. He argued that the N-K model is equivalent to directed percolation. More specifically, Obukhov and Stauffer [1989] argued that the quenched problem has a critical point which is in the same "universality class" (Kadanoff [2000]) as directed percolation. This would imply that the critical point is essentially the same as that of the directed percolation problem. The qualitative properties of both the ordered and the frozen phases would also be essentially similar in the percolation case and the N-K lattice. In the latter situation, the preferred direction would be that of the "time" axis. The structure of both would vary with dimensionality and become like that of mean field theory above four dimensions. This is the same mean field theory which describes the Kauffman net. Thus, the behavior of information transfer in N-K problems was mostly understood in terms of percolation.

2.3 Lattice versus Kauffman net

We can now point to an important difference between systems in which all elements are coupled to all others, as in the Kauffman net, and lattice systems in which the elements which are "close" to one another are likely to be mutually coupled. "Closeness" is a reciprocal relation. If a is close to b, then b is also close to a. Therefore, close elements are likely to be coupled to one another and thereby form a closed *linkage loop*. Any large-N lattice system might be expected to form many such loops. When K is small, different spatial regions tend to be unconnected and so many different modules will form[3]. The dynamics of the elements in different modules are independent. In contrast, in a Kauffman net, influence is not a reciprocal relation. If element σ_j appears in the coupling function f_i associated with element σ_i, there is only a small chance, proportional to (K/N), that σ_i will

[3]A module is a loop or more complex topology of dependencies in which all the functions are non-constant, plus all other elements that are influenced by that structure. See Section 3.1.

appear in f_j. For large N and small K, the probability that a given element will participate in a linkage loop will be quite small, so there will then be a small number of modules. When K is small, the number of modules in uniformly coupled systems grows more slowly than the system size, while in lattice systems the number of modules is proportional to the size of the system. This distinction will make a large difference in the cycle structure.[4] For the flow of information the difference between the two kinds of nets is more quantitative than qualitative. One can see the similarity between them by comparing the curves shown in Figures 2.1 and 2.2.

2.4 Calculations of overlap and divergence

Before coming to a careful description of the phases, we should describe more fully the kinds of calculation of overlap that can be performed. Equation (2.4) is just the beginning of what can be done with the trajectories of states in this problem. In fact, exactly the same logic which leads to that equation can give a much more general result. If the overlap between two states at time t is $a(t)$, and if the elements which are different arise at random, then the probability that the arguments of the function f_i will be the same for the two configurations is

$$\rho = [a(t)]^K . \tag{2.5}$$

If all arguments are the same, then the contribution to the overlap at time $t+1$ is $1/N$. (The N arises from the normalization of the overlap.) If one or more arguments of the coupling function are different in the two configurations, and the functions f_i are picked at random, then the chance of having the same functional output is $1/2$ and the contribution to the overlap is $1/(2N)$. Since there are N of such contributions, weighted by ρ and $1-\rho$ respectively, the equation for the overlap defined by equation (2.3) is

$$a(t+1) = \tfrac{1}{2}\left[1 + [a(t)]^K\right] . \tag{2.6}$$

There are several possible variants of this equation. For example, if the different outcomes 0 and 1 of the function f_i are weighted with probabilities p and $1-p$ respectively, to produce a sort of magnetization bias, then equation (2.6) is replaced by (Derrida and Pomeau [1986]; Derrida and Stauffer [1986])

$$a(t+1) = 1 - \left[1 - [a(t)]^K\right]/K_c , \tag{2.7}$$

---

[4] The interested reader will recall that in quantum field theory and statistical mechanics, mean field theory appears in a situation in which fluctuations are relatively unimportant. This will arise when the analog of linkage loops make a small contribution. Then, the physicist puts the effect of loops back into the problem by doing what is called a *loop expansion*. Since they both expand in loops, the percolation mean field theory and the mean field theory of the Kauffman net are essentially the same.

where K_c is given in terms of p as

$$K_c = 1/[2p(1-p)]. \tag{2.8}$$

In the limit $t \to \infty$, $a(t)$ asymptotically approaches the fixed point a^*,

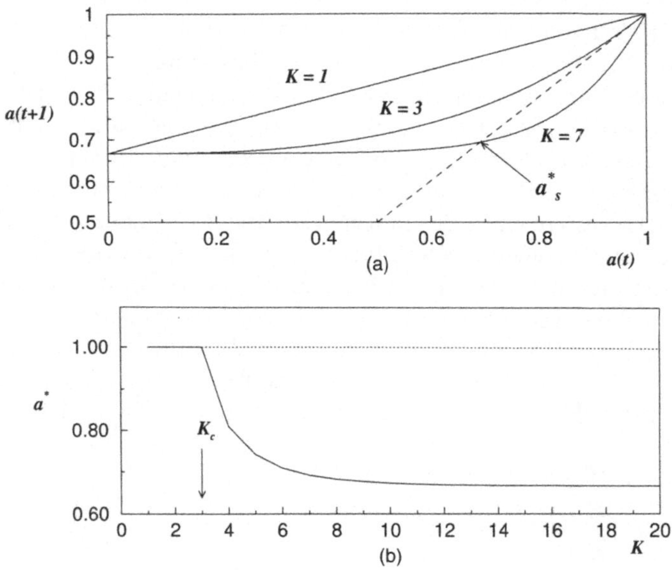

FIGURE 2.3. (a) The mapping $F(a) = 1 - [1 - a^K]/K_c$ (see Eq. (2.7)) for $K_c = 3$ and three different values of K (solid curves), corresponding to the three different phases of the system. The dotted line is the identity mapping. (b) Bifurcation diagram of equation (2.9). For $K \leq K_c$ the only fixed point is $a^* = 1$. For $K > K_c$ the previous fixed point becomes unstable and another stable fixed point a_s^* appears.

which obeys, from equation (2.7)

$$a^* = 1 - \left[1 - [a^*]^K\right]/K_c\,. \tag{2.9}$$

We might expect equation (2.9) to reflect the three-phase structure of the problem, and indeed it does. Figure 2.3a shows the graph of the mapping $F(a) = 1 - [1 - a^K]/K_c$ for different values of K, and Figure 2.3b shows the bifurcation diagram of equation (2.9). Both graphs were calculated with p chosen so that $K_c = 3$. As can be seen, if $K \leq K_c$ there is only one fixed point $a^* = 1$, whereas for $K > K_c$ the fixed point $a^* = 1$ becomes unstable as another stable fixed point, $a_s^* \neq 1$, appears. The value of the infinite time overlap a^* describes the fraction of elements whose value is sensitive to the cycle entered.

When $K > K_c$, the system is chaotic; a^* is less than one even when the starting points are quite close to one another. This reflects the fact that the system has a long-term behavior which can include some cyclic oscillations, the initial data determining the cycle entered. (We discuss the cycles in detail in the next section.) As K approaches K_c from above, a^* increases since fewer elements have final values which are sensitive to the initial data. On the other hand, for $K \leq K_c$, the infinite time overlap is exactly one and therefore the proportion of elements whose final value depends upon the starting point is precisely zero. Thus, independently of the starting point, the system is always stuck in essentially the same configuration. This surprising result pertains to the Kauffman net. In contrast, for any lattice system with $K > 0$ the final overlap is less than unity, reflecting the fact that the system can store information about initial data in some finite proportion of its elements. Such storage is impossible for the Kauffman net.

It is worth noticing that for a given value of K, the system can be put in one of the three different phases by adjusting the value of the probability bias p. The critical value of p is then obtained by inverting equation (2.8), which gives the *critical line* separating the frozen phase from the chaotic phase in the p–K parameter space, as illustrated in Figure 2.4.

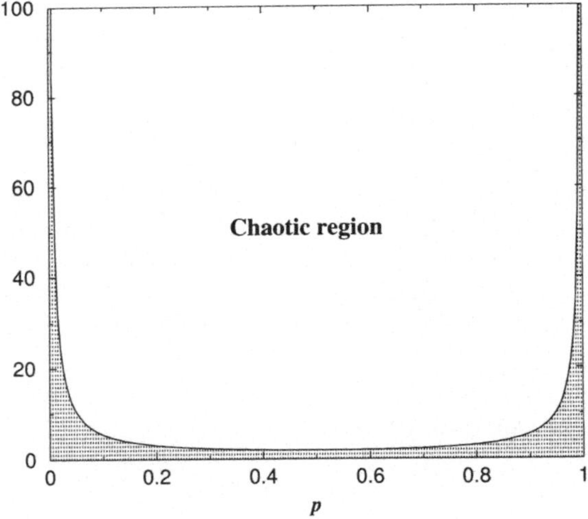

FIGURE 2.4. Phase diagram for the N-K model. The shaded area corresponds to the frozen phase, whereas the upper region corresponds to the chaotic phase. The curve separating both regions is the critical phase $K_c = [2p(1-p)]^{-1}$.

The rigidity of the Kauffman net was emphasized by Flyvbjerg (Flyvbjerg [1988, 1989]), who wished to establish the equation for the *stable core*, the set of variables whose values at time infinity do not depend upon initial data (see also Lynch [1995, 1993a,b]; Bastolla and Parisi [1998b].) He calculated the time dependence of the proportion of variables which had already settled down, and thereby found a closed form for the size of the core. He found that in the ordered state of the Kauffman net the fraction of elements in the core is unity.

Another kind of rigidity is studied by counting the *weak elements*. In a particular realization, elements are termed weak if changing their value never affects the long-term behavior of the system. Many authors have studied these elements (see for example Lynch [1995, 1993a,b]). In the Kauffman net, but not in the lattice system, the proportion of weak elements is unity throughout the frozen phase.

3 Cycle Behavior

For quenched systems the evolution functions f_i are independent of time. For any initial condition, the system eventually returns to a previously visited point and then cycles repeatedly. The state space, which consists of the $\Omega = 2^N$ configurations of the system, breaks up into different cycles and their basins of attraction, as shown schematically in Figure 3.1. Each initial condition will eventually fall in one of these cycles, which are determined by the evolution functions and the linkages.

The description of a cycle is, in some sense, much more complex than the description of orbit separation. In separation, one is dealing with a very gross property of the system: Different parts of it can behave independently and additively in the separation process. We utilized this fact in our calculations of overlaps and Hamming distances carried out in sections 2.2.1 and 2.4 above. On the other hand, to get a cycle to close, each and every element must simultaneously return to its value at a previous time-step. This is a much more delicate process and makes for much more delicate calculations. As we saw, information flow in N-K systems is closely analogous with the well-studied behavior of percolation problems. In contrast, the behavior of cycles in N-K models is something special, not closely analogous to any of the usual calculations for the usual systems of statistical mechanics. When the N-K system forms cycles, one can ask many quantitative questions. One might wish to know the number of steps the system takes before it falls into a cycle (called the *transient* time), and about the length of the cycle. For each cycle, there is a basin of attraction, which is the set of initial states which eventually fall into that cycle. Therefore, one can develop a quantitative description of the distribution of cycle lengths, transient lengths, basin sizes, etc. either within a given realization or averaged over realizations.

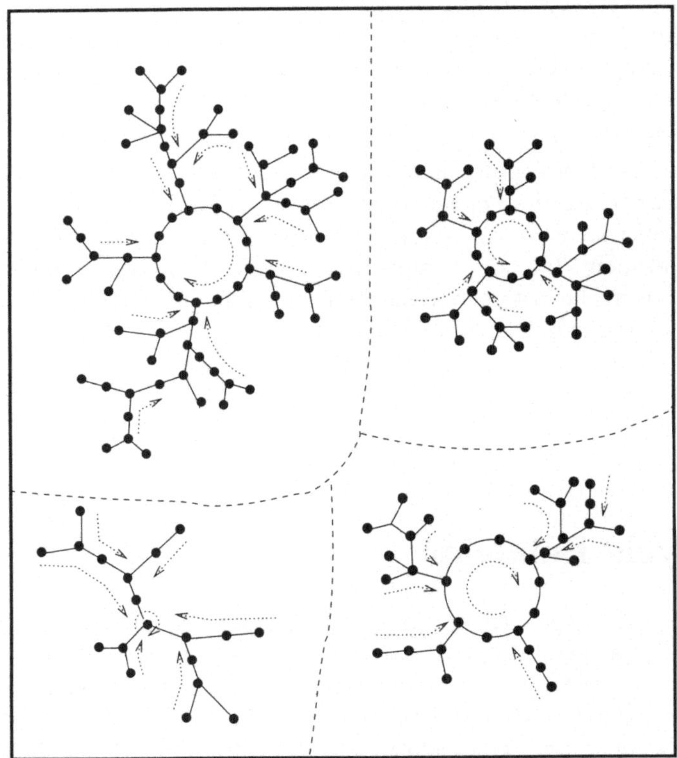

FIGURE 3.1. Schematic representation of the state space of the N-K model. Each state is represented as a bold point. Under the quenched dynamics, the state space is broken down into several cycles, or attractors, represented as circles. Each initial state eventually will end up in one of these cycles (the arrows show the direction of the flow). The totality of states which evolve towards a given cycle, is the basin of attraction of that cycle. Note that there can be attractors consisting of only one point, corresponding to a cycle of period 1.

3.1 Linkage Loops and Cycles

The structure of the linkages is an important ingredient in determining the nature of the cycles in the system. Each element can appear in the coupling functions for other elements. These in turn can appear in the couplings of others. For each element one can trace out its *descendants*, i.e., the elements it affects. Similarly, one can chain backward and find the *ancestors* for each element, i.e., the elements which affect it. For a cycle of length longer than one to exist, at least one element must be its own ancestor, and thus its own descendant. (If no such element existed, one could trace back and find elements with no ancestors. They would then have fixed values and be stable elements. The elements controlled by only them would also be stable. The line of argument would go forward until it was found that all elements were stable.) Figure 3.2 illustrates the idea of ancestors and descendants for the case $K = 1$. As can be seen, the connections between different elements of the network give rise to *linkage loops* and *linkage trees*, each tree being rooted in a loop.

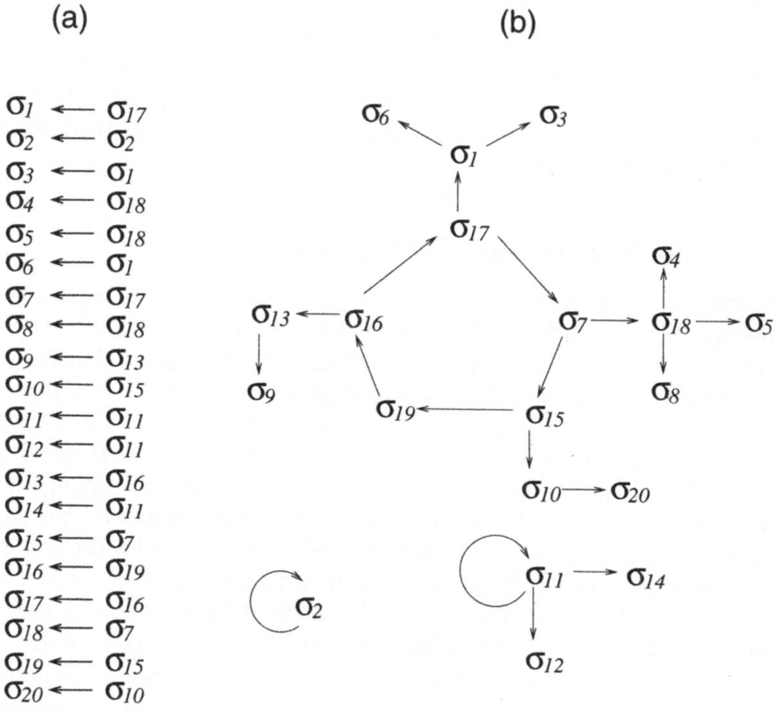

FIGURE 3.2. Linkage loops for an N-K net of $N = 20$ elements and connectivity $K = 1$. (a) Particular realization of linkage assignment in the net. The first column is the list of the N elements of the net, $\{\sigma_1, \sigma_2, \ldots, \sigma_N\}$, and the second column shows the particular linkage every element has been assigned. (b) Schematic representation of the linkage loops. Each arrow points towards the descendants of a given element. In this particular realization there are three modules, one of which consits of only one element, σ_2.

Flyvbjerg and Kjaer [1988] stress the importance of elements which are their own ancestors by pointing out that any unstable element must be influenced by at least one such element. Of particular relevance are those elements which belong to a loop in which there are only non-constant functions. These self-influencing elements, and the unstable elements that they influence, are called the *relevant elements* of the system. They naturally form themselves into groups called *modules* (Thieffry and Romero [1999]; Bastolla and Parisi [1998a]; Zawidzki [1998]). Different modules do not influence one another and fully determine the cycle structure. Notice that the chain of linkages by which an element can influence itself, namely its linkage loop, is completely defined by the linkages, i.e. by the specific assignment of the different elements in the f_i's. Linkage loops do not describe the functions themselves. Consequently, such loops are necessary but not sufficient for the existence of non-trivial cycles. Only loops formed by relevant elements (with non-constant functions) are important in determining the properties of the limit cycles of the network. The number of such cycles and their lengths depend crucially on the modular organization of the relevant variables (Flyvbjerg and Kjaer [1988]; Bastolla and Parisi [1998b,a]). Consequently, linkage loops and relevant elements will form an important part of our further discussions.

3.2 Phase Transitions for Cycles.

In the previous chapter, we saw that N-K systems fall into different phases, depending upon how effectively they transfer information. The cycles can be quite different in the different phases. The chaotic phase is characterized by very long cycles, in which the typical cycle-lengths grow as a power of the size, $\Omega = 2^N$, of the entire state space. Each cycle includes the participation of many different elements which undergo a strongly correlated dynamics. In this phase, the transients are similarly long and complex. In contrast, the frozen phase tends to have much shorter and simpler cycles, and also shorter transients in which the individual modules do not grow as the system gets larger. If the system sits on a lattice of low dimension, the different modules are spatially localized. Each module shows a strong correlation in its dynamics, but different modules move independently. The critical phase has larger modules and has a behavior intermediate between the frozen and the chaotic phase. These behaviors are known in considerable detail.

The remainder of this section describes in detail the behaviors of the different phases. We start by discussing the exactly solvable models which give a solid description of the limiting behavior of the phases. We then describe how information obtained from simulations and the exact solutions can be pieced together to give a qualitative description of the phases.

3.3 Soluble Models.

3.3.1 Independent Subsystems

Here we do a preliminary calculation which will be of use in interpreting results involving cycles. Imagine a system composed of N independent subsystems. Each subsystem has a probability ρ_l of having a cycle of length l. We imagine that ρ_l gets quite small for large l and ask what is the chance of finding a long cycle in the entire system. Notice that the chance of not finding a piece with a cycle of length l in the entire system is

$$q_l = (1 - \rho_l)^N \approx \exp(-N\rho_l)\,.$$

If then, ρ_l varies exponentially with l (the justification for this assumption will be given below), namely, if

$$\rho_l = A \exp(-\alpha l)\,, \tag{3.1}$$

then we might expect to find parts with all cycle lengths up to

$$l_{mx} = (\ln N)/\alpha \tag{3.2}$$

(so that q_l is not that small, say of order $q_l \sim e^{-A}$). To make a long cycle in the entire system, one puts together many sub-cycles of different lengths, $l_i, i = 1, 2, \ldots, N$. The total cycle length, L, is the smallest number divisible by each of the l_i's. Then L will be a product of all prime numbers, p_r, which are less than l_{mx}, each raised to a power s_r which is the largest integer for which the inequality

$$\left[\frac{l_{mx}}{p_r^{s_r}}\right] \geq 1$$

is satisfied ($[x]$ being the integer part of x). Hence, to a decent approximation, the largest cycle length L_{mx} will be

$$L_{mx} \approx (l_{mx})^{\pi(l_{mx})}\,,$$

where $\pi(l)$ is the number of primes less than l, which can be estimated in the asymptotic limit of large l as [5]

$$\pi(l) = l/\ln l\,.$$

In the end then, the longest cycle length L_{mx} obeys

$$\ln L_{mx} \approx \pi(l_{mx})(\ln l_{mx}) \approx l_{mx}\,,$$

so that

$$L_{mx} \approx N^{1/\alpha}\,. \tag{3.3}$$

We have reach the remarkable conclusion that even though the probability of long cycles in each component of the system falls exponentially, the typical maximum cycle length in the entire system depends algebraically upon the size of the system. This calculation does not apply directly to Kauffman nets because we have not accounted for the fact that the different modules have different distributions ρ_l, but nonetheless it is instructive.

3.3.2 $K = 0$

The case in which K is zero is simple and uninteresting. After the first step, each element has a value which is completely determined by its function f_i. Each element remains with the value it had at time one for all subsequent times. Thus the system is completely frozen.

[5] A better approximation is $\pi(l) = 1/(\ln l - 1)$ (Rosser and Schoenfeld [1962]). Also, in this reference it is shown that $l/\ln l < \pi(l) < 1.0423l/\ln l$ for all $l > 10$.

3.3.3 $K = 1$

In an important paper, Flyvbjerg and Kjaer [1988] analyze the structure of the case in which each coupling function depends upon the value of just one element. To analyze this case the authors focus upon those spins which are ancestors of themselves. As we have seen (see Figure 3.2), each such element forms part of a coupling loop of a length which we will denote as m. No information may pass into such a loop from other parts of the system. Each element in the loop may affect others, but the affected elements either are constant or they inherit the cycle length of the coupling loop. In a lattice system, each element is coupled with a neighboring element. In any finite number of dimensions, there is a nonzero probability that two nearest neighbor sites are inputs to each other. Therefore, a system with N sites will contain a number of loops that is proportional to N. In contrast, on the Kauffman net, the couplings are not to neighbors but randomly chosen from the whole system. When $K = 1$, the probability that two sites are inputs to each other is proportional to $1/N$, and the average number of loops grows logarithmically with N (Bastolla and Parisi [1998a]). Flyvbjerg and Kjaer calculate the probability of observing n_m loops of length m in a system of $N \to \infty$ elements. Let m_T be the total number of elements contained in loops

$$m_T = \sum_{m=1}^{\infty} m n_m \,.$$

Then the distribution of $\{n_1, n_2, \dots\}$ takes the form

$$P(n_1, n_2, \dots) = \frac{m_T}{N} \exp\left[-m_T^2/(2N)\right] \prod_{m=1}^{\infty} \frac{(1/m^{n_m})}{n_m!} \,. \qquad (3.4)$$

Thus one can have reasonably long loops with a maximum likely length of order $N^{1/2}$. The linkage loops do not determine the cycle structure. To know the number and kind of cycles, one has to know the coupling functions. For a given loop to make a non-trivial cycle, all the functions on it must be either identity or negation, for if there is a constant function in the loop, one element has a fixed value and it passes on its constancy to all the other elements in the loop. So think of a specific case. Let us have a loop, with $m = 3$, in which element 1 is the ancestor of element 2 which is the ancestor of element 3 which is the ancestor of element 1. Let all the coupling functions be identity. Then the initial data just cycles around the loop. If the initial data is (ABC) then next is (CBA), etc. There are two cycles of length one in which all elements are identical and two of length three in which they are not. On the other hand, if all the coupling functions of this three elements are negation, there is one cycle of length two and one of length six. All the different situations are similarly easy to analyze.

Despite the fact that the Kauffman net can contain very large linkage loops for $K = 1$, the cycles are reasonably short. Figure 3.3a shows the

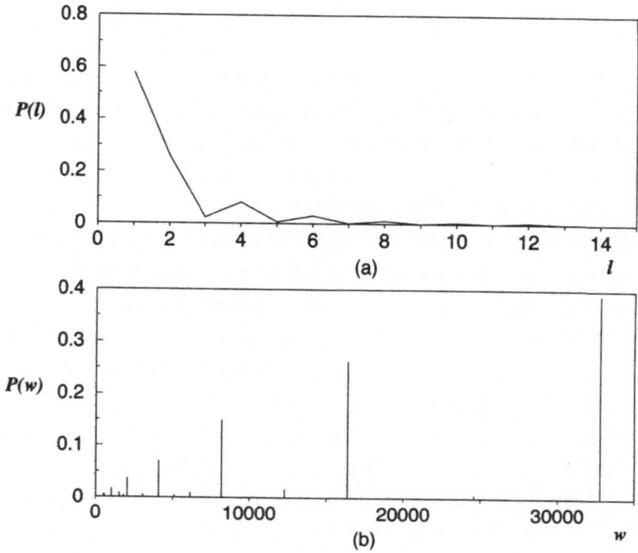

FIGURE 3.3. (a) Distribution of the probability that a cycle has length l, $P(l)$, as a function of l for a Kauffman net of $N = 15$ elements and $K = 1$. Note that the probability of having cycles with lengths larger than 10 is rather small. (b) Probability $P(w)$ that an arbitrary state belongs to a basin of attraction of size w, also for $N = 15$ and $K = 1$.

distribution of cycle lengths in a Kauffman net composed of $N = 15$ elements, and Figure 3.3b shows the probability $W(n)$ that an arbitrary state belongs to a basin of attraction of size n. For $N = 15$, there are $2^{15} = 32768$ different states. Nonetheless, as can be seen from the figure, the probability of finding a cycle with length bigger than 10 is negligibly small. The reason is that the probability that a loop is relevant (has only non-constant functions) decreases exponentially as the loop length increases. In the usual calculation one assigns the $K = 1$ functions with equal weight. If either of the two constant functions are present in the loop, the cycle length is unity. If a is the probability of assignment of the constant functions and $b = 1 - a$ is the probability of assignment of the other two functions, identity and negation, the probability of finding a cycle of length l is proportional to b^l. In most calculations b is $\frac{1}{2}$, so that the probability of really long cycles falls off exponentially in l (this is the justification of equation (3.1)). Flyvbjerg and Kjaer [1988] point out that in the special case with $b = 1$, the probability of long cycles falls algebraically. They speculate that the behavior in this limit might be, first, analyzable, and second, very similar to the behavior of the Kauffman net in the critical case ($K = K_c$) described below. As far as we know, these speculations remain unproven.

3.3.4 $K \geq N$

Another case in which it is possible to analyze the structure of the cycles in great detail is the one in which each coupling function depends upon all the values of all the different elements in the system.[6] This case has the simplifying characteristic that a change in a single spin changes the input of every coupling function. Therefore, one can analyze some features of the behavior of the system as if the system were annealed rather than quenched. In particular, one can calculate probabilities for hopping from configuration to configuration as if the system were undergoing a random walk through a space of size Ω. The quenched nature of the system only asserts itself when the hopping takes onto a configuration previously visited. After that, one can be sure that the subsequent behavior will be cyclic. There are almost classical mathematical analyses of this situation (Harris [1960]; Derrida and Flyvbjerg [1987b]). We describe this case by considering the calculation of typical distributions of cycle lengths and of transients. Imagine a Kauffman net with $K \geq N$. Imagine that we start from a random configuration Σ_0 at time $t = 0$. At subsequent times, we step by step follow the dynamics and essentially go through a random walk Σ_0, Σ_1, $\Sigma_2, \ldots,$ through the configuration space, which has size Ω. This walk continues until we land upon a point previously visited. Let us define q_t as the probability that the trajectory remains unclosed after t steps. If the trajectory is still open at time t, we have already visited $t+1$ different sites (including the sites Σ_0 and Σ_t). Therefore, there are $t+1$ ways of terminating the walk at the next time step and a relative probability of termination $\rho_t = (t+1)/\Omega$. The probability of still having an open trajectory after $t+1$ steps is

$$q_{t+1} = q_t(1-\rho_t) = q_t\left(1 - \frac{t+1}{\Omega}\right) \qquad \text{with } q_0 = 1\,,$$

while the probability p_{t+1} of terminating the excursion at time $t+1$ is

$$p_{t+1} = \frac{t+1}{\Omega} q_t\,.$$

To obtain $P(L)$, the probability that a given starting point is in the basin of attraction of a cycle of length L, we note that a closure event at time t yields with equal probability all cycle lengths up to t. Therefore,

$$P(L) = \sum_{t=L}^{\Omega} \frac{p_t}{t}\,, \tag{3.5}$$

[6] The results of this section have been known for quite some time; see references Harris [1960]; Derrida and Flyvbjerg [1987b], which study what is called the *random map model*. Here one studies systems in which one has a random map from point to point in configuration space. The form of argumentation in this section closely follows Coppersmith, Kadanoff, and Zhang [2001a].

which, in the limit of large Ω can be approximated by

$$P(L) \approx \int_L^\infty \frac{1}{\Omega} e^{-x(x-1)/(2\Omega)} dx \tag{3.6}$$

It is also important to consider the probability $P(m, L)$ of finding a cycle of length L after having gone through a precursor of length m, given by

$$P(m, L) = \frac{1}{\Omega} q_{m+L-1} \, .$$

(The factor $1/\Omega$ comes from the fact that only one point, Σ_m, of the state space can split the entire sequence $\{\Sigma_0, \Sigma_1, \ldots, \Sigma_m, \Sigma_{m+1}, \ldots, \Sigma_{m+L}\}$ of $m + L$ states into two pieces of lengths m and L respectively.) In the limit of large Ω, the previous expression can be approximated by

$$P(m, L) \approx \frac{\exp\left[-(m + L)^2/(2\Omega)\right]}{\Omega} \, . \tag{3.7}$$

The most important characteristic of the results (3.6) and (3.7) is that the typical cycle length and the typical precursor length are each of order $\Omega^{1/2}$. Thus, a very small fraction of the total configurations participate in each transient or cycle, but nonetheless the cycles and the transients may be very long. There is another, and very nice, interpretation of the results just calculated. If the precursor length is zero, equation (3.7) gives the probability that our system will contain a cycle element in a cycle of length L. Since there are Ω possible starting points, the average number of cycles of length L in our system is

$$\langle N_c(L) \rangle = \frac{\exp[-L^2/(2\Omega)]}{L} \, . \tag{3.8}$$

Here the $\langle \cdots \rangle$ represents an average over realizations. An integration then gives us the information that the average total number of cycles is proportional to $\ln \Omega$, or more precisely[7]

$$\langle N_c \rangle = \tfrac{1}{2} \ln 2 \, N + \mathcal{O}(1) \, . \tag{3.9}$$

Figure 3.4a shows the distribution $P(L)$ for a Kauffman net with $N = K = 10$, while Figure 3.4b shows the probability $P(w)$ that a given starting point belongs to a a basin of attraction of size w. Note that most of the distribution $P(w)$ is concentrated around $w = 1$ and $w = \Omega$, being nearly constant (zero) in between. The above reflects the fact that there are large

---

[7] Another result, often reported in the literature (Zoli, Guidolin, Fuxe, and Agnati [1996], or Table 1 in Kauffman [1990]) is $\langle N_c \rangle = N/e$. We do not know the justification for this, and suspect that it is wrong.

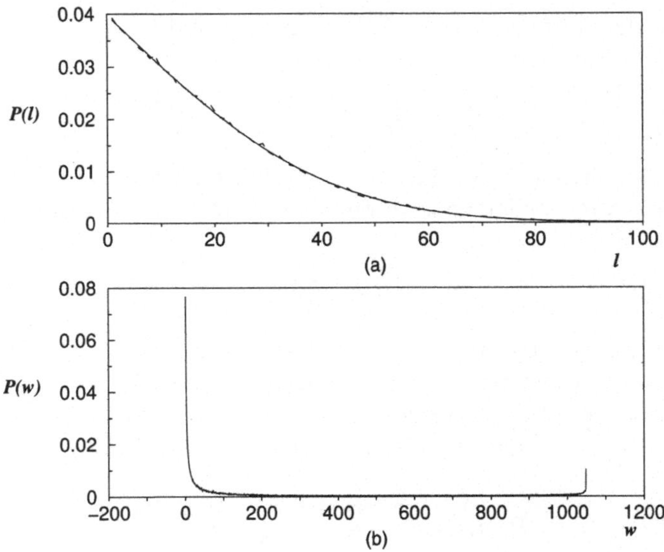

FIGURE 3.4. (a) Probability distribution of cycle lengths for a Kauffman net with $N = K = 10$. The solid curve is the theoretical result (3.5) and the broken line is the result of our numerical simulation. (b) Probability that a given starting point Σ_0 belongs to a basin of attraction of size w, also for the case $N = K = 10$.

fluctuations in the number of cycles in different realizations. If the number of cycles were a typical extensive quantity, its median would be the same as its average value. Instead, for large N its median value is half the average, indicating that, in the Kauffman net, the average is dominated by a few situations with anomalously many cycles (Harris [1960]; Derrida and Flyvbjerg [1987b]). The typical size of the largest basin is given by dividing the entire space (of size $\Omega = 2^N$), by the number of cycles (of order N, which is actually of order one, given the variance involved with the various fluctuations). To form a basin of size $\Omega = 2^N$, one starts from the cycle with its typical size $2^{N/2}$ and count backward finding the set of all first order ancestors, then backwards again to find their ancestors, which are second order ancestors of the cycle, and so on. Since these classes are non-overlapping and bounded in size, after some number of steps (perhaps of order $2^{N/2}$, the typical precursor size) the classes will shrink in size and eventually one will terminate on one or a few of the most remote ancestors. The total basin size is the sum of the number of elements in these ancestor classes. Since each configuration may have no ancestors, one ancestor or many ancestors, this backward-stepping, ancestor-counting process will have a character similar to a multiplicative random walk. As such it is an inherently highly fluctuating process. So we might expect huge fluctuations in the basin sizes (see Figure 3.4b).

3.4 Different Phases — Different Cycles

3.4.1 Frozen Phases

In the frozen phase, information typically propagates from a given element to only a very few other elements. Thus a change in initial data will typically affect the subsequent behavior of only a few elements. Similarly, damage to a single coupling function will produce changes which propagate to only a limited number of elements in the system. In these respects the lattice N-K models and the Kauffman net are very similar. However, in other respects they are very different. For example, in the lattice system, the number of relevant modules is proportional to N, so each initially different element has a nonzero chance of being in a relevant module, and each difference at time zero has a nonzero probability of developing into a difference at infinite time. In contrast, in the Kauffman net there are typically only a few short cycles and the overlap rapidly approaches unity, no matter what its initial value might have been. For the N-K model on the lattice there are many loops. In one dimension with nearest neighbor couplings all loops have length unity, but with longer range couplings and in more dimensions some loops might be quite long. Despite the fact that at any given point in the lattice it is exponentially improbable to find long loops, the argument of Section 3.3.1 indicates that the average loop size might well vary algebraically with the size of the system.

1. **Lattice system.** For small K, the lattice system falls into a phase in which there are many short cycles. The number of cycles grows as a power μ of the volume of the state space, Ω; thus, the typical basin size will be $\Omega^{1-\mu}$. The typical cycle length grows as a power of N, with multiplicative fluctuations of order $\ln N$. The growth in cycle length occurs because different modules will have sub-cycles of different lengths. The entire cycle length is the product of the prime factors coming from the modules. Large primes are exponentially unlikely in a given region, but the number of regions observed grows with N.

   Compare the time development coming from two starting configurations in which some small fraction of elements, spread out through the system, are different. These differences will each have a finite probability of causing a different cyclical behavior, localized in its particular region of the lattice. Independent of the size of the system, at long times the Hamming distance will go to a constant D^*, proportional to $D(0)$ with a constant of proportionality of order one.

2. **Kauffman net.** Because the Kauffman net has many fewer relevant modules than the lattice system, even for large N, one will be able to observe realizations which always relax into a time-independent behavior, a cycle of length one. According to Bastolla and Parisi [1998b],

the average number of cycles observed in a given realization will be independent of N. Realizations with long cycles will be exponentially unlikely. However, when one gets a long cycle, one can expect to have very many of them. The frozen phase is one in which almost all the variables end up with fixed values. That is, for a given realization, after a large number of steps most variables approach a value which is independent of the initial state of the system. Thus the system behaves, for most of the variables, as if it were highly frictional with the result that each variable comes to a stop its own 'best' value. This is the reason that the overlap goes to one as time goes to infinity in this phase.

The majority of our knowledge of the Kauffman-net behavior of the frozen phase comes from two calculations. As discussed above, the Kauffman net with $K = 1$ was solved exactly by Flyvbjerg and Kjaer (Flyvbjerg and Kjaer [1988]). For the case $K = 2$, Lynch [1993b] has proven a group of theorems which apply on the frozen side of the $K = 2$ system. To ensure the system was frozen, Lynch looked at the case in which there is unequal weight to the four classes of $K = 2$ functions, and demanded that the constant functions have a larger weight than the non-canalizing ones. This ensured that the Hamming distance, as calculated by an approach like that in equation (2.4), would decay exponentially. He then proved that the system was in a frozen phase by showing:

1. Almost all elements in the system were in their stable state.

2. Almost all gates were weak, that is, changing their value does not affect the cycle that is entered.

3. The length of the transient, i.e. the number of steps before the system enters its cycle, is bounded below by a constant times $\ln N$.

4. There is also a rather strong bound on the cycle length. The bound includes the statement that the median cycle-length is bounded by a number which is independent of N.

3.4.2 Chaotic Phases

In contrast to the frozen phase, the chaotic phase is one in which a nonzero fraction of the variables remain sensitive to initial conditions. In fact, most variables keep changing their values throughout the time development of the system. In the chaotic phase the average length of limit cycles and of the transient which occurs before the entry of the limit cycle both grow exponentially with N. Because of this explosive growth as the system is made larger, numerical investigations of orbits in the chaotic phase are limited to small system sizes (see, e.g., Bastolla and Parisi [1998b]).

As discussed above, the case when each input is coupled to all the others can be solved exactly (Derrida and Pomeau [1986]; Derrida and Flyvbjerg [1987b]), with the typical cycle length growing with system size as $2^{N/2}$. When K is finite but greater than the critical value K_c, both the Kauffman net and the lattice models have typical cycle lengths that grow exponentially with N. This behavior reflects a complex network of dependency loops in this regime. Though it is plausible that the details of the interconnected loop structures could be different for lattice models and for the Kauffman net, they do not appear to lead to marked differences in the behavior of the two types of models in the chaotic regime.

3.4.3 Critical Behavior

We have already seen that there is a close connection between the phase transition of the N-K model and the standard percolation transition of statistical physics. According to Obukhov and Stauffer [1989], random Boolean networks on nearest-neighbor d-dimensional lattices belong to the 'universality class' of directed percolation with quenched disorder in $d+1$ dimensions. However, this result is an argument, not a theorem. The sharing of the universality class means that the N-K models share many of the detailed properties of correlation and ordering with the directed percolation models of standard statistical mechanics. More particularly, the annealed N-K model based upon nearest neighbor interactions is equivalent to a directed percolation problem (Derrida and Stauffer [1986]). Moreover, when the interaction range is infinite (roughly corresponding to infinite dimensions or a simplectic geometry), the quenched (or usual) N-K model is equivalent to the corresponding directed percolation problem (Derrida and Pomeau [1986]; Derrida and Weisbuch [1986]).

Many studies have been carried out to determine the statistical properties of cycles in the critical phase. The main results can be summarized by saying that in the critical phase, both the typical cycle lengths as well as the mean number of different attractors grow algebraically with N. More explicitly, it was believed that these quantities were proportional to $\sqrt{N}$ (see for instance, Bastolla and Parisi [1998b]; Flyvbjerg [1989]; Kauffman [1990, 1969, 1995, 1993]). This was one of the most attractive results of the N-K model in that it matched one of the power-law behaviors exhibited by living organisms, as we will see in Section 6.1. However, S. Bilke and F. Sjunnesson have recently brought into question the $\sqrt{N}$ dependence of the mean number of different attractors, arguing that this result comes as a consequence of a biased *undersampling* of the whole state space (Bilke and Sjunnesson [2001]).

Because the size of the state space grows exponentially with the number of elements N, it is not feasible to enumerate completely all the different cycles in a given network realization unless N is quite small. What people usually do is to probe the state space with a small fraction of its elements

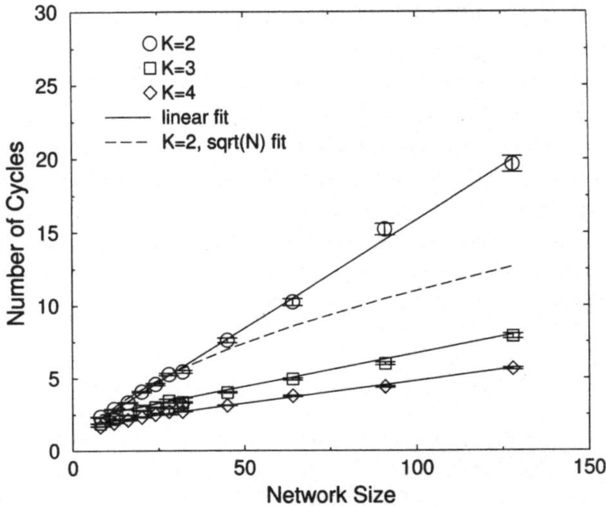

FIGURE 3.5. Mean number of different limit cycles as a function of the network size for critical Kauffman networks with connectivities $K = 2, 3, 4$. The solid curves are the best linear fits to the numerical data, whereas the dashed curve is a $\sqrt{N}$ fit for the case $K = 2$. Taken from Bilke and Sjunnesson [2001].

(which are assumed to be representative), and then to infer the statistical properties of the model from this sampling. However, this method, which has yielded the $\sqrt{N}$ dependence referred to above, has the problem that it is possible to miss a small but still important fraction of cycles.

In their work, Bilke and Sjunnesson (ibid.) use a different approach. They present a decimation method which eliminates the stable elements of the network[8], retaining only the relevant ones. Since for any realization of the network the number of different cycles as well as their lengths depend only on the relevant elements, all the statistical information related to limit cycles is preserved through this decimation procedure. The advantage in eliminating the stable elements is that the number of variables is drastically reduced. As a consequence, it is much easier to perform a *full enumeration* study of the number of different cycles for each realization of the network. Through this approach, Bilke and Sjunnesson found that, in the critical phase of the Kauffman net, the mean number of different attractors $\langle N_c \rangle$ grows linearly with N instead of as $\sqrt{N}$. Figure 3.5 (taken from Bilke and Sjunnesson [2001]) shows the mean number of different cycles $\langle N_c \rangle$ as a function of the system size N. As can be seen, the linear dependence $\langle N_c \rangle \sim N$ fits the numerical data much better than the $\sqrt{N}$ behavior.

[8]A stable element is one which evolve to the same fixed value independently of the initial state.

In addition to the exponential growth of the state space as N increases, the number of network realizations grows superexponentially with K. The evidence presented by Bilke and Sjunnesson raises the possibility that some of the results obtained so far related to cycle lengths and basins of attraction could also include a systematic bias due to undersampling. More work is needed to address these questions.

4 Reversible Models

Kauffman nets are generic models for N elements coupling K different variables. In this section we specialize this generic model to include an interesting symmetry property: *time reversal symmetry.* The discrete symmetry and quenched randomness together cause some new properties to emerge.

The standard Kauffman net that we have considered so far, whose dynamics is given by equation (1.1), is dissipative because multiple different states can map into one, so that information is lost. In this section we will refer to this system as the *dissipative Kauffman net* or simply as the *dissipative N-K model.* Not all systems are dissipative; many of the systems considered in Hamiltonian mechanics are *reversible.* Reversible systems have the property that some transformation of the coordinates (for example changing the sign of all velocities) makes the system retrace perfectly its former path. This section discusses some aspects of the behavior of discrete reversible maps. In the time-reversible Boolean network studied in Coppersmith, Kadanoff, and Zhang [2001a,b], the state of the system at time $t+1$ is governed by the equation

$$\sigma_i(t+1) = f_i(\sigma_{j_1(i)}(t), \ldots, \sigma_{j_K(i)}(t)) \oplus \sigma_i(t-1), \tag{4.1}$$

where the $\oplus$ denotes addition modulo 2. Each time-reversible network realization has a corresponding dissipative realization with the same functions and connections. In the dissipative N-K model, the state Σ_{t+1} is completely determined by the previous state Σ_t. Nevertheless, from equation (4.1) we see that in the reversible model both Σ_{t-1} and Σ_t are required to calculate Σ_{t+1}. Thus, in the reversible model the state of the system at time t is represented as

$$S_t = \begin{pmatrix} \Sigma_{t-1} \\ \Sigma_t \end{pmatrix},$$

and the state space has now 2^{2N} points. The behavior of the reversible model is in some ways closely analogous and in other ways quite different than that of the dissipative model. Both models exhibit a phase transition at K_c, a critical value of K below which the system "breaks up" into disconnected sections, and above which there is a percolating cluster of connections. The value of K_c is slightly lower for the reversible model ($K_c \approx 1.6$) than for the corresponding dissipative model ($K_c = 2$), and some details of the transition are different. But for both models the observed behavior is consistent with a percolation picture. For $K < K_c$, changing one element of a system leads to changes in only a finite number of other elements, whereas for $K > K_c$, changing one element causes a cascade of influence that spreads to a nonzero fraction of all the other elements. The typical cycle length grows slower than linearly with the size of the system when $K < K_c$, and exponentially with system size when $K > K_c$.

One big difference between the reversible and dissipative models is that they have vastly different numbers of attractors (or, in the reversible case, limit cycles). Large differences in the behavior are not entirely unexpected because in the dissipative model many different state space points end up at the same attractor, whereas in the reversible model every state space point is on exactly one limit cycle. For example, when $K = 0$ (all input functions either 1 or 0), the usual Kauffman net has only one attractor, while the number of limit cycles of a reversible Kauffman net is proportional to 2^{2N}. The reversible result can be understood by noting that for either input function, each element is in one of four different cycles, depending on its initial conditions, and that when $K = 0$ the elements are all independent. In the other limiting case $K = N$, as discussed above, the number of attractors of a dissipative Kauffman net is proportional to N (see equation (3.9)). In contrast, in the reversible model with $K = N$, and indeed throughout the regime $K > K_c$, the number of limit cycles grows as 2^N.

The number of attractors in the reversible model with $K = N$ can be understood by studying the mechanisms that lead to orbit closure. One way to close an orbit is to repeat two successive configurations, so that $\Sigma_T = \Sigma_0$ and $\Sigma_{T+1} = \Sigma_1$. Using the approximation that each successive Σ is chosen randomly from the 2^N possibilities yields a probability of an orbit closure at a given time of order $(2^{-N})^2 = 2^{-2N}$. However, there is another mechanism for orbit closure that leads to much shorter orbits and thus totally dominates the behavior. If at some time τ one has $\Sigma_\tau = \Sigma_{\tau-1}$, then time-reversibility implies that $\Sigma_{\tau+n} = \Sigma_{\tau-1-n}$ for all n. Similarly, if $\Sigma_{\tau+1} = \Sigma_{\tau-1}$, then $\Sigma_{\tau+n} = \Sigma_{\tau-n}$ for all n. Because the orbits reverse at these points, we call them "mirrors." Once two mirrors have occurred, then the orbit must close in a time that is less than the time it has taken to hit the second mirror. Again assuming that each successive Σ is chosen randomly from the 2^N possibilities, one finds that a mirror occurs at a given time with probability proportional to 2^{-N}, so the expected number of steps needed to hit two mirrors is of order 2^N. Hence, typical orbit lengths are of order 2^N. Since there are 2^{2N} points in the state space altogether, the number of limit cycles is proportional to 2^N. In the reversible Kauffman net, when K is finite but greater than the critical value K_c, the distribution of cycle lengths can be extremely broad. For example, when $N = 18$ and $K = 2$, the median cycle length $\bar{l}$ is approximately $\bar{l} \approx 140$, and yet the probability of observing an orbit of length 10^{10} is greater than 10^{-4}. This huge variability arises because of a nontrivial interplay between the discrete symmetry and the quenched randomness in the system. The occasional extremely long orbit arises because some combinations of coupling realizations and initial conditions are such that mirrors cannot occur at all. If the mirrors are not available to close the orbits, then the system must wait until two successive configurations happen to repeat. In (Coppersmith, Kadanoff, and Zhang [2001a]) it is shown that in a system with finite K in the true thermodynamic limit, almost all realizations and initial conditions yield no mirrors, and typical orbit lengths grow as $2^{2N(1-\epsilon)}$, where ϵ is of order $1/(2^{2^K})$. However, the crossover to the limiting behavior occurs only when $N \sim 2^{2^K}$, so that even for moderate K, this behavior is not accessible numerically (for example, when $K = 3$, one requires $N \sim 256$, a value at which $2^N \sim 10^{77}$ and $2^{2N} \sim 10^{154}$).

The significance of this enormous variability in the cycle lengths is not clear. One way to interpret these results is to conclude that characterizing cycles is not the right way to study the model. It would be interesting to investigate whether enormous variability arises in other random systems with discrete symmetries, and to determine whether there are possible experimental consequences. One possible starting point for comparison is with properties of random magnets, for which regions of atypical couplings lead to Griffiths singularities (Griffiths [1969]), which have stronger effects on dynamic than static properties (Randeria, Sethna, and Palmer [1985]). One must explore whether the analogy is appropriate (this is not obvious because Kauffman networks are not lattice-based), and if so, whether the results for the reversible Kauffman model indicate that dynamics far from equilibrium can be even more strongly affected by atypical coupling realizations than are the properties in spin models that have been studied to date.

5 Beyond Cycles

5.1 Non-robustness of Cycles

A cycle forms when the system returns to a configuration which is exactly the same as one it has previously visited. A demand for an exact return should be viewed not as a single constraint but in fact as N constraints upon the configuration. Such a strict demand makes the properties of cycles quite special and delicate. For this reason, *a study of cycles is probably not what one would want for understanding the possible physical or biological consequences of the models like the N-K model.* The statement just made flies in the face of a very large body of work, some of which we have just described. We should, for this reason, argue for this statement with some care.

Why don't we believe in cycles?

1. The characterizers of cycles are neither intensive nor extensive variables.
2. With exponentially short cycles in localized regions one gets power laws overall.
3. The critical situation has very many, short cycles which are not observed when one starts from randomly picked starting points.
4. Attractor basins are complex in character, being at best multi-fractal for large chaotic systems.
5. In a large system you must wait so long to see a cycle that it cannot be really important.
6. In a large chaotic system, changing one rule changes the cycles quite a bit.
7. In both glasses and biological systems one wants to characterize the system by things which are very robust.

We do believe that generic properties of cycles are important though, in that they characterize general aspects of dynamical systems. Nonetheless, the huge fluctuations throughout realizations in such important quantities as cycle lengths and number of different attractors, calls for other types of characterizations. Real networks, whether they are genetic or neural or of any other kind, are always subjected to external perturbations. The robustness in the dynamics of the network can not rely on quantities which change dramatically with perturbations. Hence, it is important to characterize the dynamical properties of the network in the presence of noise, trying to find out which kind of quantities are preserved under the influence of noise, and which ones are not.

5.2 Characterization via noise

The addition of noise to a map provides a possibility for generating additional information about the behavior of the models. Noise naturally blurs out the sharpness of behavior, making for a more "fuzzy" characterization. Noise is, then, a natural way to get away from the difficulty posed by the overly-precise characterization provided by the cycles.

Unfortunately, most previous work on the N-K model does not include noise. We do wish to point to two papers (Miranda and Parga [1989]; Golinelli and Derrida [1989]) in which noise has been used to probe N-K behavior. For reasons which will become more evident later on, we describe these papers respectively as a "crossing paper" and a "convergence paper".

These papers both break the precision of the dynamical rule of equation (1.1) by saying that the rule is broken with a probability r

$$\sigma_i(t+1) = \begin{cases} f_i(\sigma_{j_1(i)}(t), \ldots, \sigma_{j_{K_i}(i)}(t)) & \text{with probability} \quad 1-r \\ \neg f_i(\sigma_{j_1(i)}(t), \ldots, \sigma_{j_{K_i}(i)}(t)) & \text{with probability} \quad r, \end{cases} \tag{5.1}$$

which can also be written in the alternative form:

$$\sigma_i(t+1) = \begin{cases} f_i(\sigma_{j_1(i)}(t), \ldots, \sigma_{j_{K_i}(i)}(t)) & \text{with probability} \quad 1-2r \\ 1 & \text{with probability} \quad r \\ 0 & \text{with probability} \quad r. \end{cases} \tag{5.2}$$

This equation can be described as providing probabilities r for the two possible values of the outcome, independently of the value of f_i. In Golinelli and Derrida [1989] the rules are described in terms of a temperature T, related to r by

$$r = \frac{1 - \tanh(1/T)}{2}. \tag{5.3}$$

Both groups examine the development of two or more different initial configurations using the same realizations. They also apply exactly the same rules and the same probabilistic choices to the different configurations. So far, both papers are essentially similar. There are two kinds of differences, the first being the choice of measurement, and the second being the way they apply equations (5.1) and (5.2).

The convergence paper starts with two or more randomly chosen initial configurations, Σ_0^1, $\Sigma_0^2, \ldots,$ Σ_0^m, and calculates the resulting trajectories step by step:

$$\begin{array}{c} \Sigma_0^1 \to \Sigma_1^1 \to \Sigma_2^1 \to \ldots \Sigma_\tau^1 \\ \Sigma_0^2 \to \Sigma_1^2 \to \Sigma_2^2 \to \ldots \Sigma_\tau^2 \\ \ldots \\ \Sigma_0^m \to \Sigma_1^m \to \Sigma_2^m \to \ldots \Sigma_\tau^m \end{array}$$

At each step, and for each i, a choice is made among the three branches of equation (5.2), and that choice is equally applied to the m configurations.[9] The calculation continues until two of the m configurations become identical (say for example $\Sigma_\tau^1 = \Sigma_\tau^2$). The time needed to achieve the convergence is noted. We will denote this time by τ_m, stressing the fact that m configurations are being analyzed.

In some ways, the convergence calculation is more complicated than the crossing one. The noise as defined by equation (5.2) tends to produce convergence because it is applied equally to all trajectories and because makes

[9] The two-branch versus three-branch methods become inequivalent when they are applied to several configurations at once.

the values of the elements to be equal in all trajectories. In the limit of infinite N, the system shows three phases: the low noise phase in which almost always trajectories will not converge, the high noise phase in which trajectories will always converge, and a separating critical phase. These are respectively described as low temperature, high temperature, and critical phases.

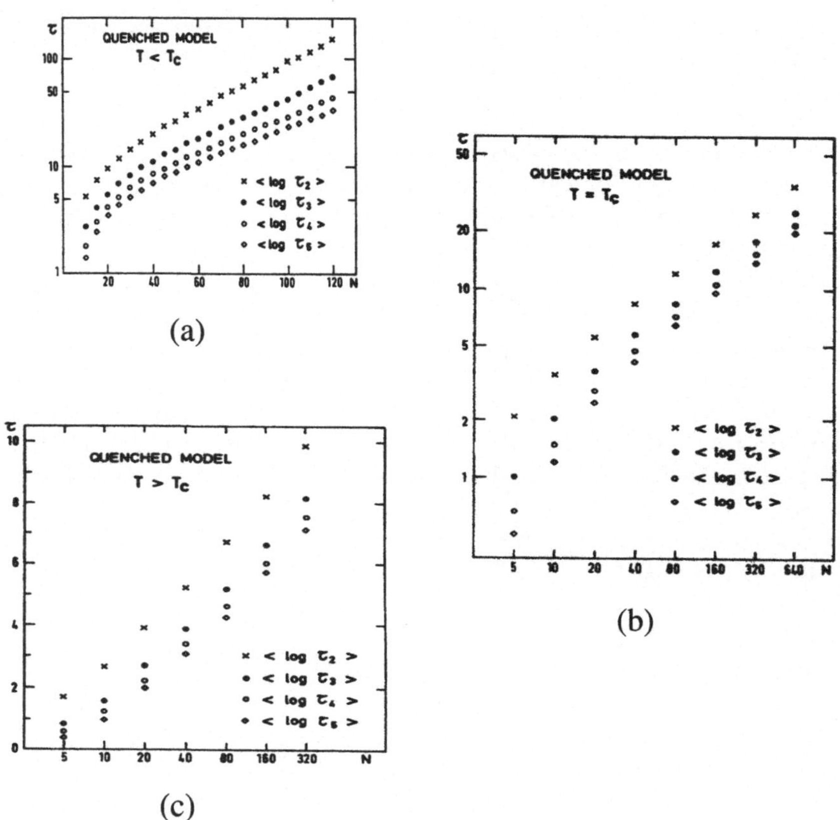

FIGURE 5.1. Plots of $\langle \ln \tau_m \rangle$ versus N, taken from reference Golinelli and Derrida [1989]. In all the cases the connectivity of the network was $K = 4$. The three graphs correspond to three values of r: (a) Low temperature phase, $r = 0.15$; (b) Critical phase, $r = 0.25$; (c) High temperature phase, $r = 0.35$.

The authors examine $K = 4$ and large-N, picking the noise-levels $r = 0.15$ (low temperature phase), $r = 0.25$ (critical phase), and $r = 0.35$ (high temperature phase). Without noise, the N-K model would show chaotic behavior, and very long cycles. In the presence of noise the average of $\ln \tau$, with τ being the convergence time, diverges as N goes to infinity. The behavior of the convergence time for two orbits, τ_2, can be calculated in the

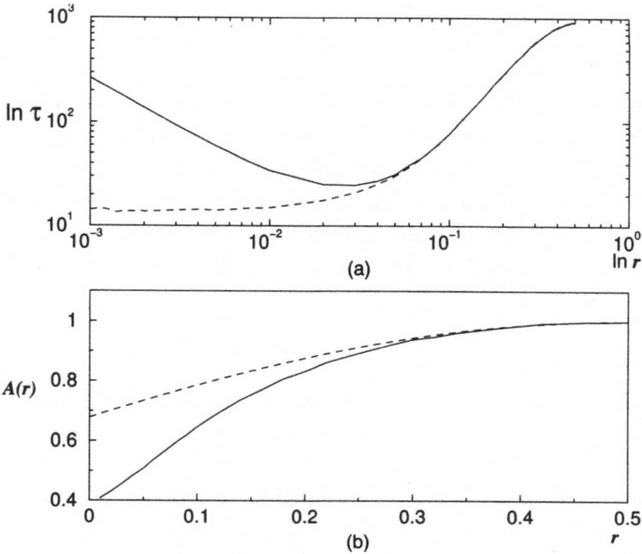

FIGURE 5.2. (a) Log-log plot of the crossing time τ as a function of the level of noise r. The solid curve is τ_d, the crossing time starting from different attractors, whereas the dashed curve corresponds to τ_s, the crossing time starting from the same attractor. (b) The index $A(r)$ describing the fractal dimension of the attractor plotted as a function of r. The solid curve is the result of the numerical simulation, obtained as $A(r) = 2\ln\tau/\ln\Omega$, and the dashed line is the theoretical prediction given in (5.7). Note that these curves become identical when $r \to 0.5$. Both graphs (a) and (b), calculated for a situation with $K = 2$ and $N = 20$, were taken from reference Qu, Kadanoff, and Aldana [2002].

annealed sequential update case, and the results compared with simulations for both annealed and quenched system. In both simulation and theory, the three phases are characterized by having different forms of divergence of $\langle \ln \tau_2 \rangle$ with N (see Figure 5.1). At low temperatures the divergence is linear: $\langle \ln \tau_2 \rangle \sim N$. In the critical phase $\langle \ln \tau_2 \rangle$ diverges linearly with $\ln N$ as $\langle \ln \tau_2 \rangle = 0.5 \ln N$. The weakest divergence occurs in the high temperature phase in which $\langle \ln \tau_2 \rangle$ varies as $\ln \ln N$. These N-dependencies describe the variation of the number of elements forming the barrier to having two configurations merge into one another. The noise causes these elements to be identical and then the merge occurs. Thus, for example, in the low temperature phase, we must bring to equality a finite fraction of all the elements in the system in order to have the convergence.

In some loose sense, these numbers measure the size of the barriers which hold together the attractors for this system. But it is hard to know what the attractors themselves might be. Because r is relatively large, these presumed attractors are probably not the cycles of the original system. In fact, the behavior of τ_2 is much the same for the annealed system (which has no cycles) as for the quenched system, which does have cycles at $r = 0$. Nobody has yet investigated the limiting case as r goes to zero. It might be most interesting to look at this limit, particularly in association with a limit which keeps the system critical (say K goes to 2).

Let us consider the case in which more than two trajectories are analyzed. Start with m trajectories and let τ_m measure the first time when any two of these have converged. If there are only M large or important basins, one might well expect $\langle \ln \tau_m \rangle \ll \langle \ln \tau_M \rangle$ for $m > M$. Instead one observes that

$$\langle \ln \tau_m \rangle = \langle \ln \tau_2 \rangle - \ln \frac{m(m-1)}{2},$$

in both theory (in the annealed approximation) and simulation. This form indicates an indefinitely large number of attractors, all with basins of comparable size.

The crossing paper considers two different configurations Σ_0^1 and Σ_0^2, which can belong either to different attraction basins or to the same basin of attraction, and then iterates forward, noting all configurations $\{\Sigma_0^1, \Sigma_1^1, \Sigma_2^1, \ldots\}$ and $\{\Sigma_0^2, \Sigma_1^2, \Sigma_2^2, \ldots\}$ they produce. In each step of iteration, and for each i-value, a choice is made between the two branches of equation (5.1), and that choice is applied to both configurations. This continues until the time τ in which one of the two trajectories attains a configuration previously entered by the other one (for example, if Σ_τ^2 is equal to one of the configurations $\{\Sigma_0^1, \Sigma_1^1, \ldots, \Sigma_\tau^1\}$). The calculation is terminated at this crossing event. The measured quantity is the "time" τ needed to produce the crossing.

In the absence of noise, if two initial configurations belong to different basins of attraction, the time for the two subsequent trajectories to cross is infinite. In the presence of noise, there is a chance for each trajectory to "jump out" of its basin of attraction, exploring a bigger part of the state space. The two trajectories will have a number of opportunities equal to τ^2, to cross one another before τ steps have elapsed. If the size of the space being explored by the trajectories is $\Omega(r)$, then the typical time for the crossing will be

$$\tau \approx \Omega(r)^{1/2}. \tag{5.4}$$

Miranda and Parga simulated the system and measured τ as a function of r and N for the critical Kauffman net in which $K = 2$. Their result for large N may be summarized as

$$\Omega(r) = \Omega^{A(r)}, \tag{5.5}$$

where $\Omega = 2^N$ is the volume of the state space in the system.

In the work by Miranda and Parga the "fractal" exponent $A(r)$ was not estimated accurately. This work was recently extended by X. Qu et al., who consider larger values of noise and different connectivities of the network (Qu, Kadanoff, and Aldana [2002]). The authors analyze two cases to compute the crossing time, when the two initial configurations belong to the same basin of attraction, and when they belong to different basins. We will denote these two crossing times by τ_s and τ_d respectively. Figure 5.2a shows the average crossing times τ_d and τ_s as funcions of r, for a net with $N = 20$ and $K = 2$. As can be seen, when r is close to its maximum value 0.5, both times are practically the same. In fact, Qu et al. have shown that for large values of r both τ_s and τ_d behave as

$$\tau_{d,s} \approx \frac{\sqrt{\pi}}{2} 2^{N/2} \left\{ 1 + \frac{(1-2r)^2}{2^K} \right\}^{-N/2} . \tag{5.6}$$

The above expression agrees with equations (5.4) and (5.5) by identifying

$$A(r) = 1 - \frac{\ln\left[1 + (1-2r)^2/2^K\right]}{\ln 2} . \tag{5.7}$$

In contrast, when r is close to 0 the behavior of τ_d and τ_s differ substantially. For $r \longrightarrow 0$ the divergence of τ_d is given simply by

$$\tau_d \approx C_1/r + \tau_0 \tag{5.8}$$

where C_1 only depends on K and N, and τ_0 is the value of τ_s at $r = 0$.

The complete analytical expresion of $A(r)$, valid in the whole interval $[0, 1/2]$ is not known yet. Figure 5.2b shows a plot of $A(r)$ obtained by numerical simulations. It is interesting to note that the attractor has a fractal volume which depends upon r. Once again, one is frustrated because one does not know what the attractor might be. It is once again probably not anything directly related to a cycle, since starting points in the same cycle or in different cycles both give the same τ-values for the higher values of N and r. Here too one might guess that studies with smaller values of r might shed light on the N-K model attractors.

6 Applications

6.1 Genetic networks and cell differentiation

N-K models have been widely used in the modeling of gene regulation and control (Somogyi and Sniegoski [1996]; De Sales, Martins, and Stariolo [1997]; Kauffman [1993]; Huang and Ingber [2000]). A very remarkable characteristic of multicellular organisms is that all the different cells of which they are made have the same genetic information. What makes the difference between the different cell types are the genes which are *expressed* in every cell at every moment. In a given cell type, some particular genes are turned off and others are turned on. So, in a liver cell, only the "liver genes" are being expressed while all the other genes are turned off, whereas in a neuron the "neuron genes" are the only ones which are expressed.

The physical and chemical mechanisms by which the cell determines which genes are to be expressed and which are not are not yet fully understood. There is evidence that gene regulation and control can occur at every stage during the metabolic pathways leading up from the genetic information contained in the DNA, to the translation of this information into proteins. Nevertheless, most of the gene regulation and control seems to occur at the level of transcription of the genetic information. At this level, one gene of DNA is transcribed into a molecule of messenger ARN (mARN) only if the conditions for this transcription are present. In the most simple model (applicable to bacteria), for the transcription of one gene into a molecule of mARN, it is necessary a protein, called *activator*, which attaches to the beginning of the gene indicating that this gene is ready to be transcribed (see Figure 6.1a). On the other hand, there also exist *repressor* proteins which, when attached to the beginning of the gene, inhibit its transcription, turning the gene off (see Figure 6.1b).

In eucaryotic cells the situation is more complicated in that many activator or repressor proteins might be needed to activate or to inhibit the expression of a single gene. For example, it is known that the human β-globine gene (expressed in red blood cells) is regulated by more than 20 different proteins. Some of these proteins may function as both repressors or activators, depending on how they are assembled.

The activator or repressor proteins of a given gene are themselves specified by some other genes, whose expression, is in turn controlled by other proteins codified by other genes and so on. Genes interact each other through the proteins they specify: the product protein of one activated gene can influence the activation or deactivation of other genes. Similarly, the *absence* of the product protein of a deactivated gene can influence the activation or deactivation of several other genes. In some particular cases, it is known which gene controls which other one, but in most of the cases the interactions between genes in a given cell type are completely unknown. A picture emerges in which genes are mingled together forming a network of interacting elements which are connected in a very complicated way.

The N-K model was first suggested by Kauffman as a way of modeling the dynamics and evolution of this complicated network of interacting genes (Kauffman [1969]). Within this model, two genes are linked if the product protein of one gene influences the expression of the other one. In real cells, genes are not randomly linked. Nevertheless, the web of linkages and connectivities among the genes in a living organism is so complicated (and mostly unknown), that to model these genes as randomly linked seems to be a reasonable approximation. By this means, the results coming out of the model are not restricted to a particular set of linkages and connectivities, but reflect instead the generic properties of this class of networks.

The state of the cell at every moment is determined then by the state of its genes $\sigma_1, \ldots, \sigma_N$. Before proceeding further, it is important to recall the main assumptions that are usually made to describe the dynamics of

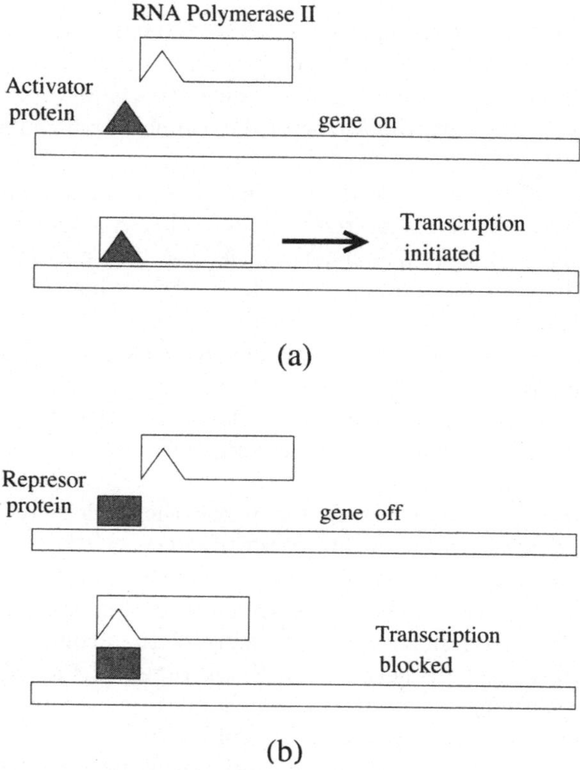

FIGURE 6.1. Schematic representation of gene regulatory proteins. (a) An activator protein attaches to the gene at a specific binding site, activating the polymerase II which then can transcript the gene into a molecule of mRNA. (b) When a repressor protein is attached to the gene, the polymerase II is blocked up and therefore unable to transcript the information contained in the gene.

genetic networks by means of the N-K model:

- Every gene σ_i can be only in one of two states, turned on (1) or turned off (0).
- Every gene is randomly linked with exactly K other genes.
- The evolution rule f_i associated to gene σ_i is a weighted function which acquires the value 1 with probability p ant the value 0 with probability $1 - p$.
- The updating of the network is synchronous.

As we have seen, under these assumptions the phase space breaks up into different cycles or attractors whose properties in the frozen phase and the chaotic phase are substantially different. According to Kauffman's interpretation, *each attractor represents a cell type or a cell fate*, whereas a single state of the system represents just a temporal state which the cell is passing through. From this point of view, cell reproduction processes start with an initial configuration of genes, which eventually evolves towards its corresponding attractor; the attractor determines a particular cell type or cell fate.

For this mechanism of cell differentiation to be meaningful, the length of the attractors must not be too long, for otherwise the cell would never reach its stable cycle. In the chaotic phase the length of the cycles grows exponentially with the system's size ($l \sim 2^{\alpha N}$, where α is of order 1). Therefore, the system has to go through very many states before reaching a stable set of configurations. In addition, systems in the chaotic phase are very sensitive to perturbations (mutations), partially because the number of relevant elements is comparable to the size of the system. The above prompts the thought that genetic networks of living organisms *are not* in the chaotic phase.

On the other hand, in the frozen phase the cycles are much shorter than in the chaotic phase: an initial set of genes swiftly reaches its stable configurations. The fraction of relevant elements in the frozen phase is close to zero. As a consequence, the system is extremely resistant to point mutations [10] or to damage in one or more of the evolution rules. But in order to evolve, genetic networks of living organisms should allow some degree of sensitivity to mutations, which rules out the frozen phase as a physical state which living organisms could be in.

Kauffman suggested that gene networks of living organisms operate *at the edge of chaos* (Kauffman [1993]), meaning that the parameters have been adjusted through evolution so that these networks are at or near the critical phase. In the critical phase, both the number of different attractors as well as their lengths are proportional to a power of N, so the cell can reach very quickly its stable configurations. Also, the fraction of relevant nodes in critical networks, even though is small, is not zero, which means that these kinds of networks present some degree of sensitivity to changes in the initial conditions. In other words, critical networks exhibit *homeostatic stability*, a term which we will come to in the next section.

A very remarkable observation supporting the idea of life at the edge of chaos consists in the fact that the number of different cell types in an organism is roughly proportional to the square root of its DNA content. Furthermore, the mitotic cycle period, which can be considered as a measure of the time required for a cell to reproduce, seems also to be proportional to the square root of the cell's DNA content (Kauffman [1993]). Thus, random networks in the critical phase seem to satisfy the requirements of order, evolvability and stability found in living organisms.

Even though this idea is very attractive, there are some problems yet to

---

[10] A point mutation is a change in the value of one gene.

be solved. As we have mentioned, for unbiased evolution rules f_i, the critical phase is characterized by the low connectivity $K = 2$. This implies that genetic networks of real organisms are restricted to have very low connectivities in order to be at "the edge of chaos". As soon as the connectivity grows, the system becomes more and more chaotic. But it is well known that the connectivity in real genetic networks is rather high. For example, the expression of the *even-skipped* gene in *Drosophila* is controlled by more than 20 regulatory proteins, also the Human β-globine gene we have referred to before (Alberts, Bray, Lewis, Raff, Roberts, and Watson [1994]). In eucaryotes it is common to find that one single gene is regulated by a bunch of proteins acting in association. On the other hand, sometimes when a single signaling receptor protein is activated, it can influence directly the activation of a very large array of genes. Let us consider for instance the activation of the platelet-derived growth factor β receptor (PDGFRβ), which induces the expression of over 60 genes (Fambrough, Mcclure, Kazlauskas, and Lander [1999]). These examples, among many others, suggest that the connectivity in real genetic networks is not low, but on the contrary, it is very high. Nonetheless, cells do not seem to operate in the chaotic phase.

There are two ways to increase the connectivity in the N-K model without going out of the critical or ordered phase: (a) by the use of weighted evolution functions f_i, or (b) by the use of canalizing functions. When evolution functions f_i are weighted with the probability parameter p, the critical line is given by equation (2.8). Figure 2.4 shows the graph of K_c as a function of p. Even if we suppose that genetic networks can be either in the frozen phase or in the critical phase, they would be restricted to remain within the shaded area of the figure.

On the other hand, it is known that the fraction of canalizing functions becomes very small as K increases (Kauffman [1984]). An upper bound for this fraction is

$$\frac{4K}{2^{2^{K-1}}}, \tag{6.1}$$

which tends to zero as $K \to \infty$. Consequently, canalizing functions are extremely rare when K is large[11]. If the apparent order seen in living cells relies on either weighted or canalizing functions, it has still to be solved what kind of mechanisms drove, through evolution, the genetic networks towards the generalized use of such type of functions.

Another problem lies in the assumption that genes can be in only two states. It is true that a given gene is expressed or is not. But the product protein of a gene can participate in a variety of metabolic functions, producing even opposite effects depending on the physical and chemical context in which it acts. Such behavior can be modeled by assuming that every gene can acquire more than two states. But in such a case the con-

[11] Even for K as small as $K = 10$, equation (6.1) gives a fraction of canalizing functions of the order 10^{-153}.

nectivity of the network must be even smaller to keep the system within the ordered phase. For if we assume that every gene can be, on average, in one of m possible states, and if every one of these states is activated with the same probability, the critical connectivity K_c is then given by

$$K_c = \frac{m}{m-1}. \tag{6.2}$$

The critical connectivity decreases monotonically when $m > 2$, approaching 1 as $m \to \infty$. The moral is that for this kind of multi-state networks to be in the ordered phase, the connectivity has to be very small, contrary to what is observed in real genetic networks.

Partially to overcome these difficulties, De Sales, Martins, and Stariolo [1997] have proposed a model which is slightly different from the original N-K model. In their model, the dynamics of the system is governed by the equation

$$\sigma_i(t+1) = \text{Sign}\Big\{ J_{ii}\sigma_i(t) + \sum_{l=1}^{K-1} J_{ij_l(i)}\sigma_{j_l(i)}(t) \Big\}, \tag{6.3}$$

where $J_{ij_l(i)}$ is the coupling constant representing the regulatory action of the $j_l(i)$ input of gene i, ($l = 1, 2, \ldots, K-1$). There is also an autogenic regulation, given by J_{ii}. The set of coupling constants J_{ij} represents the very complicated set of biochemical interactions between genes. Since these interactions are mostly unknown, the authors assign the coupling constants in a random way, according to the following criteria:

- The product protein of a given gene can activate, inhibit, or not affect at all the transcription of another gene.
- Interactions between genes are not necessarily symmetrical, namely, $J_{ij} \neq J_{ji}$ in general.
- Autogenic or self-regulation is allowed.

The coupling constants are then assigned according to the probability function given by:

$$P(J_{ij}) = \frac{1-p_1}{2}\big[\delta(J_{ij}-1) + \delta(J_{ij}-1)\big] + p_1\delta(J_{ij}). \tag{6.4}$$

In this way, the couplings J_{ij} can be activating (+1) or inhibitory (−1), each with probability $\frac{1}{2}(1-p_1)$, or neutral (0) with probability p_1. Furthermore, the linkages for every gene are chosen either at random among the whole set of genes, with probability p_2, or only among the nearest neighbors, with probability $1 - p_2$. By varying the parameter p_2 one can go from lattices where local interactions are the most important, to random nets where all ranges of interaction are present.

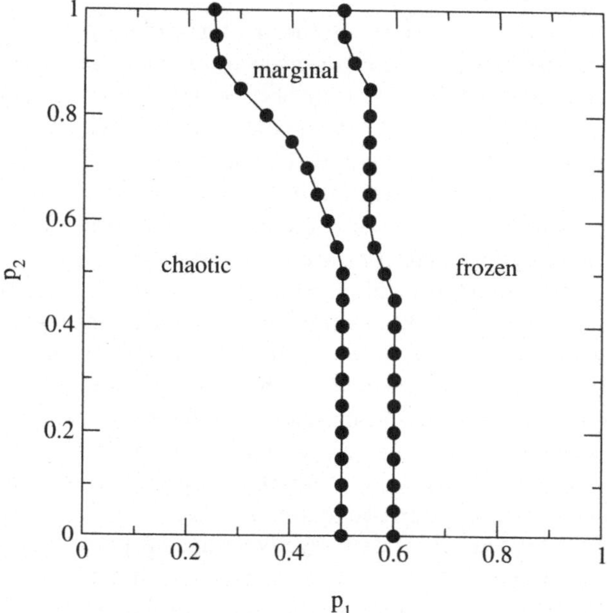

FIGURE 6.2. Phase space for the N-K model with the dynamics given by equation (6.4). As can be seen, under this dynamics the system exhibits three different "phases": frozen, marginal and chaotic (taken from De Sales, Martins, and Stariolo [1997]).

By analyzing the Hamming distance between two initial configurations, the authors show that there are three phases in the dynamics of the system: a frozen phase, a marginal phase and a chaotic phase, as shown in Figure 6.2. The frozen and chaotic phases are as in the traditional N-K model. The marginal phase is characterized by the fact that all the attractors are stable since a small change in the initial configuration neither vanishes nor grows. Also, in this phase both the number of different attractors and their lengths grow as a power of N. The above results were obtained for networks with connectivity $K = 9$ and $N = 625$. It is probable that the marginal phase is only a finite-size effect, vanishing for an infinite system. More work has to be done to explore the whole space of parameters of this model, but these preliminary results show that it is possible to obtain ordered and marginal behaviors even in the case of high connectivities by changing the type of interaction between the genes.

6.2 Evolution

In a traditional framework of the theory of evolution (Simon [1969]; Dawkins [1989, 1986]), changes in the phenotype of organisms are accumulated gradually, yielding a gradual increase in complexity of form and function. Simple structures slowly assemble together to form more complex structures, which in turn assemble to build up even more elaborate systems, and so on. At every stage in the formation of a complex system (organism) out of simpler elements, many sub-systems are created, which represent temporal stable states along the way in the construction of the whole system. As T. D. Zawidzki has pointed out in Zawidzki [1998], from this point of view the problem of evolution of complex systems translates into a search problem. For the construction of a complex system, evolution searches in the space of the possible configurations leading to the whole system. If no stable intermediate configurations were found in this search, the construction of the whole system would be extremely improbable, since the number of possible configurations in which the parts can be arranged increases exponentially with the number of different parts. Evolution does not search in one step for the "final" configuration of a complex system. Instead, the search is carried out gradually, finding first intermediate stable configurations of sub-systems that are then assembled, giving rise to the whole complex system. Every sub-system solves a particular problem (or set of problems), facilitating the construction and functioning of the whole organism. Furthermore, once an evolutionary problem has been solved by primitive organisms, more complex species which evolve from them still continue solving this problem in the same way. New species are faced with new problems, but still preserving the old solutions to the old problems (the genetic code, for example, was "invented" only once). According to this thesis, genomes of complex organisms are made up of functional modules of genes, each module encoding the solution of a given biological problem encountered by the species at some point through evolution.

The study of Boolean networks has suggested new mechanisms for evolutionary processes. Kauffman has stressed that many evolutionary changes involve reorganizing of the same genetic material rather than making it more complex (Kauffman [1993, 1995]). In Kauffman's approach, genes are organized in complex genetic networks provided with some given dynamics. The "search" of evolution consists in searching for the most stable organization of genes, which in terms of the N-K model means stable cycles. From this point of view, there is not an increasing complexification of the evolving system, but a inherent complex organizational dynamics which settles down in a finite number of stable attractors. The "role" of evolution then is to look for the more stable attractors which the system can fall into.

A Boolean network made up of N genes has 2^N states, but the system organizes itself into a much smaller number of attractors. Depending on the parameter values being used (K and p), these attractors are stable or they are not. Stability is defined according to the response of the network to perturbations, which can be of three different kinds:

- changes in the states of a few genes by flipping the value of some randomly chosen ones;

- permanent changes in the linkages of some genes;
- permanent changes in the values of the evolution functions f_i associated with some genes.

As we have already mentioned, only networks in the critical phase have the stability required to constitute evolvable systems, in that these networks are able to recover to most of the mutations described above. In the chaotic phase, the attractors are extremely unstable since any kind of perturbation would shift the system to another attractor. In the frozen phase, even though the majority of the genes are motionless, small changes in the linkages or in the evolution functions would make the system jump to a very different attractor if these damages are carried out on the relevant elements of the net. But networks operating in the critical phase show a very high homeostatic stability, which means that after some perturbation, such networks are very likely to fall again in the same attractor.

Consequently, in this approach, evolution is also interpreted as a sort of search for gene networks possessing stable dynamics and not merely as a searching for stable sub-systems out of which more complicated systems can be built up. The fact that more complex living organisms have bigger amounts of genetic material, together with the fact that real genetic networks actually exhibit high homeostatic stability, suggest that evolutionary processes consist of both kinds of "searching," hierarchical-modular complexification and dynamical stability.

So far we have considered processes occurring at the level of genomes, but evolution also acts at the level of populations, making the organisms of a given species become better adapted to their environment. Living beings are subjected to all kinds of external fluctuations, and the survival of the species depends on the capability of its members to recover from those random perturbations. Evolutionary processes produce organisms with a high degree of homeostatic stability and of adaptability to the environment, even in the presence of external fluctuations. When thinking of evolutionary processes, one usually supposes that living organisms tend to exclude noise since stability is more conductive to functioning than chaos, and also that evolutionary processes are able to recognize and favor such stability. Nevertheless, Michael D. Stern has pointed out that none of these assumptions has been rigorously proven yet (Stern [1999]). Furthermore, he has shown that, under certain circumstances, noise not only is not excluded from an adaptative system, but it is *required* for the adaptation of the system to the environment; without noise, such adaptation would not be possible.

In this work Stern considers a population of M organisms, each of which is a N-K model composed of $N = 100$ elements. Each organism is in the critical phase ($K = 2$ and $p = 0.5$) and therefore, according to Kauffman, they are in the state of highest homeostatic stability. The linkages among the elements and the evolution rules f_i are assigned in the usual way, but additionally, a noisy signal $\eta(t)$ is applied to m randomly chosen elements of every organism. The noise is introduced through only one of the two inputs of each one of the m elements. Suppose for example that σ_{i_1} and σ_{i_2} are the two inputs of σ_i. With no noise, the value of σ_i at time $t+1$ would be given, as usual, by

$$\sigma_i(t+1) = f_i\big(\sigma_{i_1}(t), \sigma_{i_2}(t)\big)\,.$$

But if σ_i is one of the m elements to which the noisy signal $\eta(t)$ is being applied, then the value of σ_i at time $t+1$ is given now by

$$\sigma_i(t+1) = f_i\big(\sigma_{i_1}(t), \eta(t)\big)\,.$$

Note that the noisy signal $\eta(t)$ is the same for the m elements of every one of the M organisms. This takes into account the fact that in a real population, every organism in the population is subjected to the same (noisy) environment.

The evolution of the population is now determined by an external criteria which has to be fulfilled. This is the *phenotype* of the population on which natural selection will be acting. In Stern's work, the phenotype to be selected is an integer time series, obtained by counting the number of positive states (+1) occurring in a given subset of elements (output elements) in each organism. This time series is to fit a predefined time function $F(t)$ (the target function), and in each generation the organisms better adapted to $F(t)$ are selected.

Selection of the organisms is made as follows. At the beginning of each generation, every organism is replicated R times allowing some mutations which consist mainly of randomly moving one of the input connections of a randomly chosen element, and randomly changing the Boolean function of a randomly chosen element. So, at the beginning of every generation the population actually consists of $R \times M$ elements. The system is then evolved during 100 time steps, after which the M elements which best fit the external criteria $F(t)$ are selected, starting another generation.

The results of the simulation are shown in Figure 6.3, in which three output series are compared with the target function $F(t)$ (dotted line). Figure 6.3C shows the output signal of the initial generation, which has not yet passed through any selection process. In the presence of noise $\eta(t)$, after 2500 generations the target function is very well approximated by the output signal of the evolving network, as shown if Figure 6.3D. However, in the absence of noise the same network fails completely to approximate the target function (Figure 6.3E). This surprising result implies that noisy perturbations may be essential for the adaptation of organisms to their environments. Stern has called this phenomenon *noise imprinting* in evolution, and claims that it "may be a prototype of a general form of symmetry breaking that leads to the evolution of structures that are suboptimal and unnecessarily dependent on fortuitous features of the environment".

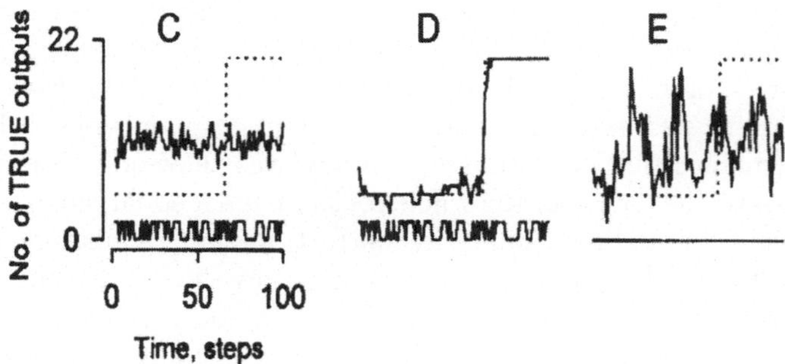

FIGURE 6.3. (C) Random output of the starting network (solid), the target function (dotted), and the fixed binary noise added to the system (lower) in Stern's model. (D) The evolved network generates a good approximation of the target function after 2500 generations. (E) The same network as in D but operating in the absence of noise fails completely in fitting the target function, indicating that the evolved network has been "imprinted" by the arbitrary noise sequence present during its evolution and requires it to function (taken from Stern [1999]).

6.3 Social Systems

It is always difficult to extrapolate the models and techniques of physics to the realm of social sciences. To begin with, it is not yet clear how to define a social system in a way that is suitable for the kind of analysis that physicists are used to. The complexity we see in human societies is of a quite different nature than the one present in physical systems, so one must be extremely cautious when talking about "complexity" and "chaos" in social sciences. Nonetheless, some parallels can be established between the complex behaviors of Boolean networks and human organizations.

Boolean networks were first introduced in social sciences by Schelling in 1971 (Shelling [1971]). Following Klüver and Schmidt [1999], the main aspect in a social system is that *social actors act according to certain social rules; the intended or unintended consequences of their actions generate other actions by the same or other actors and so forth.* The result of such interactions is reflected in the dynamics of the system.

A key difficulty in the modeling of social networks is determining the quantities that convey enough information to describe a given system, and which have some degree of predictability. One must also identify the key parameters governing the behavior. Two kind of parameters can be distinguished: those which determine the topology of the system, namely, how individuals interact, and the ones which determine the social rules, i.e., the consequences of the interactions. To illustrate these two kind of parameters, we can focus our attention on the connectivity K of Boolean networks and the weight parameter p associated to the evolution functions f_i.

In ancient societies, characterized by very rigid dictatorial or monarchic regimes, the connectivity among people was very small since the majority of individuals were restricted to interact only with the few people in their communities. Mobility of individuals among different social classes was also minimal. Poor people were condemned to remain poor while rich people usually remained rich; social mobility, if any, was allow only from upper to lower classes. The course of the whole society depended largely on very few people (the King and his clique). In contrast, modern democratic societies are characterized by much bigger connectivities since individuals have a broader spectrum of interactions. In addition, social mobility is greater in democratic societies since, at least in principle, anybody can become president or actor or any other thing.

A binary network can be used for the modeling of a society of interacting individuals, in which poverty is represented by 1 while richness is represented by 0. Of course, in real societies the connectivity among individuals varies from one individual to another, and so does the range of influence of their decisions; but it is clear that even if we assign the same connectivity to each member of the society, the out-degree will not be the same for all individuals[12], and therefore the range of influence that each member of the society has on the other members varies among them.

With this setup, a rigid feudal society would be characterized by low values of the parameter K and by evolution functions f_i whose weight parameter p is very low, reflecting the fact that the majority of the individuals are poor no matter what they do or whom they interact with. This kind of society presents a very simple dynamics in a frozen regime; no changes are expected nor unpredictable behavior. The dynamics is governed by a very reduced fraction of relevant elements, while the rest of the individuals remain in a frozen state with nearly zero mobility.

On the other hand, a perfect democratic society would be rather characterized by high connectivities and weighted functions f_i whose parameter p is close to 0.5. In this situations, we have seen that the system exhibits chaotic behavior, with an apparently random dynamics. The society would never reach a stable state or a stable attractor, for in this regime the length of the cycles is extremely large.

If these results reflect some of the fundamental aspects behind the dynamics of social organizations, the conclusion would be that the complexity that we see in modern democratic societies is inherent to the democratic principles (parameter values) on which these societies have been constructed. Apparently, democracy and complexity are tied together, and it is a political decision whether or not it is worthwhile to sacrifice democracy in the interest of predictability. We should stress, though, that Boolean

---

[12]The out-degree of node σ_i is the number of different nodes which are affected by σ_i, and can run from 0 up to N.

networks are far from accurate representations of a human society, so we should consider the previous results only as indicative trends and not as the matter of the facts.

6.4 Neural Networks

The first neural network model was introduced nearly 60 years ago by McCulloch and Pitts [1943] in an attempt to understand some of the cognitive processes of the brain. The topic of neural networks has since grown enormously, and presently it covers a great variety of fields, ranging from neurophysiology to computational algorithms for pattern recognition and non-parametric optimization processes (Cheng and Titterington [1994]). Nonetheless, it is somehow accepted that the two major applications of neural networks consist in the understanding of real neural systems (the brain), and in the development of machine-learning algorithms (Farmer [1990]). In this section, we will address the subject from a point of view based on statistical physics.

As in the N-K model, a typical neural network consists of a set of N nodes, each of which receives inputs from K other nodes. The difference from the N-K model is that in a neural network the dynamics are given by[13]

$$\sigma_i(t+1) = \mathrm{Sign}\Big(\sum_{j=1}^{K} w_{ij}\sigma_{i_j}(t) + h\Big), \qquad (6.5)$$

where the *synapse weights* $\{w_{ij}\}$ and the *activation threshold* h are random variables that can adapt (in learning processes) or stay fixed (in recall or association processes). Note that with the above definitions, the nodes have the values $\sigma_i = \pm 1$.

K. E. Kürten has considered the case in which the synapse weights are independent random variables distributed according to a symmetric probability function $\mathrm{P}(w)$. By analyzing the Hamming distance, he has shown that this kind of neural network has behavior similar to that of the N-K model, in that the three phases (frozen, critical and chaotic) are also

[13]There are other choices for the dynamics, such as

$$\sigma_i(t+1) = \tanh\Big(\sum_{j=1}^{K} w_{ij}\sigma_{i_j}(t) + h\Big),$$

which produces an output between -1 and $+1$, or

$$\sigma_i(t+1) = \Big(1 + \exp\Big\{\sum_{j=1}^{K} w_{ij}\sigma_{i_j}(t) + h\Big\}\Big)^{-1},$$

with an output between 0 and 1. Possible choices abound, depending on the particular task for which the network has been designed. We will focus on the particular class of neural networks obeying the dynamics given by equation (6.5).

present (Kürten [1988a,b]), depending on the connectivity K and the dilution of the network (dilution is a measure of the amount of nodes for which the synapse weight is 0). The main result is that for low connectivity and dilution, the neural network and the N-K model exhibit exactly the same dynamics in the Hamming distance.

Another important application of neural networks consists in investigating the tolerance to the influence of external noise in the dynamical organization of the network. Huepe and Aldana-González [2002] have analyzed a neural network whose dynamics is given by

$$\sigma_i(t+1) = \begin{cases} \text{Sign}\Big(\sum_{j=1}^{K} w_{ij}\sigma_{i_j}(t)\Big) & \text{with probability } 1-\eta\,, \\ -\,\text{Sign}\Big(\sum_{j=1}^{K} w_{ij}\sigma_{i_j}(t)\Big) & \text{with probability } \eta\,. \end{cases} \tag{6.6}$$

The dynamics can thus be changed from purely deterministic to purely random by varying the noise intensity η between 0 and $\frac{1}{2}$. In the case $\eta = 0$ the network will typically converge to an ordered state in which all the σ_i acquire the same value. For other values of η, the instantaneous amount of order in the networks is determined by the parameter

$$s(t) = \lim_{N\to\infty} \frac{1}{N}\sum_{i=1}^{N} \sigma_i(t)\,, \tag{6.7}$$

so that $|s(t)| \approx 1$ for an "ordered" system in which most elements take the same value, while $|s(t)| \approx 0$ for a "disordered" system in which the elements randomly take values +1 or −1. The order parameter of the network is defined then as the time-independent quantity

$$\Psi = \lim_{T\to\infty} \frac{1}{T}\int_0^T |s(t)|\, dt\,. \tag{6.8}$$

Huepe and Aldana have shown that, under very general conditions, the neural network model given in (6.6) undergoes a dynamical second order phase transition, as it is illustrated in Figure 6.4. More specifically, if the probability density $\mathrm{P}(w)$ of the synapse weights is a non-symmetric, but otherwise arbitrary function, and if the linkages of the network are chosen randomly, then there exists a critical value η_c of the noise in the vicinity of which the order parameter behaves as

$$\Psi = \begin{cases} C\big(\eta_c - \eta\big)^{\frac{1}{2}} & \text{for } \eta < \eta_c\,, \\ 0 & \text{for } \eta > \eta_c\,, \end{cases} \tag{6.9}$$

where C and η_c are constants whose values depend on the connectivity K and on the particular form of $\mathrm{P}(w)$. It is worth mentioning that the dynamics of the network are exactly the same if both the linkages and the synapse weights w_{ij} are time-independent, or if they are re-assigned at

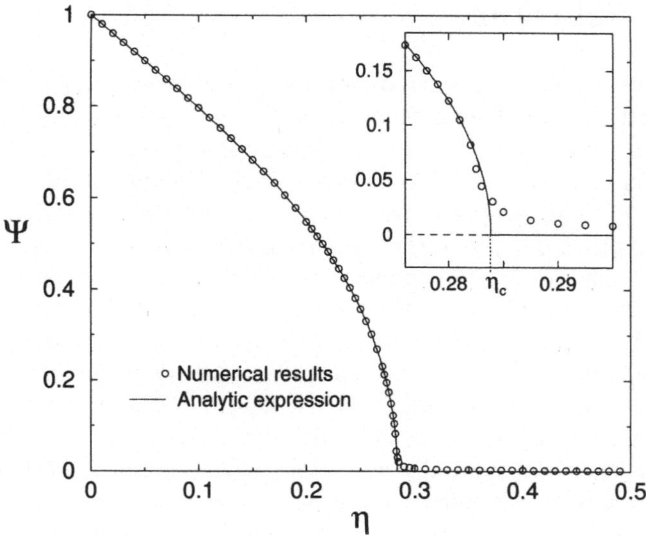

FIGURE 6.4. Bifurcation diagram of the order parameter Ψ as a function of the noise intensity η, for the neural network model in (6.6). In the case shown, the network was composed of $N = 10^5$ elements with connectivity $K = 11$, and the synapse weights w_{ij} were uniformly distributed between 0 and 1. The inset shows the finite-size effects in the numerical results near the critical noise $\eta_c \simeq 0.2838$. (Taken from Huepe and Aldana-González [2002].)

every time step. As far as we know, this is the first network model in which it has been shown that the quenched and annealed approaches coincide exactly.

Other models of particular interest are those in which the synapse weights w_{ij} are in turn given by "sub-layers" of connections, such as in the Little-Hopfield model (Hopfield [1982]; Little [1974]),

$$w_{ij} = C_{ij} \sum_{\mu=1}^{p} \xi_i^\mu \xi_j^\mu ,$$

where $\xi_i^\mu = \pm 1$ is the value of site i in the sub-layer μ, and the coefficients C_{ij} are independent random variables distributed according to a given probability distribution $\mathrm{P}(C)$. These networks are important because by suitably changing coefficients C_{ij}, the network can be either synchronized or "taught" some particular task (Hopfield [1999]; Abarbanel, Rabinovich, Selverston, and Bazhenov [1996]; Derrida, Gardner, and Zippelius [1987]).

7 Conclusions

N-K models can be considered as prototypical dynamical systems. They exhibit ordered and chaotic behaviors as well as a phase transition between these two regimes. The properties of this phase transition, first measured by Derrida and Pomeau [1986] in terms of the Hamming distance between two distinct configurations, are very well known by now. The important parameters that determine the dynamical behavior of the system (frozen, critical or chaotic) are the connectivity K of the network and the probability p that a coupling function has a given output. For a network with fully random topology, the critical value K_c of the connectivity at which the phase transition occurs is given in terms of the probability p through the equation $K_c = [2p(1-p)]^{-1}$. When $K < K_c$ (frozen phase), the Hamming distance decays exponentially with time, whereas for $K > K_c$ (chaotic phase) it grows exponentially with time, asymptotically reaching a value which depends on the initial condition. For $K = K_c$ (critical phase), the temporal evolution of the Hamming distance is determined by fluctuations.

The statistical properties of the limit cycles and the basins of attraction also change from one phase to another. In the frozen phase the mean number $\langle N_c \rangle$ of different attractors remains constant even in the limit $N \to \infty$, whereas in the chaotic phase $\langle N_c \rangle \sim N$. With regard to the critical phase, it was long believed that $\langle N_c \rangle \sim N^{1/2}$. This was the main result supporting the biological interpretation of the Kauffman model that identifies each limit cycle with a given cell type (Kauffman [1969, 1990, 1993]). However, recent studies have shown that the $N^{1/2}$ behavior of the mean number of different cycles in the critical phase could just be a consequence of an under-sampling of the configuration space (Bilke and Sjunnesson [2001]). These authors have found instead that in this phase $\langle N_c \rangle \sim N$. In any event, the fact that the $\Omega = 2^N$ configurations of the network organize in a small number of different attractors (of order N), is a very remarkable property of the system. This organization reflects the modular structure of the network produced by the clusterization of the relevant elements (Bastolla and Parisi [1998b,a]). The phase transition in the N-K model can thus be described at the level of cycles by considering the clusterization of the relevant elements as a percolation problem (Stauffer [1987b]; Hansen [1988a]; Lam [1988]): the clusters, very sparse in the frozen phase, spread out in the chaotic phase, filling up most of the network.

Another important factor determining the dynamical properties of the boolean network is the topology of the linkages. For example, in N-K lattices there are very many modules of relevant variables in both the frozen and the chaotic phases. This is reflected in the distribution of the number of different cycles. Stauffer [1987b] have shown numerically that in the two-dimensional N-K lattice with first neighbor linkages ($K = 4$ and $K = 5$), the mean number of different cycles in both the frozen and the chaotic phases seems to grow exponentially with N. Much more work is called for in order to fully characterize the statistical and dynamical properties of N-K lattices. Other boolean networks with different linkage topologies might also be worthy of study. For example, dynamical boolean networks in which the number of linkages per node is a random variable following a power law distribution (scale free networks), remain basically unexplored despite their ubiquity (Albert and Barabási [2002]; Barabási [2002]).

The study of the dynamics of N-K models during the last 30 years has contributed to our understanding of some important aspects of dynamical and complex systems. In particular, the fact that very many states in an essentially random system organize themselves into a small number of different attractors, has been a major breakthrough in the understanding of self-organization and the emergence of order in complex systems. Probably this is the reason why most of this study has been focused on the statistical characterization of limit cycles and basins of attraction. However, the enormous fluctuations (from one realization to another) in quantities such as the mean cycle length or the mean number of different attractors, lead us to conclude that these are not the most relevant quantities for a description of physical or biological systems. The limit cycles exhibit stable dynamics only when the connectivities are much lower than those in real networks. One way around this problem is to make use of forcing or canalizing functions, or by adjusting the value of p. But even when this is done, the cycle dynamics of the network change drastically between realizations. This problem is even more serious if we take into account that real networks are always subjected to random perturbations (external noise). The identification of the robust dynamical properties of the N-K models, as well as their statistical characterization, are still to be done.

Acknowledgments: We would like to thank Xiaohui Qu for her valuable help in the elaboration of this article. This work was partially supported by the MRSEC Program of the NSF under Award Number 9808595, and by the NSF-DMR 0094569. We also thank the Santa Fe Institute of Complex Systems for partial support through the David and Lucile Packard Foundation Program in the Study of Robustness. M. Aldana also acknowledges CONACyT-México for a postdoctoral grant.

References

Abarbanel, H. D. I., M. I. Rabinovich, A. Selverston, and M. V. Bazhenov [1996], Synchronization in Neural Networks, *Physics-Uspeki* **39**, 337–362.

Albert, R. and A.-L. Barabási [2000], Dynamics of Complex Systems: Scaling Laws for the Period of Boolean Networks, *Physical Review Letters* **84**, 5660–5663.

Albert, R. and A.-L. Barabási [2002], Statistical Mechanics of Complex Networks, *Reviews of Modern Physics* **74**, 47–97.

Alberts, B., D. Bray, J. Lewis, M. Raff, K. Roberts, and J. D. Watson [1994], *Molecular Biology of the Cell, Third Edition.* Garland Publishing, New York.

Andrecut, M. and M. K. Ali [2001], Chaos in a Simple Boolean Network, *International Journal of Modern Physics B* **15**, 17–23.

Atlan, H., F. Fogelman-Soulie, J. Salomon, and G. Weisbuch [1981], Random Boolean Networks, *Cybernetics and Systems* **12**, 103–121.

Bagley, R. J. and L. Glass [1996], Counting and Classifying Attractors in High Dimensional Dynamical Systems, *Journal of Theoretical Biology* **183**, 269–284.

Baillie, C. F. and D. A. Johnston [1994], Damaging 2D Quantum Gravity, *Physics Letters B* **326**, 51–56.

Bak, P., H. Flyvbjerg, and B. Lautrup [1992], Coevolution in a Rugged Fitness Landscape, *Physical Review A* **46**, 6724–6730.

Barabási, A.-L. [2002], *Linked: The New Science of Networks.* Perseus Publising, Cambridge, Massachusetts.

Bastolla, U. and G. Parisi [1996], Closing Probabilities in the Kauffman Model: an Annealed Computation, *Physica D* **98**, 1–25.

Bastolla, U. and G. Parisi [1997], A Numerical Study of the Critical Line of Kauffman Networks, *Journal of Theoretical Biology* **187**, 117–133.

Bastolla, U. and G. Parisi [1998a], The Modular Structure of Kauffman Networks, *Physica D* **115**, 219–233.

Bastolla, U. and G. Parisi [1998b], Relevant Elements, Magnetization and Dynamical Properties in Kauffman Networks: a Numerical Study, *Physica D* **115**, 203–218.

Bhattacharjya, A. and S. Liang [1996a], Median Attractor and Transients in Random Boolean Nets, *Physica D* **95**, 29–34.

Bhattacharjya, A. and S. Liang [1996b], Power-Law Distributions in Some Random Boolean Networks, *Physical Review Letters* **77**, 1644–1647.

Bilke, S. and F. Sjunnesson [2001], Stability of the Kauffman Model, *Physical Review E* **65**, 016129.

Bornholdt, S. [1998], Genetic Algorithm Dynamics on a Rugged Landscape, *Physical Review E* **57**, 3853–3860.

Bornholdt, S. and T. Rohlf [2000], Topological Evolution of Dynamical Networks: Global Criticality From Local Dynamics, *Physical Review Letters* **84**, 6114–6117.

Bornholdt, S. and K. Sneppen [1998], Neutral Mutations and Punctuated Equilibrium in Evolving Genetic Networks, *Physical Review Letters* **81**, 236–239.

Bornholdt, S. and K. Sneppen [2000], Robustness as an Evolutionary Principle, *Proc. Royal Soc. Lond. B* **266**, 2281–2286.

Bull, L. [1999], On the Baldwin Effect, *Artificial Life* **5**, 241–246.

Burda, Z., J. Jurkiewicz, and H. Flyvbjerg [1990], Classification of Networks of Automata By Dynamic Mean-Field Theory, *Journal of Physics A: Mathematical and General* **23**, 3073–3081.

Castagnoli, G. [1998], Merging Quantum Annealing Computation and Particle Statistics: A Prospect in the Search of Efficient Solutions to Intractable Problems, *International Journal of Theoretical Physics* **37**, 457–462.

Cheng, B. and D. M. Titterington [1994], Neural networks: a review from a statistical perspective, *Statistical Science* **9**, 2–54.

Coppersmith, S. N., L. P. Kadanoff, and Z. Zhang [2001a], Reversible Boolean Networks I: Distribution of Cycle Lengths, *Physica D* **149**, 11–29.

Coppersmith, S. N., L. P. Kadanoff, and Z. Zhang [2001b], Reversible Boolean Networks II: Phase Transitions, Oscilations and Local Structures., *Physica D* **157**, 54–74.

Corsten, M. and P. Poole [1988], Initiation of Damage in the Kauffman Model, *Journal of Statistical Physics* **50**, 461–463.

Dawkins, R. [1986], *The Blind Watchmaker.* W.W. Norton and Company, USA.

Dawkins, R. [1989], *The Selfish Gene.* Oxford University Press, Oxford, second edition.

De Sales, J. A., M. L. Martins, and D. A. Stariolo [1997], Cellular Automata Model for Gene Networks, *Physical Review E* **55**, 3262–3270.

Derrida, B. [1980], Random-Energy Model: Limit of a Family of Disordered Models, *Physical Review Letters* **45**, 79–82.

Derrida, B. [1987a], Dynamical Phase Transitions in Non-Symmetric Spin Glasses, *Journal of Physics A: Mathematical and General* **20**, L721–L725.

Derrida, B. [1987b], Valleys and Overlaps in Kauffman Model, *Philosophical Magazine B: Physics of Condensed Matter, Statistical Mechanics, Electronic, Optical and Magnetic Properties* **56**, 917–923.

Derrida, B. and D. Bessis [1988], Statistical Properties of Valleys in the Annealed Random Map Model, *Journal of Physics A: Mathematical and General* **21**, L509–L515.

Derrida, B. and H. Flyvbjerg [1986], Multivalley Structure in Kauffman Model - Analogy With Spin-Glasses, *Journal of Physics A: Mathematical and General* **19**, 1003–1008.

Derrida, B. and H. Flyvbjerg [1987a], Distribution of Local Magnetizations in Random Networks of Automata, *Journal of Physics A: Mathematical and General* **20**, L1107–L1112.

Derrida, B. and H. Flyvbjerg [1987b], The Random Map Model: a Disordered Model With Deterministic Dynamics, *Journal De Physique* **48**, 971–978.

Derrida, B., E. Gardner, and A. Zippelius [1987], An Exactly Solvable Asymmetric Neural Network Model, *Europhysics Letters* **4**, 167–173.

Derrida, B. and Y. Pomeau [1986], Random Networks of Automata - a Simple Annealed Approximation, *Europhysics Letters* **1**, 45–49.

Derrida, B. and D. Stauffer [1986], Phase-Transitions in Two-Dimensional Kauffman Cellular Automata, *Europhysics Letters* **2**, 739–745.

Derrida, B. and G. Weisbuch [1986], Evolution of Overlaps Between Configurations in Random Boolean Networks, *Journal De Physique* **47**, 1297–1303.

Domany, E. and W. Kinzel [1984], Equivalence of Cellular Automata to Ising Models and Directed Percolation, *Physical Review Letters* **53**, 311–314.

Fambrough, D., K. Mcclure, A. Kazlauskas, and E. S. Lander [1999], Diverse Signaling Pathways Activated By Growth Factor Receptors Induce Broadly Overlapping, Rather That Independent, Sets of Genes, *Cell* **97**, 727–741.

Farmer, J. D. [1990], A Roseta Stone for Connectionism, *Physica D* **42**, 153–187.

Flyvbjerg, H. [1988], An Order Parameter for Networks of Automata, *Journal of Physics A: Mathematical and General* **21**, L955–L960.

Flyvbjerg, H. [1989], Recent Results for Random Networks of Automata, *Acta Physica Polonica B* **20**, 321–349.

Flyvbjerg, H. and N. J. Kjaer [1988], Exact Solution of Kauffman Model with Connectivity One, *Journal of Physics A: Mathematical and General* **21**, 1695–1718.

Flyvbjerg, H. and B. Lautrup [1992], Evolution in a Rugged Fitness Landscape, *Physical Review A* **46**, 6714–6723.

Fogelman-Soulie, F. [1984], Frustration and Stability in Random Boolean Networks, *Discrete Applied Mathematic* **9**, 139–156.

Fogelman-Soulie, F. [1985], Parallel And Sequential Computation On Boolean Networks, *Theor. Comp. Sci.* **40**, 275–300.

Genoud, T. and J.-P. Metraux [1999], Crosstalk in Plant Cell Signaling: Structure and Function of the Genetic Network, *Trends in Plant Science* **4**, 503–507.

Glass, L. and C. Hill [1998], Ordered and Disordered Dynamics in Random Networks, *Europhysics Letters* **41**, 599–604.

Golinelli, O. and B. Derrida [1989], Barrier Heights in the Kauffman Model, *Journal De Physique* **50**, 1587–1601.

Griffiths, R. [1969], Nonanalytic Behavior Above the Critical Point in a Random Ising Ferromagnet, *Physical Review Letters* **23**, 17–19.

Hansen, A. [1988a], A Connection Between the Percolation Transition and the Onset of Chaos In the Kauffman Model, *Journal of Physics A: Mathematical and General* **21**, 2481–2486.

Hansen, A. [1988b], Percolation and Spreading of Damage in a Simplified Kauffman Model, *Physica A* **153**, 47–56.

Harris, B. [1960], Probability Distributions Related to Random Mappings, *Annals of Mathematical Statistics* **31**, 1045–1062.

Herrmann, H. J. [1992], Simulation of Random Growth-Processes, *Topics in Applied Physics* **71**, 93–120.

Hilhorst, H. J. and M. Nijmeijer [1987], On the Approach of the Stationary State in Kauffmans Random Boolean Network, *Journal De Physique* **48**, 185–191.

Hopfield, J. J. [1982], Neural Networks and Physical Systems with Emergent Collective Computational Abilities, *Proceedings of the National Academy of Sciences* **79**, 2554–2558.

Hopfield, J. J. [1999], Brain, Neural Networks and Computation, *Reviews of Modern Physics* **71**, S431–S437.

Huang, S. and D. E. Ingber [2000], Shape-Dependent Control of Cell Growth, Differentiation, and Apoptosis: Switching Between Attractors in Cell Regulatory Networks, *Experimental Cell Research* **261**, 91–103.

Huepe, C. and M. Aldana-González [2002], Dynamical Phase Transition in a Neural Network Model with Noise: An Exact Solution, *Journal of Statistical Physics* **108**, (3/4), 527–540.

Ito, K. and Y.-P. Gunji [1994], Self-Organization of Living Systems Towards Criticality at the Edge of Chaos, *Biosystems* **33**, 17–24.

Jan, N. [1988], Multifractality and the Kauffman Model, *Journal of Physics A: Mathematical and General* **21**, L899–L902.

Kadanoff, L. P. [2000], *Statistical Physics: Statics Dynamics and Renormalization.* World Scientific, Singapore.

Kauffman, S. [1984], Emergent Properties in Random Complex Automata, *Physica D* **10**, 145–156.

Kauffman, S. A. [1969], Metabolic Stability and Epigenesis in Randomly Constructed Nets, *Journal of Theoretical Biology* **22**, 437–467.

Kauffman, S. A. [1974], The Large Scale Structure and Dynamics of Genetic Control Circuits: an Ensemble Approach, *Journal of Theoretical Biology* **44**, 167–190.

Kauffman, S. A. [1990], Requirements for Evolvability in Complex Systems - Orderly Dynamics and Frozen Components, *Physica D* **42**, 135–152.

Kauffman, S. A. [1993], *The Origins of Order: Self-Organization and Selection in Evolution.* Oxford University Press, Oxford.

Kauffman, S. A. [1995], *At Home in the Universe: the Search for Laws of Self-Organization and Complexity.* Oxford University Press, Oxford.

Kauffman, S. A. and W. G. Macready [1995], Search Strategies for Applied Molecular Evolution, *Journal of Theoretical Biology* **173**, 427–440.

Kauffman, S. A. and E. D. Weinberger [1989], The NK Model of Rugged Fitness Landscapes and Its Application To Maturation of the Immune Response, *Journal of Theoretical Biology* **141**, 211–245.

Kaufman, J. H., D. Brodbeck, and O. M. Melroy [1998], Critical Biodiversity, *Conservation Biology* **12**, 521–532.

Kirillova, O. V. [1999], Influence of a Structure on Systems Dynamics on Example of Boolean Networks, *International Journal of Modern Physics C* **10**, 1247–1260.

Klüver, J. and J. Schmidt [1999], Control Parameters in Boolean Networks and Cellular Automata Revisited from a Logical and Sociological Point of View, *Complexity* **5**, 45–52.

Krapivsky, P. L., S. Redner, and F. Leyvraz [2000], Connectivity of Growing Random Networks, *Physical Review Letters* **85**, 4629–4632.

Kulakowski, K. [1995], Relaxation and Limit-Cycles in a Global Version of the Quenched Kauffman Model, *Physica A* **216**, 120–127.

Kürten, K. E. [1988a], Correspondence Between Neural Threshold Networks and Kauffman Boolean Cellular Automata, *Journal of Physics A: Mathematical and General* **21**, L615–L619.

Kürten, K. E. [1988b], Critical Phenomena in Model Neural Netwoks, *Physics Letters A* **129**, 157–160.

Kürten, K. E. and H. Beer [1997], Inhomogeneous Kauffman Models at the Borderline Between Order and Chaos, *Journal of Statistical Physics* **87**, 929–935.

Lam, P. M. [1988], A Percolation Approach to the Kauffman Model, *Journal of Statistical Physics* **50**, 1263–1269.

Langton, C. G. [1990], Computations at the Edge of Chaos: Phase Transitions and Emergent Computation, *Physica D* **42**, 12–37.

Lee, C.-Y. and S. K. Han [1998], Evolutionary Optimization Algorithm By Entropic Sampling, *Physical Review E* **57**, 3611–3617.

Levitan, B. and S. Kauffman [1995], Adaptive Walks With Noisy Fitness Measurements, *Molecular Diversity* **1**, 53–68.

Little, W. A. [1974], The Existence of Persistent States in the Brain, *Mathematical Bioscience* **19**, 101–120.

Luczak, T. and J. E. Cohen [1991], Stability of Vertices in Random Boolean Cellular Automata, *Random Structures and Algorithms* **2**, 327–334. reference from Lynch.

Luque, B. and R. V. Solé [1997a], Controlling Chaos in Random Boolean Networks, *Europhysics Letters* **37**, 597–602.

Luque, B. and R. V. Solé [1997b], Phase Transitions in Random Networks: Simple Analytic Determination of Critical Points, *Physical Review E* **55**, 257–260.

Luque, B. and R. V. Solé [1998], Stable Core and Chaos Control in Random Boolean Networks, *Journal of Physics A: Mathematical and General* **31**, 1533–1537.

Luque, B. and R. V. Solé [2000], Lyapunov Exponents in Random Boolean Networks, *Physica A* **284**, 33–45.

Lynch, J. F. [1993a], Antichaos in a Class of Random Boolean Cellular-Automata, *Physica D* **69**, 201–208.

Lynch, J. F. [1993b], A Criterion for Stability in Random Boolean Cellular-Automata, *Los Alamos Data Base* **http://arXiv.org/abs/adap-org/9305001**.

Lynch, J. F. [1995], On the Threshold of Chaos in Random Boolean Cellular-Automata, *Random Structures and Algorithms* **6**, 239–260.

Ma, S. [1976], *Modern Theory of Critical Phenomena.* Benjamin, Reading Pa.

Macisaac, A. B., D. L. Hunter, M. J. Corsten, and N. Jan [1991], Determinism and Thermodynamics - Ising Cellular Automata, *Physical Review A* **43**, 3190–319.

Manrubia, S. C. and A. S. Mikhailov [1999], Mutual Synchronization and Clustering in Randomly Coupled Chaotic Dynamical Networks, *Physical Review E* **60**, 1579–1589.

McCulloch, W. S. and W. Pitts [1943], A Logical Calculus of Ideas Immanent in Nervous Activity, *Bulletin of Mathematical Biophysics* **5**, 115–133.

Mestl, T., R. J. Bagley, and L. Glass [1997], Common Chaos in Arbitrarily Complex Feedback Networks, *Physicl Review Letters* **79**, 653–656.

Metropolis, N. and S. Ulam [1953], A Property of Randomness of an Arithmetical Function, *American Mathematical Monthly* **60**, 252–253.

Mezard, M., G. Parisi, and M. A. Virasoro [1987], *Spin Glass Theory and Beyond.* World Scientific, Singapore.

Miranda, E. N. and N. Parga [1988], Ultrametricity in the Kauffman Model - a Numerical Test, *Journal of Physics A: Mathematical and General* **21**, L357–L361.

Miranda, E. N. and N. Parga [1989], Noise Effects in the Kauffman Model, *Europhysics Letter* **10**, 293–298.

Nirei, M. [1999], Critical Fluctuations in a Random Network Model, *Physica A* **269**, 16–23.

Obukhov, S. P. and D. Stauffer [1989], Upper Critical Dimension of Kauffman Cellular Automata, *Journal of Physics A: Mathematical and General* **22**, 1715–1718.

Ohta, T. [1997a], The Meaning of Near-Neutrality at Coding and Non-Coding Regions, *Gene* **205**, 261–267.

Ohta, T. [1997b], Role of Random Genetic Drift in the Evolution of Interactive Systems, *Journal of Molecular Evolution* **44**, S9–S14.

Ohta, T. [1998], Evolution By Nearly Neutral Mutations, *Genetica* **103**, 83–90.

Owezarek, A., A. Rechnitzer, and A. J. Guttmann [1997], On the Hulls of Directed Percolation Clusters, *Journal of Physics A: Mathematical and General* **30**, 6679–6691.

Petters, D. [1997], Patch Algorithms in Spin Glasses, *International Journal of Modern Physics C* **8**, 595–600.

Preisler, H. D. and S. Kauffman [1999], A Proposal Regarding the Mechanism Which Underlies Lineage Choice During Hematopoietic Differentiation, *Leukemia Research* **23**, 685–694.

Qu, X., L. Kadanoff, and M. Aldana [2002], Numerical and Theoretical Studies of Noise Effects in Kauffman Model, *Journal of Statistical Physics*, **109**, (5/6), 967–986.

Randeria, M., J. Sethna, and R. Palmer [1985], Low-Frequency Relaxation in Ising Spin-Glasses, *Physical Review Letters* **54**, 1321–1324.

Rosser, J. B. and L. Schoenfeld [1962], Approximate Formulas for some Functions of Prime Numbers, *Illinois Journal of Mathematics* **6**, 64–94.

Sakai, K. and Y. Miyashita [1991], Neural Organization for the Long-Term-Memory of Paired Associates, *Nature* **354**, 152–155.

Serra, R. and M. Villani [1997], Modelling Bacterial Degradation of Organic Compounds With Genetic Networks, *Journal of Theoretical Biology* **189**, 107–119.

Shelling, T. C. [1971], Dynamic Models of Segregation, *Journal of Mathematical Sociology* **1**, 143–186.

Sherrington, D. and K. Y. M. Wong [1989], Random Boolean Networks for Autoassociative Memory, *Physics Reports: Review Section of Physics Letters* **184**, 293–299.

Sherrington, D. and S. Kirkpatrick [1975], Solvable Model of a Spin-Glass, *Physical Review Letters* **35**, 1792–1796.

Sibani, P. and A. Pedersen [1999], Evolution Dynamics in Terraced NK Landscapes, *Europhysics Letters* **48**, 346–352.

Simon, H. A. [1969], *The Sciences of the Artificial.* The MIT Press, Cambridge, MA.

Solov, D., A. Burnetas, and M.-C. Tsai [1999], Understanding and Attenuating the Complexity Catastrophe In Kauffman's NK Model of Genome Evolution, *Complexity* **5**, 53–66.

Solow, D., A. Burnetas, T. Roeder, and N. S. Greenspan [1999], Evolutionary Consequences of Selected Locus-Specific Variations In Epistasis and Fitness Contribution in Kauffman's NK Model, *Journal of Theoretical Biology* **196**, 181–196.

Somogyi, R. and C. A. Sniegoski [1996], Modeling the Complexity of Genetic Networks: Understanding Multigenetic and Pleiotropic Regulation, *Complexity* **1**, 45–63.

Somogyvári, Z. and S. Payrits [2000], Length of State Cycles of Random Boolean Networks: an Analytic Study, *Journal of Physics A: Mathematical and General* **33**, 6699–6706.

Stadler, P. F. and R. Happel [1999], Random Field Models for Fitness Landscapes, *Journal of Mathematical Biology* **38**, 435–478.

Stauffer, D. [1985], *Introduction to Percolation Theory.* Taylor and Francis, London.

Stauffer, D. [1994], Evolution By Damage Spreading in Kauffman Model, *Journal of Statistical Physics* **74**, 1293–1299.

Stauffer, D. [1987a], On Forcing Functions in Kauffman Random Boolean Networks, *Journal of Statistical Physics* **46**, 789–794.

Stauffer, D. [1987b], Random Boolean Networks - Analogy With Percolation, *Philosophical Magazine B: Physics of Condensed Matter, Statistical Mechanics, Electronic, Optical and Magnetic Properties* **56**, 901–916.

Stauffer, D. [1988], Percolation Thresholds in Square-Lattice Kauffman Model, *Journal of Theoretical Biolog* **135**, 255–261.

Stauffer, D. [1989], Hunting for the Fractal Dimension of the Kauffman Model, *Physica D* **38**, 341–344.

Stauffer, D. [1991], Computer Simulations of Cellular Automata, *Journal of Physics A: Mathematical and General* **24**, 909–927.

Stern, M. D. [1999], Emergence of Homeostasis and Noise Imprinting in an Evolution Model, *Proceedings of the National Academy of Sciences of the U.S.A.* **96**, 10746–10751.

Stölzle, S. [1988], Universality Two-Dimensional Kauffman Model for Parallel and Sequential Updating, *Journal of Statistical Physics* **53**, 995–1004.

Strogatz, S. H. [2001], Exploring Complex Networks, *Nature* **410**, 268–276.

Thieffry, D. and D. Romero [1999], The Modularity of Biological Regulatory Networks, *Biosystems* **50**, 49–59.

Toffoli, T. and N. H. Margolus [1990], Invertible Cellular Automata: a Review, *Physica D* **45**, 229–253.

Volkert, L. G. and M. Conrad [1998], The Role of Weak Interactions in Biological Systems: the Dual Dynamics Model, *Journal of Theoretical Biology* **193**, 287–306.

Waelbroeck, H. and F. Zertuche [1999], Discrete Chaos, *Journal of Physics A: Mathematical and General* **32**, 175–189.

Wang, L., E. E. Pichler, and J. Ross [1990], Oscillations and Chaos in Neural Networks — an Exactly Solvable Model, *Proceedings of the National Academy of Sciences of the United States of America* **87**, 9467–9471.

Weinberger, E. D. [1991], Local Properties of Kauffman NK Model - a Tunably Rugged Energy Landscape, *Physical Review A* **44**, 6399–6413.

Weisbuch, G. and D. Stauffer [1987], Phase Transitions in Cellular Random Boolean Networks, *Jounal De Physique* **48**, 11–18.

Wilke, C. O., C. Ronnenwinkel, and T. Martinetz [2001], Dynamic Fitness Landscapes in Molecular Evolution, *Physics Reports* **349**, 395–446.

Wolfram, S. [1983], Statistical Mechanics of Cellular Automata, *Reviews of Modern Physics* **55**, 601–644.

Wuensche, A. [1999], Discrete Dynamical Networks and their Attractor Basins, *Complexity International* **6**, http://www.csu.edu.au/ci/idx–volume.html.

Zawidzki, T. W. [1998], Competing Models of Stability in Complex, Evolving Systems: Kauffman Vs. Simon, *Biology and Philosophy* **13**, 541–554.

Zoli, M., D. Guidolin, K. Fuxe, and L. F. Agnati [1996], The Receptor Mosaic Hypothesis of the Engram: Possible Relevance of Boolean Network Modeling, *International Journal of Neural Systems* **7**, 363–368.

3

Oscillatory Binary Fluid Convection in Finite Containers

Oriol Batiste
Edgar Knobloch

To Larry Sirovich, on the occasion of his 70th birthday.

ABSTRACT Linear and weakly nonlinear theory of overstable convection in large but bounded containers is reviewed and the results compared with detailed numerical simulations of binary fluid convection in a two-dimensional domain with realistic boundary conditions. For sufficiently negative separation ratios convection sets in as growing oscillations; the corresponding eigenfunctions take the form of 'chevrons' of either odd or even parity. These may bifurcate sub- or supercritically. Simulations of ^{3}He–^{4}He and water-ethanol mixtures show that the oscillations may equilibrate in finite amplitude chevron states, or that these states are unstable to blinking or repeated transient states. The results compare favorably with available experiments.

Contents

1 Introduction

Binary fluid mixtures with a negative separation ratio exhibit a wide variety of behavior when heated from below. Particular attention has focused on the transition to various types of traveling waves with increasing Rayleigh number, hereafter R. The experimental situation is summarized by Sullivan and Ahlers [1988], Kolodner, Surko, and Williams [1989], Steinberg, Fineberg, Moses, and Rehberg [1989] and Kolodner [1993], and sample data are reproduced in fig. 1.1. These experiments have either been carried out in narrow gap annular containers, or in extended rectangular boxes. The two experimental arrangements differ in a fundamental way. In the former the system is periodic and consequently the initial instability can develop into a uniform pattern of traveling waves. This is no longer so when sidewalls are present: the presence of sidewalls destroys the translation invariance present in the annular (or unbounded) system, with the result that the finite system has only a left-right reflection symmetry. Consequently, the eigenfunctions of the latter system are either odd or even under left-right reflection, but are otherwise unconstrained by the symmetries [Dangelmayr and Knobloch, 1987, 1991; Dangelmayr, Knobloch, and Wegelin, 1991]. In contrast, in the annular (or unbounded) case the presence of translation invariance with periodic boundary conditions forces the eigenfunctions to be sinusoidal functions with a single wavenumber in the horizontal. Such eigenfunctions take the form of left- and right-traveling waves. In systems with up-down reflection symmetry the additional symmetry may also affect the eigenfunctions and constrain the dynamics.

The difference in symmetry between the bounded and unbounded systems is crucial, and is present regardless of the aspect ratio of the system. It suggests that while unbounded systems are best described in terms of amplitude equations for the amplitudes of left- and right-traveling waves, bounded systems should be described in terms of odd and even modes, cf. Landsberg and Knobloch [1996]. As shown by Batiste, Mercader, Net, and Knobloch [1999] these modes typically have a complex spatial structure. We summarize here the properties of these eigenfunctions for the parameter values used in experiments and relate them to two classes of weakly nonlinear theories developed for the onset of oscillatory instability in large aspect ratio domains. In particular we show that for large values of the aspect ratio Γ the differences between the growth rates and frequencies of the first two modes that set in both scale as Γ^{-2}. This result supports the description of the system in terms of an interaction between the first even and odd modes [Landsberg and Knobloch, 1996]. Direct numerical simulations of the governing partial differential equations for both ^{3}He-^{4}He and water-ethanol mixtures confirm the important role played by these pure parity modes, and shed light on the presence of two classes of dynamical behavior observed in the experiments, referred to as *blinking states* and *repeated transients*. We show that both of these states are fundamentally finite-dimensional and that they may occur even in extended systems, provided only that these are not too large in the sense that $1 \ll \Gamma \lesssim |\epsilon|^{-1/2}$, where $\epsilon \equiv (R - R_c)/R_c$ measures the fractional distance from onset of the primary instability. Throughout this article we focus almost exclusively on this regime, since it is amenable to both theory and direct numerical simulation.

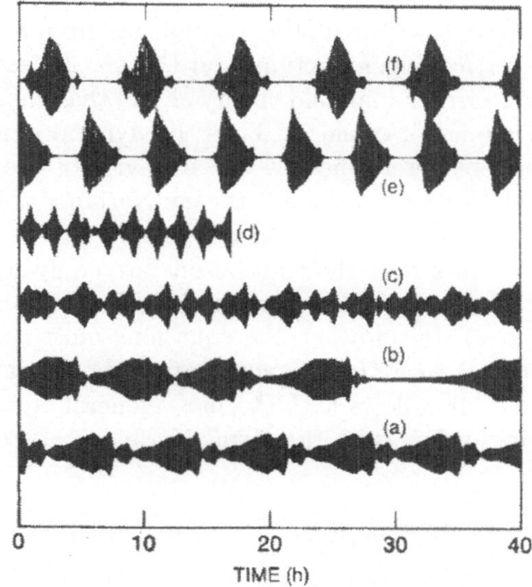

FIG. 3. The image intensity measured at a single spatial point in a cell of dimensions 4.90×16.75 is plotted as a function of time for different values of the Rayleigh number: (a) ϵ=0.0006; (b) ϵ=0.0008; (c) ϵ=0.0015; (d) ϵ=0.0030; (e) ϵ=0.0105; (f) ϵ=0.0111. The repeated-transient states seen at the lowest values of ϵ [(a), (b)] give way at higher ϵ to blinking states [(c), (d)] whose period grows quite long at the highest values of ϵ shown [(e), (f)] and diverges at ϵ~0.012.

FIGURE 1.1. Figure 3 of Kolodner [1993] with $\epsilon \equiv (R - R_c)/R_c$, reproduced with permission.

2 Abstract Considerations

We consider first an annular translation-invariant domain with left-right reflection symmetry and length Γ. Systems of this type necessarily possess a trivial (i.e., O(2)-symmetric) basic state, here the conduction state. An oscillatory instability of this state with a finite azimuthal wave number breaks the O(2) symmetry of the system. As a result the multiplicity of the purely imaginary eigenvalues at criticality is doubled. This is because the reflection-related eigenfunctions $\exp \pm ikx$ ($k \equiv 2\pi n/\Gamma \neq 0$) represent two *independent* eigenfunctions of the eigenvalue $i\omega_c$. The dynamics at such a bifurcation (called a Hopf bifurcation) can be described in terms of the (complex) amplitudes of these eigenfunctions, viz.,

$$\theta(x,z,t) = \mathrm{Re}[v(t)\,e^{ikx} + w(t)\,e^{-ikx}]\,f(z) + \cdots \tag{2.1}$$

Here θ represents the departure of the temperature from its conduction profile, $f(z)$ is the *vertical* eigenfunction, and the $\cdots$ represent higher order spatial harmonics. Center manifold theory shows that the latter are slaved to the (slow) dynamics of v and w. Thus the dynamics near the Hopf bifurcation are entirely determined by the behavior of the two amplitudes (v,w). The equations for these inherit the equivariance of the physical system under translations $x \to x + \ell$ and reflection $x \to -x$. An examination of equation (2.1) shows that the translations are equivalent to the operation $R_\ell : (v,w) \to (ve^{ik\ell}, we^{-ik\ell})$, while the reflections are equivalent to $R_x : (v,w) \to (w,v)$. In addition, the equations must commute with the operation $T_\varphi : (v,w) \to e^{i\varphi}(v,w)$ representing the effect of time translation, $t \to t + \varphi/\omega_c$. It follows that the most general equations for (v,w), truncated at third order, take the form [Knobloch, 1986; Crawford and Knobloch, 1991]

$$\begin{aligned} \dot{v} &= (\lambda + i\omega)v + b|v|^2 v + (a+b)|w|^2 v\,, \\ \dot{w} &= (\lambda + i\omega)w + b|w|^2 w + (a+b)|v|^2 w\,. \end{aligned} \tag{2.2}$$

In writing these equations we have included the small linear terms $(\lambda + i(\omega - \omega_c))(v,w)$, proportional to the bifurcation parameter $\epsilon \equiv (R - R_c)/R_c \ll 1$. In the following we refer to v, w as the amplitudes of left- and right-traveling waves. The truncated equations are valid under the nondegeneracy conditions $a_R \neq 0$, $b_R \neq 0$ and $a_R + 2b_R \neq 0$, where the subscript R indicates the real part.

Analysis of equations (2.2) shows that at $\lambda = 0$ two branches of solutions bifurcate simultaneously from the trivial state $(v,w) = 0$. These are the traveling waves $(v,w) = (v,0)$ and the standing waves $(v,w) = (v,v)$. Of course, the symmetries of the problem can be used to generate new solutions from these. Thus the reflection R_x shows that if $(v,0)$ is a solution so is $(0,v)$ while the translations R_ℓ generate $(e^{ik\ell}v, e^{-ik\ell}v)$ from (v,v), i.e., spatial translates of the standing waves. Note that the traveling waves have a spatio-temporal symmetry: $(v,0)$ is invariant under $T_{-k\ell} \circ R_\ell$, a spatial translation followed by an appropriate time translation, while the standing waves form a circle foliated by periodic orbits. Figure 2.1 summarizes the results of the analysis in the (a_R, b_R) plane and shows that at most one of the solutions can be stable, and that this requires that both branches bifurcate supercritically. Moreover, the stable branch is the one with the larger amplitude $A \equiv \sqrt{|v|^2 + |w|^2}$.

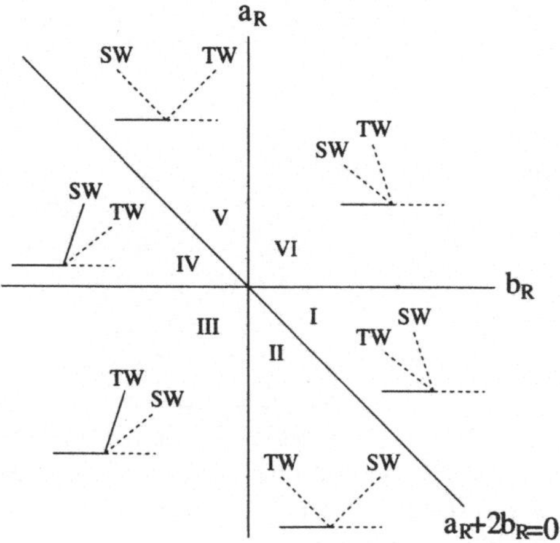

FIGURE 2.1. The solutions of eqs. (2.2) under the nondegeneracy conditions $a_R \neq 0$, $b_R \neq 0$, $a_R + 2b_R \neq 0$.

In a laterally bounded domain translation invariance is absent. Thus only the reflection symmetry remains, and this symmetry is a reflection about a particular point, the center $x = \Gamma/2$ of the container, rather than about an arbitrary point. The boundaries select from the circle of standing waves (SW) two representatives, with odd or even parity with respect to $x = \Gamma/2$, and these bifurcate in succession rather than simultaneously. Moreover, the loss of translation invariance implies that traveling waves (TW) can no longer bifurcate from the trivial state, and must therefore turn into secondary branches. From a symmetry point of view the resulting solutions are periodic in time but lack pure parity, i.e., they are neither odd nor even with respect to $x = \Gamma/2$. Consequently there are two such solutions, related by reflection, neither of which has spatio-temporal symmetry. These results follow from group-theoretic considerations [Dangelmayr and Knobloch, 1987, 1991]. But the loss of translation symmetry has other important consequences: it permits complex dynamics. The reason is simple. The two continuous symmetries of equations (2.2), originating in invariance under translations in space and time, imply that the phases $\arg(v)$, $\arg(w)$ decouple from the amplitudes $|v|$, $|w|$. The equations for the latter are therefore two-dimensional; it is these that lead to fig. 2.1. The loss of translation invariance implies that only one overall phase will decouple, and hence that the dynamics will become three-dimensional, permitting complex dynamics.

These notions can be made explicit [Dangelmayr and Knobloch, 1987, 1991]. To this end we include in equations (2.2) the dominant terms that break translation invariance (i.e., the symmetry R_ℓ) while preserving the reflection and normal form symmetries R_x, T_φ. The most general such equations with a linear symmetry-breaking term are the equations [Dangelmayr

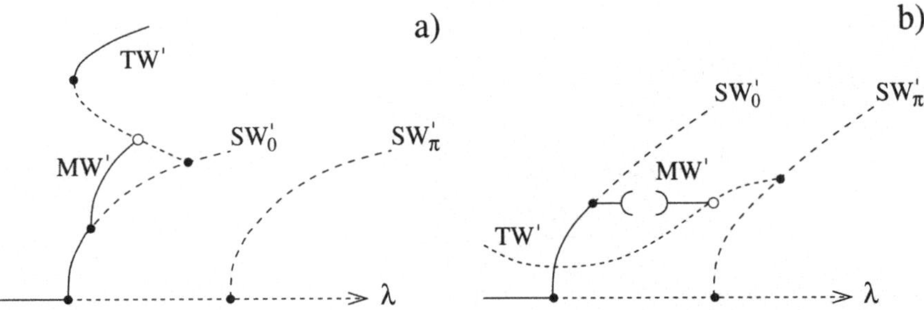

FIGURE 2.2. Schematic bifurcation diagrams from eqs. (2.3) in (a) region $B3_0$ of region III, and (b) region $C2_0$ of region II. The break in the MW′ curve in (b) indicates complicated behavior inaccessible to local analysis. The open circles indicate global bifurcations, while the filled circles correspond to local bifurcations. After Hirschberg and Knobloch [1996].

and Knobloch, 1987, 1991]

$$\begin{aligned} \dot{v} &= (\lambda + i\omega)v + ew + b|v|^2 v + (a+b)|w|^2 v\,, \\ \dot{w} &= (\lambda + i\omega)w + ev + b|w|^2 w + (a+b)|v|^2 w\,. \end{aligned} \tag{2.3}$$

Here e is a small but complex coefficient measuring the strength of the symmetry breaking. In writing eqs. (2.3) we have implicitly assumed that small symmetry-breaking cubic terms do not qualitatively affect the resulting bifurcation diagrams (but see Nagata [1991]).

Equations (2.3) have been studied in detail by Dangelmayr and Knobloch [1991], who showed that when $e \neq 0$ the double multiplicity Hopf bifurcation splits into two successive simple Hopf bifurcations each of which gives rise to a standing wave. Following Dangelmayr and Knobloch [1987, 1991] we refer to these as SW_0' and SW_π', the prime denoting solutions of equations (2.3) when $e \neq 0$. The SW_0' is an even mode while SW_π' is odd. Both solutions are characterized by time-independent amplitudes with $|v| = |w|$. There are no other primary bifurcations. Analogues of traveling waves, hereafter TW′, exist but only appear through secondary (pitchfork) bifurcations from the standing wave branches. The TW′ are single frequency solutions with $|v| \neq |w|$, $|vw| > 0$, with $|v| > |w|$ representing a state dominated by left-traveling waves and vice versa. More surprisingly perhaps, depending on the argument of e, an additional secondary Hopf bifurcation can also take place (fig. 2.2). This bifurcation gives rise to a two-frequency state MW′ in which the new frequency describes oscillations in the amplitudes $|v|, |w|$. Solutions of this type have been seen in experiments on water-ethanol mixtures [Kolodner and Surko, 1988] as well as in numerical simulations [Cross, 1986, 1988; Ning, Harada, and Yahata, 1997; Batiste, Knobloch, Mercader, and Net, 2001a], where they are referred to as *blinking* states. These states persist only in an interval of λ, and in the context of equations (2.3) terminate on the TW′ branch in either a Hopf or a global bifurcation (fig. 2.2). The appearance of such global bifurcations once translation invariance is broken provides the key to the origin of chaotic dynamics in the perturbed system.

An important additional effect of nonzero e is the appearance of stable behavior in parameter regions where the unperturbed problem has no stable solutions. For example, in region II of fig. 2.1 the SW branch bifurcates forwards and TW backwards, and hence neither is stable. In such a case for $\lambda > 0$ all solutions escape to large amplitude, and a study of the saturation of the instability therefore requires numerical simulations of model equations. This is in contrast to the perturbed problem in which the SW′ continue to bifurcate supercritically but the first SW′ branch is initially stable (Clune and Knobloch [1992]; see fig. 2.2b). Consequently the solutions MW′ and TW′ which bifurcate from it in secondary bifurcations also enjoy limited parameter windows of stability. Other examples of the effect of the perturbation $e \neq 0$ are described by Hirschberg and Knobloch [1996].

Equations (2.3) arise naturally as the small amplitude limit of the coupled complex Ginzburg-Landau (CCGL) equations that have been used to describe the dynamics of traveling waves in large aspect ratio containers. These equations, defined on the (long) interval $0 \leq x \leq \Gamma/\varepsilon$, take the following form

$$\begin{aligned} \mathcal{A}_T - s\mathcal{A}_X &= \Lambda\mathcal{A} + \tilde{b}|\mathcal{A}|^2\mathcal{A} + (\tilde{a}+\tilde{b})|\mathcal{B}|^2\mathcal{A} + D\mathcal{A}_{XX}\,, \\ \mathcal{B}_T + s\mathcal{B}_X &= \Lambda\mathcal{B} + \tilde{b}|\mathcal{B}|^2\mathcal{B} + (\tilde{a}+\tilde{b})|\mathcal{A}|^2\mathcal{B} + D\mathcal{B}_{XX}\,, \end{aligned} \tag{2.4}$$

together with the model boundary conditions

$$\begin{aligned} &\mathcal{A} + \varepsilon(\mu_1\mathcal{A}_X + \nu_1\mathcal{B}_X) = 0\,, \quad \mathcal{B} + \varepsilon(\mu_2\mathcal{B}_X + \nu_2\mathcal{A}_X) = 0\,, \quad X = 0\,, \\ &\mathcal{A} - \varepsilon(\mu_2\mathcal{A}_X + \nu_2\mathcal{B}_X) = 0\,, \quad \mathcal{B} - \varepsilon(\mu_1\mathcal{B}_X + \nu_1\mathcal{A}_X) = 0\,, \quad X = \Gamma\,, \end{aligned} \tag{2.5}$$

cf. Cross [1986]. Here $\mathcal{A}(X,T)$, $\mathcal{B}(X,T)$ denote the envelopes of left- and right-traveling waves with the slow variables $X \equiv \varepsilon x$, $T \equiv \varepsilon^2 t$, and $\tilde{a}$, $\tilde{b}$, Λ are complex coefficients. The scaled group velocity $s \equiv \frac{1}{\varepsilon}\frac{d\omega}{dk}$ is assumed to be of order unity. A straightforward center manifold reduction yields a solution in the form [Dangelmayr, Knobloch, and Wegelin, 1991]

$$\begin{aligned} \theta(x,t) =& \mathrm{Re}\,\varepsilon^{3/2}\left\{ v(T')\exp\left(\frac{s(\Gamma-2X)}{4D}\right) e^{ik_\infty x} \right. \\ &\left. + w(T')\exp\left(-\frac{s(\Gamma-2X)}{4D}\right) e^{-ik_\infty x} \right\} e^{i\omega_c t}\sin\frac{\pi X}{\Gamma} + \mathcal{O}(\varepsilon^{5/2})\,, \end{aligned} \tag{2.6}$$

where $T' \equiv \varepsilon T$ and the amplitudes $v(T')$, $w(T')$ satisfy (2.3), with the frequency ω_c factored out and $\lambda \equiv \Lambda - \Lambda_c$. Here $\Lambda_c = (s^2/4D) + (\pi^2 D/\Gamma^2)$, and ω_∞ and k_∞ are the onset frequency and wavenumber in the unbounded system [Knobloch and Moore, 1988], with $\omega_c - \omega_\infty = \mathcal{O}(\varepsilon)$. Note that the presence of the boundaries at $X = 0, \Gamma$ shifts the threshold of the instability by an $\mathcal{O}(1)$ amount even in the limit $\Gamma \to \infty$. This is because at leading order the boundary conditions (2.5) are *absorbing*.

The coefficients in (2.3) are computable functions of the parameters appearing in equations (2.4,2.5). Specifically,

$$e = -\left(\frac{2\pi^2 D}{\Gamma^3}\right)[\nu_1 e^{-s\Gamma/2D} + \nu_2 e^{s\Gamma/2D}], \qquad (2.7)$$

cf. Dangelmayr, Knobloch, and Wegelin [1991], indicating that in general the perturbation depends *exponentially* on the aspect ratio. However, since the amplitude of e can be scaled out the different possible bifurcation diagrams depend only on $\arg(e)$, a quantity that depends on Γ approximately linearly. Since a change of sign of e is equivalent to a change of sign of either v or w we see that increasing $\arg(e)$ by π is equivalent to replacing an even mode by an odd mode or vice versa. Equation (2.7) suggests therefore that the dynamics depend sensitively on Γ for $n\pi < \arg(e) < (n+1)\pi$, but more or less repeat for $(n+1)\pi < \arg(e) < (n+2)\pi$ except for a change in the parity of the primary modes SW$'$. This prediction is confirmed by both experiments and direct numerical simulations, as discussed further in section 5.

A striking consequence of the nonzero group velocity is the fact that the two constituent eigenfunctions peak in opposite halves of the container. As a result the waves near one end of the container travel in the opposite direction to those at the other end. Patterns obtained by the substitution of the simple solutions SW$'$, TW$'$ and MW$'$ for v and w in the wavefunction (2.6) resemble, respectively, the 'chevron', confined state, and blinking patterns observed in the experiments [Kolodner, Surko, and Williams, 1989; Steinberg, Fineberg, Moses, and Rehberg, 1989; Andereck, Liu, and Swinney, 1986; Croquette and Williams, 1989]. In fact equations (2.3) have chaotic solutions as well [Hirschberg and Knobloch, 1996] and these may be responsible for the irregularly reversing waves observed both in binary fluid convection [Kolodner, Surko, and Williams, 1989; Steinberg, Fineberg, Moses, and Rehberg, 1989] and in the counter-rotating Taylor-Couette system [Andereck, Liu, and Swinney, 1986]. These are typically the consequence of a global bifurcation. For example, depending on $\arg(e)$ or equivalently on the aspect ratio Γ, the heteroclinic connection with which the MW$'$ branch terminates in region III (see fig. 2.2a and Knobloch [1996]) may connect two *saddle-foci* (the TW$'$) with eigenvalues λ_u, $-\lambda_s \pm i\Omega$, $\lambda_u > 0$, $\lambda_s > 0$, $\Omega > 0$, satisfying the Šil'nikov inequality $0 < \delta < 1$. Here $\delta \equiv \lambda_s/\lambda_u$. The formation of a global connection of this type thus results in Šil'nikov dynamics [Glendinning and Sparrow, 1984], and so provides a natural mechanism for the origin of chaotic reversals.

While promising, a comparison of the above approach with the CCGL equations suggests that the distant boundaries may only be treated perturbatively when propagative effects are weak and ε is finite. However, this requirement prevents us from proceeding to the asymptotic limit $\varepsilon \to 0$ required by the theory. Consequently we develop below a theory based on the even and odd modes characteristic of a bounded container. We let

(z_+, z_-) be the (complex) amplitudes of these modes, and note that due to the parabolic nature of the neutral stability curve in unbounded systems the first even and odd modes will typically come in in close succession, well separated from the next mode, cf. Jacqmin and Heminger [1994]. If this is so it makes sense to project the partial differential equations onto these two modes. The resulting equations must be equivariant with respect to

$$R_x : (z_+, z_-) \to (z_+, -z_-), \tag{2.8}$$

owing to the reflection symmetry $x \to \Gamma - x$. However, in the large aspect ratio limit one must expect that there is in addition an *interchange* symmetry between the odd and even modes since these are effectively indistinguishable throughout most of the domain. Thus the normal form should also be equivariant under [Landsberg and Knobloch, 1996]

$$I : (z_+, z_-) \to (z_-, z_+). \tag{2.9}$$

These two reflections generate the symmetry D_4. However, since the interchange symmetry is not exact for any finite Γ (at any finite Γ the first mode is either odd or even, except for a discrete set of Γ) this D_4 symmetry is broken. The dominant interchange-breaking terms are once again linear, and hence the system should be described by amplitude equations of the form [Landsberg and Knobloch, 1996]

$$\begin{aligned} \dot{z}_+ &= (\mu_+ + i\omega_+)z_+ + A|z_+|^2 z_+ + B|z_-|^2 z_+ + C\bar{z}_+ z_-^2 \,, \\ \dot{z}_- &= (\mu_- + i\omega_-)z_- + A|z_-|^2 z_- + B|z_+|^2 z_- + C\bar{z}_- z_+^2 \,. \end{aligned} \tag{2.10}$$

Here A, B, C are complex $\mathcal{O}(1)$ coefficients and the small interchange-breaking parameters $\Delta\mu \equiv \mu_+ - \mu_-$, $\Delta\omega \equiv \omega_+ - \omega_-$ capture the effects of a finite aspect ratio. As shown below both $\Delta\mu$ and $\Delta\omega$ are $\mathcal{O}(\Gamma^{-2})$ for large Γ; consequently the two modes remain close to 1:1 resonance for all Γ and the resonant terms $(\bar{z}_+ z_-^2, \bar{z}_- z_+^2)$ must be retained.

It is of interest to rewrite equations (2.10) in terms of the traveling wave coordinates (v, w), where $z_+ = v + w$, $z_- = v - w$:

$$\begin{aligned} \dot{v} &= (\mu + i\omega)v + \frac{1}{2}(\Delta\mu + i\Delta\omega)w + b|v|^2 v + (a+b)|w|^2 v + cw^2\bar{v} \,, \\ \dot{w} &= (\mu + i\omega)w + \frac{1}{2}(\Delta\mu + i\Delta\omega)v + b|w|^2 w + (a+b)|v|^2 w + cv^2\bar{w} \,, \end{aligned} \tag{2.11}$$

where $\mu \equiv \frac{1}{2}(\mu_+ + \mu_-)$, $\omega = \frac{1}{2}(\omega_+ + \omega_-)$, and $a = A - B - 3C$, $b = A + B + C$, $c = A - B + C$. There is a notable difference between these equations and (2.3): the presence of the terms $(w^2\bar{v}, v^2\bar{w})$ is evidently *nonperturbative* and indicates that the sidewalls play an important role in the near-onset behavior of the system, however large the system! Indeed, as demonstrated by Renardy [1999], equations of the form (2.11), and hence (2.10), can also be derived via center manifold reduction of a pair of (local) coupled complex Ginzburg-Landau equations (2.4) on a periodic domain with the (generic) boundary conditions

$$\begin{aligned} \mathcal{A}(0,t) + i\mathcal{B}(0,t) &= \mathcal{A}(\Gamma,t) - i\mathcal{B}(\Gamma,t) = 0, \\ \mathcal{A}_X(0,t) - i\mathcal{B}_X(0,t) &= \mathcal{A}_X(\Gamma,t) + i\mathcal{B}_X(\Gamma,t) = 0. \end{aligned} \tag{2.12}$$

Since the two approaches just summarized lead to qualitatively different predictions in the large aspect ratio limit, we focus in the following on some of the assumptions behind them, and confront the predictions with direct numerical simulations in two dimensions.

In the same spirit one can examine the effect of breaking of translation invariance on the transition between oscillatory and steady convection that occurs when ω_c is *small*, i.e., $S \approx S_{TB}$. In the absence of sidewalls the different possibilities are captured by the normal form for the Takens-Bogdanov bifurcation with O(2) symmetry [Knobloch, 1986]. Breaking this symmetry at leading order while preserving reflection symmetry leads to the equation

$$\ddot{v} = \mu v + \nu\dot{v} + A|v|^2 v + C(\bar{v}\dot{v} + v\dot{\bar{v}}) + D|v|^2\dot{v} + E\bar{v} + F\dot{\bar{v}}. \tag{2.13}$$

Here μ, ν are two real unfolding parameters, and we have written $\theta(x,z,t) = \mathrm{Re}\, v(t)\exp ikx + \cdots$. The coefficients $A, \ldots, F$ are also real, with E, F breaking the translation symmetry $v \to v\exp ik\ell$. Unfortunately the properties of this equation for $EF \neq 0$ remain unknown.

3 Convection in Binary Mixtures

Binary fluid mixtures are characterized by the presence of cross-diffusion terms in the diffusion matrix. In liquids the dominant cross-diffusion term is the Soret term, and the sign of this term determines the behavior of the mixture in response to an applied temperature gradient. For mixtures with a negative Soret coefficient the heavier component migrates towards the hotter boundary, i.e., a concentration gradient is set up that opposes the destabilizing temperature gradient that produced it. Under these conditions the onset of convection may take the form of growing oscillations. This is the situation that is of interest here.

We consider a binary mixture in a two-dimensional rectangular container $D \equiv \{x,z | 0 \le x \le \Gamma, -\frac{1}{2} \le z \le \frac{1}{2}\}$ heated uniformly from below, and nondimensionalize the equations using the depth of the layer as the unit of length and t_d, the thermal diffusion time in the vertical, as the unit of time. In the Boussinesq approximation appropriate to the experiments the resulting equations take the form [Clune and Knobloch, 1992]

$$\mathbf{u}_t + (\mathbf{u}\cdot\boldsymbol{\nabla})\mathbf{u} = -\boldsymbol{\nabla}P + \sigma R[\theta(1+S) - S\eta]\hat{\mathbf{z}} + \sigma\nabla^2\mathbf{u}, \tag{3.1}$$

$$\theta_t + (\mathbf{u}\cdot\boldsymbol{\nabla})\theta = w + \nabla^2\theta, \tag{3.2}$$

$$\eta_t + (\mathbf{u}\cdot\boldsymbol{\nabla})\eta = \tau\nabla^2\eta + \nabla^2\theta, \tag{3.3}$$

together with the incompressibility condition

$$\boldsymbol{\nabla}\cdot\mathbf{u} = 0. \tag{3.4}$$

Here $\mathbf{u} \equiv (u, w)$ is the velocity field in (x, z) coordinates, P is the pressure, and θ denotes the departure of the temperature from its conduction profile, in units of the imposed temperature difference ΔT. The variable η is defined such that its gradient represents the dimensionless mass flux. Thus $\eta \equiv \theta - C$, where C denotes the concentration of the heavier component relative to its conduction profile in units of the concentration difference that develops across the layer as a result of the Soret effect. The system is specified by four dimensionless parameters: the Rayleigh number R providing a dimensionless measure of the imposed temperature difference ΔT, the separation ratio S that measures the resulting concentration contribution to the buoyancy force due to the Soret effect, and the Prandtl and Lewis numbers σ, τ, in addition to the aspect ratio Γ.

To model the experiments we take the boundaries to be no-slip everywhere, with the temperature fixed at the top and bottom and no sideways heat flux. The final set of boundary conditions is provided by the requirement that there is no mass flux through any of the boundaries. The boundary conditions are thus

$$\mathbf{u} = \mathbf{n}\cdot\boldsymbol{\nabla}\eta = 0 \text{ on } \partial D, \tag{3.5}$$

and

$$\theta = 0 \text{ at } z = \pm 1/2, \qquad \theta_x = 0 \text{ at } x = 0,\, \Gamma. \tag{3.6}$$

Here ∂D denotes the boundary of D.

Equations (3.1-3.6) are equivariant with respect to the operations

$$R_x: \qquad (x,z) \to (\Gamma - x, z), \qquad (\psi, \theta, C) \to (-\psi, \theta, C), \tag{3.7}$$

$$\kappa: \qquad (x,z) \to (x, -z), \qquad (\psi, \theta, C) \to (-\psi, -\theta, -C), \tag{3.8}$$

where $\psi(x, z, t)$ is the streamfunction, defined by $(u, w) = (-\psi_z, \psi_x)$. These two operations generate the symmetry group D_2 of a rectangle. It follows that even solutions, i.e., solutions invariant under R_x, satisfy $(\psi(x,z), \theta(x,z), C(x,z)) = (-\psi(\Gamma - x, z), \theta(\Gamma - x, z), C(\Gamma - x, z))$ at each instant in time, while odd solutions are invariant under κR_x and satisfy $(\psi(x,z), \theta(x,z), C(x,z)) = (\psi(\Gamma - x, -z), -\theta(\Gamma - x, -z), -C(\Gamma - x, -z))$, again at each instant of time. At midlevel, $z = 0$, these solutions therefore satisfy $(\psi(x,0), \theta(x,0), C(x,0)) = (\psi(\Gamma - x, 0), -\theta(\Gamma - x, 0), -C(\Gamma - x, 0))$.

4 Linear Theory

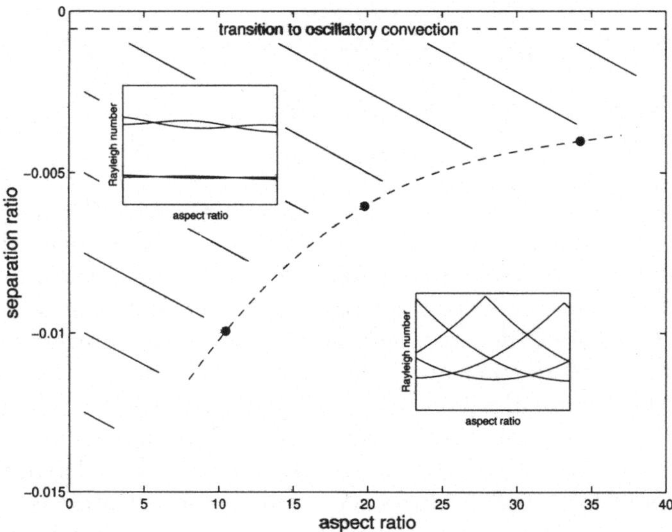

FIGURE 4.1. The (Γ, S) plane for ^{3}He-^{4}He parameters showing the approximate location of mode avoidance (hatched region) and of mode crossing (unhatched region).

To determine the critical value of the Rayleigh number R at which the conduction state $\psi = \theta = C = 0$ loses stability to overstable convection and the corresponding frequency ω_c we look for solutions to the linearized equations of the form $f(x, z)e^{(s+i\omega)t}$. The condition $s = 0$ defines the onset of instability and yields a complex condition that can be solved for $R \equiv R_c$ and $\omega \equiv \omega_c$ as a function of the aspect ratio Γ for various values of σ, τ and the separation ratio S. In the following we summarize our results for the parameters $\sigma = 0.6$, $\tau = 0.03$ (typical of ^{3}He-^{4}He mixtures) and $\sigma = 6.97$, $\tau = 0.0077$ (typical of water-ethanol mixtures). These choices are motivated by the experiments of Sullivan and Ahlers [1988] and Kolodner [1993], respectively.

The solution of this problem [Batiste, Mercader, Net, and Knobloch, 1999] indicates that the competition between odd and even modes in this system takes one of two basic forms, depending on the separation and aspect ratios. When $|S|$ is small (i.e., close to $|S_{TB}|$, the Takens-Bogdanov point) and Γ not too large the mode interaction takes the form familiar from Rayleigh-Bénard convection with non-Neumann boundary conditions: the neutral curves $R_c(\Gamma)$ divide neatly between different families with no intermingling among them. Each family consists of a pair of braided neutral curves, one for an odd mode and the other for an even mode, with each family well separated from the next, at least for the low-lying families. The crossings between odd and even modes within each family are structurally stable because of their different parity. For the case of interest here, i.e., Γ and $|S|$ large enough, the situation is quite different. There are now no distinct families of neutral curves and all modes (including like-parity modes) cross. In general these mode crossings are all structurally stable, either because the modes have opposite parity, or because their frequencies at the mode crossing are nonresonant. Figure 4.1 shows the location of the transition between these types of behavior in the (Γ, S) plane.

4.1 ^{3}He–^{4}He Mixtures

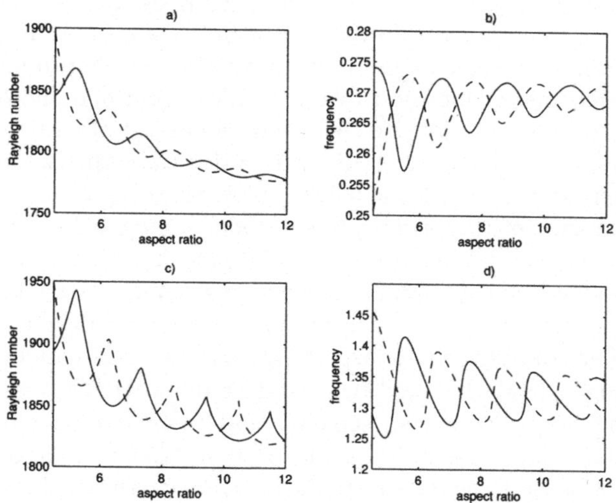

FIGURE 4.2. Onset of convection in ^{3}He-^{4}He mixtures ($\sigma = 0.6$, $\tau = 0.03$) in moderate aspect ratio containers. (a) The critical Rayleigh number R_c and (b) the corresponding frequency ω_c for the first even (solid line) and odd (dashed line) mode as a function of the aspect ratio Γ for $S = -0.001$. (c,d) The same but for $S = -0.01$.

We describe first the results for typical ^{3}He-^{4}He parameters, $\sigma = 0.6$, $\tau = 0.03$, and modest aspect ratios, $4 \leq \Gamma \leq 12$. Figure 4.2 shows the eigenvalues $R_c(\Gamma)$ and $\omega_c(\Gamma)$ for two values of the separation ratio, $S = -0.001$ and $S = -0.01$, in each case for the first even (solid line) and the first odd (dashed line) mode. For $S = -0.001$ fig. 4.2 reveals an oscillatory approach of both sets of curves towards the critical Rayleigh number R_∞ and ω_∞ for an unbounded domain. The braiding of the neutral stability curves $R(\Gamma)$ seen in fig. 4.2a is familiar from studies of stationary Rayleigh-Bénard convection [Daniels, 1977; Nagata, 1990; Hirschberg and Knobloch, 1997]. Because of the braiding the neutral stability curves for the first odd and even modes cross repeatedly. Such mode crossings indicate the presence of codimension two bifurcations. Figure 4.2b shows that at these points the frequencies of the competing modes are distinct. Consequently these mode interaction points correspond to nonresonant double Hopf bifurcations. However, when $S = -0.01$ the situation changes: for large enough aspect ratios ($\Gamma > 10$) the two neutral curves develop cusps (fig. 4.2c). The presence of these cusps is reflected in the discontinuous jumps in the corresponding frequency curves (fig. 4.2d). Figure 4.3 shows the development of these cusps with decreasing S, focusing on the range $10 \leq \Gamma \leq 12$. The figure shows the first *two* even (solid lines) and odd (dashed lines) modes at (a,b) $S = -0.005$, (c,d) $S = -0.008$ and (e,f) $S = -0.01$. In fig. 4.3a thick (thin) lines are used to indicate the first (second) mode of each parity and this coding is used to identify the corresponding modes in figs. 4.3c,e. In the latter the dotted and dashed-dotted curves indicate even and odd modes originating from yet higher modes in fig. 4.3a. Observe that in fig. 4.3a the neutral stability curves for the two odd modes avoid one another, as do the corresponding curves for the two even modes. At the same time the two sets of frequency curves intertwine. As S decreases the two odd modes come together near $\Gamma = 10.5$ and their frequencies coalesce, apparently with the frequency of the primary even mode; similar behavior involving the two even modes takes place near $\Gamma = 11.5$ (see fig. 4.3d). At somewhat smaller S the two Rayleigh number curves cross transversely (fig. 4.3e) as the first and second modes of each parity trade place, forming the cusps seen in fig. 4.2c. In the range of aspect ratios shown this happens first for the even modes, closely followed by the odd modes. At the same time the corresponding frequency curves separate and thereafter no longer cross. The same interchange mechanism is also responsible for the appearance of the cusp in the second even mode neutral stability curve near $\Gamma = 10.3$ with a yet higher order even mode involved (dotted line), with similar behavior occurring for the second odd mode near $\Gamma = 11.3$ as well (fig. 4.3c). These results suggest (and more detailed calculations [Batiste, Mercader, Net, and Knobloch, 1999] confirm) that the necessary crossings between modes of like parity originating in adjacent families are mediated by double Hopf bifurcations with 1:1 resonance located at discrete points $(R_c^{(3)}, \Gamma_c^{(3)}, S_c^{(3)})$ in the three-dimensional parameter space (R, Γ, S) (cf. fig. 4.1).

In fig. 4.4 we show the bifurcating modes for $\Gamma = 10$ when $S = -0.001$ (top panels) and $S = -0.01$ (lower panels). Both modes are even and are represented in the form of space-time diagrams, with time increasing upwards. When $S = -0.001$ the eigenvector takes the form of a standing wave, with the dynamics at the two sidewalls in phase. The amplitude of the standing oscillations peaks in the middle of the container and decreases towards the sidewalls. There is a considerable phase lag between the temperature and concentration oscillation, a consequence of the small value of τ. As S decreases the eigenvector gradually develops into a 'chevron'-like

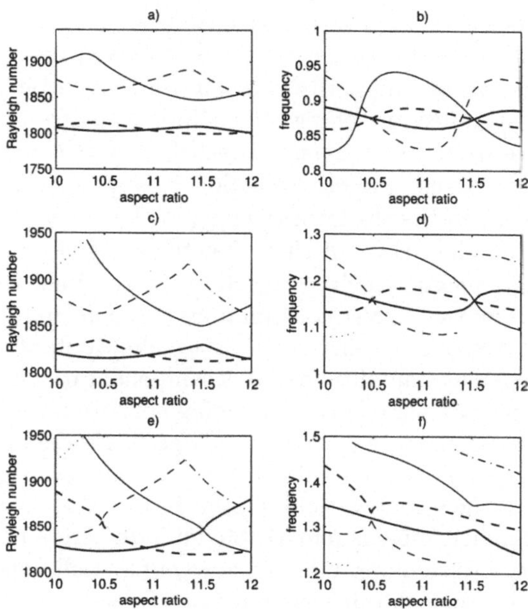

FIGURE 4.3. Details of the reconnection process between two modes of like parity for (a,b) $S = -0.005$, (c,d) $S = -0.008$, and (e,f) $S = -0.01$ when $\sigma = 0.6$, $\tau = 0.03$. In (a) solid (dashed) lines denote even (odd) modes while thick (thin) lines denote first (second) modes of each type.

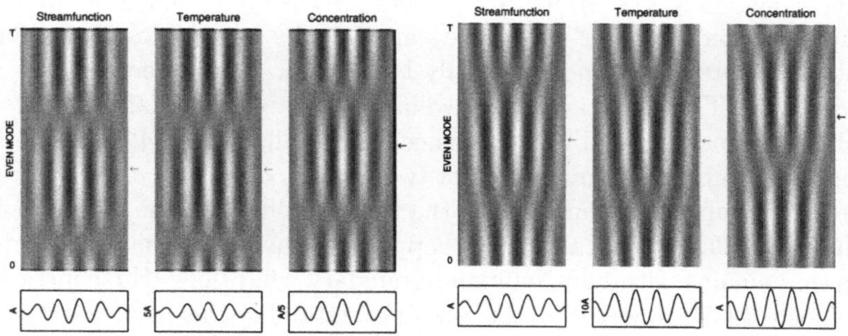

FIGURE 4.4. The eigenfunction (ψ, θ, C) of the linear stability problem at $z = 0$ for $\Gamma = 10$, $\sigma = 0.6$, $\tau = 0.03$, $S = -0.001$, $R_c = 1784.09$, $\omega_c = 0.267$ (left panels), and $S = -0.01$, $R_c = 1827.72$, $\omega_c = 1.35$ (right panels), shown in the form of space-time diagrams with $0 \leq x \leq \Gamma$ horizontally and time increasing upwards. The solutions are sinusoidal with period $T \equiv 2\pi/\omega_c$. In the former case the eigenfunctions are standing waves but become 'chevron'-like in the latter. The amplitude variation across the domain is indicated below each space-time plot.

state, consisting of waves propagating outward from the container center in such a way that the reflection symmetry in $x = \Gamma/2$ is preserved at all times (fig. 4.4b). In contrast an odd parity chevron (at $z = 0$) is at all times odd with respect to this reflection, cf. fig. 4.8b. Note that despite appearances these solutions are *strictly sinusoidal* in time: the periodic defect formation at $x = \Gamma/2$ arises because the eigenfunction ψ is a superposition of four functions each of which has the form $\psi_j \exp i(\omega_c t \pm k_j x)$, $j = 1, \ldots, 4$, with the k_j possibly complex. Each of these functions describes waves propagating with (local) phase velocity $\pm\omega_c/\mathrm{Re}k_j$. In the bulk the eigenfunction is dominated by the largest contribution; this contribution has a real wavenumber and describes oscillations that are almost standing. However, when the time-dependent amplitude of this component passes through zero (which occurs twice per period) the remaining contributions briefly reveal themselves. In the eigenfunction shown the largest of these has a relatively large phase velocity, and is responsible for the episodic propagation that is so characteristic of the eigenfunctions shown. In both fig. 4.4a,b the dominant local wavenumber remains remarkably uniform across the cell despite the nonuniformity of the amplitude of the eigenfunction. The sources (or sinks) described by these *linear* eigenfunctions persist into the nonlinear regime (section 5), indicating that in the present system these defects have a nontopological origin.

It is of interest to compare the above results with those for $\Gamma = 34$, the aspect ratio used by Sullivan and Ahlers [1988]. In fig. 4.5 we show the neutral stability curves and corresponding frequencies for the first four modes in the range $33 \leq \Gamma \leq 35$ for $S = -0.001$, $S = -0.004$ and $S = -0.01$. A comparison with figs. 4.2 and 4.5 reveals that the frequencies of the dominant modes are determined primarily by the fluid parameters and not the aspect ratio. This is because the oscillations are bulk oscillations that are modified but not caused by the presence of sidewalls. Figure 4.5a shows that when $|S|$ is sufficiently small the first two families of neutral curves are separated by a gap that is much larger than the amplitude of the braids within each family. This is typical of what happens in Rayleigh-Bénard convection in large domains with non-Neumann boundary conditions [Hirschberg and Knobloch, 1997]. However, in the case of overstability this behavior changes as $|S|$ increases (fig. 4.5c) and begins to look like that shown in fig. 4.5e. This figure shows the corresponding neutral curves for $S = -0.01$ and reveals the crossing of adjacent even modes. This mode crossing involves a *nonresonant* double Hopf bifurcation (fig. 4.5f) and is the result of a resonant 1:1 mode crossing at $S = -0.00403$ (see figs. 4.5c,d), i.e., it is formed by the same process as that leading to the nonresonant crossings shown in figs. 4.3e,f. The fact that the frequency curves in fig. 4.5f are essentially parallel "straight lines" confirms that this mode crossing is "far" from the 1:1 resonance at $S_c^{(3)} = -0.00403$. The figure is also in agreement with the plausible hypothesis that in large aspect ratio systems the frequencies of

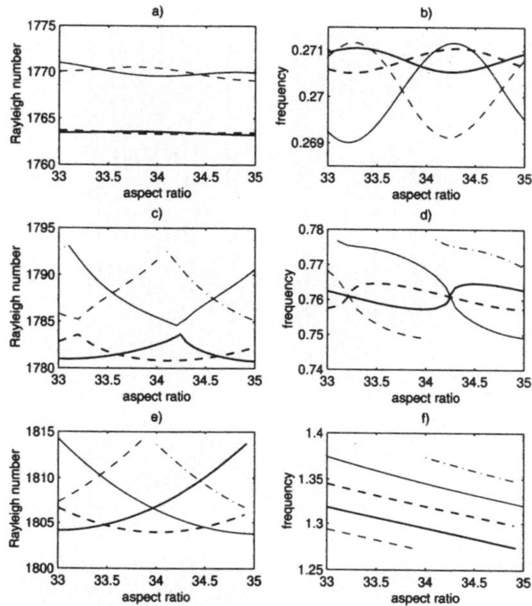

FIGURE 4.5. Critical Rayleigh numbers R_c and onset frequencies ω_c for ^{3}He-^{4}He mixtures ($\sigma = 0.6$, $\tau = 0.03$) in large aspect ratio containers when (a,b) $S = -0.001$, (c,d) $S = -0.004$ and (e,f) $S = -0.01$. In (a) solid (dashed) lines denote even (odd) modes while thick (thin) lines denote first (second) modes of each type.

the first few modes must take the form

$$\omega_n \sim \omega_\infty + c_{1n}\Gamma^{-1} + c_{2n}\Gamma^{-2} + \cdots, \qquad n = 1, 2, \ldots, \tag{4.1}$$

where the index n specifies the order in which the modes become primary as Γ increases. Thus the nth mode is primary in the interval $\Gamma_{n-1} \leq \Gamma \leq \Gamma_n$, etc. It follows that $n = \mathcal{O}(\Gamma)$. The results of fig. 4.5 suggest that $c_{1n} \sim c_1$, $c_{2n} \sim nc_2$, where c_1 and c_2 are $\mathcal{O}(1)$ constants independent of n and Γ. Then at any (large) Γ there is an $\mathcal{O}(\Gamma^{-1})$ correction to ω_∞, while $\Delta_n\omega \equiv \omega_{n+1} - \omega_n = \mathcal{O}(\Gamma^{-2})$. Thus near any particular Γ the quantity $\Delta_n\omega$ takes the form, as a function of Γ, of a set of equally spaced almost horizontal lines. We have checked that similar behavior occurs for $S = -0.021$ as well. Figure 4.5 therefore suggests that for large Γ the splitting $\Delta\omega$ in frequency and ΔR in Rayleigh number between the first odd and even modes are both of the *same* order as $\Gamma \to \infty$. This result supports the hypothesis [Landsberg and Knobloch, 1996] that the normal form describing the interaction between odd and even modes in the large aspect ratio limit has approximate D_4 symmetry, as discussed in section 2.

Figure 4.6 shows the first odd temperature eigenfunctions for $S = -0.001$ and $S = -0.021$, again in the form of space-time diagrams. In the former

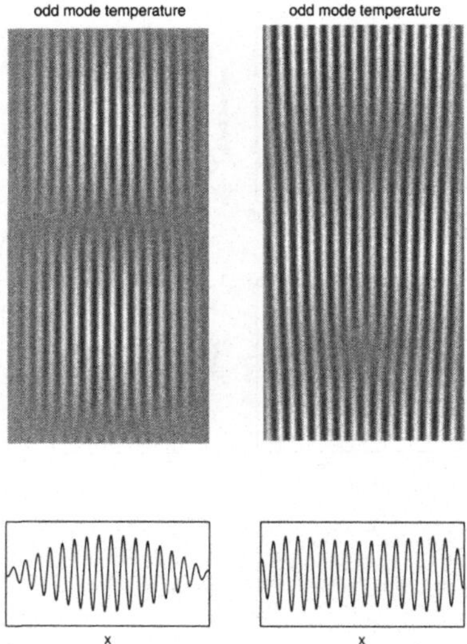

FIGURE 4.6. Odd parity eigenfunctions $\theta(x, 0, t)$ of the linear stability problem for $\Gamma = 34$, $\sigma = 0.6$, $\tau = 0.03$, $S = -0.001$, $R_c = 1768.12$, $\omega_c = 0.272$ (left panel), and $S = -0.021$, $R_c = 1833.86$, $\omega_c = 2.025$ (right panel), shown in the form of space-time diagrams with $0 \leq x \leq \Gamma$ horizontally and time increasing upwards. The solutions are sinusoidal with period $T \equiv 2\pi/\omega_c$. The eigenfunctions in the former case are standing waves but become 'chevron'-like in the latter. The lower panels show the corresponding midplane temperature profiles at $t = 0.3T$, $t = 0.5T$, respectively.

case the eigenfunction is essentially a standing oscillation with a wavelength that is once again very uniform across the container despite the fact that the amplitude varies substantially (lower left panel). When $S = -0.021$ the direction of propagation is outwards with the center of the container having become a source. The amplitude now has a local minimum at the center and increases outwards, peaking near the sidewalls (lower right panel), cf. eq. (2.6) with $s > 0$. For aspect ratios this large the odd and even (not shown) eigenfunctions are essentially indistinguishable [Batiste, Mercader, Net, and Knobloch, 1999].

4.2 Water–Ethanol Mixtures

In fig. 4.7 we show the linear theory results for $S = -0.021$, $\sigma = 6.97$, $\tau = 0.0077$, corresponding to the experimental mixture used by Kolodner [1993]. The corresponding critical eigenfunctions are shown in fig. 4.8 in the form of space-time diagrams for the three fields (ψ, θ, C) evaluated at $z = 0$. As shown in fig. 4.7a the first odd and even modes cross near $\Gamma = 16.8$. Unlike the odd-odd crossing near $\Gamma = 16.25$ this mode interaction is accessible from the conduction state and hence plays an important role in the dynamics of the system (see below).

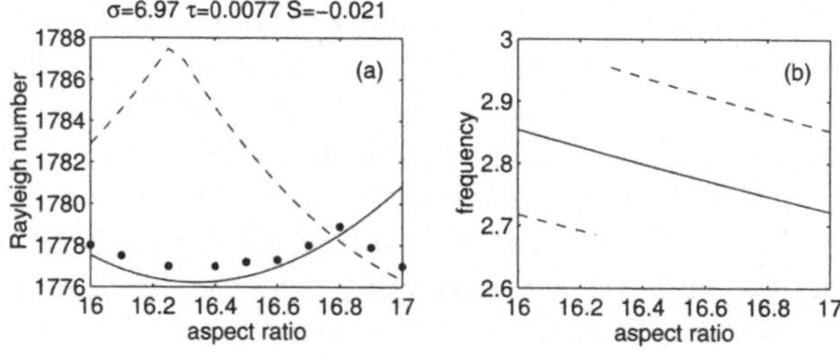

FIGURE 4.7. (a) The critical Rayleigh number R_c and (b) the corresponding frequency ω_c for $S = -0.021$, $\sigma = 6.97$, $\tau = 0.0077$ as a function of the aspect ratio Γ. Solid (broken) lines indicate even (odd) parity chevrons. The solid dots correspond to the solutions shown in fig. 5.20.

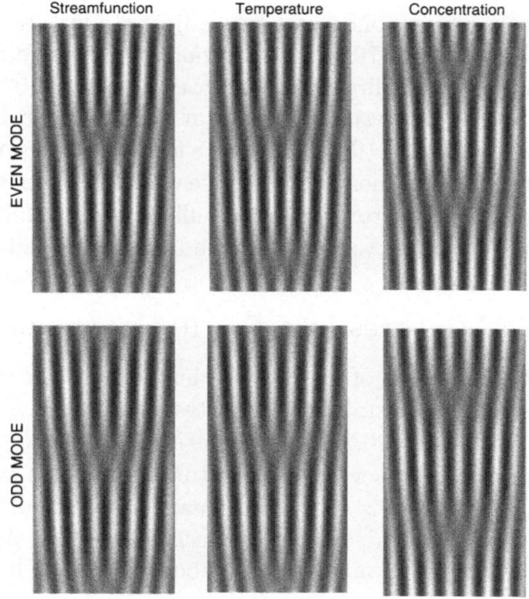

FIGURE 4.8. The eigenfunction (ψ, θ, C) of the linear stability problem at $z = 0$ for $\Gamma = 16.25$, $S = -0.021$, $\sigma = 6.97$, $\tau = 0.0077$ and (a) $R_c = 1776.30$, $\omega_c = 2.819$ (even chevron), (b) $R_c = 1787.47$, $\omega_c = 2.686$ (odd chevron), shown in the form of space-time diagrams with $0 \leq x \leq \Gamma$ horizontally and time increasing upwards. The solutions are sinusoidal with period $2\pi/\omega_c$.

5 Simulations

Direct numerical simulations of equations of the form (3.1-3.4) in two dimensions with idealized boundary conditions have revealed the presence of complex dynamics associated with standing waves in small aspect ratio systems [Knobloch, Moore, Toomre, and Weiss, 1986] and of chevron-like states in larger domains [Deane, Knobloch, and Toomre, 1988], both of which are related to finite-dimensional dynamics [Knobloch, Proctor, and Weiss, 1993]. Three-dimensional simulations of Rayleigh-Bénard convection in order one domains likewise yield evidence for finite-dimensional behavior [Sirovich and Deane, 1991]. Unfortunately for the parameter values and boundary conditions relevant to experiments equations (3.1-3.6) possess very long transients, even in two dimensions, requiring considerable patience in order to obtain reliable results. We solve these equations using a time-splitting method with an improved boundary condition for the pressure and a second order accurate time integration scheme based on a modified Adams-Bashforth formula [Hugues and Randriamampianina, 1998]. For the spatial discretization we use a Chebyshev collocation pseudospectral method [Zhao and Yedlin, 1994]. In all cases the time step and the number of collocation points used was adjusted until the solutions converged. Typically we used 170 collocation points in the x-direction and 30 collocation points in the z-direction, with a time step of $10^{-3}t_d$.

Throughout we use the vertical velocity at the points $(x, z) = (0.13\Gamma, 0)$ (near the left sidewall) and $(0.87\Gamma, 0)$ (near the right sidewall) as a proxy for shadowgraph intensity measurements (see fig. 1.1). Moreover, monitoring point quantities at mirror locations suffices to determine the spatial symmetry properties of the various possible time-dependent states.

5.1 ^{3}He–^{4}He Mixtures in a $\Gamma = 10$ Container

For $S = -0.001$ simulations of the growing instability at $R = 1785 > R_c \approx 1784.088$ show that the instability saturates in an even parity standing wave with frequency $\omega_1 = 0.25$ near the critical frequency $\omega_c \approx 0.2675$. Figure 5.1 shows the time series of the saturated vertical velocity $w(x = 0.87\Gamma, z = 0, t)$, and indicates that the primary bifurcation is a supercritical Hopf bifurcation; no evidence of hysteresis was found. With increasing Rayleigh number this state undergoes a (subcritical) Hopf bifurcation that introduces a new frequency ω_2 into the dynamics. Strictly speaking this bifurcation is a torus bifurcation. However, in the following we do not distinguish between Hopf bifurcations of equilibria and of periodic orbits (or tori), since resonance phenomena appear to play little role in the observed dynamics. Figure 5.1 shows that stable single frequency and two-frequency states coexist at $R = 1785.5$.

The two-frequency state can be identified with the *blinking* states predicted by the abstract theory. Figure 5.2 shows that this state has the

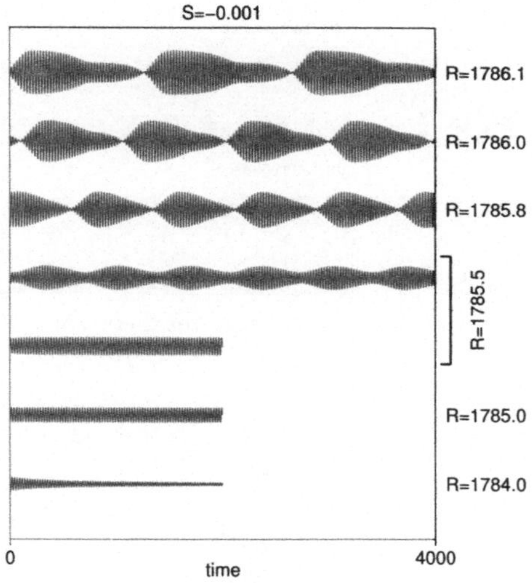

FIGURE 5.1. Time series $w(x = 0.87\Gamma, z = 0, t)$ in a ^{3}He-^{4}He mixture with $S = -0.001$, $\sigma = 0.6$, $\tau = 0.03$ for several different values of the Rayleigh number.

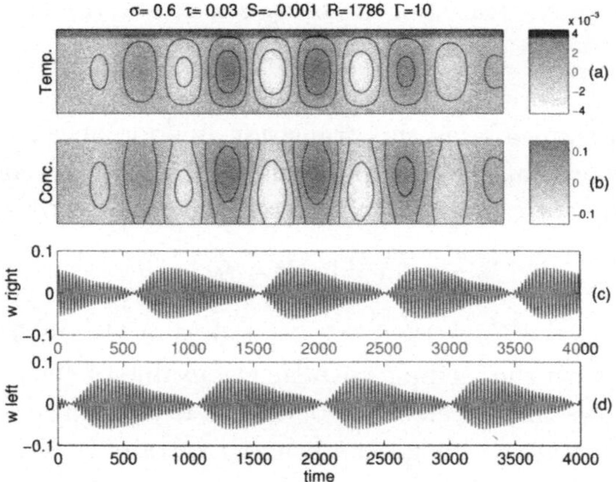

FIGURE 5.2. As for fig. 5.1 but showing a symmetric periodic blinking state at $R = 1786$. (a,b) Contours of $\theta(x, z, t)$ and $C(x, z, t)$ at $t = 4000$. (c,d) $w(x = 0.87\Gamma, z = 0, t)$ and $w(x = 0.13\Gamma, z = 0, t)$.

required symmetry: if we ignore for the moment the fast frequency ω_1 the blinking state has the symmetry $R_x w(x, 0, t) = w(x, 0, t + T_2/2)$, where $T_2 \equiv 2\pi/\omega_2$ is the blinking period. In the following we refer to states of this type as *symmetric periodic blinking states*. When R is increased to $R = 1786.2$ this state loses stability and the system jumps to a large amplitude even parity steady state (fig. 5.3). The modulation period T_2 appears

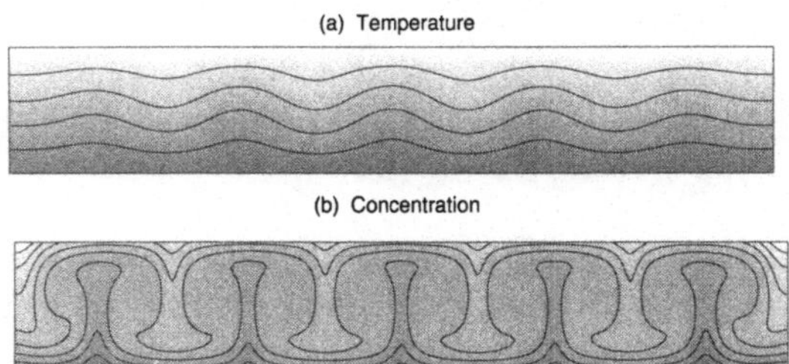

FIGURE 5.3. The large amplitude steady state reached when R is increased from $R = 1786$ to $R = 1786.2$.

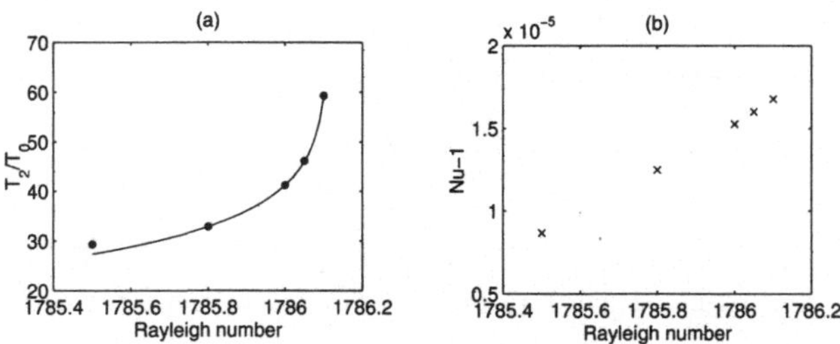

FIGURE 5.4. (a) The blinking period T_2 in units of the Hopf period $T_0 \equiv 2\pi/\omega_c$ as a function of R near the transition to steady convection. (b) The corresponding Nusselt number.

to diverge logarithmically as this transition is approached (fig. 5.4), suggesting that the oscillations disappear when the two-torus collides with an (unstable) steady state branch. A fit to the theoretical prediction

$$T_2 = -2\lambda_u^{-1} \ln |R - R_h| + d \tag{5.1}$$

leads to the estimates $R_h \approx 1786.112$, $\lambda_u \approx 0.2452$, $d \approx 8.1537$. Here λ_u is to be identified with the leading *unstable* eigenvalue of the steady state.

Figure 5.5 shows that when $S = -0.021$ the transition to steady convection is quite different. Although the primary instability is still to an even mode ($R_c = 1855.75$, $\omega_c = 2.076$) the bifurcation is now slightly hysteretic, so that stable even chevrons are present even for $R < R_c$. As R is raised a second frequency appears in the time series, corresponding to the onset of a symmetric blinking state. This bifurcation is apparently also subcritical. This behavior is qualitatively similar to that described for $S = -0.001$. However, with increasing R the modulation period T_2 begins to increase but then apparently saturates at a finite value (fig. 5.5d). At the same time the period $T_1 \equiv 2\pi/\omega_1$ appears to diverge (fig. 5.5c). These properties are reflected in the time series presented in fig. 5.5a, which show that at $R = 1862.0$ the blinking state loses stability to steady convection from the lowest frequency portion of the wavetrain, and indicate that the transition to steady convection now occurs via a radically different mechanism (see

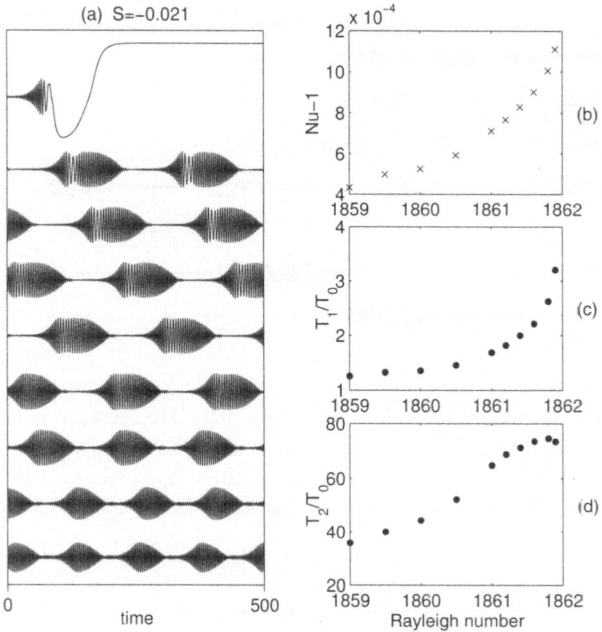

FIGURE 5.5. A ^{3}He-^{4}He mixture with $S = -0.021$, $\sigma = 0.6$, $\tau = 0.03$ in a $\Gamma = 10$ container. (a) Time series $w(x = 0.87\Gamma, z = 0, t)$ for different values of R increasing upwards ($1859.5 < R < 1862.0$), showing a hysteretic transition to steady convection at $R \approx 1862.0$. (b) The corresponding Nusselt number as a function of R. (c) The chevron period $T_1 \equiv 2\pi/\omega_1$ as a function of R. (d) The blinking period $T_2 \equiv 2\pi/\omega_2$ as a function of R. The state at $R = 1861.8$ resembles the "fish state" observed in experiments.

below). In the following we refer to the state just prior to this transition as a *fish state*, cf. Kolodner and Surko [1988].

The case $S = -0.1$ is even more interesting. Here $R_c = 1972.13$, $\omega_c = 4.918$. In this case the primary bifurcation to the (even) chevron state is substantially subcritical, and the first stable nonlinear state takes the form of a *repeated transient* (fig. 5.6). States of this type were studied in detail by Kolodner [1993] in experiments on water-ethanol mixtures, and their origin is discussed in detail in section 6. Figure 5.6 suggests that these states are *three-frequency* states, in which ω_1 is the fast chevron frequency, ω_2 represents the blinking frequency, while the third frequency ω_3 represents the slow modulation frequency. Figure 5.6 shows that these states consist of long intervals consisting of a slowly growing (even) chevron state. Instead of saturating this state becomes unstable to the onset of blinking, which leads to a collapse of the state to small amplitude, followed by a slow regrowth. In the time series shown these collapse events are periodic with period $T_3 = 2\pi/\omega_3$. In section 6 we show that states of this type come about via

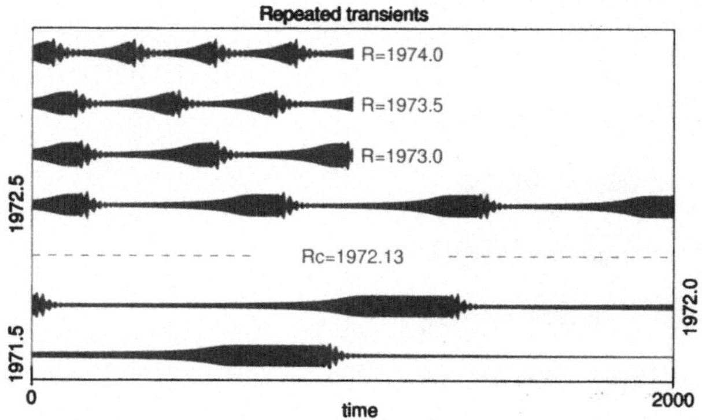

FIGURE 5.6. The repeated transient state in a ^{3}He-^{4}He mixture with $S = -0.1$, $\sigma = 0.6$, $\tau = 0.03$ and $\Gamma = 10$ at several values of R. The time series $w(x = 0.87\Gamma, z = 0, t)$ suggest the presence of three frequencies, with the lowest frequency (ω_3) increasing from zero as R increases from $R \approx 1971.5$.

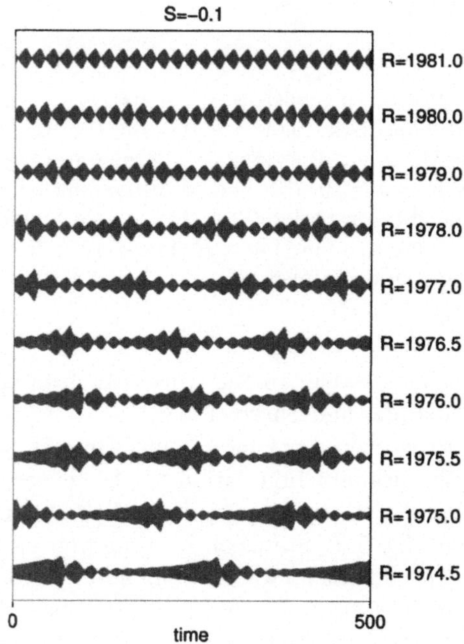

FIGURE 5.7. The time series $w(x = 0.87\Gamma, z = 0, t)$ for a ^{3}He-^{4}He mixture with $S = -0.1$, $\sigma = 0.6$, $\tau = 0.03$ and $\Gamma = 10$ at several values of R showing the transition from the repeated transient state in fig. 5.6 to a symmetric periodic blinking state at $R = 1981$.

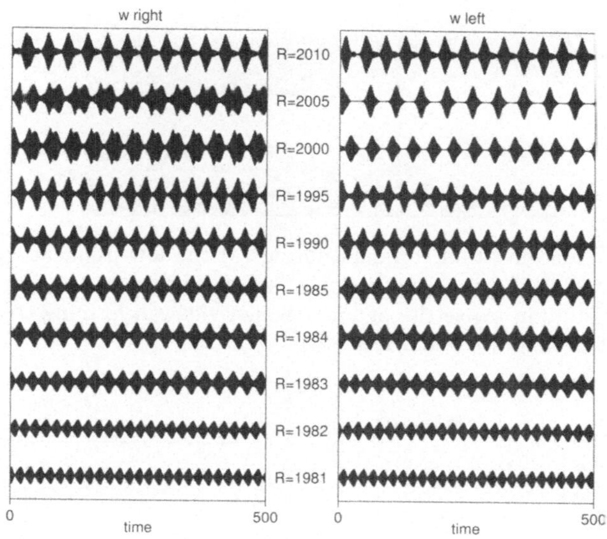

FIGURE 5.8. Time series $w(x = 0.87\Gamma, z = 0, t)$ (left) and $w(x = 0.13\Gamma, z = 0, t)$ (right) for different values of R increasing upwards, and $S = -0.1$, $\sigma = 0.6$, $\tau = 0.03$, $\Gamma = 10$.

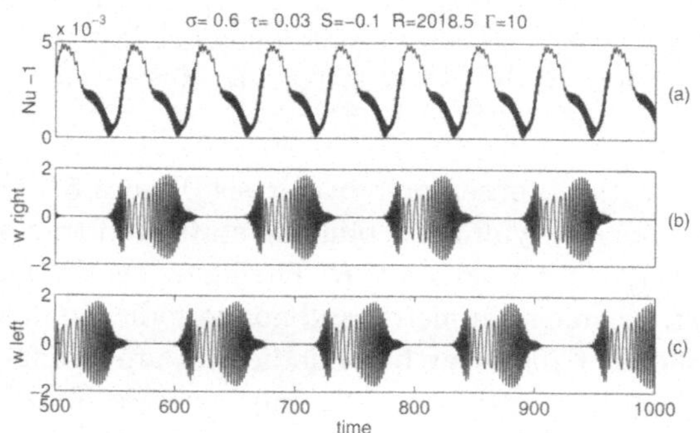

FIGURE 5.9. (a) The Nusselt number $\mathrm{Nu}(t)$, (b) $w(x = 0.87\Gamma, z = 0, t)$, (c) $w(x = 0.13\Gamma, z = 0, t)$ for $R = 2018.5$, and $S = -0.1$, $\sigma = 0.6$, $\tau = 0.03$, $\Gamma = 10$.

a very natural mechanism, and argue that quasiperiodic states with three frequencies should not be thought of as rare even in the absence of any symmetries that in other systems prevent frequency locking.

As R is increased ω_3 gradually increases, but drops out from the time series between $R = 1980.0$ and $R = 1981.0$ (fig. 5.7). We identify this transition with a Hopf bifurcation, and note that fig. 5.7 indicates that this bifurcation is *supercritical*, i.e., viewed in the direction of decreasing R this bifurcation creates a stable three-frequency state from a stable two-

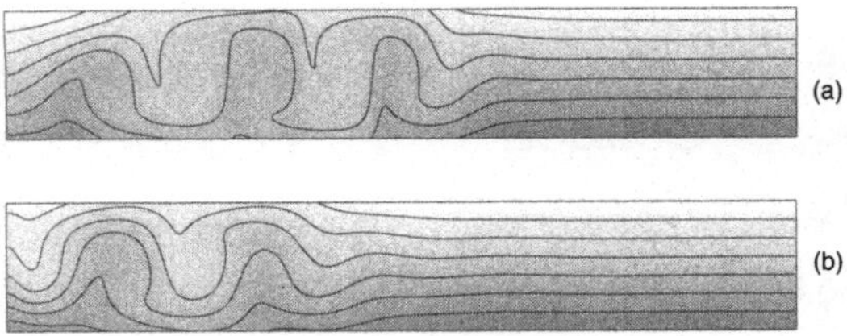

FIGURE 5.10. Concentration contours during (a) the low frequency part of the 'fish' state in fig. 5.9, and (b) during the high frequency phase that follows it. The localized state in (a) settles against the left sidewall forming temporarily a confined left-traveling wave shown in (b).

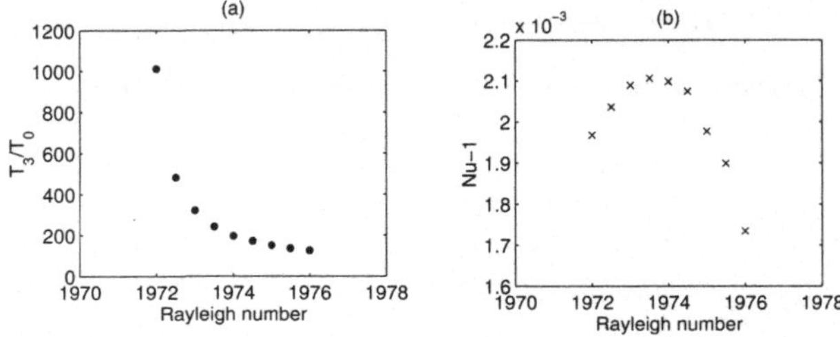

FIGURE 5.11. (a) The modulation period $T_3 \equiv 2\pi/\omega_3$ of the repeated transients when $S = -0.1$, $\sigma = 0.6$, $\tau = 0.03$, $\Gamma = 10$ in units of $T_0 \equiv 2\pi/\omega_c$ as a function of the Rayleigh number R. (b) The corresponding Nusselt number.

frequency state, with no observable hysteresis. Figure 5.8 shows that this two-frequency state is a symmetric blinking state, and traces the evolution of this state towards larger values of R. The figure shows that at $R = 1995$ the blinking has become asymmetric and nonperiodic, while the time series for $R = 2000$ and $R = 2005$ may be periodic, but are strongly asymmetric. In contrast, when $R = 2010$ the blinking becomes once again periodic and symmetric. None of these transitions appear to be hysteretic. At yet larger values of R these states again evolve into the fish state, followed by a transition to stationary convection. Figure 5.9 shows an example of the fish state just prior to this transition. An examination of the spatial structure of the waves shows that there is an instant during which the waves takes the form of a small amplitude chevron state filling the container. This state is unstable, however, with the waves at one end growing at the expense of those at the other. At this point the system shifts into a new state, one in which the waves are spatially confined towards one side, with *no* waves at the other (see fig. 5.10a). This transition is marked by a dramatic drop in the frequency ω_1. This change in frequency in turn increases the Nusselt number, and does so despite the fact that the waves no longer fill the whole domain. As time proceeds this confined state settles next to the boundary (see fig. 5.10b) and during this process both $w(x = 0.87\Gamma, z = 0, t)$ and ω_1

increase rapidly, although the Nusselt number falls, developing a prominent shoulder. The state is now so confined that waves start to regrow at the other sidewall, restoring the small amplitude extended chevron state. As this occurs the amplitude of the confined state falls and the state begins to expand, the frequency ω_1 increasing towards the Hopf frequency ω_c. The appearance of a dynamically confined state implies that the time series shown in fig. 5.9 cannot be understood in terms of the type of theory summarized in section 2. Indeed the value of ϵ corresponding to fig. 5.9, $\epsilon = 2.35 \times 10^{-2}$, indicates that for these values of R the dynamics of the system are no longer necessarily dominated by the sidewalls (since $\Gamma \gtrsim \epsilon^{-1/2}$). Indeed our calculations are consistent with the suggestion that the fish states first become possible when $\Gamma \sim \epsilon^{-1/2}$, a scaling suggested by Ginzburg-Landau theory.

Figure 5.11 reveals a fundamental property of the repeated transient state: when R is decreased from $R = 1974$ the period T_3 increases rapidly and apparently diverges at $R \approx 1971.5$; no stable solutions are present for smaller values of R. Thus the three-frequency repeated transient state represents the *first* nontrivial state of this system.

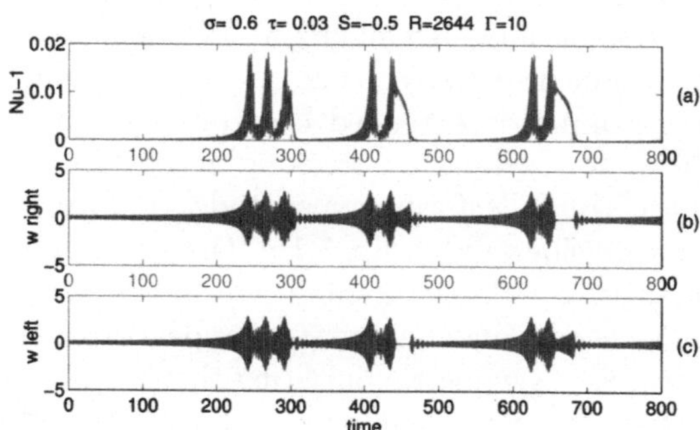

FIGURE 5.12. (a) The Nusselt number Nu(t), (b) $w(x = 0.87\Gamma, z = 0, t)$, (c) $w(x = 0.13\Gamma, z = 0, t)$ for $R = 2644$ ($\epsilon = 2.1 \times 10^{-4}$), and $S = -0.5$, $\sigma = 0.6$, $\tau = 0.03$, $\Gamma = 10$, showing irregular bursts.

The final case we have considered is $S = -0.5$. Here $R_c = 2643.43$, $\omega_c = 12.836$. The primary instability is again to an even chevron, but this time we find strongly irregular dynamics already quite close to onset (fig. 5.12). The time series in fig. 5.12 is best described as intermittent repeated transients, in which the final collapse event may be preceded by several spatially symmetric bounces before the onset of the symmetry-breaking instability that disrupts the state and leads to the temporary formation of a growing confined state at one of the sidewalls, much as already described for $S = -0.1$. As this state grows it shrinks in lateral extent, until it triggers another collapse event that permits waves to regrow at the other sidewall. The decaying symmetric blinking state that results reestablishes a small amplitude chevron state that then regrows on a much longer timescale. The spatially symmetric bounces are associated with relatively sharp peaks in the Nusselt number, while the symmetry-breaking collapse events produce bursts in the Nusselt number that are markedly asymmetric, much as in the fish state discussed above. It should be noted that the burst state coexists with a stable *odd* parity chevron state with a superposed small amplitude temporal modulation that is exactly out of phase in the two halves of the domain. This modulation has a complex waveform but appears to be periodic in time. This state is therefore a symmetric blinking state, and indeed with decreasing R one finds that the modulation disappears and (at $R = 2606$) a stable odd parity chevron is recovered. This multistability greatly complicates the behavior of the system, and appears to be a consequence of increasing subcriticality of the primary bifurcations with decreasing separation ratio.

The frequency of the burst-like events in the Nusselt number increases with R (fig. 5.13). A periodic sequence of bursts, produced by a symmetric albeit complex state, is shown in fig. 5.14. Perhaps the most remarkable time series of all is shown in fig. 5.15 for $R = 2750$. This series apparently shows an irregular switching between two states, a large amplitude state with a relatively low ω_1, and a small amplitude state with a large ω_1. The former is a wave confined spatially to one or other sidewall, while the latter is a more-or-less symmetric extended chevron-like state. This pulsating state loses stability to a symmetry-breaking instability which kills off the waves at one of the sidewalls, forming a spatially confined state that slowly drifts to the opposite sidewall (cf. fig. 5.10). Once in contact with the wall the increased dissipation reduces its amplitude and permits waves to regrow at the other wall. The growing blinking state appears to glue with its mirror image, briefly forming a symmetric pulsating chevron, which is unstable in turn to the original symmetry-breaking instability. These gluing transitions can be seen in figs. 5.7 and 5.8 as well.

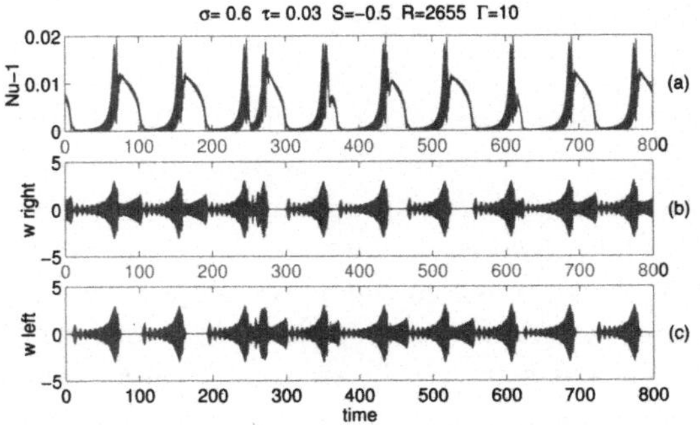

FIGURE 5.13. As for fig. 5.12 but showing the time series corresponding to $R = 2655$.

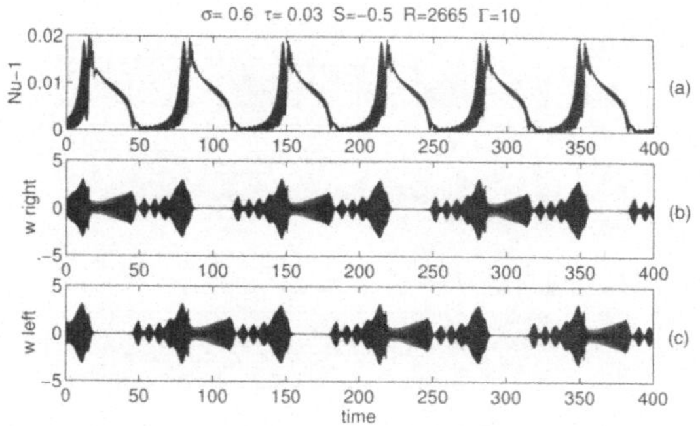

FIGURE 5.14. As for fig. 5.12 but showing periodic bursts at $R = 2665$.

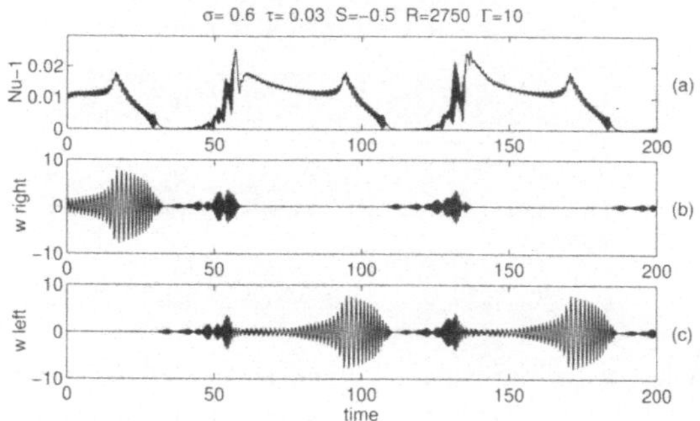

FIGURE 5.15. Irregular switching between the "fish state" and a "blinking state" at $R = 2750$.

In fig. 5.16 we show a strongly confined traveling wave found at $R = 2900$. A wave of this type evolves from the fish state when the localized low frequency state comes to rest against a sidewall, but does not collapse. The resulting wave should be interpreted in terms of a stationary *front* separating an exponentially small amplitude wave throughout most of the domain from a finite amplitude wavetrain next to the left sidewall. Consequently fig. 5.16 represents a *dynamically* localized wave, instead of the kinematic localization described by steady states of equations (2.3) with $|v| \gg |w| > 0$. In the latter the amplitude of the counterpropagating is never zero, and a small amplitude right-traveling wave is always visible near the right sidewall; the resulting state is nothing but a strongly asymmetric chevron. In contrast fig. 5.16 shows no evidence of a right-traveling wave in any part of the container except perhaps right next to the left sidewall. Such waves are described well by a *single* complex Ginzburg-Landau equation with a drift (see eqs. (2.4) with $\mathcal{B} \equiv 0$), and hence become possible once $|\epsilon| \gtrsim \Gamma^{-2}$. Theory based on this equation predicts [Cross, 1986, 1988; Tobias, Proctor, and

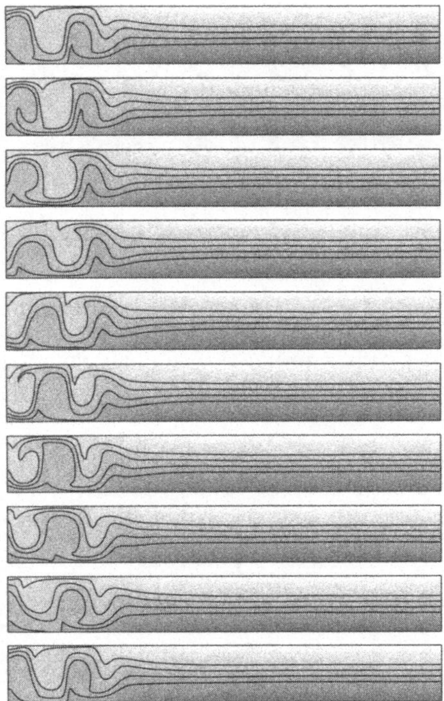

FIGURE 5.16. A left-traveling wave confined to the left sidewall when $R = 2900$, and $S = -0.5$, $\sigma = 0.6$, $\tau = 0.03$, $\Gamma = 10$, shown in terms of the total concentration contours, with time increasing upward.

Knobloch, 1998] that with increasing R the front gradually moves towards the right, but in the present case the strong nonlinear dispersion forces the frequency towards zero; once this occurs the resulting non-oscillatory state begins to expand towards the right by adding (steady) rolls and thereby expelling the lateral concentration gradient set up by the confined traveling wave. The process of adding rolls terminates once this lateral gradient is sufficiently strong, and results in the formation of a spatially confined but *steady* state (see fig. 5.17). Such confined steady states have also been found for other values of R (fig. 5.18) indicating that the confined states created by this process are in general non-unique. All of these states are numerically stable. However, to our knowledge no comparable states have been observed in any experiment.

Figure 5.19 shows the chaotic blinking state at $R = 2665$ that coexists stably with the bursts shown in fig. 5.14. As already mentioned this state is based on an odd parity chevron, in contrast to fig. 5.14 which is based on an even chevron.

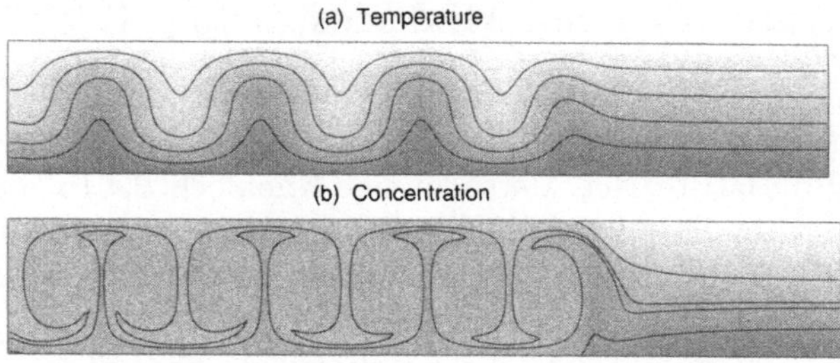

FIGURE 5.17. A stable spatially confined steady state at $R = 3050$ when $S = -0.5$, $\sigma = 0.6$, $\tau = 0.03$, $\Gamma = 10$.

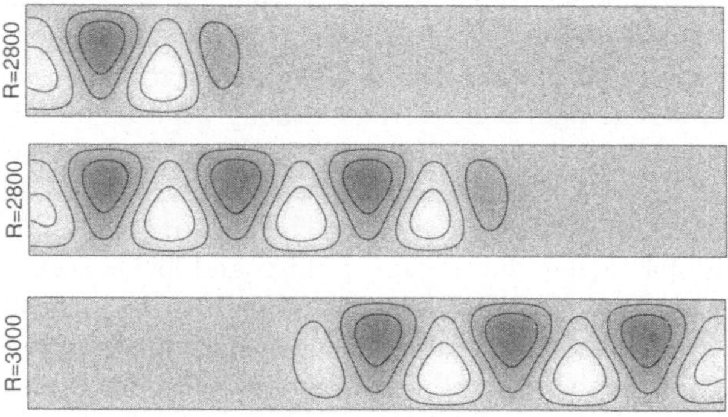

FIGURE 5.18. Other stable spatially confined steady states when $S = -0.5$, $\sigma = 0.6$, $\tau = 0.03$, $\Gamma = 10$. The corresponding value of R is indicated at the right of each panel.

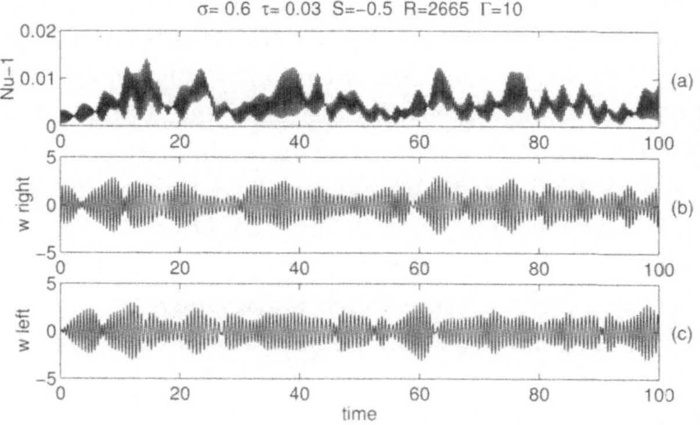

FIGURE 5.19. A stable chaotic blinking state at $R = 2665$ based on an odd parity chevron that coexists with the bursts shown in fig. 5.14.

5.2 Water-Ethanol Mixtures in $16 \leq \Gamma \leq 17.25$ Containers

In this section we describe our results [Batiste, Knobloch, Mercader, and Net, 2001a; Batiste, Net, Mercader, and Knobloch, 2001b] for the parameter values used by Kolodner [1993] in his experiments, $S = -0.021$, $\sigma = 6.97$, $\tau = 0.0077$, focusing on Rayleigh numbers near threshold (i.e., $|\epsilon| \ll 1$) and on their dependence on the aspect ratio Γ.

Figure 5.20 summarizes the evolution for different values of Γ of the midplane vertical velocity $w(x = 0.87\Gamma, z = 0, t)$ obtained by integration over $2000t_d$ after an initial transient has (almost) died out. The figure illustrates the sensitive dependence on the aspect ratio for comparison with figure 2 of Kolodner [1993], as well as the long integration times required to get reliable results. The high frequency uniform amplitude states correspond to nonlinear time-periodic chevron states such as the one shown in fig. 5.21. The figure shows that although the temperature departure θ from the conduction profile remains sinusoidal in space at this value of ϵ this is not so for the concentration departure C. As explained by Barten, Lücke, Hort, and Kamps [1989] this is a consequence of the small value of τ. Note in particular that regions of high and low concentration departure are separated by open contours, in contrast to the temperature field. The temporary straightening of these meandering concentration contours in the cell center every half period accompanies the splitting of the central concentration roll into two. Both these properties of the concentration field are absent from the Ginzburg-Landau description of this system.

In order to understand the origin and character of the nonperiodic states seen in fig. 5.20 we show in figs. 5.22 and 5.23 detailed results for $\Gamma = 16.25$ and $\Gamma = 16$, and ϵ in the vicinity of $\epsilon = 0$. These plots show that the initial bifurcation to the chevron state is subcritical, in agreement with the prediction for *standing waves* in a horizontally unbounded layer [Clune and Knobloch, 1992; Schöpf and Zimmermann, 1989]. Figure 5.22 for $\Gamma = 16.25$ shows that the (even) chevron state can equilibrate at finite amplitude ($R = 1775.5$). However, with increasing ϵ the stable chevrons lose stability in a supercritical Hopf bifurcation. This bifurcation introduces a new frequency into the system, seen in fig. 5.22 ($R = 1776.0$) as an oscillation in the amplitude of $w(x = 0.87\Gamma, z = 0, t)$; the resulting state is a symmetric periodic blinking state. Such states may set in already for $\epsilon < 0$; as ϵ increases the blinking amplitude increases and the blinking becomes nonperiodic.

The results for $\Gamma = 16$ (fig. 5.23) are quite different. In this case no subcritical stable chevrons are observed, and instead the first nontrivial state of the system appears to be a three-frequency state ($R = 1777.2$). We shall see in section 6 that such states are entirely natural in systems of this type. Figure 5.23 also shows that with increasing ϵ this state evolves into one increasingly like Kolodner's repeated transients, with a growth

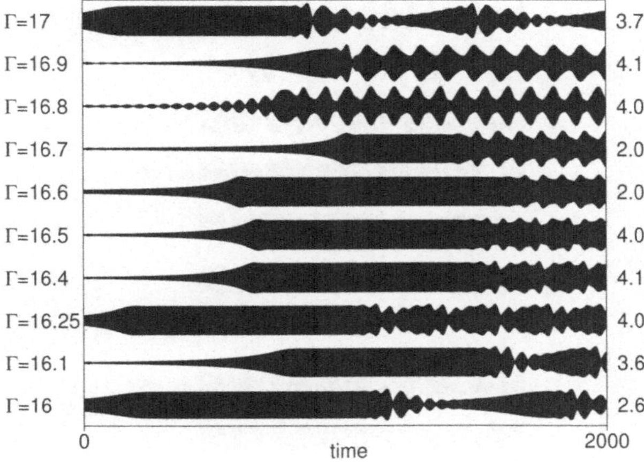

FIGURE 5.20. Water-ethanol mixtures with $S = -0.021$, $\sigma = 6.97$, $\tau = 0.0077$: an overview of the aspect ratio dependence of the equilibrated states near onset, in terms of the vertical velocity $w(x = 0.87\Gamma, z = 0, t)$ for comparison with figure 2 of Ref. [8]. The numbers at the right give the values of $10^4\epsilon$ and correspond to the solid dots in fig. 4.7.

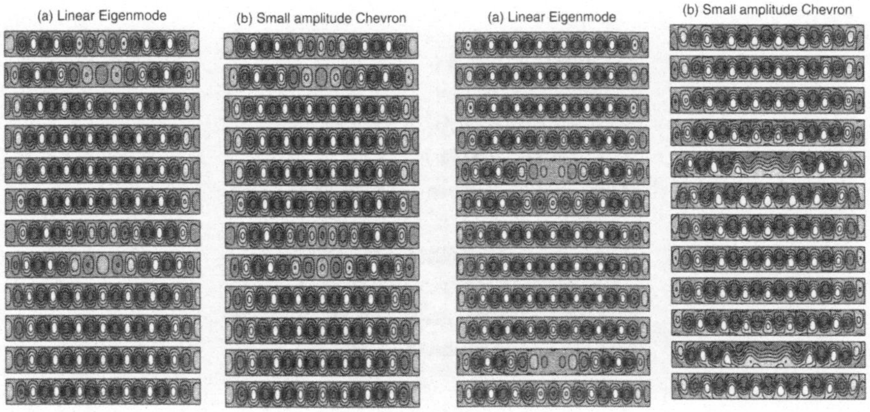

(a) Linear Eigenmode (b) Small amplitude Chevron (a) Linear Eigenmode (b) Small amplitude Chevron

FIGURE 5.21. Periodic even-parity chevron for $\Gamma = 16.25$ and (a,c) $R \equiv R_c = 1776.3$, (b,d) $R = 1775.5$ ($\epsilon = -4.5 \times 10^{-4}$) in terms of the contours of (a,b) the temperature perturbation $\theta(x, z, t)$ and (c,d) the concentration perturbation $C(x, z, t)$ of the denser component, with time increasing upwards in intervals of $0.2t_d$. The temperature contours at $R = 1775.5$ resemble the eigenfunction but the concentration contours do not.

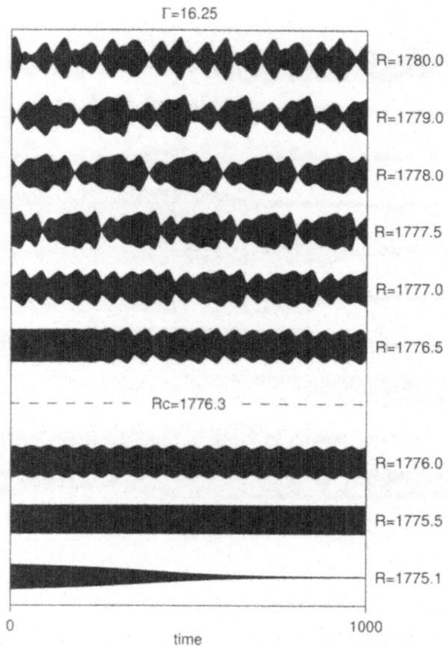

FIGURE 5.22. Time series $w(x = 0.87\Gamma, z = 0, t)$ for $\Gamma = 16.25$ and different values of the Rayleigh number R. Stable chevrons are present for $R = 1775.5$ ($\epsilon = -4.5 \times 10^{-4}$), but give way to symmetric periodic blinking states when $R = 1776$ ($\epsilon = -1.7 \times 10^{-4}$) with no detectable hysteresis. At larger R (e.g., $R = 1778$) the blinking states are asymmetric.

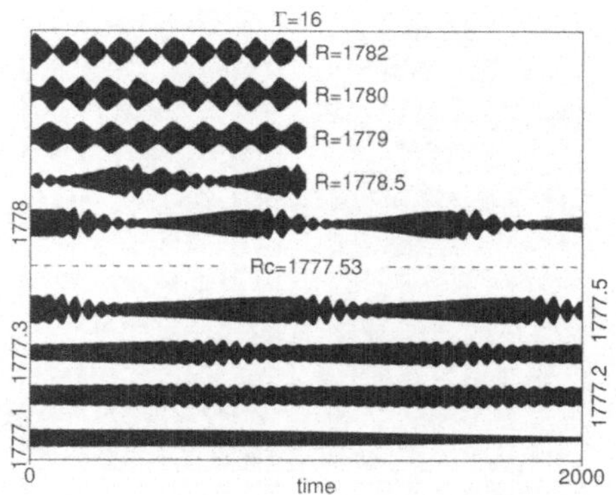

FIGURE 5.23. Time series $w(x = 0.87\Gamma, z = 0, t)$ for $\Gamma = 16$ and different values of the Rayleigh number R. The first finite amplitude state is a three-frequency state at $R = 1777.2$ ($\epsilon = -1.8 \times 10^{-4}$). This state gives way gradually and without detectable hysteresis to repeated transient states near $\epsilon = 0$ and then to symmetric periodic blinking states in a hysteretic transition between $R = 1778.5$ ($\epsilon = 5.5 \times 10^{-4}$) and $R = 1779$ ($\epsilon = 8.3 \times 10^{-4}$). The state at $R = 1782$ ($\epsilon = 2.5 \times 10^{-3}$) appears to have period-two modulation.

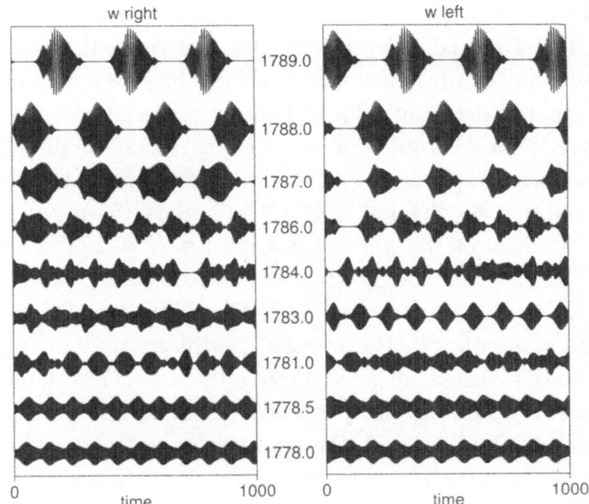

FIGURE 5.24. Time series $w(x = 0.87\Gamma, z = 0, t)$ and $w(x = 0.13\Gamma, z = 0, t)$ for $\Gamma = 16.8$ and different values of the Rayleigh number $R > R_c$, showing successive transitions from asymmetric blinking states to symmetric ones and back again. The solutions at $R = 1781$ and $R = 1784$ appear to be chaotic. Pauses, such as the one in the time series for $R = 1784$, were observed in the experiments as well (see fig. 1.1(b)).

phase that becomes progressively shorter, before a (hysteretic) transition to the symmetric periodic blinking state takes place ($1778.5 < R < 1779$). With further increase in ϵ this state gradually evolves into a chaotically blinking state much as in fig. 5.22, as anticipated in section 2 and observed in experiments [Kolodner, Surko, and Williams, 1989; Steinberg, Fineberg, Moses, and Rehberg, 1989].

Stable subcritical chevrons were found for all other values of Γ explored except $\Gamma = 16$ and $\Gamma = 17$, but their range of stability is usually narrow and may lie in $\epsilon < 0$. This may explain why stable chevrons have only rarely been found in experiments [Steinberg, Fineberg, Moses, and Rehberg, 1989]. In any case these lose stability almost immediately to a symmetric periodic blinking state, which then becomes asymmetric and/or chaotic (fig. 5.22). Thus the first state encountered on raising ϵ through $\epsilon = 0$ is typically a complex state even when $\Gamma \neq 16, 17$. However, symmetric periodic blinking states are present at onset near the mode crossing point at $\Gamma \approx 16.8$ (see figs. 5.20, 5.24), in accord with theoretical expectation: the onset of blinking is a consequence of the interaction between odd and even chevrons in the nonlinear regime. The blinking period we find, $\approx 90t_d$, is comparable to the period measured in the experiments when $\Gamma = 17.63$, viz. 8000s, since $t_d = 84.3$s [Kolodner, 1993] and the $\Gamma = 17.63$ case behaves much like $\Gamma = 16.63$ (section 2).

Figure 5.24 explores the evolution of the blinking states at larger values of ϵ when $\Gamma = 16.8$ using time series for the vertical velocity at mirror points in the two halves of the container. The solutions are in general asymmetric with respect to the middle of the container, and may be periodic (as for $R = 1783$, for example) or chaotic (as for $R = 1784$). The figure also shows that with increasing ϵ the left (right) traveling waves become increasingly confined to the left (right) half of the container, leading to characteristic time series of the type shown for $R = 1787$ and $R = 1788$. Note that the former is strongly spatially asymmetric, while the latter is symmetric, with both states being periodic. Thus transitions that break and restore the symmetry in the vertical midplane may occur repeatedly, with the symmetric states at $R = 1786$ and $R = 1788$ separated by an asymmetric one, suggesting the presence of a cascade of gluing bifurcations, cf. Arnéodo, Coullet, and Tresser [1981]. Note that the time series for $R = 1789$ ($\epsilon = 0.006$) is very similar to the fish state observed by Kolodner for $\Gamma = 16.75$ and $\epsilon = 0.0111$, even to the extent of capturing the strong amplitude dependence of the chevron frequency within this state (cf. section 5.1). Moreover, the computed period of this state, $305t_d$ (see fig. 5.24) corresponds almost exactly to that measured by Kolodner in the experiment (see fig. 1.1(f)).

Figures 5.25 and 5.26 show the corresponding results for $\Gamma = 17.25$ and $\Gamma = 17$, for comparison with figs. 5.22 and 5.23, respectively. In both cases the results are very similar. Thus when $\Gamma = 17.25$ the first stable state is an (odd) chevron, which undergoes a supercritical bifurcation to a blinking state when ϵ is still negative. With increasing ϵ the time series gradually become more complex, and begin to resemble the states we have called repeated transients (see fig. 5.25). However, these states develop from small amplitude blinking states with *increasing* R and hence can be unambiguously distinguished from them. In fact it is likely that the states shown in fig. 5.25 (at $R = 1775.5$, for example) are the result of another Hopf bifurcation from the blinking state, but this time with R increasing, not decreasing. Figure 5.25 shows that as R increases further these states gradually become chaotic, although at $R = 1779$ the time series is again periodic. The results for $\Gamma = 17$ likewise resemble those for $\Gamma = 16$. In particular the first nontrivial state is again a repeated transient, this time consisting of a slowly but exponentially growing *odd* parity chevron, followed by the characteristic oscillatory collapse. In fact the results for $\Gamma = 17$ provide a somewhat clearer illustration of the origin of the three-frequency state, since they suggest that the oscillatory collapse phase connects a larger amplitude chevron state with a smaller amplitude chevron which then regrows again into the larger amplitude state. The transition from this state to the blinking state appears to be again hysteretic, while the largest Rayleigh number solution ($R = 1780$) appears to be periodic but with a period that is three times the basic blinking period. The only substantive difference between the results of figs. 5.23 and 5.26 is that the parity of the chevron

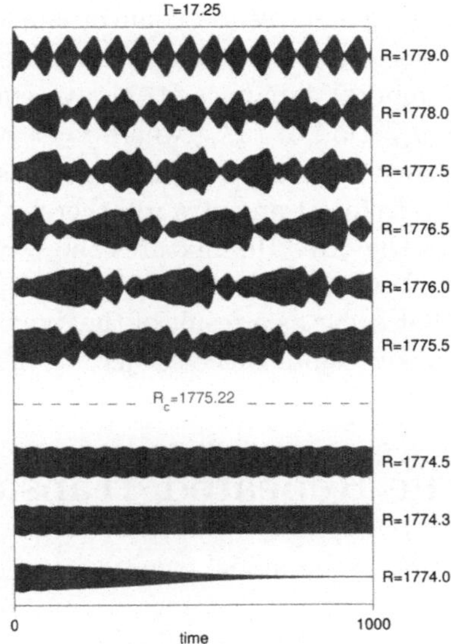

FIGURE 5.25. Time series $w(x = 0.87\Gamma, z = 0, t)$ for $\Gamma = 17.25$ and different values of the Rayleigh number R. Stable chevrons are present for $R = 1774.3$ ($\epsilon = -5.2 \times 10^{-4}$), but give way to symmetric periodic blinking states when $R = 1774.5$ ($\epsilon = -4.1 \times 10^{-4}$) with no detectable hysteresis. At larger R (e.g., $R = 1776$) the blinking state evolves to a three-frequency state superficially resembling a repeated transient.

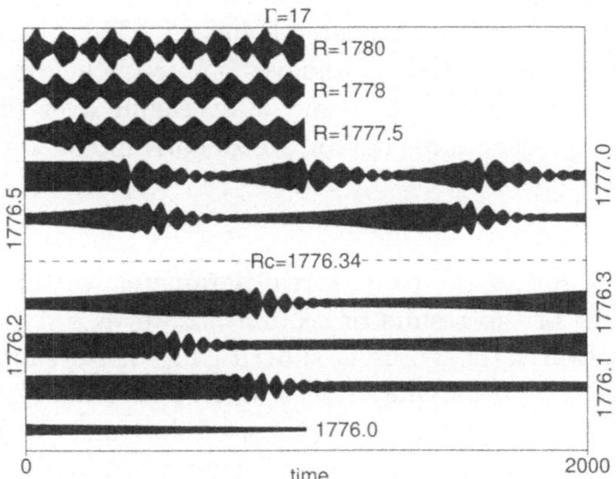

FIGURE 5.26. Time series $w(x = 0.87\Gamma, z = 0, t)$ for $\Gamma = 17$ and different values of the Rayleigh number R. The repeated transient state gives way to a periodic blinking state in a hysteretic transition between $R = 1777$ ($\epsilon = 3.7 \times 10^{-4}$) and $R = 1777.5$ ($\epsilon = 6.5 \times 10^{-4}$). The state at $R = 1776.1$ ($\epsilon = -1.4 \times 10^{-4}$) eventually decays, while that at $R = 1780$ ($\epsilon = 2.1 \times 10^{-3}$) appears to have period-three modulation. The results follow closely the sequence shown in fig. 5.23.

state involved is different. These results confirm Kolodner's experimental results and the theoretical prediction [Dangelmayr, Knobloch, and Wegelin, 1991] that the spatio-temporal dynamics of this system should be periodic with respect to Γ with a period of π/k_c, where $k_c \approx \pi$ is the wavenumber obtained from linear theory (cf. section 2). Thus increasing the aspect ratio by one allows the system to insert an extra roll thereby changing the parity of the basic state. As in the ^{3}He-^{4}He mixtures the wavelength of the rolls remains remarkably uniform across the container despite the substantial changes in amplitude that occur as a result of the dynamics of these states.

The explanation of these results is the subject of the next section.

6 Origin of the Repeated Transients

In this section we investigate the model equations

$$\begin{aligned} \dot{v} &= (-\nu + cz^2)v - \delta|v|^2 v\,, \\ \dot{z} &= (\mu + az^2 - z^4)z - |v|^2 z\,, \end{aligned} \tag{6.1}$$

constructed to retain the main properties of the partial differential equations. Here z refers to the amplitude of the chevron state (either even or odd) and is taken to be a real quantity despite the fact that the chevron states are in fact time-periodic. We justify this approximation using fig. 5.20 which shows that, for the parameter values considered, the chevron frequency ω_1 is high compared to the blinking frequency ω_2 or the slow frequency ω_3 associated with the repeated transients. This assumption has the considerable advantage in that it removes one frequency from the system, and can be considered to be the result of averaging over the fast chevron frequency. Consequently pure chevron states correspond to the solutions of the equation

$$\dot{z} = (\mu + az^2 - z^4)z, \tag{6.2}$$

and we take $\mu \propto R - R_c(\Gamma)$ to be a real parameter, with the coefficient a also real. In view of the results of section 5.2 we take $a > 0$ so that the primary bifurcation to chevrons is subcritical, with a saddle-node bifurcation (hereafter SN) occurring at $z^2 = a/2$. The stability of these states with respect to perturbations in the form of chevrons of the same parity is therefore given by the linearization of equation (6.2) about the solution $z = z_0$ satisfying $\mu + az_0^2 - z_0^4 = 0$. We denote this eigenvalue by λ. It follows that when $a < 0$ this eigenvalue is always negative (stable) while if $a > 0$ it is positive (unstable) on the subcritical branch and becomes negative above the saddle-node bifurcation.

The variable v represents perturbations transverse to the chevron invariant subspace, and is complex because these perturbations are destabilized at a secondary Hopf bifurcation, hereafter H_2. As a result the coefficients ν,

c and δ are all complex. This Hopf bifurcation is responsible for the onset of blinking. In the model the amplitude of the blinking is given by

$$\dot{y} = (-\nu_R + c_R z^2)y - \delta_R y^3 \tag{6.3}$$

and its frequency by the decoupled equation

$$\dot{\theta} = -\nu_I + c_I z^2 - \delta_I y^2. \tag{6.4}$$

Here $v \equiv ye^{i\theta}$ and the subscripts R and I denote real and imaginary parts, respectively. In these equations the important parameter is $\nu_R \equiv \nu_R(\Gamma) > 0$ and we take $c_R > 0$. Because of the decoupling of θ from the equations for y and z the resulting model is simple to analyze. Within the model the symmetry $y \to -y$ represents evolution in time by half the blinking period so that solutions with opposite signs of y are in fact identical modulo time translation. The pure chevrons $(y, z) = (0, z_0)$ begin to blink when $z_0^2 = \nu_R/c_R$ and do so with frequency $-\nu_I + c_I z_0^2$; the resulting blinking states take the form $(y, z) = (y_0, z_0)$, provided $y_0^2 > 0$, $z_0^2 > 0$. The stability of these states is described by a quadratic dispersion relation. This relation shows that the blinking states either set in (supercritically) from the larger amplitude chevron branch (hereafter A), or from the smaller amplitude branch (hereafter B). In the former case the chevrons acquire stability at the saddle-node bifurcation before losing it again at larger amplitude to stable blinking states. In the (y, z) variables this bifurcation looks like a pitchfork bifurcation, although it is of course a Hopf bifurcation. In the latter case the blinking states are initially unstable but acquire stability at a tertiary Hopf bifurcation H_3. This Hopf bifurcation is of vital importance in what follows since it introduces a third frequency, ω_3, into the dynamics of the partial differential equations. Its presence is a direct consequence of the passage of the Hopf bifurcation H_2 through the saddle-node bifurcation SN on the chevron branch when $a > 0$, $c_R > 0$, as originally noted by Guckenheimer [1981]. For a related analysis, also arising in the binary fluid context, see Knobloch and Moore [1990]. In the following we present the corresponding results for the full model equations (6.1). These are summarized in fig. 6.1 for the case in which the three-frequency state created from the blinking state branch is stable. This is always the case when $c_R = 1$, $\delta_R = 0$ and $a > 0$, and hence for sufficiently small positive values of δ_R as well. The figure shows the loci of the primary (H_1), secondary (H_2) and tertiary (H_3) Hopf bifurcations, as well as the locus of the saddle-node bifurcations (SN) on the chevron branch. It should be remembered that in the (y, z) variables only the bifurcation H_3 remains a Hopf bifurcation, with H_1 and H_2 represented by pitchfork bifurcations. In addition the figure shows the curve γ of global bifurcations at which the limit cycle (corresponding to the three-frequency states) created at H_3 disappears by simultaneous collision with the pure chevron states A and B. The location of this line must be determined numerically. An asymptotic calculation of this curve near the codimension-two point yields the heavy broken line; this line is tangent to γ at the codimension-two point, as it must.

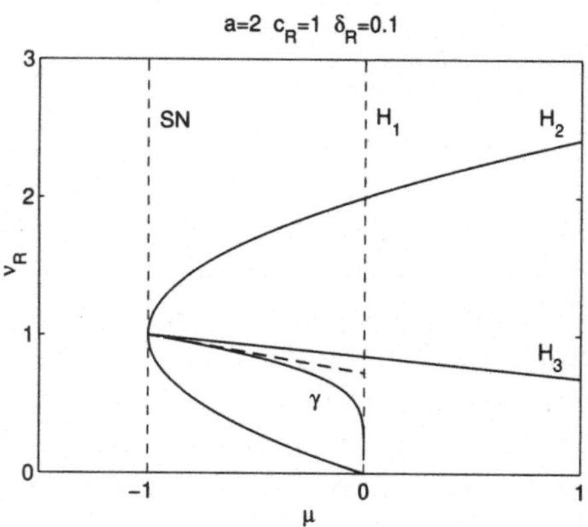

FIGURE 6.1. Codimension-one bifurcation surfaces in the (μ, ν_R) plane for equations (6.1) with $a = 2.0$, $c_R = 1.0$ and $\delta_R = 0.1$. H_1: primary (Hopf) bifurcation to the chevron state $(v, z) = (0, z_0)$, SN: saddle-node bifurcation on the chevron state, H_2: (secondary) Hopf bifurcation to blinking states $(v, z) = (v_0, z_0)$, H_3: (tertiary) Hopf bifurcation from (v_0, z_0) responsible for the appearance of the three-frequency states, and γ: global bifurcation at which these states disappear. The heavy broken line represents the asymptote to γ.

Figure 6.2 shows the bifurcation diagrams obtained by traversing the (μ, ν_R) plane in fig. 6.1 along the lines $\nu_R = 1.6$ and $\nu_R = 0.7$. These capture the two fundamentally different bifurcation diagrams characterizing the binary mixture. Figure 6.2a shows a small interval of subcritical but stable chevrons, followed by a supercritical pitchfork bifurcation to a state with $y_0 \neq 0$ that represents a blinking state in the physical variables. In the example shown this bifurcation occurs at $\mu < 0$ so that the first stable state just above onset ($\mu = 0$) is a finite amplitude blinking state. In contrast, in the case shown in fig. 6.2b the first stable state encountered beyond $\mu = 0$ is a finite amplitude periodic state which we identify with the three-frequency repeated transient state discovered by Kolodner. Figure 6.2c shows the time series corresponding to this state when $\mu = -0.21$. These oscillations represent the low frequency component of the three-frequency state, i.e., the repeated transient state with the frequencies ω_1 and ω_2 filtered out. Observe that during the growth phase of the variable z the variable y vanishes, indicating that the growing state is a pure chevron; y becomes nonzero only during the collapse phase, indicating that the collapse is triggered by a symmetry-breaking instability (i.e., the loss of stability of the growing chevron). Figure 6.3 shows similar behavior obtained from the partial differential equations when $\Gamma = 16$: during the growth phase $w(0.13\Gamma, 0, t) = w(0.87\Gamma, 0, t)$, indicating a growing

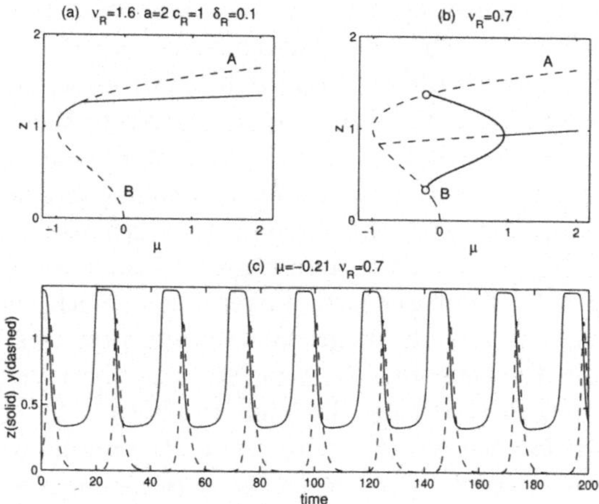

FIGURE 6.2. (a,b) The bifurcation diagrams along the lines $\nu_R = 1.6$ and $\nu_R = 0.7$ in fig. 6.1. Scenario (a) corresponds to that observed in fig. 5.22 for $\Gamma = 16.25$, while (b) corresponds to that observed in fig. 5.23 for $\Gamma = 16$. In (b) the open circles indicate the global bifurcation with which the oscillations terminate as μ decreases, with the states A and B labeled as in the text. Solid (dashed) lines indicate stable (unstable) solutions. (c) The time series $y = |v(t)|$ (dashed) and $z(t)$ (solid) for a stable repeated transient when $\mu = -0.21$, $\nu_R = 0.7$, $a = 2.0$, $c_R = 1.0$ and $\delta_R = 0.1$.

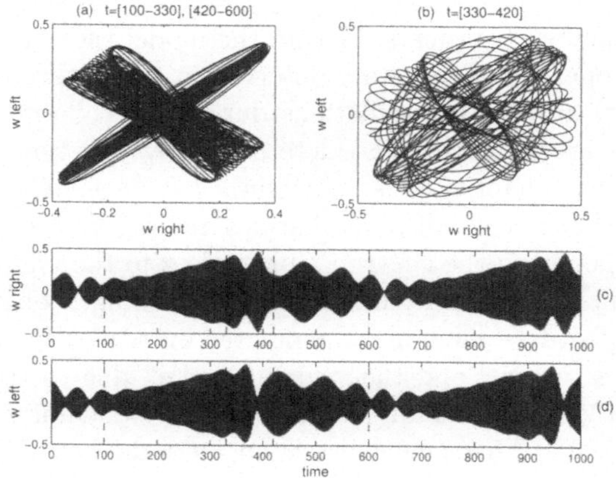

FIGURE 6.3. Plots of $w(x = 0.13\Gamma, z = 0, t)$ vs $w(x = 0.87\Gamma, z = 0, t)$ for $\Gamma = 16$, $R = 1778.5$ during (a) the growth and the collapse phases, and (b) during the start of the collapse phase, together with the time series (c) $w(x = 0.87\Gamma, z = 0, t)$ and (d) $w(x = 0.13\Gamma, z = 0, t)$. Growing symmetric chevrons evolve along the 45° line in (a) but evolve along the orthogonal direction during the collapse phase (cf. figure 8 of Kolodner [1993]). The growth, transition and collapse phases used to construct figs. (a,b) are indicated by vertical dashed lines in figs. (c,d).

chevron, a fact confirmed in fig. 6.3a by the evolution of the system away from the origin along the 45° line. The collapse phase is initiated when the difference between $w(0.13\Gamma, 0, t)$ and $w(0.87\Gamma, 0, t)$ begins to grow and the system begins to evolve in a direction orthogonal to the 45° line, much as shown in figure 8a of Kolodner [1993]. With the beginning of the collapse phase one starts to notice the onset of blinking as evidenced in the 180° phase difference between the decaying oscillations in $w(0.87\Gamma, 0, t)$ and $w(0.13\Gamma, 0, t)$ (fig. 6.3c,d). The amplitude and the period $2\pi/\omega_3$ of the limit cycle in fig. 6.2c decreases with increasing μ, with the oscillations disappearing at H_3. As already mentioned we interpret this transition as the transition from the repeated transient state to the (symmetric) periodic blinking state with increasing Rayleigh number (cf. fig. 4 of Kolodner [1993], where H_3 is located at $\epsilon \approx 2 \times 10^{-3}$, i.e., the minimum of the measured "modulation" period). For the model parameters this transition is supercritical, indicating absence of hysteresis. As μ approaches closer to the global bifurcation at $\mu = \mu^* < 0$, indicated by open circles in fig. 6.2b, the time series remains similar to that shown in fig. 6.2c but the oscillation period $2\pi/\omega_3$ becomes longer, diverging as $-\ln|\mu - \mu^*|$ for $\mu \sim \mu^*$, cf. figure 4 of Kolodner [1993]. In fig. 6.4 we show another case, in which the global bifurcation at μ^* occurs very close to $\mu = 0$. As a result the chevron state grows from almost zero amplitude, and so resembles more closely the repeated transient state discovered by Kolodner. In this case there is almost no hysteresis between this state and the conduction state, and the system behaves *as if* the primary instability at $\mu = 0$ were directly responsible for generating repeated transients. Within the model the corresponding state has all the properties of this state observed in the experiments, except for the (apparent) absence of oscillations during the collapse phase. In fact, if the frequencies ω_1 and $\dot{\theta} \equiv \omega_2$ determined by equation (6.4) are incorporated, and the quantity $[z + v_R(t)] \sin \omega_1 t \equiv [z + y(t) \cos \omega_2 t] \sin \omega_1 t$ is plotted instead of z or $|v|$, these oscillations *are* present (fig. 6.5b), and their amplitude depends on the chevron amplitude z in the manner observed in the experiments. In fact, the time series shown in fig. 6.5b bears a number of qualitative features, including the pointed overshoot at maximum as the mode v begins to grow and the "ringing down" due to the fact that the variable z decays more rapidly than v, that are documented in figure 6a of Kolodner [1993]. This time series is not periodic because in general the two (nonlinear) frequencies ω_2 and ω_3 are incommensurate.

Several remarks are in order.

1. The coefficient δ can be zero without qualitative effect on the above scenarios. However, we have chosen $\delta_R > 0$ to assure that the solutions remain bounded for all time, and to move the secondary bifurcations away from the saddle-node on the primary chevron branch.

2. The invariance of the plane $z = 0$ in the two-dimensional model prevents the formation of a connection between the large amplitude

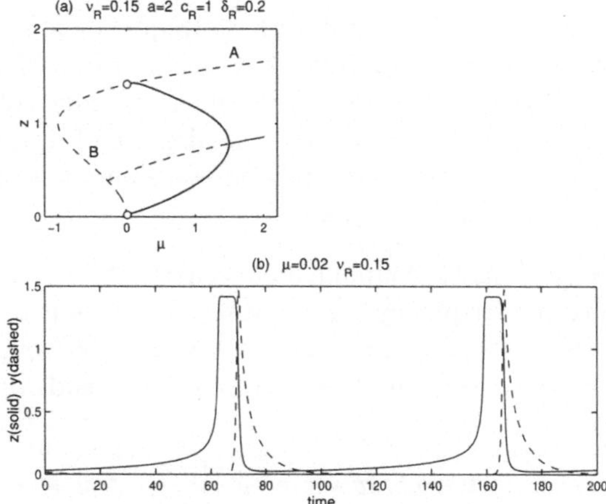

FIGURE 6.4. (a) As for fig. 6.2b but with $\nu_R = 0.15$, $a = 2.0$, $c_R = 1.0$ and $\delta_R = 0.2$, showing a global bifurcation very close to $\mu = 0$. (b) The corresponding time series when $\mu = 0.02$.

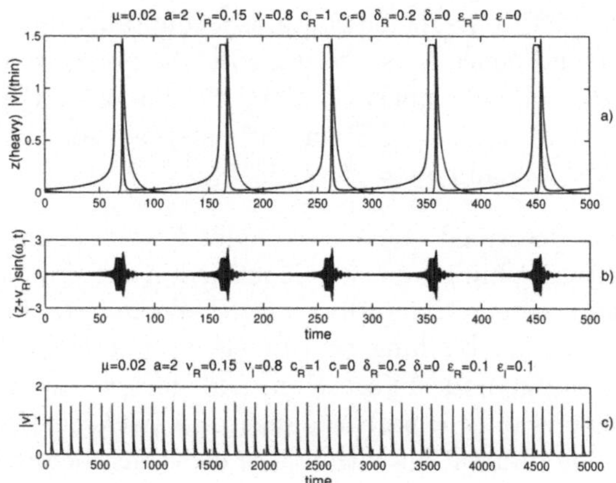

FIGURE 6.5. (a) As for fig. 6.4b but over a longer time interval. (b) The time series for $[z + v_R(t)] \sin \omega_1 t$ when $\nu_I = 0.8$, $c_I = 0$, $\delta_I = 0$ and $\omega_1 = 20$. Note the exponential growth during the chevron phase, followed by an overshoot when the blinking instability sets in, and the ringing down during the subsequent collapse phase. The time series resembles closely that in fig. 6a of Kolodner [1993]. (c) $|v(t)|$ for chaotic repeated transients from equations (6.5) with $\epsilon_1 = 0.1 + 0.1i$, $f_1 = z^2$.

chevron state A and the origin when $\mu > 0$. Consequently, the global bifurcation with which the stable three-frequency states first appear must involve the small amplitude chevron state B, and μ^* is necessarily negative. However, there is a large range of values of ν_R for which $\mu^* \approx 0$ (see fig. 6.1). Consequently, the absence of hysteresis between the conduction state and the repeated transients noted by Kolodner finds a ready explanation in fig. 6.1.

3. The model (6.1) lacks coupling between the blinking frequency ω_2 and the tertiary frequency ω_3 because of the built-in normal form symmetry. Since this symmetry is not exact the coupling between these two frequencies should be restored. This leads one to consider the model

$$
\begin{aligned}
\dot{v} &= (-\nu + cz^2)v - \delta|v|^2 v + \epsilon_1 f_1(z^2)\bar{v} + \cdots, \\
\dot{z} &= (\mu + az^2 - z^4)z - |v|^2 z,
\end{aligned}
\tag{6.5}
$$

where the coefficients $\epsilon_1, \ldots$ represent the coupling and are assumed to be small, cf. Kirk [1991, 1993]. When this is done one finds that the invariant sphere with A and B at its poles breaks down due to the transversal intersection of the unstable manifold of A and the stable manifold of B. These intersections occur in a heteroclinic *region* in the (μ, ν_R) plane whose width increases with ϵ_1 [Guckenheimer, 1981; Kirk, 1993]. This region contains a countably infinite number of 'horseshoes' and hence is associated with the presence of chaos. Note that the heteroclinic connection along the diameter BA is preserved by the perturbation $\epsilon_1 \neq 0$. Figure 6.5c shows a solution of this type for $\epsilon_1 = 0.1 + 0.1i$ and $f_1 = z^2$.

The model (6.1) described above is completely consistent with the two scenarios for generating blinking states identified in the simulations of both ^{3}He–HeliumIV and water–ethanol mixtures. For example, in the scenario depicted in fig. 5.22 the blinking sets in via a supercritical Hopf bifurcation above the saddle-node bifurcation, and does so already for $\epsilon < 0$. Consequently there is only a narrow range of ϵ between this bifurcation and the saddle-node bifurcation with stable chevrons, before blinking sets in. The blinking frequency ω_2 is quite small because the chevron amplitude at which the Hopf bifurcation takes place is small [Dangelmayr and Knobloch, 1987, 1991]. In contrast, the results for $\Gamma = 16$ (fig. 5.23) and $\Gamma = 17$ (fig. 5.26) are entirely consistent with the second scenario, i.e., that the secondary Hopf bifurcation to blinking states now occurs below the saddle-node bifurcation, thereby eliminating the stable chevrons entirely. Moreover, our results for, say, $R = 1777.2$ (fig. 5.23) and $R = 1776.5$ (fig. 5.26) are suggestive of a quasiperiodic state with three independent frequencies such as might be expected from the tertiary Hopf bifurcation H_3 on the branch of blinking states identified in the model. Indeed, our calculations are consistent with the conjecture that the bifurcations SN and H_2 on the chevron branch coincide at an aspect ratio Γ somewhere between 16 and 16.25. Note that the observed period associated with the third frequency is about $1000t_d$. Such low frequencies are characteristic of the scenario proposed in fig. 6.1. Moreover, this scenario predicts that the corresponding modulation period should increase rapidly with decreasing ϵ, i.e., as $\epsilon \downarrow \epsilon^*$, in accord with the experimental observations (see figure 4 of Kolodner [1993]). Figure 5.23 also suggests that the repeated transients observed by Kolodner evolve from this three-frequency state as ϵ increases from ϵ^*, a suggestion that is confirmed in fig. 5.26 where the three-frequency states look like Kolodner's repeated transients from the very beginning. In both cases periodic blinking states are observed only after a (hysteretic) transition from the three-frequency repeated transients. Consequently the branch of blinking states only acquires stability at H_3 and these states therefore blink with finite amplitude when they first appear, resulting in a longer blinking period than at H_2, typically $100t_d$ (compare fig. 5.23 at $R = 1779$ with fig. 5.22 at $R = 1776$). This period is also comparable to the period observed in the experiments. With further increase in ϵ the blinking state appears to undergo period-doubling as suggested by the time series for $R = 1782$ in fig. 5.23 (cf. Hirschberg and Knobloch [1996]; Knobloch [1996]), and gradually becomes more and more chaotic. Indeed, the time series for $R = 1780$ in fig. 5.26 suggests a period-three blinking state. Available theory predicts (section 2) that the blinking states terminate in another global bifurcation at which a hysteretic transition to a single frequency localized state takes place. This state consists of waves that travel under a stationary envelope attached to one or other lateral wall. It is likely that the period-doubling transitions etc. are associated with this global bifurcation. Finally, the fact that we have found the repeated transients only in the vicinity of $\Gamma = 16$ and $\Gamma = 17$, i.e., for aspect ratios differing by ≈ 1, is also consistent with theoretical expectation (section 2), and indeed the experiments as well [Kolodner, 1993].

The presence of the global bifurcation in which the repeated transients first appear, established in figs. 6.2b and 6.4a, suggests that under appropriate conditions the repeated transients may be chaotic. As already noted the frequency ω_3 decreases to zero as $\epsilon \downarrow \epsilon^*$. As this occurs the three-frequency states approach simultaneously the unstable large and small amplitude chevron states A and B. The character of the repeated transient when $\epsilon \approx \epsilon^*$ is determined by the leading eigenvalues of A and B in the chevron fixed point subspace, hereafter $-\lambda_A < 0$ and $\lambda_B > 0$, and the leading eigenvalues in the perpendicular direction. If the latter are real, $\alpha_A > 0$ and $-\alpha_B < 0$, say, and $\rho \equiv \alpha_B\lambda_A/\alpha_A\lambda_B > 1$, the repeated transients will remain periodic and stable all the way to ϵ^*, where the period diverges and the global bifurcation takes place [Batiste, Knobloch, Mercader, and Net, 2001a]. In contrast, when $0 < \rho < 1$, the periodic oscillations necessarily lose stability before the global bifurcation at ϵ^*. Similar results obtain in the case where the leading stable symmetry-breaking eigenvalue at B is complex, viz. $-\alpha_B + i\omega_B$, $\alpha_B > 0$, as suggested by the simulations. In this case stable periodic oscillations will persist down to ϵ^* if $\rho > 1$, but if $0 < \rho < 1$ complex dynamics of Shil'nikov type will be present. Possible *chaotic* repeated transients resulting from this mechanism are shown in fig. 6.6. In fact figs. 5.23 and 5.26 suggest that the leading unstable eigenvalues α_A and λ_B are also complex; this is to be expected since the bifurcations at H_1 and H_2 are in fact both Hopf bifurcations. In the following we do not consider the resulting complications further.

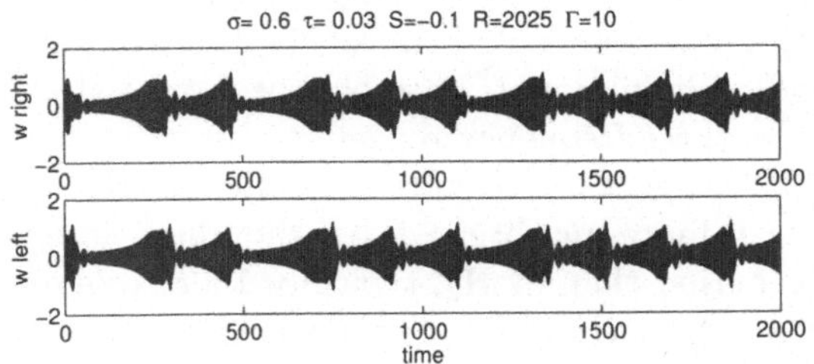

FIGURE 6.6. A chaotic repeated transient in a ^{3}He-^{4}He mixture with stress-free and fixed temperature boundary conditions at $x = 0, \Gamma$ and $R = 2025$, $S = -0.1$, $\sigma = 0.6$, $\tau = 0.03$, $\Gamma = 10$.

When λ_B is real a trajectory escaping from B describes an exponentially growing chevron state. This growth phase, including the states A and B, is clearly visible in the time series for $R = 1776.2$ (fig. 5.26). When the growing chevron reaches the vicinity of A it becomes unstable to symmetry-breaking oscillations which take it back near B. This is the collapse phase of the repeated transient state (compare figs. 6.2c and 6.4b with fig. 6.3). The frequency of the decaying oscillations observed in the time series in figs. 6.3c,d is given by ω_B. This frequency will in general be of the same order as the blinking frequency associated with the branch of blinking states when these bifurcate from the small amplitude chevron B, but quite different from (and in general larger than) the blinking frequency of the *stable* blinking states beyond H_3, cf. Kolodner [1993]. This observation explains the coincidence of the period of the blinking states and of the oscillations during the collapse phase of the repeated transient also noted by Kolodner. Note also that since the repeated transient state visits the states A and B whose amplitude decreases (resp., increases) as ϵ becomes more negative the modulation amplitude along the branch of three-frequency states should decrease towards the end of the branch. This is seen quite dramatically in fig. 5.23. Moreover, since α_B decreases as ϵ decreases (it passes through zero at H_2, i.e., at $\epsilon = \epsilon_2$) the collapse becomes slower and slower, as also seen in fig. 5.23, but is still finite when the three-frequency states disappear in the global bifurcation at ϵ^* (since $\epsilon_2 < \epsilon^* < 0$) and the system makes a hysteretic transition to the conduction state. The fact that α_B decreases with ϵ makes it likely that the Shil'nikov condition $0 < \rho < 1$ holds at ϵ^*, resulting in *chaotic* repeated transients prior to their disappearance [Knobloch, 1996]. This possibility apparently does not occur in figs. 5.23 and 5.26 although it does in fig. 6.6 and may occur in the experiments. In any case even longer time series would be required to test this prediction. Note that since $\lambda_B \propto |\epsilon|$ only periodic repeated transients will occur if $\epsilon^* \approx 0$, although even in this case there may be a few bifurcation bubbles with chaotic dynamics, as in Knobloch, Moore, Toomre, and Weiss [1986]. In fact Kolodner [1993] notes that the growth phase of the repeated transient is inversely proportional to ϵ, in accord with the above scenario. It is of interest that the repeated transients are most likely to be chaotic just prior to their extinction, as ϵ decreases.

We note, finally, that despite the fact that the repeated transients are three-frequency states they are not necessarily structurally unstable. In the model this is because we have averaged out the chevron frequency ω_1. However, even if we had not done so, genuinely three-frequency states can be observed in open parameter regions [Grebogi, Ott, and Yorke, 1983] despite the Ruelle-Takens theorem [Ruelle and Takens, 1971a,b].

7 Discussion

In this article we have described a theoretical approach to understanding the dynamics due to an oscillatory instability in moderately large domains, and confronted the predictions of this theory with the results of direct numerical simulations of oscillatory convection in binary mixtures in two dimensions. For the latter we have used the experimentally relevant boundary conditions and explored in detail the parameter values characteristic of ^{3}He-^{4}He and water-ethanol mixtures. We have seen that the competition between odd and even modes in systems of this type takes one of two basic forms, depending on the separation and aspect ratios (see fig. 4.1). When $|S|$ is small (i.e., close to S_{TB}) and Γ not too large the mode interaction takes the form familiar from Rayleigh-Bénard convection with non-Neumann boundary conditions: the neutral curves $R(\Gamma)$ divide neatly between different families and there is no intermingling among them (see fig. 4.5a). Each family consists of a pair of braided neutral curves, one for an odd mode and the other for an even mode, with each family well separated from the next, at least for the low-lying families. The crossings between odd and even modes within each family are structurally stable because of their different parity. At fixed Γ and large enough $|S|$ the situation is quite different (see fig. 4.5c). There are now no distinct families of neutral curves and all modes (including like-parity modes) cross. These mode crossings are all structurally stable, either because the modes have opposite parity, or because their frequencies at the mode crossing are nonresonant. The transition between these two types of behavior occurs via 1:1 resonant mode interactions as illustrated in fig. 4.3. These interactions allow mode crossings between like-parity modes belonging to different families and hence are responsible for the transition between the neutral stability curves in figs. 4.5a,c. Likewise at fixed S the neutral stability curves are braided when the aspect ratio Γ is not too large but with increasing Γ nonresonant crossings between like-parity modes appear (cf. fig. 4.3), as anticipated by Hirschberg and Knobloch [1997]. On the basis of our calculations we have made several conjectures about the behavior of the neutral stability curves and corresponding frequencies and eigenfunctions for large aspect ratios. These bear out the picture of large aspect ratio systems put forward by Landsberg and Knobloch [1996] with two significant clarifications. First, we have found that if the separation ratio is small there is a substantial range of aspect ratios within which the first odd and even modes to set in are separated from the next pair by a significant gap in Rayleigh number. In this range the existence of the gap justifies the reduction of the partial differential equations to amplitude equations for the first odd and even modes using a procedure called center-unstable manifold reduction [Armbruster, Guckenheimer, and Holmes, 1989]. We have presented examples of what these modes look like. However, for sufficiently large aspect ratios or sufficiently large separation ratios this gap disappears, and odd and even modes from different families are selected in succession. In this case the reduction to a pair of amplitude equations continues to be valid near all crossing points between odd and even modes, but it is no longer clear whether such a description captures the behavior of the system for all intervening aspect ratios. Second, we have found that the frequency difference between the competing odd and even parity modes scales like Γ^{-2} for large Γ instead of the expected Γ^{-1} behavior. This observation strengthens the argument in favor of the theory put forward by Landsberg and Knobloch [1996] and summarized in section 2.

The results of the direct numerical simulations reveal a complex sequence of transitions among states we have called chevrons, blinking states and repeated transients as the aspect ratio or the applied Rayleigh number varies. We have seen that

- The primary bifurcation to convection is subcritical, in agreement with the prediction for *standing waves* in a horizontally unbounded layer [Clune and Knobloch, 1992]. As a result for some aspect ratios stable chevrons can be present, but are most likely observed for negative values of ϵ.
- Stable small amplitude blinking states set in when stable chevrons lose stability at a secondary Hopf bifurcation.
- For many (though not most) aspect ratios the first nontrivial state of

the system is a three-frequency repeated transient state. The transition to this state is hysteretic since $\epsilon^* < 0$ but for many parameter values ϵ^* is so close to zero as to make the detection of hysteresis highly unlikely. Indeed, in some experiments no hysteresis was found [Kolodner and Surko, 1988]. The three-frequency states appear via a global bifurcation and acquire the characteristic behavior associated with repeated transients only as ϵ increases. Consequently the modulation period $2\pi/\omega_3$ is infinite when the repeated transients first appear, but drops rapidly with increasing ϵ, as observed in figure 4 of Kolodner [1993]. When the condition $\rho < 1$ on the eigenvalues at A and B holds the resulting repeated transients are expected to be chaotic. The repeated transients persist only over a narrow interval of ϵ close to onset, of order 10^{-3} (see figs. 5.23,5.26), which compares well with the observed range 2×10^{-3} for $\Gamma = 16.75$ [Kolodner, 1993], and give way to large amplitude blinking states in a (slightly) hysteretic transition when $\epsilon \approx 10^{-3}$. Our calculations suggest that dispersive effects are not of fundamental importance in this behavior (in contrast to Kolodner's conjecture), but that the primary chevron state must bifurcate subcritically. In particular we have been unable to identify any fundamental role of the local wavenumber changes observed in the experiments.

- Stable large amplitude blinking states set in via a (typically hysteretic) bifurcation from the repeated transient state. This bifurcation eliminates the slowest frequency from the time trace.

- In the experiments the blinking states become irregular with increasing ϵ and the blinking period gradually increases, until an abrupt transition to a localized traveling wave state attached to one boundary (or to steady overturning convection) takes place. The observed chaotic blinking is likely associated with the global bifurcation studied by Dangelmayr, Knobloch, and Wegelin [1991], and Knobloch [1996].

- Although the chaotic asymmetric blinking states are easily confused with the repeated transients because of their somewhat similar appearance, cf. Kolodner [1993], such states develop from symmetric blinking states with *increasing* ϵ and do so either via spatial symmetry-breaking followed by period doubling, or via a further Hopf bifurcation.

- Regular blinking states are observed near onset only for aspect ratios differing roughly by unity, as in the experiments. Our simulations suggest (fig. 5.20) that these special aspect ratios are in fact nothing but the mode interaction points (compare fig. 4.7(a) with fig. 5.20). The location of these points depends quite sensitively on the system parameters and in particular on the additional dissipation due to the neglected no-slip walls in the third (transverse) direction. Consequently, differences between the experimental results and our calculations may be primarily due to differences in the location of these points. For example, Kolodner finds that blinking states persist down to small amplitudes for $\Gamma = 16.63$ and $\Gamma = 17.63$ when the (dimensionless) width $\Gamma_y = 3.0$, and for $\Gamma = 16.25$ and $\Gamma = 17.25$ when $\Gamma_y = 4.9$. In contrast our strictly two-dimensional calculations ($\Gamma_y = \infty$) show that blinking states are most easily found near $\Gamma \approx 16.8$. In contrast to the experiments [Kolodner, 1993] we do not find that repeated transients predominate at all aspect ratios away from the mode interaction points. For example, at $\Gamma \approx 16.25$ we found subcritical stable chevrons that bifurcate into blinking states with increasing ϵ (fig. 5.22).

- The numerical simulations support much of the theory summarized in section 2 as to the origin of the different types of states, and provide essential information supporting the interpretation of the repeated transients as a three-frequency state. The simulations also confirm the significance of the parameter Γ mod π/k_c proposed in Dangelmayr, Knobloch, and Wegelin [1991] and confirmed so dramatically in Kolodner's experiments [Kolodner, 1993]. This parameter reflects the fact that when Γ changes by π/k_c, where k_c is the "wavenumber" of the waves, an additional roll fits into the container, thereby changing the parity of the most unstable state. Detailed calculations [Batiste, Mercader, Net, and Knobloch, 1999] indicate that the wavenumber at onset is in fact remarkably uniform across the cell (cf. Kolodner [1993]), despite the variation in amplitude, and that $k \approx \pi$. Thus when the neutral curves of both modes cross, the system feels comfortable with both, and oscillates regularly even in the nonlinear regime. This is not so for other aspect ratios for which one mode is preferred, and when that loses stability at finite amplitude to a symmetry-breaking perturbation the competing state does not fit well into the container and the system oscillates irregularly.

These results and the accompanying interpretation account quantitatively for almost all of the observations of Kolodner [1993] on water-ethanol mixtures, and suggest related experiments on ^{3}He-^{4}He mixtures. Recent improvements in experimental technique make the latter viable [Woodcraft, Lucas, Matley, and Wong, 1999]. In particular the quantitative success of the numerical simulations confirms that the restriction to two dimensions is not fatal, and helps in identification of the different dynamical states of the system. It is our view that the remaining quantitative discrepancies can all be attributed to the sensitive dependence of the mode interaction point on the width of the container, because of its effect on dissipative processes in the cell.

The results reported here indicate that for sufficiently small $|\epsilon|$ the sidewalls necessarily exert a critical influence on the dynamics of the system. This is to be expected since for such ϵ the behavior of the system is dominated by one (or at most two) unstable modes of the system, whose spatial structure is determined by the lateral boundary conditions. Order of magnitude estimates [Dangelmayr, Knobloch, and Wegelin, 1991; Renardy, 1999] suggest that this will be the case whenever $|\epsilon|\Gamma^2 \lesssim 1$, i.e., with increasing $|\epsilon|$ the sidewall influence becomes *smaller*, and indeed one may reach the situation in which the collapse described by the subcritical Ginzburg-Landau equation on an unbounded domain becomes a more and more appropriate description of the dynamics [Kaplan, Kuznetsov, and Steinberg, 1994]. The experiments of Kaplan, Kuznetsov, and Steinberg [1994] suggest that this is in fact so once $|\epsilon|\Gamma^2 \gtrsim 10$. The results of Niemela, Ahlers, and Cannell [1990], performed for slightly larger $|S|$ and subcritical values of ϵ, can likewise be interpreted as showing that $|\epsilon|\Gamma^2 \approx 5$ describes the transition between these two regimes. However, the water-ethanol simulations reported here all satisfy the condition $|\epsilon|\Gamma^2 \lesssim 1$, and hence are always dominated by the sidewalls.

The model put forward in section 6 suggests that equations (2.10) can only describe repeated transients if the primary bifurcation to the chevron state is subcritical, i.e., if $A_R > 0$, and if fifth order terms are included. However, as discussed in detail by Moehlis and Knobloch [1998, 2000] the cubic truncation may suffice to describe the type of bursts observed by Sullivan and Ahlers [1988] in ^{3}He-^{4}He mixtures. These experiments are performed in containers extended in one dimension, with $\Gamma = 34$, and the bursts located at $\epsilon = 3 \times 10^{-4} < \Gamma^{-2}$, i.e., the reported bursts should be amenable to a finite-dimensional description. Although these bursts require the presence of at least one subcritical branch fifth order terms are not necessarily required to provide saturation. Unfortunately, despite much effort we have been unable to reproduce these bursts in our simulations.

Acknowledgments: Preparation of this article was supported by EPSRC grant GR/R52879/01 and the National Science Foundation, under grant DMS-0072444. Computer time was provided by CEPBA. We also thank I. Mercader and M. Net for assistance.

References

Andereck, C. D., S. S. Liu, and H. L. Swinney [1986], Flow regimes in a circular Couette system with independently rotating cylinders, *J. Fluid Mech.* **164**, 155–183.

Armbruster, D., J. Guckenheimer, and P. Holmes [1989], Kuramoto-Sivashinsky dynamics on the center-unstable manifold, *SIAM J. Appl. Math.* **49**, 676–691.

Arnéodo, A., P. Coullet, and C. Tresser [1981], A possible new mechanism for the onset of turbulence, *Phys. Lett. A* **81**, 197–201.

Barten, W., M. Lücke, W. Hort, and M. Kamps [1989], Fully developed traveling-wave convection in binary fluid mixtures, *Phys. Rev. Lett.* **63**, 376–379.

Batiste, O., E. Knobloch, I. Mercader, and M. Net [2001a], Simulations of oscillatory binary fluid convection in large aspect ratio containers, *Phys. Rev. E* **65**, 016303 (19pp).

Batiste, O., I. Mercader, M. Net, and E. Knobloch [1999], Onset of oscillatory binary fluid convection in finite containers, *Phys. Rev. E* **59**, 6730–6741.

Batiste, O., M. Net, I. Mercader, and E. Knobloch [2001b], Oscillatory binary fluid convection in large aspect-ratio containers, *Phys. Rev. Lett.* **86**, 2309–2312.

Clune, T. and E. Knobloch [1992], Mean flow suppression by endwalls in oscillatory binary fluid convection, *Physica D* **61**, 106–112.

Crawford, J. D. and E. Knobloch [1991], Symmetry and symmetry-breaking bifurcations in fluid dynamics, *Annu. Rev. Fluid Mech.* **23**, 341–387.

Croquette, V. and H. Williams [1989], Nonlinear waves of oscillatory instability of finite convection rolls, *Physica D* **37**, 300–314.

Cross, M. C. [1986], Traveling and standing waves in binary-fluid convection in finite geometries, *Phys. Rev. Lett.* **57**, 2935–2938.

Cross, M. C. [1988], Structure of nonlinear traveling-wave states in finite geometries, *Phys. Rev. A* **38**, 3593–3600.

Dangelmayr, G. and E. Knobloch [1987], On the Hopf bifurcation with broken O(2) symmetry. In Güttinger, W. and G. Dangelmayr, editors, *The Physics of Structure Formation: Theory and Simulation*, pages 387–393. Springer-Verlag, Berlin.

Dangelmayr, G. and E. Knobloch [1991], Hopf bifurcation with broken circular symmetry, *Nonlinearity* **4**, 399–427.

Dangelmayr, G., E. Knobloch, and M. Wegelin [1991], Travelling wave convection in finite containers, *Europhys. Lett.* **16**, 723–729.

Daniels, P. G. [1977], Asymptotic sidewall effects in rotating Bénard convection, *Z. Angew. Math. Phys.* **28**, 577–584.

Deane, A. E., E. Knobloch, and J. Toomre [1988], Travelling waves in large-aspect-ratio thermosolutal convection, *Phys. Rev. A* **37**, 1817–1820.

Glendinning, P. and C. Sparrow [1984], Local and global behavior near homoclinic orbits, *J. Stat. Phys.* **35**, 645–696.

Grebogi, C., E. Ott, and J. A. Yorke [1983], Are 3-frequency quasi-periodic orbits to be expected in typical non-linear dynamical systems?, *Phys. Rev. Lett.* **51**, 339–342.

Guckenheimer, J. [1981], On a codimension-two bifurcation. In Rand, D. A. and L. S. Young, editors, *Dynamical Systems and Turbulence. Lecture Notes in Mathematics 898*, pages 99–142. Springer-Verlag, New York.

Hirschberg, P. and E. Knobloch [1996], Complex dynamics in the Hopf bifurcation with broken translation symmetry, *Physica D* **90**, 56–78.

Hirschberg, P. and E. Knobloch [1997], Mode interactions in large aspect ratio convection, *J. Nonlinear Sci.* **7**, 537–556.

Hugues, S. and A. Randriamampianina [1998], An improved projection scheme applied to pseudospectral methods for the incompressible Navier-Stokes equations, *Int. J. Numer. Meth. Fluids* **28**, 501–521.

Jacqmin, D. and J. Heminger [1994], *Double-diffusion with Soret effect in a rectangular geometry: linear and nonlinear traveling wave instabilities*, unpublished, 1994.

Kaplan, E., E. Kuznetsov, and V. Steinberg [1994], Burst and collapse in traveling-wave convection of a binary fluid, *Phys. Rev. E* **50**, 3712–3722.

Kirk, V. [1991], Breaking of symmetry in the saddle-node–Hopf bifurcation, *Phys. Lett. A* **154**, 243–248.

Kirk, V. [1993], Merging of resonance tongues, *Physica D* **66**, 267–281.

Knobloch, E. [1986], Oscillatory convection in binary mixtures, *Phys. Rev. A* **34**, 1538–1549.

Knobloch, E. [1996], System symmetry breaking and Shil'nikov dynamics. In Chadam, J., M. Golubitsky, W. F. Langford, and B. Wetton, editors, *Pattern Formation: Symmetry Methods and Applications. Fields Inst. Comm. 5*, pages 271–279. Amer. Math. Soc., Providence.

Knobloch, E. and D. R. Moore [1988], Linear stability of experimental Soret convection, *Phys. Rev. A* **37**, 860–870.

Knobloch, E. and D. R. Moore [1990], A minimal model of binary fluid convection, *Phys. Rev. A* **42**, 4693–4709.

Knobloch, E., D. R. Moore, J. Toomre, and N. O. Weiss [1986], Transitions to chaos in two-dimensional double-diffusive convection, *J. Fluid Mech.* **166**, 409–448.

Knobloch, E., M. R. E. Proctor, and N. O. Weiss [1993], Finite-dimensional description of doubly diffusive convection. In Sell, G. R., C. Foias, and R. Temam, editors, *Turbulence in Fluid Flows: A Dynamical Systems Approach. IMA Volumes in Mathematics and its Applications 55*, pages 59–72. Springer-Verlag, New York.

Kolodner, P. [1993], Repeated transients of weakly nonlinear traveling-wave convection, *Phys. Rev. E* **47**, 1038–1048.

Kolodner, P. and C. M. Surko [1988], Weakly nonlinear traveling-wave convection, *Phys. Rev. Lett.* **61**, 842–845.

Kolodner, P., C. M. Surko, and H. Williams [1989], Dynamics of traveling waves near the onset of convection in binary fluid mixtures, *Physica D* **37**, 319–333.

Landsberg, A. S. and E. Knobloch [1996], Oscillatory bifurcation with broken translation symmetry, *Phys. Rev. E* **53**, 3579–3600.

Moehlis, J. and E. Knobloch [1998], Forced symmetry breaking as a mechanism for bursting, *Phys. Rev. Lett.* **80**, 5329–5332.

Moehlis, J. and E. Knobloch [2000], Bursts in oscillatory systems with broken D_4 symmetry, *Physica D* **135**, 263–304.

Nagata, W. [1990], Convection in a layer with sidewalls – bifurcation with reflection symmetries, *Z. Angew. Math. Phys.* **41**, 812–828.

Nagata, W. [1991], A perturbed Hopf bifurcation with reflection symmetry, *Proc. R. Soc. Edinburgh A* **117**, 1–20.

Niemela, J. J., G. Ahlers, and D. S. Cannell [1990], Localized traveling-wave states in binary-fluid convection, *Phys. Rev. Lett.* **64**, 1365–1368.

Ning, L., Y. Harada, and H. Yahata [1997], Modulated traveling waves in binary fluid convection in an intermediate-aspect-ratio rectangular cell, *Prog. Theor. Phys.* **97**, 831–848.

Renardy, M. [1999], A note on bifurcation problems in large containers, *Fluid Dyn. Research* **24**, 189–199.

Ruelle, D. and F. Takens [1971a], Nature of turbulence, *Comm. Math. Phys.* **20**, 167–192.

Ruelle, D. and F. Takens [1971b], Nature of turbulence, *Comm. Math. Phys.* **23**, 343.

Schöpf, W. and W. Zimmermann [1989], Multicritical behaviour in binary fluid convection, *Europhys. Lett.* **8**, 41–46.

Sirovich, L. and A. E. Deane [1991], A computational study of Rayleigh-Bénard convection. 2. Dimension considerations, *J. Fluid Mech.* **222**, 251–265.

Steinberg, V., J. Fineberg, E. Moses, and I. Rehberg [1989], Pattern selection and transition to turbulence in propagating waves, *Physica D* **37**, 359–383.

Sullivan, T. S. and G. Ahlers [1988], Nonperiodic time dependence at the onset of convection in a binary liquid mixture, *Phys. Rev. A* **38**, 3143–3146.

Tobias, S., M. R. E. Proctor, and E. Knobloch [1998], Convective and absolute instabilities of fluid flows in finite geometry, *Physica D* **113**, 43–72.

Woodcraft, A. L., P. G. J. Lucas, R. G. Matley, and W. Y. T. Wong [1999], Visualisation of convective flow patterns in liquid helium, *J. Low Temp. Phys.* **114**, 109–134.

Zhao, S. and M. J. Yedlin [1994], A new iterative Chebyshev spectral method for solving the elliptic equation $\nabla \cdot (\sigma \nabla u) = f$, *J. Comput. Phys.* **113**, 215–223.

4

Solid Flame Waves

Alvin Bayliss
Bernard J. Matkowsky
Anatoly P. Aldushin

To Larry Sirovich, on the occasion of his 70th birthday.

ABSTRACT We consider the gasless combustion model of the SHS (Self-Propagating High Temperature Synthesis) process in which combustion waves are employed to synthesize desired materials. In this case the combustion phenomenon is referred to as a "solid flame". Specifically, we consider the combustion of a solid sample in which combustion occurs on the surface of a cylinder of radius R. In addition to uniformly propagating planar waves there are many other types of waves. The study of different waves is important since the nature of the wave determines the structure of the product material. For the fixed value of the Zeldovich number Z which we employ the uniformly propagating planar wave is unstable. We describe stable waves only. We consider solution behavior as R is increased. For R sufficiently small, slowly propagating planar pulsating flames are the only modes observed. As R is increased, transitions to more complex modes of combustion occur, including (i) spin modes in which one or several symmetrically spaced hot spots rotate around the cylinder as the flame propagates along the cylindrical axis, thus following a helical path, (ii) counterpropagating (CP) modes, in which spots propagate in opposite angular directions around the cylinder, executing various types of dynamics. These include spots which pass through each other essentially unchanged, much the same as solitons, and spots which appear to be annihilated, only to be regenerated further along the sample in one of a variety of ways, (iii) alternating spin CP modes (ASCP), where rotation of a spot around the cylinder is interrupted by periodic events in which a new spot is spontaneously created ahead of the rotating spot. The new spot splits into counterpropagating daughter spots, one of which collides with the original spot leading to their mutual annihilation, while the other continues to spin, (iv) modulated spin waves consisting of either one or two symmetrically located rotating spots which exhibit a periodic modulation in speed and temperature as they rotate, (v) asymmetric spin waves in which two spots of unequal strength and not separated by angle π, rotate together as a bound state, (vi) modulated asymmetric spin waves in which the two asymmetric spots oscillate in a periodic manner as they rotate, alternately approaching each other and then moving apart periodically in time, (vii) asymmetric ASCP modes in which a slowly varying bound state of two spots rotates around the cylinder with the leading spot, and subsequently the trailing spot, exhibiting episodes of ASCP behavior, and (viii) 3-headed spins in which three spots rotate around the cylinder in a nonuniform fashion so that each cell alternately approaches one of its neighbors and then the other. In one case the motion is quasiperiodic, with neighboring spots approaching and departing from each other periodically in time as they rotate. In another case the motion is apparently chaotic. Two neighboring spots nearly collide, after which one spot is rapidly propelled away from the other as they rotate. Finally, for a slightly higher value of R, two neighboring spots collide, leading to annihilation of one spot and collapse of the 3-headed spin to a 2-headed spin mode.

Contents

1 Introduction

In the SHS (Self Propagating High Temperature Synthesis) process of materials synthesis reactants are ground into a powder, cold pressed and ignited at one end. A high temperature combustion wave then propagates through the sample converting reactants into products. When gas plays no significant role in the process, the resulting gasless combustion wave is referred to as a "solid flame" (Merzhanov [1994]). The process was pioneered in the former Soviet Union, and has subsequently been the focus of a great deal of research e.g., Merzhanov [1981, 1990]; Munir and Anselmi-Tamburini [1989]. The SHS process enjoys a number of advantages over conventional technology, in which the sample is placed into a furnace and "baked" until it is "well done". The advantages include (i) simpler equipment, (ii) significantly shorter synthesis times, (iii) greater economy, since the internal energy of the chemical reactions is employed rather than the costly external energy of the furnace, (iv) greater product purity, due to volatile impurities being burned off by the very high combustion temperatures of the propagating combustion wave, and (v) no intrinsic limit on the size of the sample to be synthesized, as exists in conventional technology.

The primary objective of this paper is to examine item (v) above, namely the role of the sample size in SHS. Here, we expand on the results in Bayliss, Matkowsky, and Aldushin [to appear(a)]. The characteristics of the resulting combustion wave depend significantly on the size of the sample and impact both the nature of the product synthesized and indeed the ability of combustion waves to propagate in large samples. We describe various types of combustion waves that occur for sample sizes up to an order of magnitude larger than that of the preheat zone, the natural length scale of the process. While yet larger sample sizes are of technological interest, the results presented here may serve as a first step in assessing the role of sample size in the SHS process.

We note that burning sometimes occurs throughout the sample, while under other circumstances it occurs only on the surface of the sample, though this is generally due to the effect of limited oxidizer filtration through the sides of the sample, which is not accounted for here. In this paper we nevertheless restrict consideration to surface burning so that the process can be modeled in two dimensions. Our model thus describes solid flame propagation in a thin cylindrical annulus between two coaxial cylinders, as in the synthesis of hollow tubes. It also provides insight into the behavior of combustion throughout the sample since visual observation is limited to views of the surface of the sample. Since the activation energy of the combustion reaction is typically large, reaction is restricted to a thin zone. No appreciable reaction occurs ahead of the zone since the temperature is not sufficiently high and no reaction occurs behind the zone since the reactants have been consumed. In analytical studies the thin reaction zone is often approximated by an interface separating the fresh unburned mixture from the burned product. In numerical computations there is no interface. Nevertheless, we employ the descriptive term front for the thin reaction zone. It is known that in many instances the combustion wave does not propagate in a uniform spatial and temporal manner, but rather nonuniformities can develop in the front speed and in the temperature along the front as well. Since the mode of propagation determines the microstructure of the product, i.e., the nature of the final product, a study of different modes of propagation is important for technological applications of the SHS process. In particular, it is known that for sufficiently large activation energies (more precisely Zeldovich numbers, defined below) the uniformly propagating combustion wave is unstable. In this case the only stable planar mode is the planar pulsating mode (autooscillatory combustion), in which there is no spatial structure along the front, i.e., the combustion wave remains planar. However, the front speed and temperature on the front oscillate in time in a periodic, quasiperiodic or chaotic fashion, e.g., Aldushin, Martemyanova, Merzhanov, and Shkadinsky [1973]; Bayliss and Matkowsky [1987, 1990]; Margolis [1983]; Matkowsky and Sivashinsky [1978]; Merzhanov, Fiolonenko, and Borovinskaya [1973]; Raymond, Bayliss, Matkowsky, and Volpert [2001]; Shkadinsky, Khaikin, and Merzhanov [1971].

Other modes of propagation involve spatial as well as temporal structure. The modes described in this paper involve one or more localized hot spots (temperature maxima) along the front which exhibit a variety of dynamics. The spots become highly localized as the sample size increases. We will see that, in a sense, the spots behave like particles, which execute interesting dynamics. All of the modes described here involve dynamical behavior of one or more hot spots. The modes of propagation include spin combustion, in which one or more hot spots move on a helical path along the surface of the cylinder, and multiple point combustion, in which hot spot(s) repeatedly appear, disappear and then reappear. Spinning waves were observed in Merzhanov, Fiolonenko, and Borovinskaya [1973]. Subsequent observations of spinning waves and other nonplanar modes are described in, e.g., Dvoryankin, Strunina, and Merzhanov [1982]; Maksimov, Pak, Lavrenshuck, Naiborodenko, and Merzhanov [1979]; Merzhanov [1981, 1990]; Strunina, Dvoryankin, and Merzhanov [1983]. Nonplanar modes, including spinning modes, have also been described analytically, e.g., Margolis, Kaper, Leaf, and Matkowsky [1985]; Sivashinsky [1981], and numerically, e.g., Aldushin, Bayliss and Matkowsky [2000]; Bayliss, Kuske, and Matkowsky [1990]; Bayliss and Matkowsky [1999]; Ivleva, Merzhanov, and Shkadinsky [1978]; Ivleva, Merzhanov and Shkadinsky [1980]. For the case of burning throughout the sample some spinning waves are described in Ivleva and Merzhanov [1999, 2000].

Another family of nonplanar modes of propagation involves pairs of counterpropagating (CP) hot spots along the front. Such modes of solid flame propagation have been observed experimentally in Maksimov, Merzhanov, Pak, and Kuchkin [1981]; Mukasyan, Vadchenko, and Khomenko [1997] and have been described via numerical computations in Bayliss and Matkowsky [1999].

It was found in Mukasyan, Vadchenko, and Khomenko [1997] that under low pressure conditions the desired product was formed by the high temperature spots. Thus, the spots play a crucial role in the synthesis process. The observations in Mukasyan, Vadchenko, and Khomenko [1997] indicated that the spots appeared to be annihilated at a collision, only to be regenerated further along the sample. The spots can be regenerated at the same angle where the collision occurred or on the opposite side of the cylinder, e.g., 180° away from the collision site, e.g., Bayliss and Matkowsky [1999]; Mukasyan, Vadchenko, and Khomenko [1997]. Furthermore, for some parameters the angles corresponding to creation and annihilation sites can spontaneously change as the mode propagates along the cylinder (Bayliss and Matkowsky [1999]). This behavior was described in Bayliss and Matkowsky [1999] employing the model for surface combustion described below. However, it was shown that the annihilation and creation of spots was only apparent. After collision the spots would stay at a relatively low, undetectable level until they would be amplified to a detectable level further along the sample. The results in Bayliss and Matkowsky [1999] suggest that CP modes develop as transitions from standing waves. In this paper we show that CP behavior can also be associated with spinning modes.

An important issue for both the theory and applications of SHS is the behavior of unsteady solid flames in large scale systems, e.g., cylinders whose diameter significantly exceeds the scale of the combustion wave, i.e., the size of the preheat zone, which is of the order of millimeters. Technological applications of SHS are generally associated with systems whose scale is much larger since often the desired goal is to synthesize large samples of material. In addition, the problem is of theoretical interest since it involves a much wider range of scales than is typically encountered in analysis of SHS problems. Virtually all analytical studies of solid flame propagation employ the Zeldovich number $Z = N(1-\sigma)/2$, where N is the suitably nondimensionalized activation energy and $\sigma = \widetilde{T}_u/\widetilde{T}_b$ where $\widetilde{T}_u$ and $\widetilde{T}_b$ are the unburned and burned temperatures far ahead of and far behind the front, respectively, as the control parameter. In this paper we employ the radius R of the sample as the control parameter, as in Bayliss and Matkowsky [1999]. Here, we consider that the cylindrical radius increases up to $O(10)$ times the size of the preheat zone. Furthermore, the scale of the patterns, e.g., the extent of the localized hot spots associated with spin combustion, becomes progressively narrower as R increases and can be smaller than the size of the preheat zone. Thus, combustion waves for large diameter samples involve the effect of small scale behavior on the large scale dynamics, i.e., on the scale of the sample size. One characteristic that we use to categorize these branches is the mean axial flame speed, V, which is both a readily measurable quantity and which is related to the ability of the flames to survive when heat losses are accounted for. We note that for all of our computations the wave travels in the negative z direction, so that the actual flame velocity, as opposed to the speed, is negative.

We first describe the behavior of solutions for a fixed R, with Z employed as the control parameter. Planar front solutions always exist. The planar solutions can be either uniformly propagating, where the front temperature and front speed are constant, or planar pulsations, where the temperature on the front and the front speed oscillate in time, often exhibiting complex dynamics, e.g., Bayliss and Matkowsky [1990]; Raymond, Bayliss, Matkowsky, and Volpert [2001]. It is known that as Z increases past a critical value the uniformly propagating solutions lose stability to pulsating flames if R is sufficiently small and to spinning flames if R is sufficiently large. In the case of planar pulsations, as Z increases further the pulsations become increasingly relaxational, with temporally localized large temperature spikes alternating with long periods of relatively low temperatures. A sequence of period doubling transitions is known to occur as Z increases further (Bayliss and Matkowsky [1990]). The pulsating planar solutions have a lower mean front speed than the uniformly propagating flame and the mean front speed generally decreases as the pulsating solutions become more relaxational, i.e., as Z increases.

In this paper we fix Z at a value beyond the stability boundary so that the uniformly propagating planar solution is unstable and the pulsating planar solution has become relaxational and undergone a transition from a singly periodic solution to a period doubled ($2T$) solution. Furthermore, the mean speed of the pulsating planar solution is reduced considerably from the adiabatic, uniformly propagating flame speed. While we do not consider heat losses in this paper, it is well known that slowly propagating flames are more prone to extinction via heat losses than more rapidly propagating flames.

We find that the pulsating planar solution is stable for small R, but becomes unstable to nonplanar perturbations for larger values of R. Our results show that the transition from pulsating planar, i.e., one dimensional behavior, to spot behavior is a jump transition, i.e., the spots enter with finite amplitude as R increases, consistent with the phenomenological description in Aldushin, Malomed and Zeldovich [1981]. This is not surprising since instability of the uniformly propagating solution is also expected to lead to instability of spinning solutions which differ infinitesimally from it. We describe the various stable modes that we find as R increases. Throughout the discussion transition points refer to points at which transitions from one type of behavior to another occur. We have not attempted to determine the precise numerical value of the transition points, which are very sensitive to variations in Z and our computations correspond to a fixed value of Z. Thus, the precise value is not particularly informative. Rather, for the particular value of Z that we consider we have determined the approximate ranges in which various modes occur.

We next summarize the modes that we find. These modes are described in more detail in Section 4. In all cases the spatiotemporal structure of the mode is best visualized by examining the temperature of the combustion wave on the front in a moving axial coordinate system as a function of the cylindrical angle ψ and time t. In this visualization hot spots appear as localized bright regions on a dark background. The spots trace out trajectories in the $\psi - t$ plane. For example, a spinning wave traces out a straight line whose slope is inversely proportional to the rotation rate. As we describe the modes we refer to the corresponding figures. The figures are described in detail in Section 4, but are referred to here to help clarify the present description. As R increases beyond a critical value R_{cp} the first nonplanar mode to develop is a counterpropagating (CP) mode, in this case characterized as a standing wave in a coordinate system moving axially with the flame. We show that near the transition point, the CP mode emerges with nonzero amplitude hot spots. The flame speed V associated with these modes near the transition point is larger than that of the unstable, uniformly propagating planar mode.

The CP mode is characterized by two counterpropagating hot spots which either collide and pass through each other unchanged except for a phase shift, as in the behavior of solitons, or are weakened in the collision and subsequently strengthened. The latter can occur in various ways as described in Bayliss and Matkowsky [1999]. Let ψ denote the angular coordinate. There are two collision sites, ψ_1, ψ_2 on the front where collisions occur. Near the transition point the two collision sites are symmetrical, i.e., the behavior at the two sites is identical after shifting in both time and ψ (see Figure 3.2).

As R increases, the collisions of the counterpropagating spots become asymmetrical, so that the amplitudes of the spots at one collision site differ from those at the other. The spots impinging at $\psi = \psi_1$ have a large amplitude before the collision and a small amplitude after the collision. The opposite occurs at ψ_2. This pattern leads to sites of apparent annihilation and creation as described in Bayliss and Matkowsky [1999] and observed in Mukasyan, Vadchenko, and Khomenko [1997] (see Figure 3.4). This increasingly nonlinear dynamical behavior is accompanied by a reduction in the mean flame speed V. The CP modes develop highly nonlinear behavior over a very small interval of R. We provide only a limited description of the CP modes as they were discussed in detail in Bayliss and Matkowsky [1999], albeit for different parameter values.

Spin modes enter at $R = R_{tw1} > R_{cp}$ with finite amplitude. The first spin mode is a traveling wave with one hot spot (TW1). In a coordinate system moving with the flame, a single spot rotates around the cylinder. As R increases, the spot becomes more localized and the temperature of the spot increases. This increasing localization and relaxational behavior of the spin mode is associated with a reduction in the mean flame speed V. We find a transition to modulated traveling waves (MTWs) at $R_{mtw} > R_{tw1}$, in which the intensity and speed of the hot spot oscillate in time as the wave propagates (see Figure 3.6).

We next find a transition to a family of modes which does not appear to have been previously observed. For $R_{ascp} > R_{mtw}$ we find alternating Spin CP modes (ASCP). For most of the time there is a single hot spot spinning around the cylinder. Call this spot S_1 and say that it is spinning clockwise. At certain times (periodic in an appropriately moving coordinate system), a new spot is spontaneously created ahead of S_1. Call this spot S_2. The new spot S_2 splits into two counterpropagating spots, $S_{2,c}$ and $S_{2,cc}$, which propagate in the clockwise and counterclockwise directions, respectively. The spots S_1 and $S_{2,cc}$ subsequently collide and are eventually mutually annihilated after the collision while the spot $S_{2,c}$ continues to propagate (see Figures 3.7 and 3.8). In contrast to the CP modes described above, where the annihilation is only apparent, i.e., a weak spot emanates from the collision, we find that actual annihilation occurs here. The spot $S_{2,c}$ continues rotating clockwise until a new spot is created ahead of it and the process repeats periodically in time. The rotation rate is nonuniform, although it is roughly uniform away from the CP events.

We believe that the behavior of the ASCP mode is a consequence of the increasing localization of the spots as R increases. As the spot is localized, a considerable region of the combustion front is nearly planar. Since planar fronts are unstable for the parameters that we consider, additional spots are expected to form, leading to the pattern described above. The ASCP branch evolves from a branch of 1-headed spin solutions, which is stable up to $R = O(10)$ in units of the preheat zone length.

We have not found additional transitions from this branch., However, we have found other spin modes. For any value of R for which a TW1 exists, a TW2 will exist for $2R$, simply by replicating the TW1 solution. Thus, since stable TW1 modes are found in the interval (R_{tw1}, R_{mtw}), TW2 modes exist and may be stable in the interval $(2R_{tw1}, 2R_{mtw})$. Furthermore, the mean front speed V should be the same as for the corresponding TW1 solution. Thus, rapid axially propagating TW2 modes should exist for R near $2R_{tw1}$. We find that the replicated TW2 modes are stable for a subset of the interval $(2R_{tw1}, 2R_{mtw})$. The replicated TW2 mode clearly involves two hot spots of equal intensity which are symmetric, by which we mean that the two identical spots are symmetrically located on the circle, i.e., are separated by angle $\psi = \pi$. Note, however, that the spatial profiles of these waves are not symmetric. In addition to the TW2 modes we have found a family of asymmetric traveling waves consisting of 2 rotating hot spots which are of unequal temperature and are asymmetric, i.e., separated by angle $\psi < \pi$. Nevertheless, the two spots rotate together as a traveling wave in which the two spots are bound together (see Figure 3.11). We call these modes ATW2 for asymmetric TW2 modes. A preliminary description of these modes is given in Bayliss, Matkowsky, and Aldushin [to appear(b)]. The ATW2 branch appears to emanate as a period doubling of TW2 solutions for small activation energy. We find such modes for relatively large values of R, and their axial propagation speeds are smaller than for the replicated TWs. For smaller values of R a transition occurs from ATW2 modes to modulated ATW2 modes (MATW2) in which the two spots alternately approach each other and move apart as they propagate (see Figure 3.17). As R is increased the ATW2 modes lose stability to a class of solutions in which there are still two spots which are bound together though their motion is slowly varying. However, as the spots propagate, a new spot is spontaneously created ahead of the leading spot and we observe ASCP behavior, followed by similar behavior for the trailing spot and a return to the slowly varying bound state. The process then repeats, though the dynamics is not regular, but rather either quasiperiodic or weakly chaotic (see Figure 3.16). We term this an Alternating ASCP mode (AASCP).

We have also investigated 3−headed spins. We find that replicated TW3 modes are not stable. In a limited range of R we have found stable 3-headed spin modes that are not pure TWs. Rather, each spot alternately speeds up and slows down so that it alternately approaches and separates from its neighbors. We find that for R near $3R_{tw}$ the behavior is that of a modulated traveling wave (MTW3) (see Figure 3.20). The motion is quasiperiodic. For larger values of R, the behavior is apparently chaotic. At seemingly random times two of the spots nearly touch, after which one spot is rapidly propelled away from the other as they rotate (see Figure 3.21). We refer to this mode as C3. Upon increasing R further we find that two spots collide during the transient, leading to annihilation of one spot and a collapse of the solution from three heads to two (see Figure 3.22). We find that for R near $3R_{tw}$ the 3-headed spin modes share several features with the unstable replicated TW3 modes, namely a rapid axial propagation speed and relatively low temperatures in the spot.

In Section 2 we describe the model and briefly describe the numerical method employed. In Section 3 we discuss our results in detail. Finally, our results are summarized in Section 4.

2 Mathematical Model

We consider a model which accounts for diffusion of heat and one-step, irreversible Arrhenius kinetics. Since the reactants are solid we neglect diffusion of mass. We consider the case where combustion occurs on the surface of a cylindrical sample of radius $\widetilde{R}$. We further assume that there is a deficient component of the solid mixture and that all other components are present in sufficiently large quantities that the evolution of only the deficient component needs to be followed in the model.

We let $\tilde{}$ denote dimensional quantities and let $\tilde{x}$ and ψ denote the axial and angular cylindrical coordinates, respectively, and assume that the combustion wave propagates in the $-\tilde{x}$ direction. Let $\widetilde{T}$ and $\widetilde{Y}$ denote the temperature and mass fraction of the deficient component respectively. The model is given by

$$
\begin{aligned}
\widetilde{T}_{\tilde{t}} &= \tilde{\lambda}\widetilde{T}_{\tilde{x}\tilde{x}} + \tilde{\lambda}\frac{1}{\widetilde{R}^2}\widetilde{T}_{\psi\psi} + \tilde{\beta}\tilde{A}\widetilde{Y}\,\exp\left(-\frac{\widetilde{E}}{\widetilde{R}_g\widetilde{T}}\right),\\
\widetilde{Y}_{\tilde{t}} &= -\tilde{A}\widetilde{Y}\,\exp\left(-\frac{\widetilde{E}}{\widetilde{R}_g\widetilde{T}}\right),
\end{aligned}
\tag{2.1}
$$

where $\tilde{\lambda}$ is the thermal diffusivity, $\tilde{A}$ is the frequency factor, $\tilde{\beta}$ is the scaled heat of reaction, $\widetilde{E}$ is the activation energy (all assumed constant) and $\widetilde{R}_g$ is the gas constant. The boundary conditions are

$$
\begin{aligned}
\widetilde{Y} \to \widetilde{Y}_u, \quad \widetilde{T} &\to \widetilde{T}_u, \text{ as } \tilde{x} \to -\infty,\\
\widetilde{T}_x &\to 0, \text{ as } \tilde{x} \to \infty.
\end{aligned}
$$

We note that $\widetilde{T} \to \widetilde{T}_b$ as $\tilde{x} \to \infty$ where the subscripts u and b refer to unburned and burned respectively, however, we employ the Neumann condition in our model. The solution is periodic in ψ with period 2π. We observe that $\widetilde{T}_b$ is derivable from the time-independent solution of the problem as $\widetilde{T}_b = \widetilde{T}_u + \tilde{\beta}\widetilde{Y}_u$.

We nondimensionalize as in Matkowsky and Sivashinsky [1978] by introducing

$$Y = \frac{\widetilde{Y}}{\widetilde{Y}_u}, \quad \Theta = \frac{\widetilde{T} - \widetilde{T}_u}{\widetilde{T}_b - \widetilde{T}_u}, \quad t = \frac{\tilde{t}\widetilde{U}^2}{\tilde{\lambda}}, \quad x = \frac{\tilde{x}\widetilde{U}}{\tilde{\lambda}}$$

$$\sigma = \frac{\widetilde{T}_u}{\widetilde{T}_b}, \quad N = \frac{\widetilde{E}}{\widetilde{R}_g \widetilde{T}_b}.$$

Here,

$$\widetilde{U}^2 = \frac{\tilde{\lambda}\widetilde{A}}{2Z} \exp(-N)$$

where $Z = N(1-\sigma)/2$ is the Zeldovich number, $\widetilde{U}$ is the speed of the uniformly propagating front for asymptotically large Z (Margolis [1983]), and lengths are scaled by the size of the preheat zone. We next introduce the moving coordinate

$$z = x - \phi(t)$$

where $\phi(t)$ is defined by $Y(\phi(t), \psi = 0, t) = 0.5$. Here, the choice of the angle ψ at which $Y = 0.5$ is arbitrary; $\psi = 0$ is chosen merely for convenience. The velocity ϕ_t and the location $x = \phi(t)$ do not model the front velocity and location when there are nonplanar disturbances, since they, unlike these functions, also depend on ψ. However, for all angles the reaction zone will be localized in a neighborhood of $z = 0$. Thus, ϕ_t is the approximate velocity of the wave, so that the transformation to the moving coordinate system enables us to localize the reaction zone to a neighborhood of $z = 0$.

In terms of the nondimensionalized quantities, the system (2.1) becomes

$$\begin{aligned} \Theta_t &= \phi_t \Theta_z + \Theta_{zz} + \frac{1}{R^2}\Theta_{\psi\psi} + 2ZY \exp\left(\frac{N(1-\sigma)(\Theta-1)}{\sigma + (1-\sigma)\Theta}\right), \\ Y_t &= \phi_t Y_z - 2ZY \exp\left(\frac{N(1-\sigma)(\Theta-1)}{\sigma + (1-\sigma)\Theta}\right), \end{aligned} \tag{2.2}$$

where

$$R = \frac{\widetilde{R}\widetilde{U}}{\tilde{\lambda}}.$$

The coefficient of the reaction term arises from the use of the asymptotic planar adiabatic burning speed $\widetilde{U}$ in the nondimensionalization. The use of the asymptotic value of $\widetilde{U}$ with finite activation energy affects the length and time scales of the computation (for example, the speed of uniformly propagating planar waves at finite activation energy differ slightly from unity), but does not change any of the resulting spatiotemporal patterns.

In addition, we introduce a cutoff function $g(\Theta)$ which multiplies the Arrhenius term so that the reaction term vanishes for sufficiently small temperatures. The function $g(\Theta)$ is defined by

$$g(\Theta) = 0, \ \ \Theta < \Theta_{cut}, \ \ g(\Theta) = 1, \ \ \Theta > \Theta_{cut}.$$

We have tested both discontinuous and smooth cutoff functions and find that the different possibilities have virtually no effect on the computed solution. The cutoff function is employed to avoid the "cold boundary difficulty" which arises because the Arrhenius model for the reaction term does not vanish far ahead of the front, which is incompatible with the boundary condition $\widetilde{T} = \widetilde{T}_u$ as $\tilde{x} \to -\infty$. In addition, in practice no significant reaction occurs ahead of the reaction zone. For the computations presented here $\Theta_{cut} = .03$. We have found that the results are insensitive to variations in Θ_{cut} as long as its value is of this order.

The boundary conditions are specified at finite points far from the reaction zone which is in the vicinity of $z = 0$. The computations presented here were obtained with the boundary conditions imposed at $z = \pm 12$. There is virtually no effect of further increasing the size of the computational domain. Note that no boundary condition is imposed on Y in the burned region.

Since we consider distributed Arrhenius kinetics, there is strictly speaking no combustion front but rather a narrow reaction zone in which the chemical reaction terms are significant. However, a front can be defined in a number of ways, e.g., as the curve of maximum reaction rate or as the curve where the reactant mass fraction decreases to half its initial value, etc. These procedures treat the front as a curve for each value of time, $z_f(\psi, t)$, along which temperature, mass fraction and reaction rate vary. In order to present our results, we have developed a procedure to approximate the combustion front and the temperature on the front as a function of time t and the cylindrical angle ψ. We define the front for each value of ψ as the z location where the reaction term is maximal. Thus, for each value of ψ and t we compute a value $z = z_f(\psi, t)$, where the reaction term takes a maximum and define the resulting function as the combustion front. That is, with

$$W(z, \psi, t) = Y \exp\left(\frac{N(1-\sigma)(\Theta - 1)}{\sigma + (1-\sigma)\Theta}\right),$$

we let z_i denote the axial collocation points, and compute W at all collocation points. For each value of ψ and t we find i^m, the value of i where $W(z_i, \psi, t)$ attains its maximum. We then locally fit W to a quadratic in z using the points $W(z_{i^m}, \psi, t)$, $W(z_{i^{m+1}}, \psi, t)$ and $W(z_{i^{m-1}}, \psi, t)$. We next compute the value of z which maximizes this quadratic and use this value as $z_f(\psi, t)$, our approximation to the front location. Finally, we compute $\Theta_f(\psi, t)$, the temperature at $z_f(\psi, t)$, by Chebyshev interpolation. We refer to $\Theta_f(\psi, t)$ obtained in this manner as the temperature at the front.

In our model for the front it does not follow that Θ is maximal at the front. In many instances the location of maximum temperature does in fact correlate with the front location as defined above. However, under certain circumstances the temperature can increase behind the front. Axial profiles of temperature for a one dimensional problem are shown in, e.g., Bayliss and Matkowsky [1990]; Shkadinsky, Khaikin, and Merzhanov [1971].

For the numerical computations we employ an adaptive Chebyshev pseudospectral method in z that we previously developed, e.g., Bayliss, Gottlieb, Matkowsky, and Minkoff [1989]; Bayliss and Matkowsky [1987], together with a Fourier pseudo-spectral method in ψ. In order to better resolve the reaction zone in which the solution changes rapidly, we adaptively transform the coordinate z. The transformation has the effect of expanding the reaction zone so that in the new coordinate system the solution varies more gradually and is therefore easier to compute. The method is described in detail in other references, e.g., Bayliss and Matkowsky [1999]; Bayliss, Matkowsky, and Aldushin [to appear(a)].

3 Results

Unless otherwise stated for all of the computations reported here we fixed $N = 25$ and $\sigma = 0.6$ so that $Z = 5$. For these parameters, the uniformly propagating planar solution is unstable. The stability boundary is approximately $Z = 4.2$ and is very close to the analytically predicted value which is calculated for δ-function kinetics rather than the distributed kinetics that we employ. The control parameter is R which is varied over approximately the interval $0 < R < 20$.

An overview of our results is shown in Figure 3.1 where we plot the mean axial flame speed V for the different solution branches that we have found. We compute V by first computing the z location of the front for each value of ψ, as described above. We next compute $\phi(t)$ by integrating the velocity ϕ_t in time. We then compute $x_f(\psi, t)$, the front location in the fixed frame. We note that the wave propagates toward $-\infty$. For simplicity x_f denotes the absolute value of the front location so that in the plots below it appears as a positive quantity. The mean axial flame speed V is then obtained by performing a linear least squares fit to $x_f(t)$ and taking the slope as V. We do not consider a solution to be equilibrated until we find that V does not vary with different choices of angles ψ employed in computing the average V, or if the run is continued longer in time. We note that all solution branches are described (and computed) with respect to a coordinate system moving with the flame.

Figure 3.1 incorporates several solution branches. A common characteristic of these solution branches is the formation of hot spots. The branches are characterized by different dynamics exhibited by the spots. For small values of R the flame speed is close to that of the unstable, uniformly propagating mode. However, our results show that even for small values of R the primary conversion of reactant to product occurs via the spinning of the spots rather than through axial diffusion of heat. As R increases the spots become increasingly localized in space and exhibit increasing temperature.

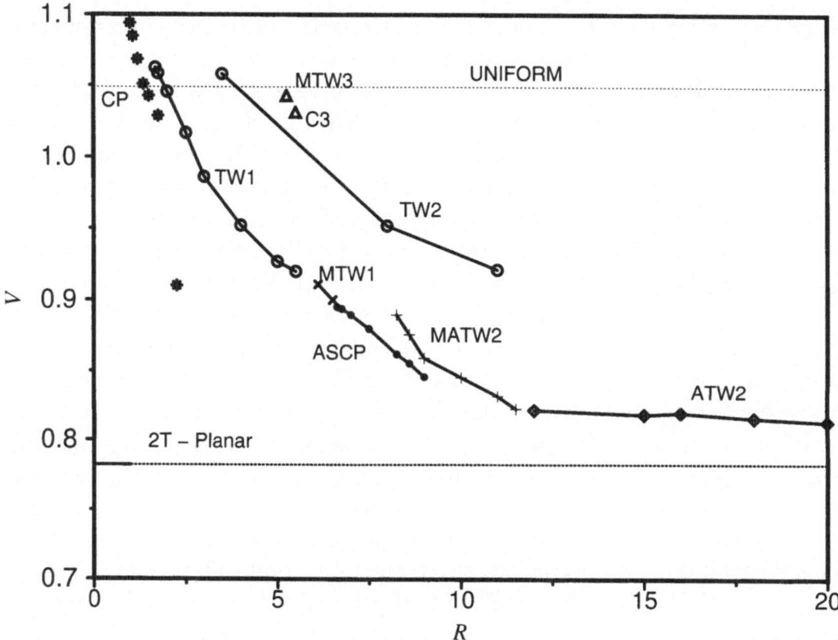

FIGURE 3.1. V as a function of R for various solution branches. Solid (dotted) lines indicate stable (unstable) solutions.

As a result large regions of the front are nearly uniform and thus prone to additional instabilities. Furthermore, as the spots become localized, the behavior at a fixed angle ψ becomes increasingly relaxational in time analogous to the behavior of the slowly propagating pulsating planar solution. Since for all solution branches the mean axial speed V is nonincreasing as R increases, the branches can be loosely thought of as providing a transition from behavior analogous to the higher speed uniform mode to behavior analogous to the lower speed planar pulsating mode as R increases. Since slowly propagating flames are more likely to be extinguished by heat losses, the modes at the left end of the branches may be more likely to survive heat loss effects than those at the right end. Due to the increased localization more of the front (in terms of ψ) attains the rather cool temperatures away from the spot, thus resulting in the lower mean speeds. This behavior is similar to that observed for planar modes of propagation, where the development of relaxation oscillations is accompanied by a decrease in the mean speed of the flame, e.g., Aldushin, Bayliss and Matkowsky [2000]; Matkowsky and Sivashinsky [1978]; Shkadinsky, Khaikin, and Merzhanov [1971].

Broadly speaking, hot spot dynamics can be classified into two categories. Counterpropagating (CP) behavior in which pairs of spots propagate in opposite directions around the cylindrical surface and spinning modes in which one or more spots rotates around the cylinder in the same direction. As seen in Figure 3.1, the nonplanar solutions which first appear, i.e., appear for the lowest values of R, are CP solutions. For values of R below those indicated in the figure, only planar pulsating solutions are found. Spin behavior develops only for larger values of R. The CP modes indicated in Figure 3.1 correspond to a variety of behaviors. An extensive discussion of different CP modes of solid flames, computed for different parameter values than those in this paper, is presented in Bayliss and Matkowsky [1999]. Here, we briefly describe various types of CP behavior that we have found.

In Figure 3.2 we show a space-time plot of a CP mode for $R = 1$. We note that there are collisions at $\psi = 0$ and $\psi = \pi$. The collision sites are determined by the initial conditions. This CP mode is periodic and symmetric in that the behavior at the two collision sites is the same, just 180° out of phase in time. In this mode, the hot spots enter and leave the collision essentially unchanged except for a phase shift, much like the behavior of solitons. Our computations indicate that the amplitude of the symmetric CP mode, i.e., the maximum temperature achieved at a collision, does not approach 0 as R approaches the transition point. Thus, the stable CP modes develop with finite amplitude. Our results also indicate that the mean speed V for these modes exceeds the speed for the unstable, uniformly propagating planar solution. Thus, near the transition point finite amplitude CP modes can propagate faster than the uniformly propagating mode. The mean propagation speed decreases as R increases.

As R increases there is a transition to CP modes which exhibit different behavior at each of the two collision sites. We illustrate such a solution in Figure 3.3 for $R = 1.2$, where we plot Θ_f as a function of t at the two collision sites $\psi = 0$ and $\psi = \pi$. Each spike in the time history of Θ_f corresponds to a collision. The solution is periodic, however, we note that had we chosen an angle different from one of the collision sites Θ_f would exhibit two spikes per period corresponding to the two counterpropagating hot spots. In contrast to the solution for $R = 1$ an asymmetry between the two collision sites has developed. Spots entering the collision at $\psi = 0$ are stronger than when they exit the collision and conversely for collisions at $\psi = \pi$. The collisions take on the character of apparent creation/annihilation sites as the asymmetry between the two sites increases. The development of nonlinear behavior for the CP solutions is accompanied by a reduction in the mean flame speed V.

The CP modes in Figures 3.2 and 3.3 are periodic in time. In Figure 3.4 we illustrate a CP mode for $R = 2.25$ in which the collisions do not occur in a periodic manner. We note that for this figure the collision sites, which are determined by initial conditions, are no longer at $\psi = 0$ and $\psi = \pi$. The problem is invariant under shifts in ψ, so that solutions which are shifted in ψ are also stable. The CP solution in Figure 3.4 exhibits apparent creation and annihilation in that at one collision site strong waves interact producing weak waves (apparent annihilation) while at the other

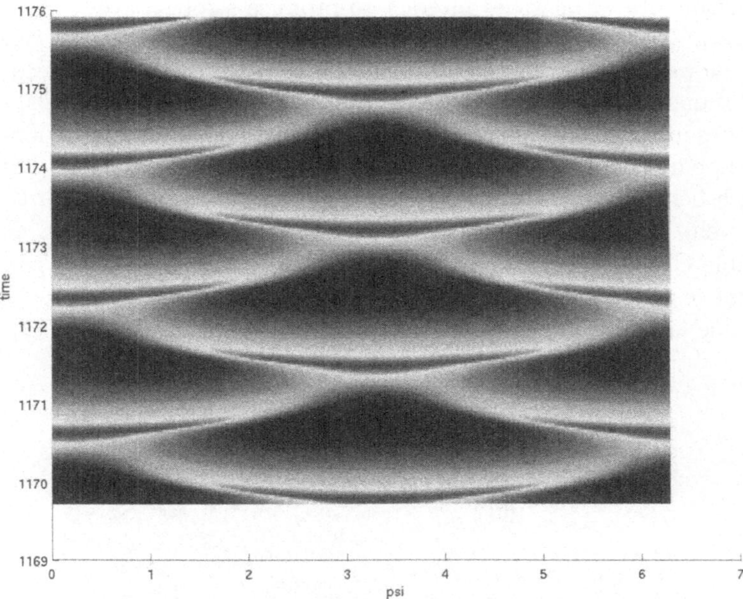

FIGURE 3.2. Θ_f as a function of ψ for a range of times for CP solution with $R = 1$. Time increases along the vertical axis.

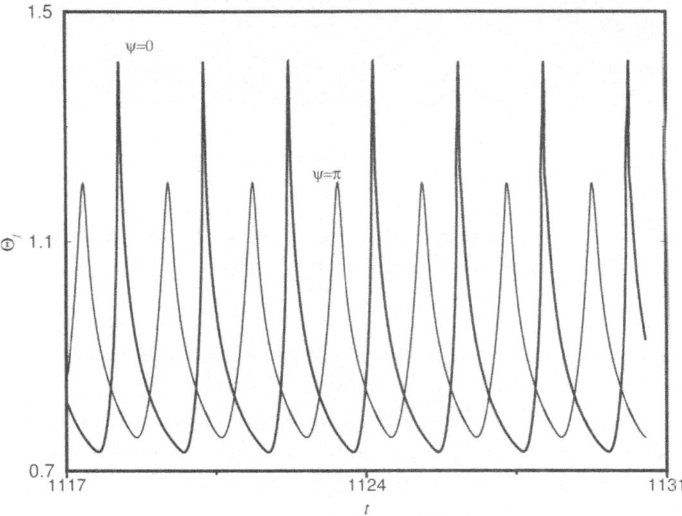

FIGURE 3.3. Θ_f as a function of t at the two collision sites $\psi = 0$ and $\psi = \pi$ for CP mode with $R = 1.2$.

collision site weak waves interact to produce strong waves (apparent creation). These and other CP modes were described in detail in Bayliss and Matkowsky [1999] for different parameters. We have not traced the CP solutions further here, however, additional dynamical behavior, e.g., CP modes in which sites alternate between being an apparent creation site and an apparent annihilation site, was found in Bayliss and Matkowsky [1999]. Such behavior is likely to occur for the parameters employed in this paper as well. In Bayliss and Matkowsky [1999] we also showed that additional stable CP solutions existed, including solutions with multiple pairs of CP spots, e.g., one pair may execute the same dynamics as the others but with a time lag.

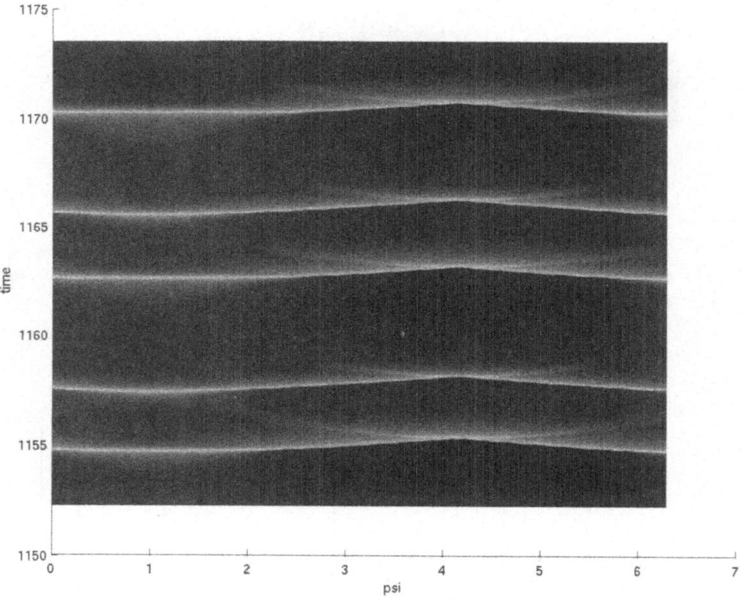

FIGURE 3.4. Θ_f as a function of ψ for a range of times for CP solution with $R = 2.25$. Time increases along the vertical axis.

We next consider the spin mode solution branches shown in Figure 3.1. The first branch we focus on corresponds to 1-headed spin modes which are traveling waves (TW1). These modes require larger radii for stability than the CP modes. Stable CP modes may be able to exist for smaller values of R as they are, in a sense, reinforced by collisions after circuiting only half the cylinder. The TW1 modes also appear to enter with nonzero amplitude of the rotating hot spot. The smallest value of R for which we can find a stable TW1 mode is R=1.68. For this value of R the temperature of the hot spot is approximately 1.21. Unstable TW1 modes may well exist for smaller values of R. If we decrease R further solutions with TW1 initial data evolve to CP modes. We note that the mean axial speeds V for the TW1 modes near onset are below those of the CP modes near onset, probably due to the fact that R is larger.

We find that the TW1 modes are stable for $1.68 \leq R \leq 5.5$. Over this interval V decreases with R as indicated in Figure 3.1. In addition, the spinning spot becomes hotter and more localized as R increases. In Figure 3.5 we plot Θ_f as a function of the distance $R\psi$ around the cylinder at specific times for TW1 modes with $R = 1.75$ and $R = 5.5$. The figure shows that as R increases the temperature on the front becomes nearly uniform over most of the front away from the hot spot.

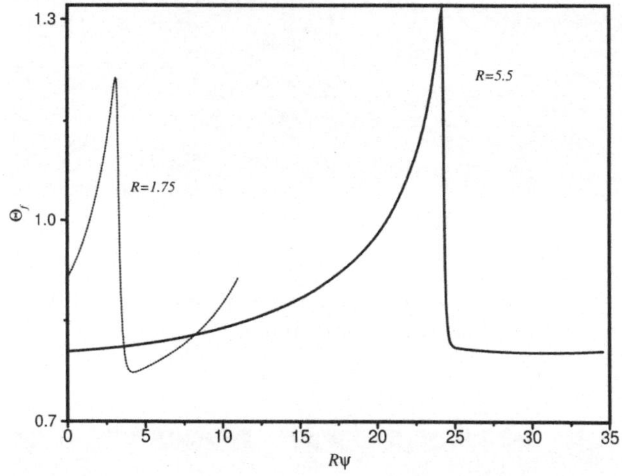

FIGURE 3.5. Θ_f as a function of distance $R\psi$ around the cylinder for TW1 solutions with $R = 1.75$ and $R = 5.5$.

For $R \geq 6.1$ we are unable to compute stable TW1 modes. Rather, we find modulated traveling waves consisting of a single hot spot which alternately speeds up and slows down and its intensity alternately increases and decreases as it rotates around the cylinder, thus describing a modulated traveling wave for a 1-headed spin (MTW1). A space time plot of such a mode is shown in Figure 3.6 for $R = 6.1$.

More complex behavior occurs for larger values of R. In Figure 3.7 we show a space time plot for $R = 7.5$. We note that there are periods of time where the trajectory of the spot (streak in the figures) is nearly horizontal. In order to analyze the nature of these solutions and relate them to the modes that occur for larger values of R we refer to Figure 3.8 which shows a space time plot of the front temperature over a small time interval around one of these events. A small spot develops spontaneously ahead of the main propagating spot. The new spot splits into two counterpropagating spots, one of which collides with the original spot leading to their eventual mutual

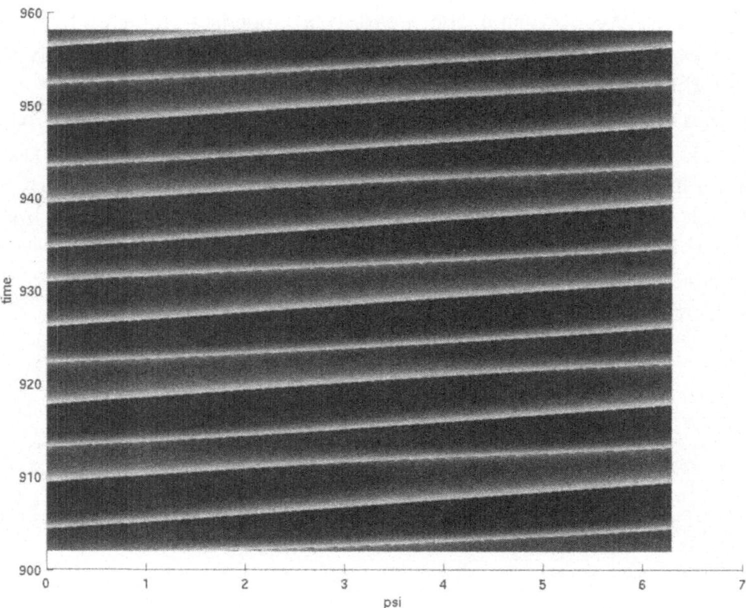

FIGURE 3.6. Θ_f as a function of ψ for a range of times for MTW1 solution with $R = 6.1$. Time increases along the vertical axis.

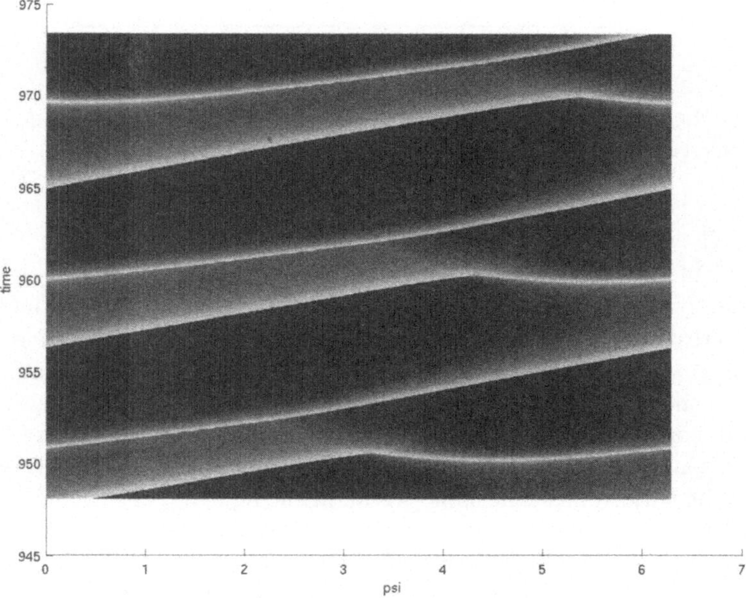

FIGURE 3.7. Θ_f as a function of ψ for a range of times for ASCP solution with $R = 7.5$. Time increases along the vertical axis.

annihilation, while the other spot continues to rotate. Thus, for most of the time there is only one spot, but at times there can be either two or three spots on the front. We refer to these modes as Alternating Spin CP (ASCP) modes.

As can be seen from Figure 3.7 these events occur periodically in time so that the mode is quasiperiodic. In Figure 3.9 we illustrate the CP event by showing Θ as a function of z and ψ at selected times around a CP event. In this figure the lateral direction is ψ so that propagation around the cylinder corresponds to lateral movement of a hot spot. The upper part of each figure corresponds to burned material (lighter shading) and the lower part (darker shading) is unburned material. The shading scale is the same for all frames in the figure. The first frame ($t = 948.8$) shows a single spot propagating counterclockwise (to the right in the figure). At $t = 950.1$ a new spot forms ahead of the main spot. At $t = 950.45$ the new spot splits into two daughter spots, one of which propagates to the right and the other to the left. At $t = 950.58$ the leftward propagating spot collides with the original spot, leading to the highest temperatures in the sequence of frames, while the rightward moving spot continues to propagate. At $t = 950.88$ the temperature at the collision site has decreased. Ultimately the two colliding spots completely annihilate each other. At $t = 957.8$ there is again a single spot as in the first frame. The large temperatures at the collision illustrated at $t = 950.58$ together with the subsequent annihilation can be explained by examining the degree of conversion $\eta = 1-Y$. In Figure 3.10 we plot η in grayscale at $t = 950.45$ and $t = 950.58$. In this plot light corresponds to complete burning ($\eta = 1$) and dark corresponds to unburned reactant ($\eta = 0$). For this problem the internal layer connecting $\eta = 0$ and $\eta = 1$ is an extremely thin strip. At $t = 950.48$ the new spot has split and one of the daughter spots is counterpropagating toward the original spot. It can be seen from Figure 3.10 that a pocket of unburned reactant penetrates the burned region between these two counterpropagating spots. At $t = 950.58$ when the two spots have collided, the pocket of unburned reactant has been essentially consumed in the high temperatures at the resulting collision. After this time the remnant of the collision decays due to thermal diffusion as there is no longer any unburned reactant to sustain the high temperature. The ASCP mode demonstrates that both spin and CP behavior can occur in a periodic fashion in the same solution. We note that the spontaneous creation of new spots ahead of rotating spots also occurs in gaseous combustion stabilized on a rotating burner (Bayliss and Matkowsky [1996]).

We conjecture that the mechanism for ASCP behavior is the increasing localization of the spot as R increases. When R is sufficiently large, the portion of the front excluding the hot spot (topologically a circle in the moving coordinate) has a nearly uniform temperature (cf. Figure 3.5). Since the uniformly propagating planar mode is unstable additional spots are expected to form in the region away from the propagating spot. These

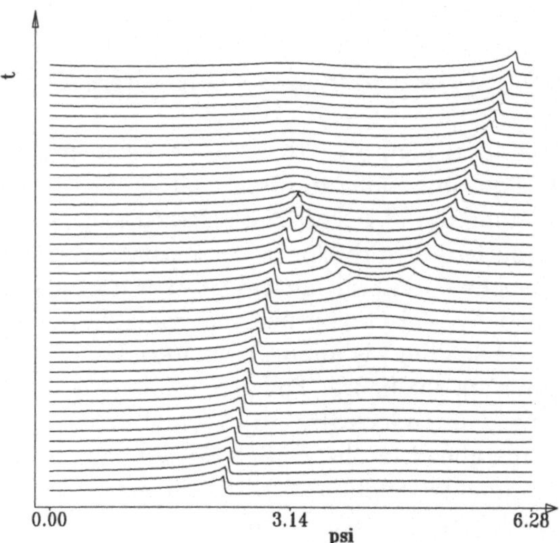

FIGURE 3.8. Θ_f as a function of ψ for a range of times for ASCP solution with $R = 7.5$. Solution is shown over a small interval of time around an event in which a new spot is spontaneously created. Time increases along the vertical axis.

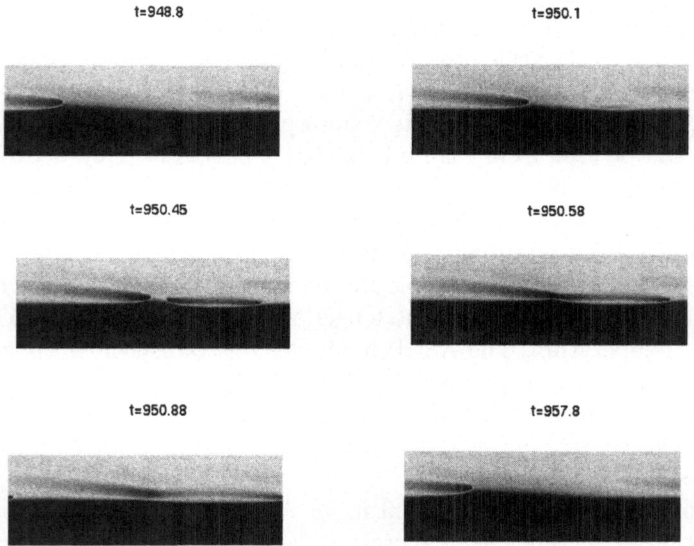

FIGURE 3.9. Θ as a function of ψ and z for selected times for ASCP solution with $R = 7.5$. Axial direction is vertical.

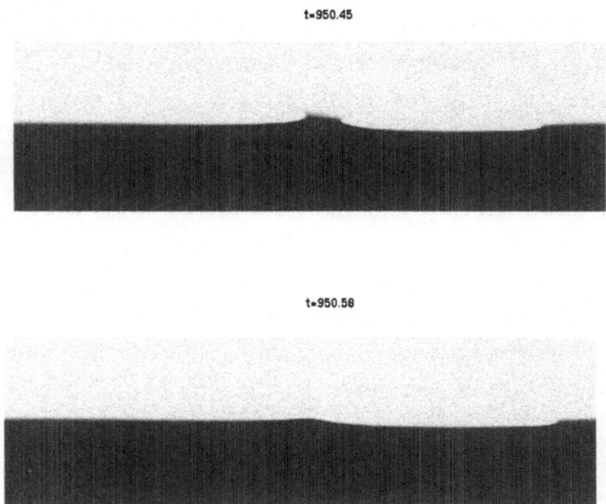

FIGURE 3.10. η as a function of ψ and z for selected times for ASCP solution with $R = 7.5$. Axial direction is vertical.

spots can either exhibit spin or CP behavior. In the ASCP mode the new spot exhibits CP behavior.

We believe that the ASCP mode evolves continuously from the MTW1 branch which evolved from the TW1 branch. This mode, while exhibiting both spin and CP behavior, is essentially a one-headed spin except for brief intervals of time. As R increases V decreases and the instantaneous temperature at collision increases. We have computed ASCP modes from $R = 6.625$ up to $R = 9$. We have not attempted to compute this mode for larger values of R.

We next consider multi-headed spins. We first refer to the branch labeled TW2 in Figure 3.1. These solutions are replicates of analogous TW1 modes. Thus, for any value of R the TW2 solution consists of two replicates of the TW1 mode at $R/2$. The two spots are of equal intensity and are located symmetrically over the interval $(0, 2\pi)$, i.e., are separated by angle π. We have tested the stability of these solutions by imposing perturbations in the initial conditions which do not correspond to replicated perturbations. (We note that replicated conditions correspond in Fourier space to even order modes. Thus, we impose perturbations which have odd Fourier modes.) We have found stable replicated TW2 modes for $3.5 \leq R \leq 11$, corresponding to TW1 modes for $1.75 \leq R \leq 5.5$. We have not computed the TW2 branch for all values of R corresponding to TW1 computations, but only for the three values of R shown in the figure. We assume that the replicated branch is stable for values of R between the extreme points that we have computed. We note that for $R = 3.36$, corresponding to the TW1 at $R = 1.68$, the smallest value of R for which we found stable TW1 modes, we were unable to compute a TW2 solution. Rather, the perturbed replicated initial conditions evolved to a TW1 solution. Thus, our computations indicate that the replicated TW2 branch is stable, but not for all values of R for which a stable TW1 solution exists for $R/2$. The replicated TW2 solutions near $R = 3.5$ have the same large mean speed as the TW1 mode near $R = 1.75$, suggesting that these modes would be less prone to extinction and thus they may be more readily observed in experiments.

In the region of bistability between the TW1 and TW2 branches, the two spin modes are differentiated by lower axial flame speeds for the TW1 solutions, along with more localized, higher temperature spots than for the TW2 mode. The more localized spots result in a large region of the front being at a relatively low temperature thus possibly explaining the slower axial flame speed as compared to the TW2 solution. Ultimately, the localization is such that a new spot is spontaneously created and ASCP behavior ensues. For the TW2 mode the spots are at lower temperature and less localized since they are replicates of TW1 spots corresponding to lower values of R. Thus, the resulting front speed V is larger and closer to that of the uniform mode.

For $R > 11$ we are unable to compute stable TW2 modes. We note that the MTW1 branch described above appears to evolve continuously from the TW1 branch in that near the transition point the solution consists of a small modulation of the TW1, with the mean axial speed V apparently continuous across the transition. We were unable to find any such branch evolving from the TW2 branch. When we tried to compute 2-headed spins for $R = 12$ using continuation, i.e., employing equilibrated TW2 data for R=11 as initial conditions, we found that the solution evolved to the mode shown in Figure 3.11, an asymmetric 2-headed spin, which we denote by ATW2, accompanied by a jump in V. This solution is a 2-headed traveling wave in that two spots rotate around the cylinder at a uniform rate and without any change in shape. However, the two spots are not symmetrically placed, i.e., are not separated by $\psi = \pi$ and they are of unequal strength. The stronger (hotter) spot leads the weaker (cooler) spot. Another perspective on the ATW2 mode can be seen in Figure 3.12 where we plot the front location $x_f(t)$ as a function of time for a fixed angle ($\psi = 0$). The time interval is the same as in Figure 3.11. The mean axial speed V of this mode is obtained from the slope of the least squares fit of this data (taken over a longer time interval) to a line and is indicated in Figure 3.1. For the instantaneous front location $x_f(t)$, nearly vertical slopes correspond to rapid front speeds at $\psi = 0$ over short time intervals. These occur when the rotating hot spots pass the angle $\psi = 0$, as an increase in temperature causes the front to propagate faster. It can be seen from Figure 3.12 that the front speedups occur at periodic intervals, since the ATW2 mode is periodic in a moving coordinate system. Furthermore, over each period there are two asymmetric spot crossings, corresponding to the two asymmetric spots which rotate together as a traveling wave. The spot crossings shown in Figure 3.12 correlate directly with a vertical slice at $\psi = 0$ in Figure 3.11. The two-headed asymmetric spatial structure of the ATW2 mode can

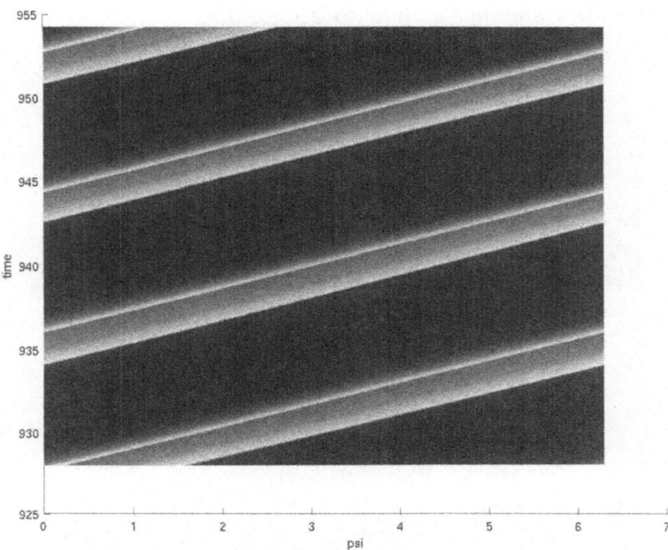

FIGURE 3.11. Θ_f as a function of ψ for a range of times for ATW2 solution with $R = 12$. Time increases along the vertical axis.

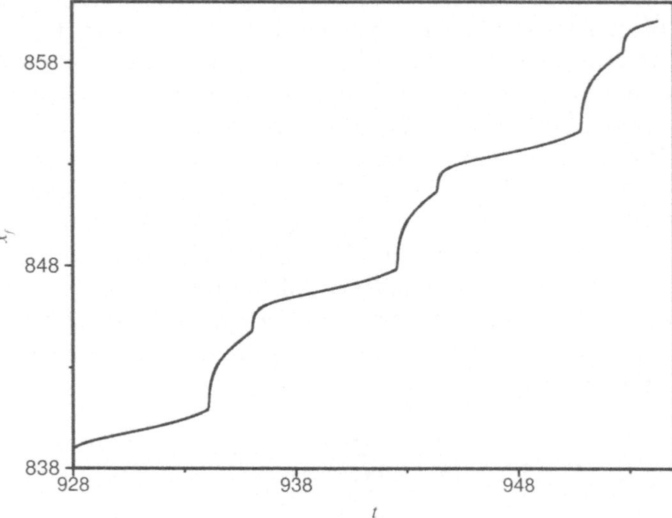

FIGURE 3.12. x_f as a function of t at $\psi = 0$ for ATW2 solution with $R = 12$.

be seen in Figure 3.13 where Θ is plotted as a function of z and ψ at two different times in a manner similar to Figure 3.9.

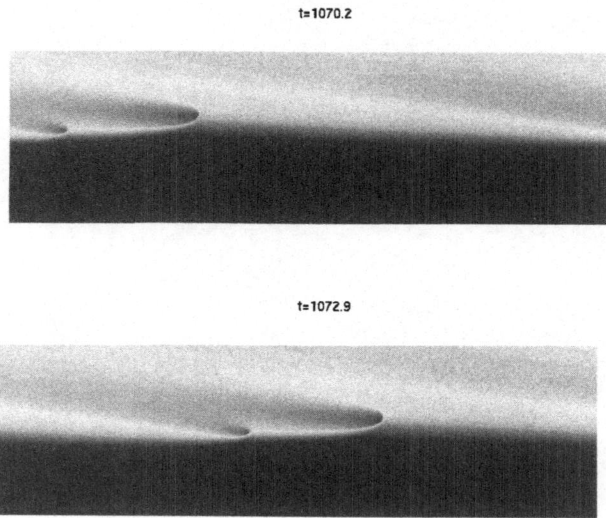

FIGURE 3.13. Θ as a function of ψ and z for selected times for ATW2 solution with $R = 12$. Axial direction is vertical.

These frames appear to suggest that the rotation of the bound state provides the mechanism for the flame propagation. The leading spot propagates directly into the cold unreacted material ahead of it (dark in the figures), thus burning the bulk of the material and advancing the front both laterally and downward in z. The weak trailing spot then advances into a small ridge of unreacted material, again propelling the front downward. We have found this to be characteristic of spin modes with similar behavior exhibited even for the TW1 mode for the smallest value of R that we compute ($R = 1.68$). The conclusion which appears to follow from the figure is that all the conversion occurs directly due to the propagation of the hot spot into virgin material. However, more detailed consideration shows that some conversion occurs ahead of the spot, due to axial diffusion of heat from the burned region to the unburned region. In order to illustrate this fact we compute the quantity

$$\kappa(\psi) = \int_0^{\psi} d\psi' \int_{z_L}^{z_U} W(z, \psi') dz/(2\pi V),$$

where $z_{L,U}$ denote the boundaries of the computational domain. This quantity is the normalized, integrated conversion rate at a fixed time over the sector of the cylinder between the angles 0 and ψ. In Figure 3.14 we plot this quantity for the TW1 mode with $R = 1.68$ and for two ATW2 modes,

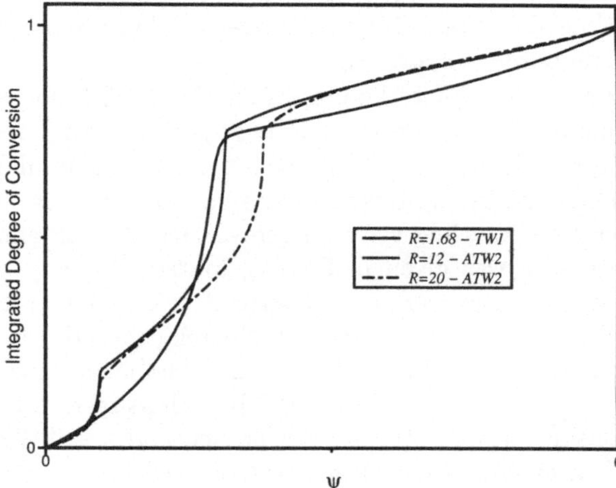

FIGURE 3.14. $\kappa(\psi)$ for a TW1 mode with $R = 1.68$ and ATW2 modes with $R = 12$ and $R = 20$.

$R = 12$ and $R = 20$ (the largest value of R for which we compute ATW2 modes). In this figure we have aligned the three solutions so that the locations of the spots are roughly the same. We note that for an axially symmetric mode we have $\kappa(\psi) = \psi/(2\pi)$. For non-axially symmetric modes regions of large gradient in the curve correspond to rapid burning, which occurs in the vicinity of the hot spot. Inflection points correspond to maxima of the reaction rate and represent the lateral edge of the hot spot. We note that for all three cases significant conversion occurs ahead of the spot. In fact, approximately 25% of the total conversion occurs ahead of the spot. The fact that the region ahead of the front does not consist of virgin unreacted material but rather, has been partially heated and therefore partially reacted, should be taken into account in the analysis of spin combustion because the propagation mechanisms in unreacted and in partially reacted mixtures differ from one another. In the former, the propagation speed is determined by the reaction rate at the combustion temperature $\tilde{T}_b$ as described by Zeldovich and Frank Kamenetskii, whereas in the latter it is determined by the reaction rate at the unburned temperature $\tilde{T}_u$ as in Kolmogorov, Petrovsky, Piskunov type problems. A discussion of the two different mechanisms is given in Aldushin, Khudayev, Zeldovich [1979]. We note that for small values of R, e.g., $R = 1.68$, this figure demonstrates that the nature of the conversion is very different from that of the uniformly propagating mode though the mean front speeds are very close (see Figure 3.1). Thus, the mean front speed, while a convenient way of characterizing solution branches and a readily measured quantity, does not really distinguish different modes of front propagation. In principle changes in the slope of V as R varies should detect transitions between solution branches (see Aldushin, Bayliss and Matkowsky [2000]). However, due to the cost of the computations the data in Figure 3.1 does not have sufficient resolution to accurately determine changes in slope.

We further note that while the graphs for the ATW2 modes differ from the TW1 mode in that the presence of two spots is clearly indicated, the overall structure of the conversion graphs is very similar for all three modes considered. These modes are referred to as bound states, as the two spots move together as a traveling wave on the surface. A preliminary description of these modes was given in Bayliss, Matkowsky, and Aldushin [to appear(b)]. Bound states were also described in one dimensional computations of an excitable system (Or-Guil, Kevrekidis, and Bär [1999]). We have computed ATW2 modes up to $R = 20$. The mean axial speed V is nearly independent of R along the ATW2 branch as seen in Figure 3.1.

The relationship between the ATW2 branch and the TW2 and TW1-MTW1-ASCP branches is not clear since, as seen in Figure 3.1, there is no apparent connection between these different manifestations of spin combustion. There are three parameters associated with this problem, R, the activation energy N and the temperature ratio σ. Figure 3.1 shows that there is no clear connection between these different manifestations of spin combustion when R is varied with the other parameters remaining fixed. We have also computed solutions for fixed R and σ with only N varying. These computations suggest that ATW2 solutions may evolve as a period doubling bifurcation from TW2 solutions corresponding to low activation energies. In order to quantify this we have computed the asymmetry ratio k for ATW2 solutions as N is varied. We define k by determining the difference $\psi_{leading} - \psi_{trailing}$ between the leading and trailing spots and normalizing by 2π. A value of k close to .5 indicates nearly symmetrically placed spots. In Figure 3.15 we plot k as a function of N for fixed R and σ. Clearly as N decreases k approaches .5. Although not indicated in the figure, the amplitudes of the two spots become comparable as well. These results suggest that the ATW2 mode can emerge as a period doubling bifurcation of the TW2 mode. We have not determined the evolution of the ATW2 branch for larger values of N than those shown in the figure.

There are transitions from the ATW2 branch to non-TW modes still exhibiting leading and trailing spot behavior at both its extremities, i.e., for smaller R and larger R. Upon increasing R beyond $R = 20$ we find slowly varying bound states in which a new spot is spontaneously created ahead of the leading spot, and subsequently ahead of the trailing spot in the bound state. The new spot, together with the leading (trailing) spot, exhibits episodes of ASCP behavior, An example of such a state is shown in Figure 3.16 where we plot Θ_f as a function of ψ and t for a solution with $R = 20.5$. Denote the strong leading spot in the bound state by S and the trailing weak spot by s. As the slowly varying bound state propagates, events occur in which a new spot forms ahead of S. Call this new spot SS. The new spot SS together with the spot S then exhibits ASCP behavior. That is, SS splits into a pair of counterpropagating spots, one of which

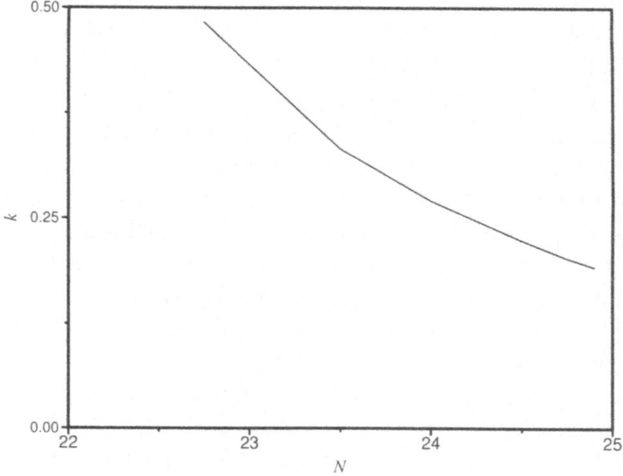

FIGURE 3.15. Asymmetry ratio $k(N)$ for a ATW2 branch as N is varied.

annihilates with S when they collide while the other continues to propagate. The effect is a speed up of the leading spot. Subsequently, similar behavior occurs for the trailing spot. We believe that the latter event is an attempt to maintain the bound state nature of the mode. We term this an Asymmetric Alternating Spin CounterPropagating (AASCP) mode. We note that the time interval between these events does not appear to be regular, but have not determined whether the motion is quasiperiodic or weakly chaotic.

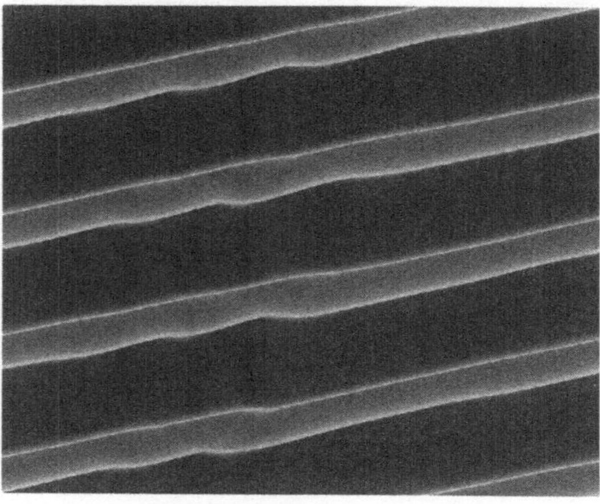

FIGURE 3.16. Θ_f as a function of ψ for a range of times for a slowly varying bound state with $R = 20.5$ exhibiting AASCP behavior. Time increases along the vertical axis.

Upon decreasing R along the ATW2 branch we find a family of modulated ATW2 modes (MATW2), where the two bound spots alternately approach one another and then separate in an oscillatory manner. The solution is quasiperiodic, as the approaches and separations occur periodically. An example of such a mode is shown in Figure 3.17 for $R = 9$. Another perspective of this mode can be seen in Figure 3.18 where we plot $x_f(t)$ for $\psi = 0$ over the same time interval as shown in Figure 3.17. The front speedups at $\psi = 0$ are now no longer periodic. There are regions where the speedups (corresponding to spots crossing the vertical line $\psi = 0$ in Figure 3.17) are closely spaced, i.e., the spots are approaching each other, alternating with regions where the front speedups are more widely spaced in time, e.g., around $t = 800$ and $t = 820$, indicating that the spots have separated.

We note that this is an example of increasing spatiotemporal complexity accompanied by decreasing R. This is in contrast to the behavior of the 1-headed spins. We were unable to compute MATW2 modes for $R < 7.5$, where we found that MATW2 initial data evolved to a 1-headed spin solution.

We have found other examples of 2-headed spin behavior, although for different parameters than those considered above. In particular, we have found symmetric 2-headed MTWs. These solutions are characterized by two spinning hot spots which undergo a modulation in phase. Thus, they can be thought of as replicated MTW1 modes. An example of a symmetric MTW2

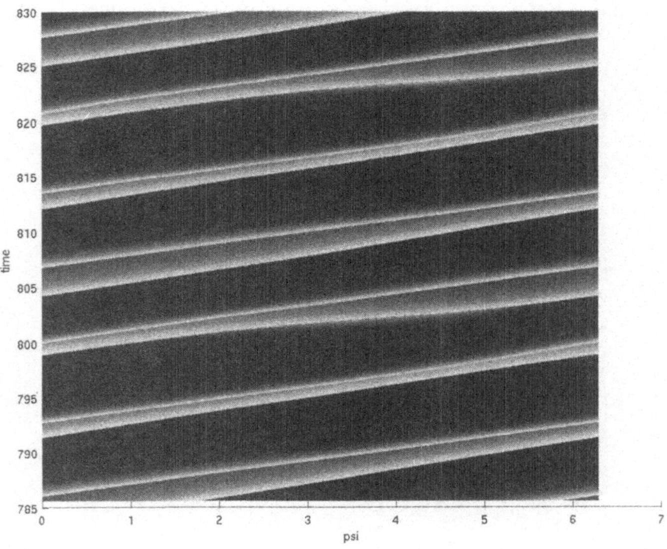

FIGURE 3.17. Θ_f as a function of ψ for a range of times for MATW2 solution with $R = 9$. Time increases along the vertical axis.

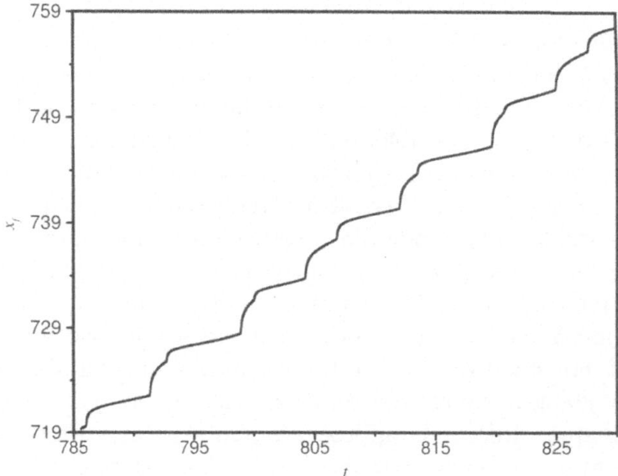

FIGURE 3.18. x_f as a function of t at $\psi = 0$ for MATW2 solution with $R = 9$.

solution is shown in Figure 3.19. The solution corresponds to $N = 22.5$ and $R = 14$ as we were unable to find such a mode for $N = 25$.

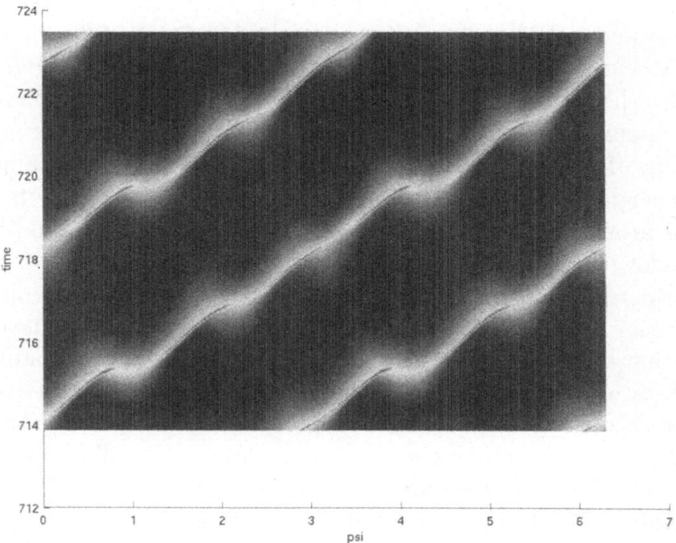

FIGURE 3.19. Θ_f as a function of ψ for a range of times for MTW2 solution with $N = 22.5$ and $R = 14$. Time increases along the vertical axis.

Finally, we have investigated 3-headed spin modes. We find that replicated 3-headed spin modes, obtained by replicating a TW1 mode three times, are unstable. We note that for replicated 3-headed spins, replicated perturbations correspond to perturbations whose Fourier wave numbers are divisible by 3. This is in contrast to 2-headed spins, where replicated perturbations correspond to Fourier wave numbers which are divisible by 2. Since the TW1 mode is stable, replicated 2-headed spin modes are necessarily stable to all even $(2n)$ perturbations and potentially unstable only to odd $(2n+1)$ perturbations. In contrast, replicated 3-headed spin modes are necessarily stable to all mode $(3n)$ perturbations and potentially unstable to mode $(3n+1)$ and $(3n+2)$ perturbations. This may be why replicated 3-headed spins are less stable than replicated 2-headed spins.

We do find 3-headed spins modes that are stable over a small range of values of R. For small values of R, these solutions are characterized by three spots undergoing a modulated rotation around the cylinder (MTW3), in which each spot alternately approaches and separates from its neighbors. These events do not occur periodically, in contrast to the MATW2 mode described above. We illustrate such a mode in Figure 3.20 for $R = 5.25$, which corresponds to three times $R = 1.75$, a value near the onset of stable TW1 modes. We note that while the replicated TW3 mode is unstable, the stable MTW3 mode for $R = 5.25$ has features similar to the replicated mode, in that the front speed is close to that of the TW1 at $R = 1.75$ (but somewhat smaller) and the amplitude of the spots is of the same order as the TW1 mode. The MTW3 solutions have a small domain of attraction and unless very small steps are taken in R, MTW3 initial data evolves to either 1-headed or 2-headed spins. Upon increasing R, we find that the 3-headed spin modes exhibit apparently chaotic behavior. For certain, apparently random, times two spots come close together and nearly touch. One of the spots is then nearly extinguished and moves rapidly away from its neighbor. The spot then becomes hotter and continues its spinning behavior around the cylinder. We refer to this mode as C3 and illustrate its behavior in Figure 3.21 for $R = 5.5$.

Upon increasing R further, we find it difficult to obtain equilibrated 3-headed spins. Starting with initial data corresponding to a 3-headed spin we find that during the transient, two spots actually collide, leading to the annihilation of one spot and the mode collapses to a 2-headed spin. We illustrate this in Figure 3.22 where we plot the evolution of Θ_f for $R = 7.2$. This figure is of a transient computation. We do not illustrate the final steady state. The initial conditions were taken from a 3-headed spin for a smaller value of R. The transient initially contains three spots, however at a certain time two of the co-propagating spots collide. A weak spot emanates from the collision, but decays and eventually disappears. The solution then continues as a 2-headed spin.

The collapse from three to two heads can also be seen from Figure 3.23 where we plot $x_f(t)$ for $\psi = 0$. The data in Figure 3.23 is taken over the same time interval as Figure 3.22. As for the other similar figures, e.g., Figures 3.12 and 3.18, a spot crossing $\psi = 0$ corresponds to a rapid movement of the front or equivalently a rapid rise in $x_f(t)$. For early times

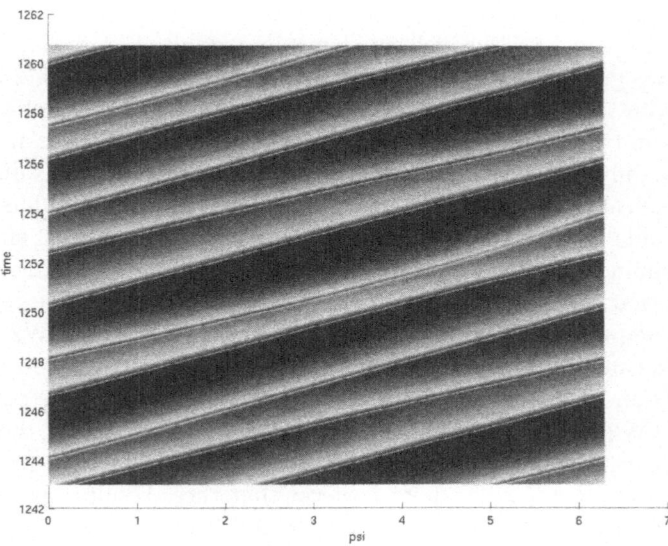

FIGURE 3.20. Θ_f as a function of ψ for a range of times for MTW3 solution with $R = 5.25$. Time increases along the vertical axis.

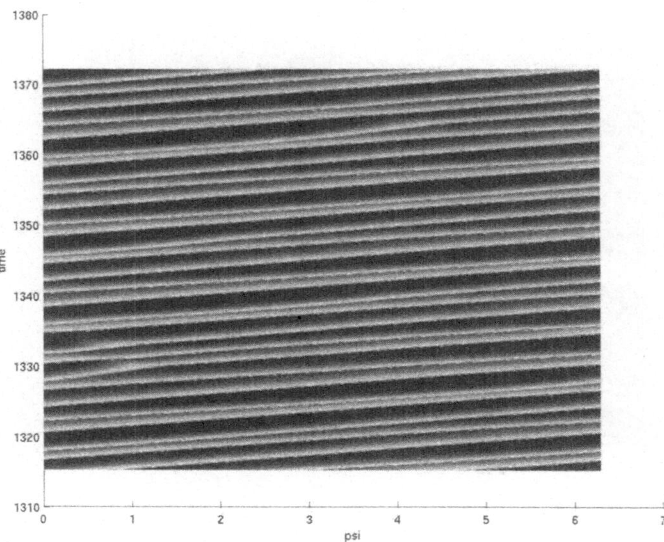

FIGURE 3.21. Θ_f as a function of ψ for a range of times for apparently chaotic 3-headed spin solution (C3) with $R = 5.5$. Time increases along the vertical axis.

in the figure there are two groups of three spot crossings, corresponding to the end of the transient 3-headed spin behavior (see also Figure 3.22). There are then two groups of two spot crossings, corresponding to the 2-headed spin behavior after the collapse. For this value of R there is yet another collapse from two heads to one head. The steady state solution is an ASCP mode. The only stable spin solutions that we can find for this value of R are the replicated TW2 and ASCP modes (see Figure 3.1). Thus, there is not a stable modulated 2-headed spin for this value of R and since the spots remaining after the collision are far from uniformly spaced it is not surprising that the steady state solution is the ASCP mode. We repeated the same computation for $R = 8.2$. For this value of R the MATW2 branch is stable as can be seen from Figure 3.1. Thus, modulated 2-headed spins can be sustained. In this case we found that the 3-headed mode collapsed to an MATW2 solution rather than to an ASCP solution, i.e., there was only a collapse from three heads to two heads, rather than two successive collapses as for $R = 7.2$. We note however that these results suggest that the domain of attraction of the replicated TW2 branch may be smaller than those of the ASCP and MATW2 branches. Thus, the TW2 spins may be difficult to observe in experiments. Heat loss, however, may change this, as discussed above.

The collapse from three to two heads is accompanied by a reduction in the mean axial speed (see Figure 3.1). The computed mean speed over the interval in which the solution is a 3-headed spin is approximately 1.0, while

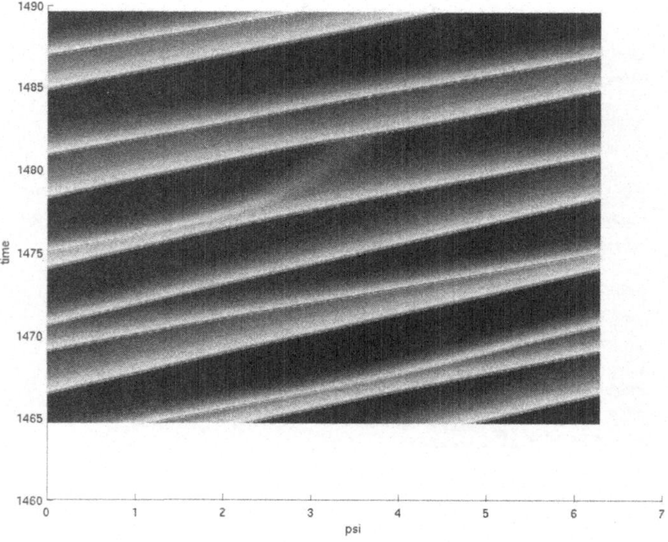

FIGURE 3.22. Θ_f as a function of ψ for a range of times for a transient 3-headed spin collapsing to a 2-headed spin with $R = 7.2$. Time increases along the vertical axis.

over the interval in which the solution is a 2-headed spin the computed mean speed is approximately 0.87. The speeds were computed by doing a least squares fit over a longer time interval than the data shown in Figure 3.23. This reduction in speed can also be seen from Figure 3.23 where it is clear that the mean slope of the curve is smaller for later times when the solution consists of only two heads.

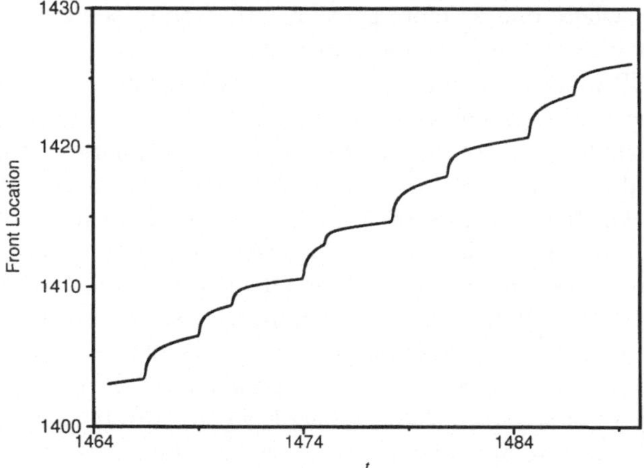

FIGURE 3.23. x_f as a function of t at $\psi = 0$ for $R = 7.2$ over the same time interval as that shown in Figure 3.22.

Thus, for increasing R, the 3-headed spins are either unstable or have a small domain of attraction. Such behavior might limit the observability of 3-headed spins in experiments. It is possible that as R increases further, collapses such as the one shown in Figure 3.22 could be avoided since there is a greater circumference for the spots to occupy without colliding, however, we have not been able to compute a steady state 3-headed spin for values of R greater than those shown in Figure 3.1.

4 Conclusion

We have examined the dynamics and spatial structure of nonplanar modes in surface solid fuel combustion employed in the SHS process of materials synthesis. The nonplanar modes involve the spatiotemporal dynamics of hot spots moving on the surface. We have focused on determining behavior as the radius of the sample R increases as an objective of this process is to synthesize large samples of product. We have characterized the different modes that we have computed in terms of the mean axial flame speed V. Although we have not considered nonadiabatic effects, generally those flames for which V is smaller are more prone to extinction from heat losses than those with larger values of V.

We consider parameters where the uniformly propagating planar flame is unstable. A pulsating, planar flame is stable for small values of R. Nonplanar behavior is characterized by the formation and motion of hot spots on the front. We find counterpropagating (CP) modes, in which two hot spots propagate in opposite angular directions around the cylinder, and spin modes, where one or more spots rotate around the cylinder, but all spots rotate in essentially the same direction.

For the CP modes, there are collisions between the counterpropagating spots. When the spots collide different interactions can occur depending on R. For small values of R, the spots enter and emerge from the collision essentially unchanged except for a phase shift, much like the behavior of solitons. For larger values of R the spots are alternately reduced and magnified in amplitude, leading to apparent annihilation and creation at the collision. These modes were observed in Mukasyan, Vadchenko, and Khomenko [1997] and described in detail in Bayliss and Matkowsky [1999].

We have found several different types of spin modes, including 2 and 3- headed spin modes, involving two or three spots. The dynamics include traveling waves of n spots (TWn), where the spots rotate around the cylinder at a uniform rate and with constant shape, modulated traveling waves (MTWs) where the spots have a nonuniform rotation rate and change shape as they rotate, and ASCP modes, where the overall spin motion is interrupted by periodic events of a newly created CP spot which splits into two CP spots, one of which annihilates with the original propagating spot while the other continues to propagate. We suggest that the newly created spots in the ASCP modes form as a result of the increasing localization of the original propagating spots as R increases, so that a considerable portion of the front is approximately uniform, a condition which is unstable. The new spots give rise to daughter spots which counterpropagate, collision by one of the daughter spots with the original propagating spot leading to their eventual mutual annihilation and continuation of the spinning behavior by the remaining daughter spot.

For the TW modes, replicated solutions exist. Thus, given a 1-headed spin at radius R_1, an $n-$headed spin consisting of n replicates of the solution exists at $R_n = nR_1$. The replicated modes are not necessarily stable. We have examined 2-headed and 3-headed replicates of the TW1 solution. We showed that TW2 modes are generally stable, while TW3 modes are not stable. This suggests that higher order replicates are unstable, possibly because of the larger class of perturbations which are potentially unstable.

We have also found 2-headed traveling waves, which are not replicates of a single TW1. In this case there are two spots, but they are of different amplitude and unequally spaced around the cylinder. The two spots are bound to each other and rotate as a single structure. We referred to these modes as asymmetric traveling waves (ATW2). These modes appear to emanate as a period doubling bifurcation from the TW2 modes for small activation energy. Our figures suggest that the spots associated with the ATW2 mode appear to be the primary impetus for the axial propagation of the combustion front. In addition to the TW modes we have found MTW and ATW2 modes, in which the spots rotate with a nonuniform rotation rate and amplitude. We have also found AASCP modes which are slowly varying bound states where the leading spot, and subsequently the trailing spot, together with a newly created spot ahead of it, exhibits periods of ASCP behavior.

For the parameters that we consider, TW3 modes are unstable, but we have found 3-headed spin modes exhibiting either quasiperiodic or apparently chaotic dynamics. The 3-headed spin modes are difficult to compute for larger valucs of R. This is due to collisions between the spots, leading to annihilation and a collapse from three spots to two spots or even one spot. This is accompanied by a reduction in the mean flame speed.

We have characterized the different modes in terms of their mean axial speed V, which is a global characteristic of the wave. As R increases the spots become increasingly localized and increase in amplitude. Most of the front is consequently at a lower temperature for a longer time. As a result, for any branch V generally decreases as R increases and the spots become more pronounced. For small values of R the first nonplanar modes that are stable are CP modes, which arise at finite amplitude and exhibit a mean propagation speed higher than that of the unstable uniformly propagating mode. As R increases a branch of 1-headed spin modes appears. These modes also enter at nonzero amplitude and are accompanied by a reduction in V. As R increases further a family of stable 1-headed spin branches evolves continuously with transitions from TW1 to MTW1 to ASCP behavior. Some form of a 1-headed spin appears to be stable up to the largest value of R considered, which is ten times larger than the value of R at which 1-headed spin modes first appear. This is in contrast to the 2-headed and 3-headed spin branches which appear to lose stability for large values of R. When a number of modes coexist at the same value of R, the one with more spots has a larger speed V. For the range of R considered V was bounded below by the speed of the (unstable) planar pulsating wave.

Acknowledgments: This research was partially supported by NASA Grant NAG3-2209, NSF grants DMS00-72491, DMS95-30937, the San Diego Supercomputer Center and RFFI Grant 01-03-32055.

References

Aldushin, A. P., A. Bayliss, and B. J. Matkowsky [2000], Dynamics of Layer Models of Solid Flame Propagation, *Physica D* **143**, 109.

Aldushin, A. P., S. I., Khudyaev, and Ya. B. Zeldovich [1979], Numerical Study of Flame Propagation in a Mixture Reacting at the Initial Temperature, in: *Selected Works of Yakov Borisovich Zeldovich, Vol. 1*, J. P. Ostriker, G. I. Barenblatt, R. A. Sunyaev, Eds., Princeton Univ. Press, 1992.

Aldushin, A. P, B. A. Malomed, and Ya. B. Zeldovich [1981], Phenomenological Theory of Spin Combustion, *Combustion and Flame* **42**, 1.

Aldushin, A. P, T. M. Martemyanova, A. G. Merzhanov, B. I. Khaikin, and K. G. Shkadinsky [1973], Autooscillatory Propagation of Combustion Front in Heterogeneous Condensed Media, *Combustion Explosion and Shock Waves* **9**, 531.

Bayliss, A., D. Gottlieb, B. J. Matkowsky, and M. Minkoff [1989], An Adaptive Pseudo-Spectral Method for Reaction Diffusion Problems, *J. Comput. Phys.* **81**, 421.

Bayliss, A., R. Kuske, and B. J. Matkowsky [1990] A Two-Dimensional Adaptive Pseudo-Spectral Method, *J. Comput. Phys.* **91**, 174.

Bayliss, A., B. J. Matkowsky, and A. P Aldushin [to appear(a)], Dynamics of Hot Spots in Solid Fuel Combustion, to appear in Physica D.

Bayliss, A., B. J. Matkowsky, and A. P Aldushin [to appear(b)], Bound States of Asymmetric Hot Spots in Solid Flame Propagation, to appear in *Proceedings of Conference on Bifurcations, Symmetry and Patterns* Eds. J. Buescu, I Labouriau, S. Castro, A. P. Dias. Birkhauser Verlag.

Bayliss, A. and B. J. Matkowsky [1987], Fronts, Relaxation Oscillations, and Period Doubling in Solid Fuel Combustion, *J. Comput. Phys.* **71**, 147.

Bayliss, A. and B. J. Matkowsky [1990], Two Routes to Chaos in Condensed Phase Combustion, *SIAM J. Appl. Math.* **50**, 437.

Bayliss, A. and B. J. Matkowsky [1996], Structure and Dynamics of Kink and Cellular Flames Stabilized on a Rotating Burner, *Physica D* **99**, 276.

Bayliss, A. and B. J. Matkowsky [1999], Interaction of Counterpropagating Hot Spots in Solid Fuel Combustion, *Physica D* **128**, 18.

Dvoryankin, A. V., A. G. Strunina, and A. G. Merzhanov [1982], Trends in the Spin Combustion of Thermites, em Combustion, Explosion and Shock Waves **18**, 134.

Ivleva, T. P. and A. G. Merzhanov [1999], Mathematical Simulation of Three Dimensional Spinning Modes of Gasless Combustion Waves, *Doklady Physics* **44**, 739.

Ivleva, T. P. and A. G. Merzhanov [2000], Three Dimensional Spinning Waves in the Case of Gas Free Combustion *Doklady Physics* **45**, 136.

Ivleva, T. P., A. G. Merzhanov, and K. G. Shkadinsky [1978], Mathematical Model of Spin Combustion, *Soviet Physics Doklady* **23**, 255.

Ivleva, T. P., A. G. Merzhanov and K. G. Shkadinsky [1980], Principles of the Spin Mode of Combustion Front Propagation, *Combustion, Explosion and Shock Waves* **16**, 133.

Maksimov, Yu. M., A. T. Pak, G. B. Lavrenshuck, Y. S. Naiborodenko and A. G. Merzhanov [1979], Spin Combustion of Gasless Systems, *Combustion, Explosion and Shock Waves* **15**, 415.

Maksimov, Yu. M., A. G. Merzhanov, A. T. Pak, and M. N. Kuchkin [1981], Unstable Combustion Modes of Gasless Systems, *Combustion, Explosion and Shock Waves* **17**, 393.

Margolis, S. B.[1983], An Asymptotic Theory of Condensed Two-Phase Flame Propagation, *SIAM J. Appl. Math.* **43**, 351.

Margolis, S. B., H. G. Kaper, G. K. Leaf, and B. J. Matkowsky[1985], Bifurcation of Pulsating and Spinning Reaction Fronts in Condensed Two- Phase Combustion, *Comb. Sci. and Tech* **43**, 127.

Matkowsky, B. J. and G. I. Sivashinsky [1978], Propagation of a Pulsating Reaction Front in Solid Fuel Combustion, *SIAM J. Appl. Math.* **35**, 465.

Merzhanov, A. G.[1981], SHS Processes: Combustion Theory and Practice, *Arch. Combustionis* **1**, 23.

Merzhanov, A. G.[1990], Self-Propagating High-Temperature Synthesis: Twenty Years of Search and Findings, in: *Combustion and Plasma Synthesis of High-Temperature Materials*, Z.A. Munir, J.B. Holt, Eds., VCH, 1.

Merzhanov, A. G.[1994], Solid Flames: Discovery, Concepts, and Horizons of Cognition, *Combustion Science and Technology* **98**, 307.

Merzhanov, A. G., A. K. Filonenko, and I. P. Borovinskaya [1973], New Phenomena in Combustion of Condensed Waves, *Soviet Phys. Dokl.* **208**, 122.

Mukasyan, A. S., S. C. Vadchenko, and L. O. Khomenko [1997], Combustion Modes in the Titanium-Nitrogen System at Low Nitrogen Pressures, *Combustion and Flame* **111**, 65.

Munir, Z. A. and U. Anselmi-Tamburini [1989], Self-Propagating Exothermic Reactions: The Synthesis of High-Temperature Materials by Combustion, *Material Science Reports, A Review Journal* **15**, 277.

Or-Guil, M., I. G. Kevrekidis, and M. Bär [1999], Stable Bound States of Pulses in an Excitable System *Physica D* **135**,157.

Raymond, C., A. Bayliss, B. J. Matkowsky, and V. Volpert [2001], Transitions to Chaos in Condensed Phase Combustion with Reactant Melting, *Int'l. J. Self-Propagating High-Temperature Synthesis* **10**, 133.

Shkadinsky, K. G., B. I. Khaikin, and A. G. Merzhanov [1971], Propagation of a Pulsating Exothermic Reaction Front in the Condensed Phase, *Combustion, Explosion and Shock Waves* **1**, 15.

Sivashinsky, G. I. [1981], On Spinning Propagation of Combustion Waves, *SIAM J. Appl. Math.* **40**, 432.

Strunina, A, G., A. V. Dvoryankin, and A. G. Merzhanov [1983], Unstable Regimes of Thermite System Combustion, *Combustion, Explosion and Shock Waves* **19**, 158.

5

Globally Coupled Oscillator Networks

Eric Brown
Philip Holmes
Jeff Moehlis

To Larry Sirovich, on the occasion of his 70th birthday.

ABSTRACT We study a class of permutation-symmetric globally-coupled, phase oscillator networks on N-dimensional tori. We focus on the effects of symmetry and of the forms of the coupling functions, derived from underlying Hodgkin-Huxley type neuron models, on the existence, stability, and degeneracy of phase-locked solutions in which subgroups of oscillators share common phases. We also estimate domains of attraction for the completely synchronized state. Implications for stochastically forced networks and ones with random natural frequencies are discussed and illustrated numerically. We indicate an application to modeling the brain structure *locus coeruleus*: an organ involved in cognitive control.

Contents

1 Introduction and Background

We consider networks of N rotator oscillators with constant forcing and pairwise phase-difference and absolute-phase 'product' coupling, described by:

$$\dot{\theta}_i = \omega_i + \frac{1}{N}\sum_{j=1}^{N} \alpha_{ij} f_{ij}(\theta_j - \theta_i) + h_i(\theta_i)\frac{1}{N-1}\sum_{j\neq i}^{N} \beta_{ij} g_j(\theta_j), \tag{1.1}$$

where $(\theta_1, \ldots, \theta_N)^{\mathrm{T}} \in \mathrm{T}^N$, α_{ij}, β_{ij} and f_{ij}, h_i, g_j are, respectively, coupling parameters and 2π-periodic functions, and ω_i are the natural frequencies of the uncoupled rotators. This paper focuses on networks with identical global (mean field) coupling, so that equation (1.1) becomes

$$\dot{\theta}_i = \omega + \frac{\alpha}{N}\sum_{j=1}^{N} f(\theta_j - \theta_i) + h(\theta_i)\frac{\beta}{N-1}\sum_{j\neq i}^{N} g(\theta_j)\ , \tag{1.2}$$

although we include some results with randomly distributed frequencies ω_i, and also with additive random noise. The denominators (N, $N-1$) are introduced to normalize coupling effects.

Rotator (phase-only) models of coupled oscillators have been widely studied, especially in the contexts of neuroscience and coupled Josephson junctions. The phase equations offer, respectively, significant simplification of more realistic neuron models of Hodgkin-Huxley or Fitzhugh-Nagumo type: see e.g. Murray [2001]; Keener and Sneyd [1998]; Hoppensteadt and Izhikevich [1997], and of the Josephson circuit equations: e.g. Watanabe and Strogatz [1994]; Watanabe and Swift [1997]; Wiesenfeld, Colet, and Strogatz [1998]. In the case that the N uncoupled oscillators have strongly attracting limit cycles in their full phase space, the persistence of normally hyperbolic invariant manifolds (Fenichel [1971]) under small perturbations (weak coupling) may be used to reduce the system to the N-torus by a suitable coordinate transformation. Two procedures for approximating the reduced system will be applied in Section 4 of this paper. The first is the 'strong attraction' (SA) method described in Ermentrout and Kopell [1990] and Hoppensteadt and Izhikevich [1997]; the second, a related 'phase response' (PR) technique, is given in e.g. Ermentrout [1996], Kuramoto [1997], and Kim and Lee [2000].

We were motivated to study systems of the form (1.1) by the study of Usher, Cohen, Servan-Schreiber, Rajkowski, and Aston-Jones [1999], which presents data showing that neurons of the brain organ *locus coeruleus* (LC) in monkeys exhibit two distinct firing patterns corresponding to different behaviors evinced in cognitive tasks, cf. Grant, Aston-Jones, and Redmond [1988]; Aston-Jones, Rajkowski, and Alexinsky [1994]. These are designated as the *phasic* and *tonic* modes. In the latter, associated with labile behavior and poor task performance, LC neurons fire at a relatively high rate, but with little synchrony; in the former, associated with good performance, their average firing rate is lower but they display greater synchrony (i.e. higher correlation among individual firing times). See Figure 1.1(a). Moreover, firing patterns are more responsive to changes in stimulus in the phasic than in the tonic mode; see Figure 1.1(b,c). Thus, the LC has been proposed as a modulator involved in cognitive control, cf. Servan-Schreiber, Printz, and Cohen [1990].

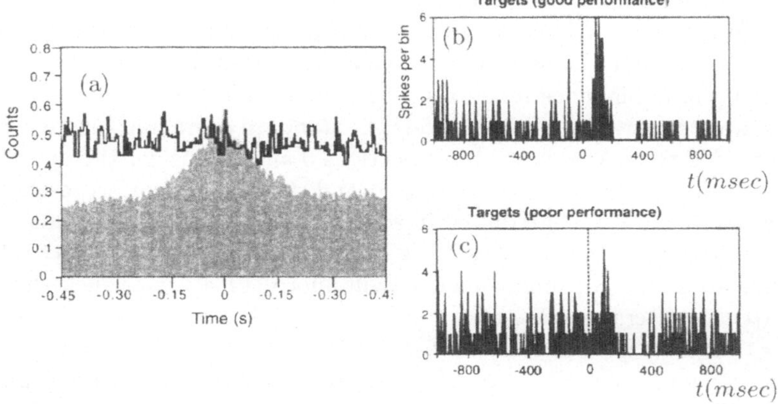

FIGURE 1.1. Reproduced from Usher, Cohen, Servan-Schreiber, Rajkowski, and Aston-Jones [1999], with adapted caption. (a) Cross correlograms for two simultaneously recorded LC neurons during phasic LC mode (filled histogram) and tonic mode (line): central peak indicates increased synchrony in phasic mode. (b,c) Histograms of LC activity in phasic (b) and tonic (c) LC modes: psychological stimulus onset (which precedes the direct stimulus $I(t)$ considered in Section 4.4) is marked by dotted line, and enhanced response in phasic versus tonic mode is apparent. Average firing rate is lower in phasic mode.

The computational model constructed in Usher, Cohen, Servan-Schreiber, Rajkowski, and Aston-Jones [1999] includes inhibitory synaptic and excitatory electrotonic coupling (Johnston and Wu [1997]), explicitly imposed refractory periods, and representations of rapid depolarization during action potentials to successfully reproduce these characteristics of the phasic and tonic modes; transitions between the two modes are affected by varying the degree of electrotonic coupling. However, the model's complexity makes analysis difficult, and we wish to develop a model that has similar behavior but is more amenable to mathematical study.

In this paper, we consider coupling functions motivated by two physically distinct mechanisms: (1) *electrotonic* or gap junction coupling, based on voltage differences between cells in electrical contact, and (2) spike-triggered *synaptic* transmission that releases a pulse of neurotransmitter across synaptic clefts. Electrotonic coupling is additive in the sense that the sum of voltage differences of all cells in contact with a given cell influence that cell; hence the first sum in equations (1.1) and (1.2). Synaptic coupling leads to absolute phase terms $\beta h_i(\theta_i)g_j(\theta_j)$ in (1.1-1.2). Intuitively, these arise because the primary effect on the post-synaptic cell occurs after the pre-synaptic cell fires, and therefore depends, via $g(\theta_j)$, on the latter's location on its phase circle. Coupling via a 'reversal potential' also depends upon the post-synaptic cell's phase through $h(\theta_i)$ (Ermentrout and Kopell [1990]; Taylor and Holmes [1998]); thus $h(\theta_i)$ multiplies the summed $g(\theta_j)$'s, leading to the product coupling form of the second term.

We furthermore assume an additional separation of scales, taking electrotonic coupling to be weaker than synaptic, so that it can be averaged to give the phase-difference functions αf_{ij} without affecting $\beta h_i g_j$ at leading order; we are currently studying when the standard averaging theorems can be extended to make this rigorous. Sections 3 and 4 consider the dynamics of equation (1.1) for various values of α and β, without a priori restricting to the $|\alpha| \ll |\beta| \ll \mathcal{O}(1)$ required in this derivation of the phase equations.

When $\beta = 0$ but frequencies differ between oscillators, equation (1.2) is referred to as the Kuramoto model (Kuramoto [1984]), on which there is an extensive literature; see the recent review of Strogatz [2000] and references therein (e.g. Crawford [1995]). Much of this work has been done in the continuum limit $N \to \infty$, and Strogatz [2000] adopts this viewpoint; specifically, stability analyses of some stationary (continuous) states are discussed. Finite-dimensional results, including a Liapunov function and dimension reduction, are found in the context of Josephson junction models in Watanabe and Strogatz [1994]. Many earlier studies take only the leading term in an odd Fourier expansion of f, so that $f(\,\cdot\,) = \sin(\,\cdot\,)$; as we shall see this is a very degenerate case for the mean field coupled system (1.2) (e.g. Nichols and Wiesenfeld [1992]; Golomb, Hansel, Shraiman, and Sompolinsky [1992]). Moreover, as shown in Izhikevich [2000], relaxation oscillators of Hodgkin-Huxley or Fitzhugh-Nagumo type lead to much richer phase difference functions than $\sin(\,\cdot\,)$. Others have recognized the importance of higher Fourier harmonics: see Daido [1994]; Golomb, Hansel, Shraiman, and Sompolinsky [1992]; Nichols and Wiesenfeld [1992]; Watanabe and Swift [1997]. Additional work on finite dimensional oscillator networks includes Kopell and Ermentrout [1990]; Kopell, Ermentrout, and Williams [1991]; Kopell and Ermentrout [1994], which consider directed coupling, Bressloff and Coombes [1998], which considers integrate-and-fire models derived from coupled spiking neurons, and Okuda [1993], which will be discussed in Section 2. Shortly before this paper was submitted, we learned of recent work of Chow and Kopell [2000], in which the effects of spike shape on electrotonically coupled integrate-and-fire networks are studied. They find that the existence and stability of splay states depends on the spike shape in a manner that would be interesting to compare with the present results.

The present paper draws on Ashwin and Swift [1992], which addresses a class of $S_N \times T^1$-equivariant oscillator networks (of which (1.2) is an example when $\beta = 0$). We now summarize the properties of symmetric dynamical systems necessary to present and apply these results; for more background, see Golubitsky, Stewart, and Schaeffer [1988].

Consider the ODE

$$\frac{dx}{dt} = f(x) \ , \ x \in \text{manifold } M \ , \tag{1.3}$$

and let Γ be a group acting on M. The ODE is said to be Γ-equivariant if f commutes with the group action, i.e.

$$f(\gamma x) = \widehat{\gamma} f(x) \ \ \forall \gamma \in \Gamma, x \in M \ , \tag{1.4}$$

where the derivative map $\widehat{\gamma}$ (Arnold [1973]) acts on the tangent space TM; for linear actions of γ, $\widehat{\gamma} = \gamma$. The symmetry of a solution $x_0 \in M$ is characterized by the isotropy subgroup $\Sigma_{x_0} = \{\gamma \in \Gamma : \gamma x_0 = x_0\}$, that is, the set of all group elements which leave the solution x_0 unchanged. Associated with an isotropy subgroup is a fixed point subspace $\text{Fix}[\Sigma_{x_0}] = \{x \in M : \sigma x = x \ \ \forall \sigma \in \Sigma_{x_0}\}$: the set of points fixed by all elements of Σ_{x_0}. Two immediate consequences of Γ-equivariance are that (1) for any solution $x(t)$ to equation (1.3), $\gamma x(t)$ is also a solution, and (2) fixed-point subspaces are invariant under the flow generated by f. We will refer to this latter property as dynamical invariance. As in Ashwin and Swift [1992], we study special classes of symmetric systems defined by the following groups: the circle group $T^1 = \{\delta : \ \delta \in [0, 2\pi)\}$ (with action on T^N, $\theta_i \mapsto \theta_i + \delta, \ \forall \ i$), the cyclic subgroups $Z_m \in T^1$ (with action $\theta_i \mapsto \theta_i + 2\pi/m$), and the subgroups of permutations on j-many coordinates, S_j.

The remainder of the paper proceeds as follows. In Section 2 we study (1.2) with $\beta = 0$ ($S_N \times T^1$ equivariant), emphasizing the influence of general coupling functions and obtaining additional results for odd functions f. In succeeding sections the symmetries are gradually relaxed. In Section 3 this is done by breaking T^1 equivariance through re-introduction of $h(\theta_i)g(\theta_j)$ terms. In Section 4 we break S_N equivariance by introducing a random distribution of frequencies as well as random excitation, and apply our model to the LC. Thus, Sections 2 and 3 are largely abstract and general, while Section 4 concerns specific 'neural' coupling. Conclusions are drawn and future directions noted in Section 5. Our major contributions to this survey on globally coupled oscillators include the implications of gradient dynamics for the existence of families of equilibria, nonlinear stability results for the synchronized state, and the analysis of a two-parameter (α, β) system, including the influence of noise, in relation to the LC model.

In a series of recent papers (e.g. Omurtag, Knight, and Sirovich [2000b]; Sirovich, Knight, and Omurtag [2000]; Omurtag, Kaplan, Knight, and Sirovich [2000a]), Larry Sirovich and his colleagues have also addressed problems in neurobiology; specifically, modeling networks of neurons in visual cortex. They employ integrate-and-fire units, and develop an evolution equation for the probability $\rho(V,t)dV$ of finding a neuron with membrane potential $V \in [V, V+dV]$ at time t. We hope that our study, which takes a different, and, we believe, a complementary viewpoint, will at once serve as a suitable tribute, a stimulus for future work, and a fond reminder of turbulent interactions past.

2 $S_N \times T^1$ Phase Difference Oscillator Systems

This section treats the system of N oscillators,

$$\dot{\theta}_i = \omega + \frac{\alpha}{N}\sum_{j=1}^{N} f\left(\theta_j - \theta_i\right), \quad i = 1, \ldots, N, \tag{2.1}$$

where $f(\,\cdot\,)$ is assumed to be continuously differentiable and 2π-periodic. Transforming to coordinates $\phi_i = \theta_i - \omega t$ rotating with the common natural frequency, equation (2.1) becomes

$$\dot{\phi}_i = \frac{\alpha}{N}\sum_{j=1}^{N} f\left(\phi_j - \phi_i\right). \tag{2.2}$$

Denoting the phase differences $\phi_j - \phi_i \equiv \phi_{ji}$, we seek 'diagonal flow' periodic solutions $\bar{\phi}$ of equation (2.2) in the form

$$\dot{\bar{\phi}}_i = \frac{1}{N}\sum_{j=1}^{N} f\left(\bar{\phi}_{ji}\right) = c, \quad i = 1, \ldots, N, \tag{2.3}$$

where c is a constant, nonzero in general. These solutions are also periodic for (2.1), and, employing a second rotating frame $\theta_i - (\omega + c)t$, they become fixed points. Since the derivatives $f'(\phi_{ji})$ are time-independent, eigenvalue calculations suffice to determine the stability of such solutions. Equation (2.3) determines the $N-1$ phase differences, without loss of generality leaving the phase ϕ_1 unspecified, as expected from the T^1 symmetry of equation (2.1).

We began this work by determining the existence and stability of diagonal flow solutions to equation (2.2). While we later found many of the following results in the literature, no unified presentation appears to exist, so we provide a summary here, including extensions and new examples of our own. Proofs are only sketched.

2.1 Gradient Property for Odd Phase-Difference Coupling

If $f(\cdot)$ is odd, we observe as in Theorem 9.15 of Hoppensteadt and Izhikevich [1997] that:

2.1 Proposition. *Equation (2.2) is a gradient dynamical system on* T^N *with potential*

$$V = \frac{\alpha}{N} \sum_{i=1}^{N-1} \sum_{j=i+1}^{N} F(\phi_j - \phi_i), \text{ where } f(\theta) = F'(\theta). \tag{2.4}$$

Proof. Note that

$$-\frac{\partial V}{\partial \phi_i} = \frac{\alpha}{N} \sum_{j<i} f(\phi_j - \phi_i) - \frac{\alpha}{N} \sum_{j>i} f(\phi_i - \phi_j); \tag{2.5}$$

the oddness of f implies that $\dot{\phi}_i = -\partial V / \partial \phi_i$. ∎

Proposition 2.1 implies that

$$\dot{V} = \sum_{i=1}^{N} \frac{\partial V}{\partial \phi_i} \dot{\phi}_i = -\sum_{i=1}^{N} \dot{\phi}_i^2 \leq 0\,, \tag{2.6}$$

with equality only at equilibria. Thus, equation (2.2) with odd $f(\cdot)$ has no periodic or homoclinic orbits or heteroclinic cycles: all solutions approach equilibria, and almost all approach stable equilibria. In particular,

2.2 Corollary. *For f odd, equation (2.3) has no solutions unless $c = 0$.*

2.2 Periodic Orbits for the Phase Difference System

We begin by stating two results from Ashwin and Swift [1992] which hold for arbitrary 2π-periodic, continuously differentiable coupling functions $f(\cdot)$:

2.3 Theorem. (Ashwin and Swift [1992]) *Every isotropy subgroup of a general $S_N \times T^1$-equivariant vector field is of the form:*

$$\Sigma_{\mathbf{k},m} \equiv (S_{k_1} \times \cdots \times S_{k_{l_B}})^m \rtimes Z_m,$$

where $N = m(k_1 + \cdots + k_{l_B})$, and $\rtimes$ denotes the semi-direct product.

The fixed-point subspace $\mathrm{Fix}\left[\Sigma_{\mathbf{k}.m}\right]$ may be thought of as being partitioned into m blocks each containing $k = (k_1 + \cdots + k_{l_B})$ oscillators. The solution is invariant under time shifts of the period divided by m, coupled with a cyclic permutation of the blocks, giving the Z_m symmetry. Each block is partitioned into clusters of k_i oscillators, and the solution is invariant under

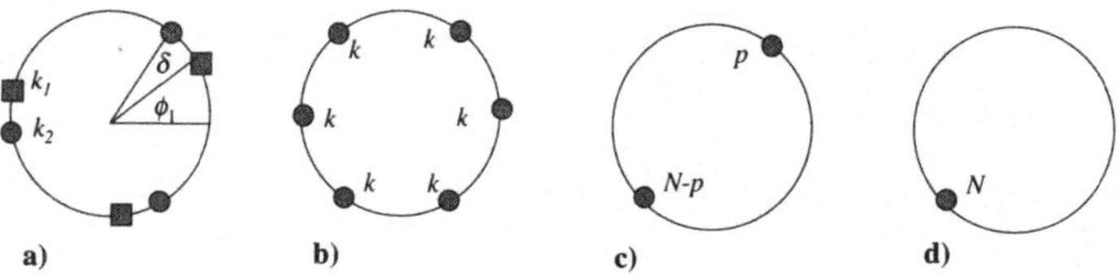

FIGURE 2.1. Illustrations of equilibria fixed under the actions of various isotropy subgroups. (a) An element of $\text{Fix}\big[(S_{k_1} \times S_{k_2})^3 \rtimes Z_3\big]$. Each square represents k_1 oscillators mutually in phase, and successive squares denote groups differing in phase by $2\pi/3$ (similar for circles). Elements of this fixed point subspace are parameterized by two angles ϕ_i and δ, so it is a 2-torus. (b) $(S_k)^m \rtimes Z_m$ (rotating block modes) with $m = 6$, (c) $S_p \times S_{N-p}$ (two-block modes), and (d) S_N (in-phase mode).

permutations of oscillators within these clusters, giving the $S_{k_1} \times \cdots \times S_{k_{l_B}}$ symmetry. These permutations all commute, hence the direct products, while the Z_m symmetry does not commute with the permutations, hence the semi-direct product. Overall, the fixed-point subspace $\text{Fix}\big[\Sigma_{\mathbf{k},m}\big]$ is an l_B-torus: there are $l_B - 1$ degrees of freedom setting the spacings between the blocks, plus an additional degree of freedom determining a 'reference' ϕ_1 (see Figure 2.1(a) for an example); this represents the T^1 group orbit and will be associated with a unit Floquet multiplier.

2.4 Theorem. (Ashwin and Swift [1992]) *Every* $\text{Fix}\big[(S_k)^m \rtimes Z_m\big]$ *with* $mk = N$ *and* $\text{Fix}\big[(S_{k_1} \times S_{k_2})^m \rtimes Z_m\big]$ *with* $m(k_1 + k_2) = N$, *generically contains a periodic orbit with diagonal flow.*

Ashwin and Swift [1992] prove this theorem by noting that, without loss of generality, the phases of the oscillators can be ordered as $\phi_1 \le \phi_2 \le \cdots \le \phi_N \le \phi_1 + 2\pi$. The oscillators retain their ordering under the dynamics, i.e., they can never 'pass' each other, since this would involve crossing an invariant fixed-point subspace. Projecting the phases onto the manifold $\phi_1 = 0$ (by subtracting the instantaneous value of ϕ_1 from each phase) gives a simplex called the 'canonical invariant region' (CIR). The intersection of $\text{Fix}\big[(S_k)^m \rtimes Z_m\big]$ with the CIR is a zero-dimensional invariant subspace, i.e., an equilibrium. In the unprojected system, this corresponds to a periodic orbit (or a circle of equilibria if $\dot\phi_1 = 0$ for all time) with isotropy $(S_k)^m \rtimes Z_m$. Furthermore, the intersection of $\text{Fix}\big[(S_{k_1} \times S_{k_2})^m \rtimes Z_m\big]$ with the CIR is a one-dimensional line segment. The end points of this segment have isotropy $(S_{k_1+k_2})^m \rtimes Z_m$, and are equilibria with stability in the direction of the line segment determined by the same eigenvalue. Provided this eigenvalue does not vanish (this is the nondegeneracy condition satisfied for generic functions f), this can only happen if there is at least one equilibrium in the interior of the line segment. In the original system, this corresponds to a periodic orbit (or one-torus of equilibria if $\dot\phi_1 = 0$ for all time) with isotropy $(S_{k_1} \times S_{k_2})^m \rtimes Z_m$. If $k_1 = k_2$, the midpoint of the line segment is an equilibrium with isotropy $(S_{k_1})^{2m} \rtimes Z_{2m}$; this can serve as the necessary equilibrium in the interior of the line segment.

Ashwin and Swift [1992] developed their proof for systems coupled with general T^1-equivariant functions. The special additive, pairwise-coupled form of the coupling in (2.1) allows a much simpler argument to prove (a restricted version of) Theorem 2.4. For a $(S_{k_1} \times S_{k_2})^m \rtimes Z_m$ solution with clusters separated by phase δ (see Figure 2.1(a)) to have diagonal flow requires $\dot{\phi}_i \equiv c(\delta)\ \forall i$, for some fixed δ. This condition reduces to

$$c_1(\delta) = c_2(\delta) , \tag{2.7}$$

where

$$c_1(\delta) = k_1 \sum_{j=0}^{m-1} f\left(\frac{2\pi j}{m}\right) + k_2 \sum_{j=0}^{m-1} f\left(\frac{2\pi j}{m} + \delta\right) \tag{2.8}$$

$$c_2(\delta) = k_2 \sum_{j=0}^{m-1} f\left(\frac{2\pi j}{m}\right) + k_1 \sum_{j=0}^{m-1} f\left(\frac{2\pi j}{m} - \delta\right) \tag{2.9}$$

are the (constant) phase velocities for oscillators in k_1 or k_2-clusters, respectively (cf. Kim and Lee [2000] for $m = 1$). A quick sketch shows that at least one $\delta \in (0, 2\pi/m)$ satisfying (2.7) must exist if $c_1'(0)/k_2 = -c_2'(0)/k_1$ is nonzero, since $c_{1,2}'(0) = c_{1,2}'(2\pi/m)$. Thus, the nondegeneracy condition becomes $\sum_{j=0}^{m-1} f'(\frac{2\pi j}{m}) \neq 0$; for $S_{N-p} \times S_p$ solutions, $m = 1$, implying $f'(0) \neq 0$. Further, for $(S_k)^m \rtimes Z_m$ rotating blocks the equality $\dot{\phi}_i \equiv \gamma$ is automatic. Finally, we note that if $k_1 = k_2$, $\delta = \pi/m$ always satisfies (2.7), so that the corresponding $(S_{k_1} \times S_{k_1})^m \rtimes Z_m$ solutions may also have symmetry $(S_{k_1})^{2m} \rtimes Z_{2m}$.

These arguments extend in a natural way to show the existence of weak solutions to the partial differential equations derived from (2.1) as $N \to \infty$ (Crawford and Davies [1999]). These are symmetrically-spaced combinations of delta distributions rotating at the frequency $c(\delta)$ found above, with the k_j-cluster distributions weighted by $k_j(N)/m(k_1(N)+k_2(N))$, $j = 1, 2$. Here, $k_j(N)$ is the number of oscillators in a cluster when the total number of oscillators is N, and the $N \to \infty$ limit is taken over a subsequence of configurations (E [2001]) with constant $k_j(N)/m(k_1(N)+k_2(N))$ such that m (fixed) divides $k_1(N)+k_2(N)$. Under the same nondegeneracy conditions as above, their existence may be shown for *any* values of $k_j/m(k_1 + k_2)$ and *any* m. Furthermore, if f lacks m-th Fourier harmonics and their multiples, families of Z_m-symmetric solutions analogous to the fixed tori of Section 2.4 also exist. A study of the stability of these solutions, their persistence under the introduction of a (diffusive) noise term, and associated convergence issues as $N \to \infty$ is in progress.

For the special case of odd phase-difference coupling, Proposition 2.1 ensures:

2.5 Theorem. *For the phase-difference system (2.2) with f odd, every* $\mathrm{Fix}\big[(S_{k_1} \times \cdots \times S_{k_{l_B}})^m \rtimes Z_m\big]$ *with* $m(k_1 + \cdots + k_{l_B}) = N$ *contains at least one equilibrium.*

Proof. Let $\widehat{V}_{\mathbf{k},m}$ denote the restriction of V from (2.4) to the l_B-torus $\Sigma_{\mathbf{k},m}$. $\widehat{V}_{\mathbf{k},m}$ is a continuous function on the (flow-invariant) compact set $\Sigma_{\mathbf{k},m}$ and thus possesses a minimum $\bar{\phi}$ on $\Sigma_{\mathbf{k},m}$. Consider a trajectory $\phi(t) \in \Sigma_{\mathbf{k},m}$ starting at $\bar{\phi}$ at $t = 0$. Since $\dot{V} = -\sum_{i=1}^{N} \dot{\phi_i}^2$, if $\phi(t) \neq \bar{\phi}$ then $V(\phi(t)) < V(\bar{\phi})$, $\forall t > 0$. This contradicts the assumption that $\bar{\phi}$ is a minimum for V, so $\bar{\phi} \in \Sigma_{\mathbf{k},m}$ must be a fixed point for equation (2.2). ■

Associated with any of the equilibria above, we expect at least one zero eigenvalue and a circle of equilibria corresponding to its T^1 group orbit.

Despite the variety of these equilibria, we can prove a result on the *non*-existence of fixed points in a region surrounding the in-phase solution. Define the open N-cylinder $\mathcal{C}_{R_1} \triangleq \big\{\theta \mid d(\theta, \theta_d(\psi)) \leq R_1 \text{ for some } \psi \in [0, 2\pi]\big\}$. Here, $d(\cdot\,,\cdot)$ is the Euclidian metric on $\mathcal{R}^N$ (and hence on T^N) and $\theta_d(\psi)$ is the N-vector with all coordinates equal to ψ (so that the axis of $\mathcal{C}_{R_1}$ is the diagonal $\mathcal{D} \triangleq \big\{\theta \mid \theta_i = \theta_j \;\forall\, i,j\big\}$; see Figure 2.2).

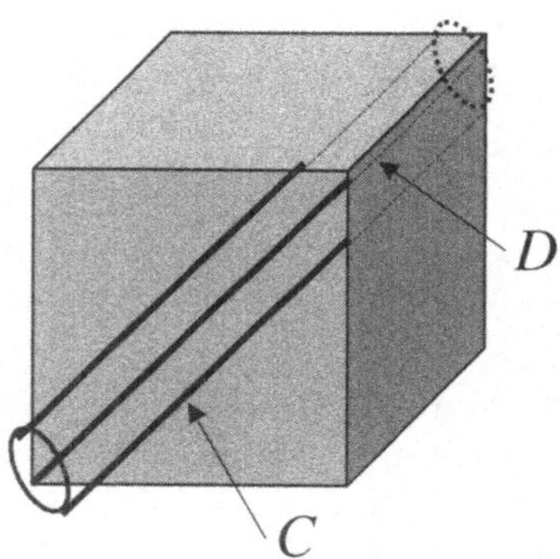

FIGURE 2.2. The diagonal $\mathcal{D}$ and cylinder $\mathcal{C}$ of Proposition 2.6. The cube represents the N-torus.

2.6 Proposition. *Let $R_1 > 0$ be such that* either *f restricted to* $(0, 2R_1)$ or *f restricted to* $(-2R_1, 0)$ *is of one sign (i.e., f is strictly negative or positive in the region). Then there are no fixed points in* $\mathcal{C}_{R_1} \setminus \mathcal{D}$.

Proof. First we note that for $\phi \in \mathcal{C}_{R_1}$, there exists a ψ s.t. $\phi \in B_{R_1}(\psi) \triangleq \big\{\phi \mid d(\phi, \theta_d(\psi)) < R_1)\big\}$. Thus in particular $|\phi_j - \psi| < R_1 \;\forall j$; summing two of these inequalities and applying the triangle inequality gives $|\phi_{ji}| < 2R_1 \;\forall\, i,j$. Next, consider an arbitrary N-vector $\phi \in \mathcal{C}_{R_1} \setminus \mathcal{D}$, where without loss of generality $\phi_1 \leq \phi_2 \leq \cdots \leq \phi_N$. If the chosen interval in the hypothesis is $(0, 2R_1)$, note that $2R_1 > \phi_j - \phi_1 \geq 0\ \forall\, j$ and $\phi_j - \phi_1 > 0$ for at least one j (since $\phi \notin \mathcal{D}$). Thus, $\dot{\phi}_1 = \frac{\alpha}{N}\sum_{j=1}^{N} f(\phi_j - \phi_1) \neq 0$, because each term in the sum is either zero or of the same sign (the continuity of f implies that $f(0)$ is either 0 or of the same sign as $f(\phi)$ for $\phi \in (0, 2R_1)$) and at least one term is nonzero. If the chosen interval is $(-2R_1, 0)$, we use the facts that $0 \geq \phi_j - \phi_N > -2R_1\ \forall\, j$ and $\phi_j - \phi_N < 0$ for at least one j. Similarly, then, $\dot{\phi}_N = \frac{\alpha}{N}\sum_{i=1}^{N} f(\phi_j - \phi_N) \neq 0$. Hence ϕ is not a fixed point. ■

This result does not exclude orbits with nonzero diagonal flow, which may exist within $\mathcal{C}_{R_1} \setminus \mathcal{D}$ if f is *not* odd.

We now consider the stability of some of the solutions found above.

2.3 Stability of Periodic Orbits

Rotating Blocks

We study solutions with isotropy $(S_k)^m \rtimes Z_m$ (rotating block modes), $S_p \times S_{N-p}$ (two-block modes), and S_N (in-phase mode). Since the Jacobian is constant along these periodic orbits with diagonal flow, this problem reduces to computation of eigenvalues (Ashwin and Swift [1992]). Stability will be discussed in terms of orbital stability, which implies asymptotic stability with respect to all perturbations transverse to the (continuous) T^1 group orbit of the solution (hence excluding the corresponding zero eigenvalue). Note that if $c \neq 0$ in equation (2.3), the group orbit and periodic orbit coincide.

To state some of the stability results, it is useful to express the coupling function in a Fourier series with coefficients b_l^o and b_l^e:

$$f(\phi_{ji}) = \sum_{l=0}^{\infty} \left(b_l^o \sin(l\phi_{ji}) + b_l^e \cos(l\phi_{ji})\right). \tag{2.10}$$

For $(S_k)^m \rtimes Z_m$ symmetric solutions we have, as in Okuda [1993] and (for $m = 1$) Watanabe and Swift [1997]:

2.7 Proposition. *Let $N = mk$ and let $\bar{\phi}$ be an $(S_k)^m \rtimes Z_m$-invariant fixed point or periodic orbit with diagonal flow. Then the eigenvalues of the Jacobian $J(\bar{\phi})$ obtained by linearization of Equation (2.2) are*

$$\left.\begin{array}{lll} \lambda = \lambda_0 = 0\,, & \text{with multiplicity 1} & \\ \lambda = \lambda_r^j\,, & j = 1, \ldots, m-1\ : \text{`rotation eigenvalues'} & \\ \lambda = \lambda_p\,, & \text{with multiplicity } m(k-1)\ : \text{`permutation eigenvalues'} & \end{array}\right\},$$

$$\lambda_r^j = \frac{\alpha}{m}\sum_{k=1}^{m-1} f'\left(\frac{2\pi k}{m}\right)\left(\exp\left(\frac{2\pi kji}{m}\right)-1\right)$$
$$= \frac{\alpha}{2}\left(\sum_{l\in\mathcal{M}(m)_1^j} l\,(b_l^o + ib_l^e) + \sum_{l\in\mathcal{M}(m)_2^j} l\,(b_l^o - ib_l^e) - 2\sum_{l\in\mathcal{M}(m)} lb_l^o\right) \tag{2.11}$$

$$\lambda_p = -\frac{\alpha}{m}\sum_{k=0}^{m-1} f'\left(\frac{2\pi k}{m}\right) = -\alpha\sum_{l\in\mathcal{M}(m)} b_l^o l\,, \tag{2.12}$$

$$\mathcal{M}(m)_1^j = \{mh - j \mid h = 1,2,...\}\,,\quad \mathcal{M}(m)_2^j = \{mh + j \mid h = 0,1,2,...\}\,,$$
$$\mathcal{M}(m) = \{mh \mid h = 1,2,...\}\,.$$

The 'rotation' and 'permutation' terminology is due to Ashwin and Swift [1992], where general formulae for eigenvalues and eigenvectors are presented. The proof of Proposition 2.7 repeatedly uses the following simple fact. Define the set $\mathcal{M} = \{l \mid l = qm \text{ for some } q \in \mathbb{Z}\}$ and let $\gamma = \exp(2\pi i/m)$; then for $\bar{l} \in \mathbb{Z}\setminus\mathcal{M}$, $\sum_{r=1}^{m-1}\gamma^{\bar{l}r} = -1$. From here, the $m\times m$-blocked structure of the Jacobian along with results on the eigenvectors of Toeplitz matrices leads to the desired conclusion.

If $m = 1$, then the proposition addresses the S_N-invariant (in-phase) solutions. Here there are no rotation eigenvalues, and the permutation eigenvalues are simply

$$\lambda_p = -\alpha f'(0)\,, \tag{2.13}$$

with multiplicity $N-1$. *Nonlinear* stability of these solutions is discussed in Section 3.2 (Proposition 3.2 with $\beta = 0$). At the other extreme, if $m = N$, the proposition addresses the Z_N-invariant solutions in which the phases of the oscillators are equally spaced; these are called 'rotating wave' solutions by Ashwin and Swift [1992], and correspond to the 'splay state' in the Josephson junction literature. In this case, there are no permutation eigenvalues, and (2.11) reduces to equation (63) of Watanabe and Swift [1997].

We now give examples to illustrate several interesting stability behaviours implied by Proposition 2.7. Including only the first harmonic in the coupling function ($f(\cdot) = b_1^o\sin(\cdot) + b_1^e\cos(\cdot)$), we have for $m > 1$:

$$\lambda_p = 0, \text{with multiplicity } m(k-1)$$
$$\lambda_r^j = \begin{cases} \frac{\alpha}{2}(b_1^o - ib_1^e) \text{ and } \frac{\alpha}{2}(b_1^o + ib_1^e) & \text{for } j = 1 \text{ and } m-1 \\ 0 & \text{otherwise (multiplicity } m-3), \end{cases}$$

in addition to λ_0. In this case the $(S_k)^m \rtimes Z_m$ solutions are *highly degenerate* and, for $\alpha b_1^o > 0$, *unstable*. For $m = N$ $(k = 0)$, there are $N-2$ zero eigenvalues. This result is well-known from the Josephson junction literature; in Watanabe and Strogatz [1994], it is shown to be related to the integrability of the equations for this choice of f.

On the other hand, we note that inclusion of higher harmonics in $f(\cdot)$ generically unfolds the degeneracy in the sense that all but one (λ_0) of the eigenvalues become nonzero, implying instability or orbital stability. For example, adding the mth harmonic ($f(\cdot) = b_1^o \sin(\cdot) + b_1^e \cos(\cdot) + b_m^o \sin(m\,\cdot) + b_m^e \cos(m\,\cdot)$; $m \neq 1$), we obtain

$$\lambda_p = -\alpha b_m^o m, \text{ with multiplicity } m(k-1)$$

$$\lambda_r^j = \begin{cases} \alpha\left[\frac{1}{2}(b_1^o - i b_1^e) - b_m^o m\right], \ c.c. & \text{if } j = 1,\ m-1 \\ -\alpha b_m^o m & \text{otherwise (multiplicity } m-3), \end{cases}$$

so that any $(S_k)^m \rtimes Z_m$ solution is *orbitally stable* if $\alpha b_m^o > \alpha b_1^o/2m$ and $\alpha b_m^o > 0$. We also note that for coupling functions whose harmonic indices belong entirely to $\mathcal{M}(l)$, any oscillator may be individually translated from a $(S_k)^m \rtimes Z_m$ solution by a multiple of $2\pi/l$ to give another equilibrium. These translations give a total of l^N fixed points, each with identical stability (due to the $2\pi/l$ periodicity of f).

The calculations proving Proposition 2.7 show which eigenvectors correspond to zero eigenvalues and hence along which directions there *may* be continuous families of equilibria. For example, with $k = 1$, $m = N = 4$ and $f(\cdot) = \sin(\cdot)$, the nondiagonal zero eigenvector is $(1, -1, 1, -1)^{\mathrm{T}}$, which reflects the fact that equilibrium is preserved if 'diametrically-opposite' pairs of oscillators are rotated independently.

Two-Block Periodic Orbits

For $f'(0) \neq 0$, equation (2.13) guarantees that the S_N-invariant solutions satisfy the nondegeneracy assumption of Theorem 2.4. Then, the Theorem (with $m = 1$) implies that for some $\delta(p) > 0$, equation (2.2) has periodic orbits with $\phi_{ji} \in \{0,\ \delta(p),\ 2\pi - \delta(p)\}$ for all $i,\ j$. This occurs when two blocks of p and $N-p$ identical-phase oscillators are mutually out of phase by δ; to avoid redundancy, we restrict $0 \leq p \leq \lfloor N/2 \rfloor$. The Jacobian from linearizing around a $S_{N-p} \times S_p$ solution has a four-blocked structure which yields:

2.8 Proposition. (Kim and Lee [2000]) *Let $\bar{\phi}$ be an $(S_p \times S_{N-p})$-invariant solution and $0 \leq p \leq \lfloor N/2 \rfloor$. Then for $p \geq 1$ the eigenvalues of the Jacobian from equation (2.2) are:*

$$\left.\begin{array}{lll} \lambda_1 = \alpha\big(b - \frac{p}{N}(a+b)\big), & \text{with multiplicity} & p-1 \\ \lambda_2 = \alpha\big(\frac{p}{N}(a+c) - a\big), & \text{with multiplicity} & N-p-1 \\ \lambda_3 = 0, & \text{with multiplicity} & 1 \\ \lambda_4 = \alpha\big(\frac{N-p}{N} b + \frac{p}{N} c\big), & \text{with multiplicity} & 1 \end{array}\right\}. \tag{2.14}$$

Here, $a = f'(0)$, $b = -f'\big(\delta(p)\big)$, $c = -f'\big(-\delta(p)\big)$.

If $f(\cdot)$ is odd, two-block states with $\delta = \pi$ exist for any p since $f(0) = f(\pi) = 0$; we write $\delta \neq \delta(p)$ to indicate this p-independence of δ. Oddness of f also implies $b = c$. This case was studied in Okuda [1993], where expressions corresponding to (2.14) are presented.

2.9 Corollary. *Assume that $b = c$, $\delta \neq \delta(p)$, and that $a, b > 0$. If $\alpha > 0$, the two-block equilibria of equation (2.2) are orbitally stable if and only if $p = 0$. If $\alpha < 0$, the equilibria are stable if and only if $p \neq 0$ and $a < bp/(N-p)$, if the equilibria are stable for $p = k$ for some $k \leq \lfloor N/2 \rfloor$, then they are stable for $p > k$.*

Proof. The results for $\alpha > 0$ are immediate from λ_4 of equation (2.14) and (2.13). For $\alpha < 0$, we note that $\lambda_{1,2} \leq 0$ implies $Na \leq p(a+b) \leq Nb$. Upon rearranging, this yields $a \leq b(N-p)/p$ and $a \leq bp/(N-p)$; for p in the given range, the latter inequality implies the first, and for fixed a and b it is clear that if the second inequality is satisfied for $p = k$, then it continues to be satisfied as p increases. In this case λ_1, λ_2 and λ_4 are all strictly negative, leading to the Corollary. ∎

We remark that if $a, b < 0$, the sign of α may be switched and the Corollary applied, and that the result that stability of equilibria for $p = k$ implies stability for $p = N/2$ is stated in Okuda [1993].

The corollary indicates that for $\alpha < 0$ and under certain conditions on a, b, and N, orbital stability of two-block fixed points can change as p is varied. For example, if $a = 1$, $b = 2$, and $N = 5$, the equilibria are unstable for $p = 0, 1$ but are stable for $p = 2$. In the special case $a = b = c$ (which occurs, for example, if $f(\cdot) = \sin(\cdot)$), note that $\lambda_1 = -\lambda_2 = \alpha(a - 2p/N)$, $\lambda_4 = \alpha a$; thus the fixed points are unstable unless $\alpha a < 0$, N is even and $p = N/2$, in which case they are neutrally stable with $N - 1$ zero eigenvalues. As above, inclusion of higher harmonics in the Fourier series for $f(\cdot)$ generically unfolds this degeneracy.

We close this subsection by remarking that techniques used to prove Propositions 2.7 and 2.8 could in principle be extended to calculate the stability of general $(S_{k_1} \times S_{k_2})^m \rtimes Z_m$ solutions for $m > 1$, where $m(k_1 + k_2) = N$. We refer the reader to Ashwin and Swift [1992] for the specific example $(S_2 \times S_1)^3 \rtimes Z_3$.

2.4 Existence of Fixed l_B-Tori

2.10 Proposition. *For ϕ contained in an invariant l_B-torus* $\mathrm{Fix}\left[(S_{k_1} \times \cdots \times S_{k_{l_B}})^m \rtimes Z_m\right]$ *with $N = m(k_1 + \cdots + k_{l_B})$, equation (2.2) reduces to:*

$$\dot{\phi}_i = \frac{\alpha}{N} \sum_{l \in \mathcal{M}(m)} \left\{ b_l^e m \sum_{q=1}^{l_B} k_q(i) \cos[l x_q(i)] + b_l^o m \sum_{q=1}^{l_B} k_q(i) \sin[l x_q(i)] \right\}, \tag{2.15}$$

where the numbers $k_q(i)$ and the angles $x_q(i)$ are as explained in Figure 2.3. In particular (as found in Ashwin and Swift [1992]*), if $b_l^{e,o} = 0$ for all $l \in \mathcal{M}(m)$, then the l_B-torus is a continuum of fixed points.*

The vector field (2.15) may be calculated directly by plugging an arbitrary point on the invariant l_B-torus (i.e., with arbitrary $\{x_1(i), ..., x_{l_B}(i)\}$ for some i) into equation (2.2) and using the relationship discussed in connection with the proof of Proposition 2.7. The existence of continua of fixed points is obvious from equation (2.15). For odd f, the fixed tori may also be found by showing that the potential (2.4) is always constant under this same condition on the Fourier coefficients of f given in Proposition 2.10.

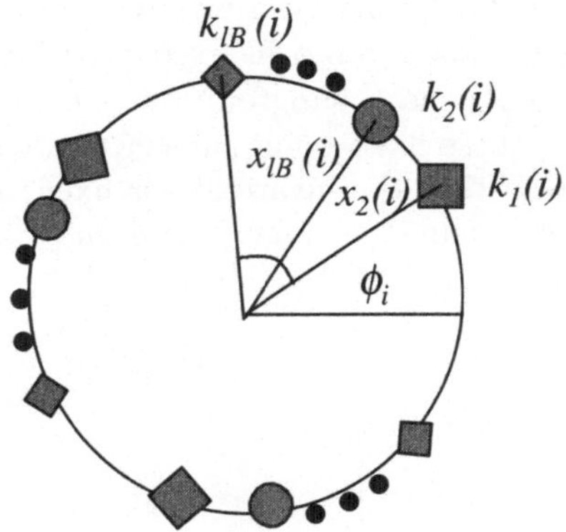

FIGURE 2.3. The labeling scheme used in Proposition 2.10. Given the reference index i corresponding to the $\dot{\phi}_i$ being computed, blocks of oscillators are numbered by the index q in a counterclockwise fashion, starting with $q = 1$ for the block containing ϕ_i itself. Each block contains $k_q(i)$ oscillators and is separated from its neighbor by the angle $x_q(i)$ (by definition $x_1(i) \equiv 0$).

If $m = N$ and $b_l^{e,o} = 0$ for all $l = 0 \pmod N$, $l_B = 1$ and Proposition 2.10 simply gives the circle of equilibria that is the T^1 group orbit of the Z_N-symmetric equilibrium of Proposition 2.7 (with $k = 1$). If $m = 1$ then Proposition 2.10 gives no new information about fixed subspaces: $b_l^{e,o} = 0$ for all $l \in \mathcal{M}(m)$ implies that the oscillators are uncoupled. Also, we note that since $S_1 \times S_1 \times \cdots \times S_1 \subseteq S_{k_1} \times \cdots \times S_{k_{l_B}}$, the l_B-tori of fixed points guaranteed by the theorem are actually contained in the (N/m)-torus $\text{Fix}\big[(S_1 \times S_1 \times \cdots \times S_1)^m \rtimes Z_m\big] = \text{Fix}[Z_m]$.

The following examples illustrate implications of Proposition 2.10.

Example. Consider $N = 4$ and suppose $b_l^{e,o} = 0$ for even l. For $m = 4$ the torus of fixed points guaranteed by the proposition is just the one-torus $\text{Fix}[Z_4]$. For $m = 2$, we get the two-torus of fixed points $\text{Fix}[Z_2]$. This describes the set of points for which two oscillators are out of phase by π, and the other two are also out of phase by π, corresponding to

$(\phi_1, \phi_2, \phi_3, \phi_4) = (\xi_1, \xi_2, \xi_1 + \pi, \xi_2 + \pi)$. Fix$[Z_2]$ contains both Fix$[Z_4] = \{(\xi, \xi + \pi/2, \xi + \pi, \xi + 3\pi/2)\}$ and Fix$[(S_2)^2 \rtimes Z_2] = \{(\xi, \xi, \xi + \pi, \xi + \pi)\}$.

Fix$[Z_2]$ also coincides with the ($N-2 = 2$)-dimensional 'incoherent manifold' found for averaged arrays of Josephson junctions (Watanabe and Swift [1997]). The incoherent manifold is defined as the set with zero centroid of phases ϕ_i on the unit circle. Because Fix$[Z_2]$ is a fixed point subspace, the two-dimensional incoherent manifold is dynamically invariant as found in Watanabe and Swift [1997]; Proposition 2.10 gives conditions under which it is also dynamically fixed as well as the expression for drift along the manifold. Watanabe and Swift [1997] also show that the $(N-2)$ dimensional incoherent manifold is not dynamically invariant when $N \geq 5$.

However, this manifold contains dynamically invariant (and perhaps dynamically fixed) submanifolds: for ϕ in fixed point subspaces of isotropy subgroups which have Z_m as a subgroup (where $m \geq 2$), the relevant centroid is zero. Thus, these fixed point subspaces are contained in the incoherent manifold. Note that the invariant (or fixed) tori have dimension $l_B \leq N/m$, which is less than $N-2$ for $N \geq 5$, $m \geq 2$. ◆

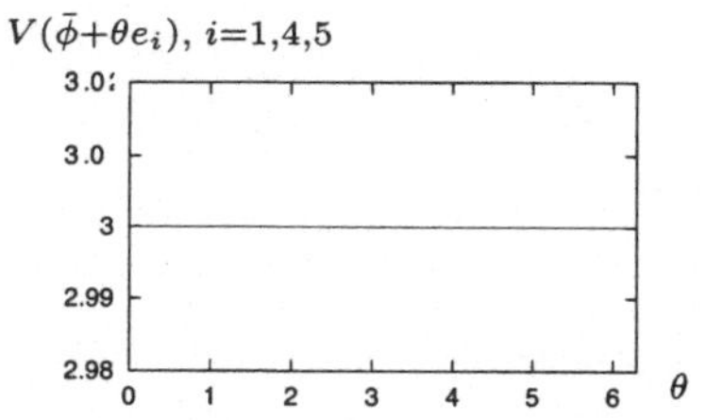

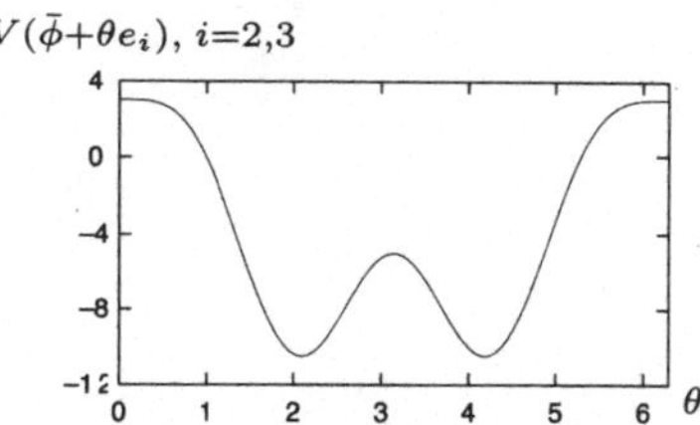

FIGURE 2.4. Potential V for perturbations to $\bar{\phi}$ in the directions of the null eigenvectors, as defined in the text.

Example. Suppose $N = 6$ and $f(\cdot) = \sin(\cdot)$, and consider the $((S_3)^2 \rtimes Z_2)$-invariant equilibria (e.g., $(\phi_1, \phi_2, \phi_3, \phi_4, \phi_5, \phi_6) = (0, 0, 0, \pi, \pi, \pi) \equiv \bar{\phi}$). From Proposition 2.7, the eigenvalues for such equilibria are 0 with multiplicity five, and 6α with multiplicity one. The null eigenvectors may be taken to be $e_1=(1,1,1,1,1,1)$, $e_2=(2,-1,-1,0,0,0)$, $e_3=(1,1,-2,0,0,0)$, $e_4 = (2,-1,-1,2,-1,-1)$, $e_5 = (1,1,-2,1,1,-2)$. Figure 2.4 shows the potential V corresponding to perturbations to $\bar{\phi}$ in the directions of these null eigenvectors. V is flat for perturbations in the e_1, e_4, and e_5 directions (each with a corresponding one-dimensional continuum of fixed points, overall giving a three-torus of equilibria), but not for perturbations in the e_2 and e_3 directions. Proposition 2.10 guarantees the existence of the three-torus of equilibria Fix$[Z_2]$ given by $(\phi_1, \phi_2, \phi_3, \phi_4, \phi_5, \phi_6) = (\xi_1, \xi_2, \xi_3, \xi_1 + \pi, \xi_2 + \pi, \xi_3 + \pi)$; note that perturbations to $\bar{\phi}$ in the e_1, e_4, and e_5 directions keep the system in the Fix$[Z_2]$ subspace. The e_2 and e_3 perturbations illustrate that every zero eigenvalue of Propositions 2.7 or 2.8 does *not* necessarily imply a corresponding one-dimensional continuum of fixed points. ♦

3 Breaking the T^1 Symmetry: Product Coupling

Reintroducing the h and g terms and going back to θ coordinates, we return to the S_N-equivariant system

$$\dot{\theta}_i = \omega + \frac{\alpha}{N}\sum_{j=1}^{N} f\left(\theta_j - \theta_i\right) + h(\theta_i)\frac{\beta}{N-1}\sum_{j\neq i}^{N} g(\theta_j)\,. \tag{3.1}$$

The results of this section are valid for arbitrary C^1 2π-periodic functions g and h; without loss of generality, we assume that g takes values in $[0,1]$. Additional assumptions on the product function $G(\theta) \triangleq h(\theta)g(\theta)$ simplify the discussion of bifurcations in Section 3.1.

3.1 Bifurcations of Fixed Points on the Diagonal $\mathrm{Fix}[S_N]$

This section concerns analysis local to the diagonal of T^N, defined by $\mathcal{D} \triangleq \{\theta \mid \theta_i = \theta_j \ \forall\ i,j\}$, which is dynamically invariant. Restricted to $\mathcal{D}$ and with $\theta_i \equiv \theta$, equation (3.1) becomes

$$\dot{\theta} = \omega + \alpha f(0) + \beta G(\theta)\,. \tag{3.2}$$

This equation has fixed points given by $\bar{\theta} = G^{-1}\big(-(\omega + \alpha f(0))/\beta\big)$. To simplify the analysis in this section, we assume that G has a single minimum θ_{min} with $G''(\theta_{min}) \neq 0$, as it does for the 'neurobiological' coupling functions to be considered in Section 4.

These conditions on G and $\omega > 0$ imply that there are no, one, or two on-diagonal fixed points, in the latter case denoted by $\bar{\theta}_1 < \bar{\theta}_2$. The eigenvalues of the Jacobian of equation (3.1) linearized about these fixed points are

$$\begin{aligned} \lambda_1^k &= -\alpha f'(0) + \beta\big(h'(\bar{\theta}_k)g(\bar{\theta}_k) - \tfrac{1}{N-1}h(\theta_k)g'(\theta_k)\big)\,, && \text{multiplicity } N-1,\\ \lambda_2^k &= \beta G'(\bar{\theta}_k)\,, && \text{multiplicity } 1. \end{aligned}$$

Stability in the transverse directions (with respect to the diagonal) is determined by λ_1, and in the axial direction by λ_2; note that our hypothesis on G implies that $\lambda_2(\bar{\theta}_1) < 0$, and $\lambda_2(\bar{\theta}_2) > 0$. As β is decreased through $\beta = (\omega + \alpha f(0))/|G(\theta_{min})|$, the two fixed points coalesce and disappear in a saddle node bifurcation (the appearance of these fixed points as β

increases represents the phenomenon of oscillator death: Ermentrout and Kopell [1990]; Taylor and Holmes [1998]). For the remaining values of β, the orbit along $\mathcal{D}$ is the (S_N symmetric) periodic orbit $\theta_D(t)$. We will investigate the period and stability of this orbit in the following sections.

3.2 Frequency and Stability of the In-Phase Periodic Orbit

If $\beta < (\omega + \alpha f(0))/|G(\theta_{min})|$, the period of the orbit on $\mathcal{D}$ is given by

$$\tau = \int_0^{2\pi} \left(\frac{d\theta}{dt}\right)^{-1} d\theta = \int_0^{2\pi} \frac{d\theta}{\omega + \alpha f(0) + \beta h(\theta) g(\theta)} . \tag{3.3}$$

Moreover, we have

3.1 Proposition. (Local stability of $\mathcal{D}$.) *The S_N-symmetric periodic solution $\theta_i(t) \equiv \theta_D(t)$ along $\mathcal{D}$ is asymptotically stable if*

$$\alpha > \frac{\beta N}{(N-1)\tau f'(0)} \int_0^{2\pi} \frac{g(\theta)h'(\theta)}{\omega + \alpha f(0) + \beta g(\theta) h(\theta)} d\theta , \tag{3.4}$$

where τ is the (generally α-dependent) period of $\theta_D(t)$ given in equation (3.3) and we assume $f'(0) > 0$.

Closely related results are found in Tsang, Mirollo, and Strogatz [1991]; Golomb, Hansel, Shraiman, and Sompolinsky [1992].

Proof. Linearized around $\theta_D(t)$, equation (3.1) becomes $\dot{\xi}_i = [A_d(t)\xi]_i$. The proof uses the fact that the (t-dependent) symmetric matrix $A_d(t)$ has a particularly simple structure, and that it can be diagonalized by a t-independent similarity transformation. Specifically, the eigenvalues of $A_d(t)$, where t is viewed as a (fixed) parameter, are:

$$\lambda_1(t) = -\alpha f'(0) + \beta\big(g(t)h'(t) - \frac{1}{N-1} h(t)g'(t)\big) , \quad \text{multiplicity } N-1 ,$$
$$\lambda_2(t) = \beta G'(t) , \quad \text{multiplicity } 1,$$

where $g(t)$ is written for $g\big(\theta_D(t)\big)$, etc. The orthogonal eigenvectors of $\lambda_1(t)$ (denoted by $\chi_1, \dots, \chi_{N-1}$) may be chosen constant and orthogonal to the eigenspace of $\lambda_2(t)$, which is spanned by the eigenvector $(1, \dots, 1)^{\mathrm{T}}$. Thus, $\chi_1, \dots, \chi_{N-1}$ span the space normal to $\theta_D(t)$. In these eigencoordinates the linearized system decouples as

$$\dot{\xi}_i = \lambda_1(t)\xi_i, \; i = 1, \dots, N-1, \;\; \dot{\xi}_N = \lambda_2(t)\xi_N .$$

Define the $(N-1)$-dimensional plane $\sum = \{\chi \mid \chi_N = 0\}$ and consider the Poincaré map $P : U \to U$ for some neighborhood $U \subset \sum$ of 0. The orbit

$\theta_D(t)$ intersects $\sum$ at 0, which is a fixed point for P. For $i = 1, \dots, N-1$, $P : \xi_i \mapsto (\exp \int_0^\tau \lambda_1(t)dt)\xi_i$ so that 0 is a stable fixed point for P if $\int_0^\tau \lambda_1(t)dt < 0$. (Due to the periodicity of $G(t)$, $\int_0^\tau \lambda_2(t)dt \equiv 0$, as it must, being the Floquet exponent along the periodic orbit). We have

$$\begin{aligned}
&\int_0^\tau \lambda_1(t)dt \\
&= \int_0^\tau \left(-\alpha f'(0) + \beta\left[g(t)h'(t) - \frac{1}{N-1}h(t)g'(t)\right]\right) dt \\
&= -\alpha f'(0)\tau + \int_0^{2\pi} \beta\left(-\frac{[h(\theta)g'(\theta) + h'(\theta)g(\theta)]}{N-1} + \frac{Ng(\theta)h'(\theta)}{N-1}\right)\dot{\theta}^{-1}d\theta \\
&= -\alpha f'(0)\tau - \frac{1}{N-1}\ln[\omega + \alpha f(0) + \beta h(\theta)g(\theta)]_0^{2\pi} \\
&\quad + \int_0^{2\pi} \frac{\beta N g(\theta)h'(\theta)}{(N-1)(\omega + \alpha f(0) + \beta h(\theta)g(\theta))}d\theta \qquad (3.5) \\
&= -\alpha f'(0)\tau + \int_0^{2\pi} \frac{\beta N g(\theta)h'(\theta)}{(N-1)(\omega + \alpha f(0) + \beta h(\theta)g(\theta))}d\theta\,, \qquad (3.6)
\end{aligned}$$

where the second term in equation (3.5) vanishes due to the 2π-periodicity of h and g. Thus, $\int_0^\tau \lambda_1(t)dt < 0$ when the inequality of Proposition 3.1 is satisfied. Since stability of the fixed point 0 under P implies stability of $\theta_D(t)$ for equation (2.1), the Proposition is proven. ■

A simple calculation using integration by parts and the 2π-periodicity of g and h shows that for $\alpha = 0$ and asymptotically small β, the right-hand side of (3.6) becomes $\frac{\beta N}{N-1}f_s'(0)$, which (cf. (2.13)) determines the stability of the in-phase solution if synaptic coupling $\frac{\beta}{N-1}h(\theta_i)\sum_{j\neq i} g(\theta_j)$ is taken to be weak and then averaged to yield $\frac{\beta}{N-1}\sum_{j\neq i} f_s(\theta_j - \theta_i)$. This agreement between the averaged and original versions of (3.1) for sufficiently small β is expected from the averaging theorem (Guckenheimer and Holmes [1983]), and reveals how (3.4) generalizes the stability result found in van Vreeswijk, Abbot, and Ermentrout [1994]; Hansel, Mato, and Meunier [1995] for $N = 2$ and averaged synaptic coupling.

Equation (3.4) may be used to estimate a critical value α_{loc} such that $\theta_D(t)$ is asymptotically stable for $\alpha > \alpha_{loc}$. Letting $\widehat{h}$ be a Lipschitz constant for h, we note that

$$\int_0^{2\pi} \frac{g(\theta)h'(\theta)}{\omega + \alpha f(0) + \beta h(\theta)g(\theta)}d\theta \leq \int_0^{2\pi} \frac{\widehat{h}}{\omega + \alpha f(0) + \beta h(\theta)g(\theta)}d\theta = \widehat{h}\tau\,, \qquad (3.7)$$

where the inequality follows from the bound on g and the definition of the Lipschitz constant. Thus, from (3.4) we have stability if $\alpha > \frac{N\beta\widehat{h}}{(N-1)f'(0)} \triangleq \alpha_{loc}$. With $f(0) = 0$ (e.g. if f is odd), this estimate can be refined: the right-hand side of equation (3.4) is independent of α, so that the (smallest) critical value $\tilde{\alpha}_{loc}$ is

$$\tilde{\alpha}_{loc} = \frac{\beta N}{(N-1)\tau f'(0)} \int_0^{2\pi} \frac{g(\theta)h'(\theta)}{\omega + \beta g(\theta)h(\theta)} d\theta \,. \tag{3.8}$$

We now turn to the nonlinear stability properties of $\mathcal{D}$.

Estimate for the Domain of Attraction of $\mathcal{D}$

3.2 Proposition. (Nonlinear stability of $\mathcal{D}$.) *For some $s > 1$, assume $f'(0) > 0$ and let $R > 0$ be the smallest value for which either $f'(2R) = f'(0)/s > 0$ or $f'(-2R) = f'(0)/s > 0$ (implying $\min_{\theta\in[-2R,2R]} f'(\theta) = f'(0)/s$). Let $\widehat{G}$ be the Lipschitz constant for $G(\cdot) = g(\cdot)h(\cdot)$, and define $\widehat{h}_1(\theta_i) = \max_\theta\{|h'(\theta_i)g(\theta)| : |\theta_i - \theta| < 2R\}$ and $\widehat{h}_1 = \max_{\theta_i}\{\widehat{h}_1(\theta_i)\}$. Then, for*

$$\alpha > \alpha_{glob} \triangleq \frac{s\beta(N\widehat{h}_1 + \widehat{G})}{(N-1)f'(0)} \,, \tag{3.9}$$

the domain of attraction for $\mathcal{D}$ includes $\mathcal{C}_R \triangleq \{\theta \mid d(\theta, \theta_d(\psi)) \leq R$ for some $\psi \in [0, 2\pi]\}$ (cf. Figure 2.2).

Proof. Fix an arbitrary $\psi \in [0, 2\pi)$. Consider the (non-orthogonal) basis $b \equiv \{x_i \mid i = 1, \ldots, N-1\}$, where $x_i \equiv \theta_i - \theta_{i+1}$. We define X_ψ, the $N-1$ dimensional space perpendicular to the axis of $\mathcal{C}_R$ at $\theta_d(\psi)$, as the copy of span b containing $\theta_d(\psi)$. In other words, X_ψ is the normal space $\mathbf{N}(\theta_d(\psi))$.

Now, define the squared 'radius' $\mathcal{R} = \sum_{i=0}^{N-1} x_i^2$. We will show that $\dot{\mathcal{R}} = 2\sum_{i=0}^{N-1} x_i \dot{x}_i \leq 0$ for all $\mathbf{x} \in \mathcal{C}_R$. The cylindrical surfaces $\{\mathbf{x} \mid \mathcal{R}(\mathbf{x}) = c\}$ will therefore be crossed 'inward' toward the axis of $\mathcal{C}_R$.

Take an arbitrary $\theta \in \mathcal{C}_R \cap X_\psi$. For such a θ, we also have $\theta \in B_R(\psi) = \{\theta \mid d(\theta, \theta_d(\psi)) < R\}$. Thus $|\theta_j - \theta_i| < 2R \;\forall\; i, j$ (and, in particular, $|x_i| < 2R \;\forall\; i$). These inequalities allow us to find a bound on each $\dot{x}_i$:

$$\begin{aligned}
\dot{x}_i &= \overline{\theta_i - \theta_{i+1}}^{\,\cdot} \\
&= \frac{\alpha}{N}\sum_{j=1}^{N} f(\theta_j - \theta_i) - \frac{\alpha}{N}\sum_{j=1}^{N} f(\theta_j - \theta_{i+1}) \\
&\quad + \frac{\beta}{N-1} h(\theta_i) \sum_{j\neq i}^{N} g(\theta_j) - \frac{\beta}{N-1} h(\theta_{i+1}) \sum_{j\neq i+1}^{N} g(\theta_j) \\
&= \frac{\alpha}{N}\sum_{j=1}^{N} [f(\theta_j - \theta_i) - f(\theta_j - \theta_i + x_i)] + \frac{\beta[h(\theta_i) - h(\theta_{i+1})]}{N-1} \sum_{j=1}^{N} g(\theta_j) \\
&\quad + \frac{\beta[h(\theta_{i+1})g(\theta_{i+1}) - h(\theta_i)g(\theta_i)]}{N-1} .
\end{aligned} \tag{3.10}$$

$$\dot{x}_i \begin{array}{ll} < -\alpha[f'(0)/s]x_i + \beta\frac{N}{N-1}\widehat{h}_1 x_i + \frac{\beta}{N-1}\widehat{G}x_i \triangleq kx_i & \text{if } x_i > 0 \\ > -\alpha[f'(0)/s]x_i + \beta\frac{N}{N-1}\widehat{h}_1 x_i + \frac{\beta}{N-1}\widehat{G}x_i \triangleq kx_i & \text{if } x_i < 0 \end{array} \Bigg\} \quad (3.11)$$

The inequalities (3.11) use the hypothesis on f', the bound $g(\theta) \leq 1$, and the definitions of $\widehat{h}$ and $\widehat{G}$. Thus, for $k < 0$ (i.e. $\alpha > \alpha_{glob}$), $\dot{\mathcal{R}} = 2\sum_{i=0}^{N-1} x_i \dot{x}_i < 0$ unless $x_i = 0$, $\forall i$. This argument may be repeated for any ψ and therefore for any arbitrary $\theta \in \mathcal{C}_R$, so the Proposition follows. ■

Since nonlinear stability implies local stability, it must follow from $\alpha > \alpha_{glob}$ that inequality (3.4) is satisfied. This may be seen from the fact that $\alpha > \alpha_{glob}$ implies $\alpha > \alpha_{loc}$ and comparing equation (3.9) with (3.7).

Finally, we note that Proposition 3.2 may be sharpened by refining the estimates in (3.11) in any manner that also implies $\text{sign}(\dot{x}_i) = -\text{sign}(x_i)$. For example, a lower value $\widehat{h}_2$ can replace $\widehat{h}_1$ above, where $\widehat{h}_2 = \max_{\theta_i} \widehat{h}_2(\theta_i)$ and $\widehat{h}_2(\theta_i) = \max_\theta\{h'(\theta_i)g(\theta) : |\theta_i - \theta| < 2R\}$ (note that although we have dropped the absolute value in the $\max_\theta$, $\widehat{h}_2 \geq 0$ since h is periodic). The bound $\widehat{h}_2$ arises as follows. If the second term in (3.10) is of opposite sign to x_i, it favors the conclusion $\text{sign}(\dot{x}_i) = -\text{sign}(x_i)$ and hence may be ignored for the purposes of bounding α such that $k < 0$. Thus the natural question is: *assuming* that it is of the same sign as x_i, can we find a smaller upper bound than $\beta\frac{N}{N-1}\widehat{h}_1 x_i$ on the magnitude of this second term? The answer is yes: since $[h(\theta_i) - h(\theta_{i+1})] = [h(x_i + \theta_{i+1}) - h(\theta_{i+1})]$, this difference cannot exceed the upper bound $\beta\frac{N}{N-1}\widehat{h}_2 x_i$, as desired.

4 Application to a Model of the *Locus Coeruleus*

Here we apply the analysis above to a model of the *locus coeruleus* brain nucleus. First, we introduce specific coupling functions f, g, and h appropriate to neuronal coupling.

4.1 Coupling Functions

The functions f and g, h, corresponding to electrotonic and synaptic coupling, were computed using both the strong attraction (SA) and phase response (PR) methods mentioned in the introduction. The Hodgkin-Huxley (HH) equations with input current 10 $\mu A/cm^2$ were used (Hodgkin and Huxley [1952]). In their original form these equations were derived from the giant axon of a squid, so their use here merely represents a proof of concept. Reduction of more realistic mammalian neuron models, which include calcium-dependent potassium channels and whose action potential spikes occupy a much smaller fraction of the period than in the rescaled HH equations, is in progress (Brown, Moehlis, Holmes, Aston-Jones, and Clayton [2002]), and leads to coupling functions somewhat different from those considered here, although the general structure of the phase equations survives.

The effect of electrotonic coupling on the time derivative $\dot{V}_i$ of neuron i's voltage was taken to be $\frac{\alpha}{N}\sum_{j=1}^{N}(V_j - V_i)$ (c.f. Johnston and Wu [1997]), and the inhibitory synaptic effect to be $(E_K - V_i)\frac{\beta}{N-1}\sum_{j\neq i} A(V_j, t)$, where E_K is the reversal potential for potassium and $A(V_j, t)$ is an 'alpha function' which takes values in $[0, \tilde{A}]$, $\tilde{A} < 1$, and represents the influence of neuron j on post-synaptic cells. Specifically, $A(V_j, t) = \big((t - t_s^j - t_d)/\tau_A\big) \cdot \exp\big(-(t - t_s^j - t_d)/\tau_A\big)$, where t_s^j is the time at which the voltage of neuron j spikes (see below), t_d is the synaptic delay, and τ_A is the synaptic time constant (e.g. Kim and Lee [2000]). The effective value of τ_A for LC neurons has been observed to be much longer than those of typical synaptic connections due to the slow dynamics of norepinephrine neurotransmitter uptake (Grant, Aston-Jones, and Redmond [1988]; Aston-Jones, Rajkowski, and Alexinsky [1994]). We parameterize the limit cycle of the uncoupled HH equations by a time scaled so that the period $T = 2\pi/\omega$ is $\frac{1}{3}$ sec (to match our estimate for LC neurons), and take $\tau_A = 0.025\ sec$ and $t_d = 0.150\ sec$. These neuron and coupling models and parameters lead to the reductions of the coupling functions to T^N displayed in Figure 4.1.

Under the idealisation that neurons in the small LC nucleus are identical (Williams, North, Shefner, Nishi, and Egan [1984]) and globally coupled, our LC model simplifies to equation (1.2). As coupling becomes stronger, modifications to f, g, and h may be required to maintain the accuracy of phase reductions; for the purpose of this paper, these effects are neglected. However, Brown, Moehlis, Holmes, Aston-Jones, and Clayton [2002] includes a careful comparison of phase-reduced and 'full' conductance-based LC models in the relevant parameter range.

4.2 Modeling Synchrony in LC Modes

This section demonstrates that cross correlograms qualitatively similar to those of Figure 1.1(a) can arise in our phase-reduced LC model due to increased coupling in the phasic LC mode relative to the tonic mode.

Cross correlograms are derived from solutions of (3.1) as follows. A spike is deemed to occur when a rotator θ_i crosses through a threshold value θ_s: the solution of $\{V(\theta_s) = V_s, V'(\theta_s) > 0\}$, where $V_s = -30mV$ is a depolarized voltage characteristic of a neuron firing an action potential and the function $V(\theta)$ is defined by $V(t) = V(\theta/\omega)$ over the period of one neuron action potential. The set of all pairwise differences between times at which distinct spike events occur is computed according to this definition, and the cross correlogram is the histogram of this set.

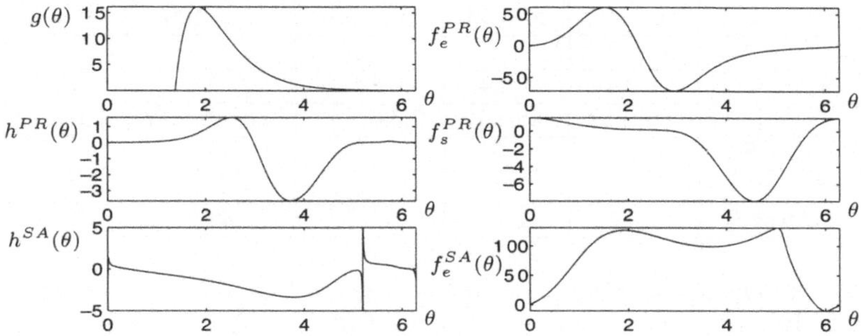

FIGURE 4.1. Coupling functions derived from the (time-reparameterized) Hodgkin–Huxley equations. The subscript e (s) refers to electrotonic (synaptic) coupling, while the superscript PR (SA) indicates that the function was derived using the phase reduction (strong attraction) method. The h's and g are calculated for the synaptic coupling described in the text: f_s^{PR} is obtained by averaging the product of g and h^{PR}, and f_e^{PR} is obtained by averaging the electrotonic coupling using a phase response method (cf. Kim and Lee [2000]). f_e^{SA} is obtained by first assuming that the limit cycle is infinitely attracting, followed by averaging (cf. Ermentrout and Kopell [1990]; Hoppensteadt and Izhikevich [1997]). The 'spikes' in $h^{SA}(\theta)$ are associated with the projection of coupling functions near turning points in the original phase variables; the tips extend to approximately ± 30 (there is also an $\mathcal{O}(1)$ spike near $\theta = 0$, not visible here).

Phase-Difference coupling

In this section we assume that synaptic coupling $\frac{\beta}{N-1} h(\theta_i) \sum_{j \neq i} g(\theta_j)$ is sufficiently weak that it can be averaged to yield $\frac{\beta}{N-1} \sum_{j \neq i} f_s(\theta_j - \theta_i)$, and we take $\beta = \kappa\alpha$, so that (3.1) becomes $\dot{\theta}_i = \omega + \frac{\alpha}{N} \sum_j f(\theta_{ji})$, where the phase difference function $f(\cdot) = f_e(\cdot) + \kappa f_s(\cdot)$. We use the methods of Section 2 to determine stability of periodic orbits. The results are shown in Table 4.1 for various values of κ and coupling functions derived with the PR and SA methods. The SA coupling functions give rise to a larger set of distinct stable periodic orbits, the consequences of which will be discussed below.

To model the phasic and tonic behavior of Figure 1.1, we consider equation (2.1) in the presence of noise represented as additive Brownian forcing on the torus, so that:

$$d\theta_i = \left[\omega + \frac{\alpha}{N} \sum_{j \neq i}^{n} f\left(\theta_j - \theta_i\right) \right] dt + \sigma dW_t^i . \tag{4.1}$$

The inclusion of random noise represents additional inputs currents to LC neurons, a common stratagem in accounting for the influence of neural subgroups neglected in the model. The stochastic averaging theorem (Zhu

	PR $\kappa = 0$	PR $\kappa = 10$	PR $\kappa = 50$	PR $\kappa = 100$	SA $\kappa = 0$
$(\mathbf{S_k})^\mathbf{m} \rtimes \mathbf{Z_m}$					
S, $m =$	1,2,4	1,2,4	1,4	1	1,3,4
U, $m =$	3,6,8,12	3,6,8,12	2,3,6,8,12	2,3,4,6,8,12	2,6,8,12
$\mathbf{S_{N-p} \times S_p}$					
S, $p =$	11,12	12	1,2	1-4	1-6
U, $p =$	1-12	1-12	1-12	3-12	7-12

TABLE 4.1. The linear stability of various 'clustered' periodic orbits of diagonal flow for different phase difference couplings $f_e(\theta_{ji}) + \kappa f_s(\theta_{ji})$, computed using the PR or SA methods. $(S_k)^m \rtimes Z_m$ stability is given for allowable m ($N = 24$), and $S_{N-p} \times S_p$ stability for $N = 24$ and $p = 1, ..., 12$; a value of p being listed twice indicates the correspondence of multiple δ's. S and U indicate asymptotic stability and instability, respectively, for the m or p values given in the subsequent columns.

[1988]; Freidlin and Wentzell [1998]) leads to the approximation above, including σ being independent of θ; in particular, $\sigma^2 = \overline{z^2}\gamma$, where γ is the variance of Brownian input currents, $z(\theta)$ is the phase response curve (PRC, cf. Kuramoto [1997]), and the overbar denotes the average $\overline{(\cdot)} = \frac{1}{2\pi}\int_0^{2\pi}(\cdot)d\theta$. Simulations of equation (4.1) to be discussed below were performed using a second order stochastic Runge–Kutta method (Honeycutt [1992]).

The stability results of Table 4.1 will persist for α sufficiently larger than σ, but solutions approach a constant-drift Brownian motion on T^N as the coupling/noise ratio α/σ decreases. These two regimes correspond to the biological hypothesis discussed in the Introduction: that tonic behavior can be expected for relatively weak coupling (α_{ton}) and phasic behavior for larger α_{ph}. For α_{ph}, the presence of appropriate stable states can lead to the phasic-mode cross correlogram via several mechanisms (see below and Figures 4.2(b-e)). Meanwhile, for the lesser value of α_{ton}, the essentially constant-drift Brownian motion on the torus leads to the flat cross correlogram characteristic of this mode (i.e., no preferred firing time difference between neurons); see Figure 4.2(a).

If a variety of diagonal flow solutions are simultaneously stable (as for the SA coupling functions) the following mechanism can produce the 'peak-shoulder' cross correlogram patterns characteristic of the phasic LC mode (Figure 1.1(a)). For every revolution (of diagonal flow), $(S_{k_1} \times S_{k_2})^m \rtimes Z_m$ states with $\delta \neq 0, \pi$ produce cross correlograms with $m(k_1^2 + k_2^2 - k_1 - k_2)$ counts at $t = 0$, $m(k_1^2 + k_2^2)$ counts at times proportional to $\pm 2\pi j/m$, $j = 1, ..., m-1$, and mk_1k_2 counts at times proportional to $\pm(2\pi j/m) \pm \delta$, $j = 0, ..., m-1$. If $N \geq 5$, this leads to a dominant central peak in the cross correlogram for (two-cluster) states with $m = 1$. Moreover, all peaks *except* for the central peak at 0 will be differently spaced for each distinct

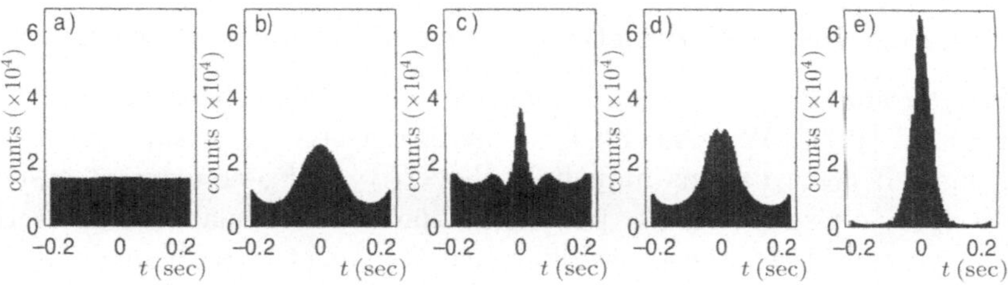

FIGURE 4.2. Cross correlograms for simulations of the tonic 'low-coupling' (a) and phasic 'high-coupling' (b-e) LC modes using the phase-difference model (4.1). To facilitate comparison, α_{ph} was chosen so that $\alpha_{ph} \max_\theta\{|f(\theta)|\} = 2.25$ for both the SA and PR coupling functions; also, $\alpha_{ton} = \alpha_{ph}/5$, $\omega = 6\pi$, and $N = 24$. (a) PR model with $\kappa = 0$, $\sigma = 0.8$, and $\alpha = \alpha_{ton}$; (b) same except with $\alpha = \alpha_{ph}$; (c) SA model with $\kappa = 0$ and $\sigma = 0.2$, $\alpha = \alpha_{ph}$. (d) As described in Section 4.2, phasic PR simulation as in (a) but with randomly distributed frequencies (drawn from the Cauchy distribution with parameter 0.2 and mean ω) and $\sigma = 0.1$. (e) as in (b), but with the addition of phase-dependent (synaptic) coupling of strength $\beta = 0.23$. The range of all histograms is $[-.7\tau, .7\tau]$, where $\tau = \frac{1}{3}$ sec is the simulated natural period; all histograms are averaged over five simulated recordings with uniformly distributed initial conditions, with each trial 2 min in duration.

$(S_{k_1} \times S_{k_2})^m \rtimes Z_m$ orbit. Thus, if the individual cross correlograms from many of these states are combined (e.g. due to stochastic switching due to random noise in (4.1)), the common central maxima would conspire to produce a central peak in the cross correlogram while the combination of many secondary maxima could give rise to the relatively flat shoulder. This is demonstrated in Figure 4.2(c).

However, the stochastic switching mechanism does not apply to the PR functions, which exhibit relatively few stable periodic orbits significantly different from the in-phase mode (in particular, $\delta < 1$ for the stable $S_p \times S_{N-p}$ modes for all κ values in Table 4.1). For the PR functions, then, the shoulders in the cross correlogram instead arise when the stable distribution around the in-phase equilibrium is broad; this was found (Figure 4.2(b)) to require high noise strengths, approximately $\sigma > \alpha \max_\theta\{|f_e(\theta)+\kappa f_s(\theta)|\}/3$. Faced with this perhaps unrealistically high value, in the next section we explore the presence of random natural frequencies as another method for creating the cross correlogram shoulders.

To check the validity of the SA and PR phase reductions, we compared a few cases of our stability predictions with numerical simulations of the full HH equations (cf. Kim and Lee [2000]); for $\kappa = 0$, stability was consistent with the full equations for the PR reductions but *not* for SA. Thus, all subsequent computations are given for the PR case; however, to illustrate properties of (2.1) we will continue to refer to the SA functions when their general form gives additional (e.g. contrasting) results.

Breaking the S_N Symmetry with Random Natural Frequencies

By the standard theorems for normally hyperbolic invariant manifolds Fenichel [1971]; Wiggins [1994], many of our stability results persist under small perturbations from fully $S_N \times T^1$- or S_N-equivariant systems. To study these effects, we performed numerical simulations: in particular, we considered the phase-difference system (4.1) with randomly distributed frequencies $\omega \to \omega_i$ (so that the noise-free system has *no* nontrivial isotropy subgroups). This is appropriate because individual LC neurons exhibit a range of natural frequencies Grant, Aston-Jones, and Redmond [1988]; Aston-Jones, Rajkowski, and Alexinsky [1994]. In Figure 4.2(d), we show a reproduction of the phasic mode cross correlogram for the averaged PR coupling function with $\kappa = 0$ and frequencies randomly distributed around ω with a Cauchy distribution with parameter 0.2. This corresponds to a tight distribution; nevertheless, the relatively low noise value $\sigma = 0.1$ is sufficient to reproduce patterns similar to those in Figure 1.1.

We note that there are some analytical results (e.g. Kuramoto [1984]; Strogatz [2000]; Crawford [1995]; Crawford and Davies [1999]) for the distributed frequency, phase-difference system as $N \to \infty$. However, a finite-N analysis of the full equation (2.1) with distributed phases and phase product coupling term remains open, to the best of our knowledge.

Phase-Dependent Coupling

We now return to consider equation (3.1) with unaveraged synaptic (product) coupling. Since $f(0) = 0$ for electrotonic coupling, we may use equation (3.8) to explicitly calculate a lower bound on α for this orbit to be stable. The resulting $\tilde{\alpha}_{loc} < 0$ (Figure 4.3), so that the in-phase state is stable for synaptic coupling and any positive α (since $f'(0) > 0$). In addition, the domain of attraction may be estimated: for example, taking $s = 2.14$ in Proposition 3.2 gives $R = 1/4$ and $\widehat{h}_2 = 28.1$ for PR coupling functions with $\kappa = 0$. Thus, the domain of attraction of the in-phase orbit includes $\mathcal{C}_R$ if $\alpha > \alpha_{glob} = 7.11\beta$ (equation (3.9) for $N \to \infty$). Figure 4.2(e) demonstrates the collapse of the cross correlogram (d) upon addition of the synchronizing (cf. Section 4.2) product coupling.

For the SA coupling functions, the following observation is useful: since the denominator in the integrand of (3.4) is always positive, if $h'(\theta) < 0$ for θ in the (essential) support of the positive function g then the integrand itself will always be *negative*, giving stability for any $\alpha > 0$. The plots in Figure 4.1 show that for delays t_d (which correspond to translations of g) taken in a wide range around 0.150 *sec*, stability holds for arbitrary g of reasonably compact essential support and *any* $\alpha, \beta > 0$ such that the in-phase periodic orbit exists. We note here that related stability conditions are derived in van Vreeswijk, Abbot, and Ermentrout [1994]; Ermentrout [1996]; Gerstner, van Hemmen, and Cowan [1996], in which the time course of g is also important.

4.3 Modeling Firing Rates in LC Modes

Here we show that the depressed firing rates characteristic of the phasic mode can be captured by our LC model if explicitly phase-dependent synaptic terms are included.

Phase-Difference Coupling

Under the assumption that the period of the in-phase state largely determines average LC firing rates, the depressed firing rate observed in the phasic mode cannot be reproduced with averaged (phase-difference), PR coupling functions. The period of the in-phase state of (4.1) is $\frac{2\pi}{\omega+\alpha f(0)}$, and Figure 4.1 shows that $f(0) > 0$ for any $\kappa > 0$, so that the period will always *decrease* as α increases. Hence, the (highly-coupled) phasic state would actually have an *increased* firing rate. In contrast, we shall see that (3.1) allows us to correctly reproduce observed firing rates.

Phase-Dependent Coupling

For small β, where averaging is valid, the period of the S_N symmetric orbit must decrease with β. However, Figure 4.3 also shows that for β sufficiently large, the period *increases* with β (the term $G(\theta)$ in (3.2) slows the flow). Since the S_N symmetric orbit is attracting, its increasing period indicates a mechanism for lower firing rates in the phasic mode. This shows the importance in this case of considering the explicit product coupling of (3.1); however, for other parameter values and neuron models phase-difference coupling may correctly capture trends in firing rates (cf. Brown, Moehlis, Holmes, Aston-Jones, and Clayton [2002]).

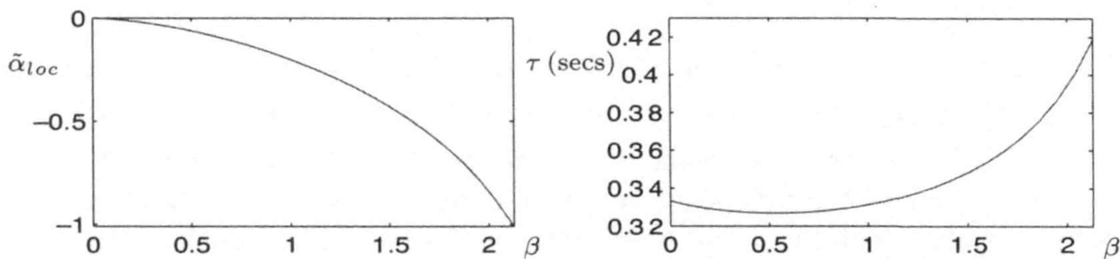

FIGURE 4.3. The β-dependence of (left) bounds for stability and (right) the period of the S_N symmetric orbit with PR coupling functions, from equations (3.8) and (3.3) in the large-N limit.

4.4 LC Response to Stimulus

Figures 1.1(b,c) show averaged histograms of the times at which spikes were recorded in LC neurons, in data accumulated from many stimulus-response trials. The phasic histogram displays a post-stimulus window in which the probability of a given LC neuron firing in an arbitrary trial is relatively high; this represents increased activity following stimulus. After this burst the histogram exhibits a 'refractory' period in which there is a lower firing probability, followed by a gradual return to equilibrium firing probability. We now indicate a mechanism by which this pattern can emerge from averages over many trials of oscillator dynamics with the appropriate initial conditions.

As noted in the Introduction, the LC response in the phasic mode is stronger than in the tonic mode. This effect can be shown to arise from the fact that phasic-mode LC neurons tend to have lower firing frequencies (Brown, Moehlis, Holmes, Aston-Jones, and Clayton [2002]). Here we only study phase-difference models based on equation (4.1) for the phasic-mode LC response. We assume that external stimuli act (via other neural groups) by increasing the input current to each LC neuron identically via a function $I(t)$, contributing a term $z(\theta)I(t)$ to the phase velocities. Approximating the phasic mode as perfectly synchronized so that the average $\frac{\alpha}{N}\sum_{j\neq i}^{n} f(\theta_j - \theta_i)$ may be replaced by the constant $f(0)$, equation (4.1) becomes

$$d\theta = \left[\omega + \alpha f(0) + z(\theta)I(t)\right] dt + \sigma dW_t \stackrel{\triangle}{=} v(\theta,t)dt + \sigma dW_t \ . \tag{4.2}$$

Initial conditions $\theta(0) = \theta_0$ are drawn from the appropriate distribution (see below).

The probability density of oscillator phases for this stochastic ODE (4.2) evolves according to the forward Kolmogorov equation

$$\frac{\partial \rho(\theta,t)}{\partial t} = -\frac{\partial}{\partial \theta}\left[v(\theta,t)\rho(\theta,t)\right] + \tfrac{1}{2}\sigma^2 \frac{\partial^2 \rho(\theta,t)}{\partial \theta^2} , \tag{4.3}$$

with $\rho(\theta, t = 0) = \delta(\theta - \theta_0)$. Due to the linearity of (4.3), histograms to be compared with Figure 1.1 may be produced from (4.3) with initial density representing an average over *many* trials, in which case $\rho(\theta,t)d\theta$ represents the probability that a given neuron observed in an arbitrary trial has phase in the interval $[\theta, \theta + d\theta)$ at time t. This 'density' approach to modeling experimental firing time histograms is also employed in Ritt and Kopell [2002]. The experimentally relevant initial condition ρ_0 may be found by reasoning as follows: if v lacks explicit t-dependence (i.e. $I(t) = 0$), then the probability that an oscillator obeying (4.2) is in $[\theta, \theta + d\theta]$ should scale with $1/T(\theta)$, where $T(\theta)$ is the time spent in this interval during one cycle. To lowest order in $d\theta$ and neglecting noise, this implies $\rho(\theta, t = 0) = C/v(\theta)$ for some constant C; normalization gives ρ_0 itself, which is simply $1/2\pi$ in this case. Histograms of firing times may be extracted from the solutions of (4.3) by noting that the firing probability (for an arbitrarily chosen neuron) at time t is proportional to the threshold probability flux $FL(t) \stackrel{\triangle}{=} v(\theta_s, t)\rho(\theta_s, t)$.

We can develop explicit solutions to (4.3) for certain classes of stimuli and PRCs in the absence of noise for comparison with experimental data; these analyses will appear in a future paper (Brown, Moehlis, Holmes, Aston-Jones, and Clayton [2002]). Figure 4.4 illustrates preliminary results relevant to the data of Figure 1.1(b,c) via numerical solutions of (4.3) and direct simulations of (4.1) (with identical stimuli), which show that $FL(t)$, derived from (4.3), provides a reasonable approximation for post-stimulus firing probabilities. We note that $FL(t)$ displays the characteristic peak and refractory period of Figures 1.1(b,c), but the subsequent return to equilibrium is slow and exhibits prolonged 'ringing' not seen in the experimental data. The ringing is due to a 'resonance' between the stimulus duration and the oscillator frequency; it is diminished or disappears entirely, giving histograms more similar to Figures 1.1(b,c), when data is averaged over a distribution of frequencies (Brown, Moehlis, Holmes, Aston-Jones, and Clayton [2002]).

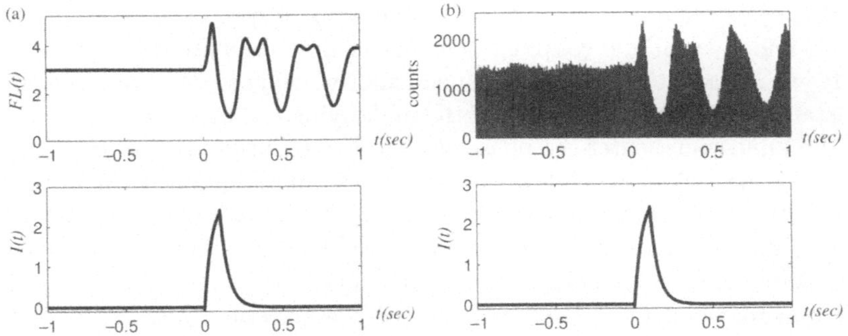

FIGURE 4.4. LC firing rates (proportional to the probability that a neuron sampled from the LC will fire at a particular time), computed using equation (4.3). Upper plot: (a) $FL(t)$, (b) firing time histogram for the corresponding finite-trial simulation, see text. Parameters: $N = 24$, $\kappa = 0$, $\alpha = 0.032$, $\sigma = 0.8$ (as in Figure 4.2(b)). Lower plot (a,b): $I(t)$ in $\mu A/cm^2$.

5 Conclusions

The dynamics of a finite set of identically (mean field) coupled oscillators were analyzed in general cases of phase-difference coupling and in specific cases of combined phase-difference and phase-dependent 'product' coupling. The existence of symmetric equilibria and fixed tori for $S_N \times T^1$-equivariant systems of coupled oscillators was demonstrated, as was the dependence of their stability and degeneracy on the Fourier content of the coupling functions. In particular, while single harmonic sine and cosine functions are degenerate in that they give rise to steady states with multiple zero eigenvalues, the inclusion of higher harmonics generically produces equilibria with only a single (necessarily) zero eigenvalue in the relative phase direction. Illustrative examples of such functions were derived from Hodgkin-Huxley neural models, and stability and domains of attraction of synchronized periodic orbits were investigated in detail.

This analysis of idealised oscillator networks guided the numerical simulations of Section 4, which demonstrate that networks of rotators can reproduce phenomena observed in firing patterns of the brain organ *locus coeruleus*. Specifically, the model suggests three mechanisms potentially responsible for broadening of the phasic cross correlogram (Figure 1.1(a)): multiple equilibria with distinct phase differences, random external effects, and randomly distributed natural frequencies (Figures 4.2(b-d)). The former is perhaps least likely, since it tends to give multi-peaked correlograms. In addition, relatively strong synaptic (phase product) coupling appears necessary to reproduce the depressed firing rates of the phasic state: averaged (phase-difference) synaptic coupling actually produced the opposite effect, and electrotonic coupling does not change in-phase firing rates. Finally, we note that these results depend subtly on synaptic delays and time constants, and thus that details of the underlying cell membrane and channel dynamics must enter the phase oscillator description. In fact in ongoing work (Brown, Moehlis, Holmes, Aston-Jones, and Clayton [2002]), we incorporate an additional slow calcium-dependent potassium current, leading to coupling functions differing from those of Figure 4.1, and, in some cases, remarkably close to the degenerate pure sinusoids.

A general point of contact between the present study and neuroscience lies in the relationship between explicitly modeling individual neurons and their couplings, and averaging the behavior of a (sub-)population into a single connectionist-type 'unit' (e.g. Rumelhart and McClelland [1986]). Domain of attraction and probability density results for synchronized states may inform conditions under which such approximations are justifiable. In the present paper, a simple probabilistic evolution (equation (4.3)) for coherent phase states produces acceptable results (Figure 4.4). Extension of this result to a true population average for systems with distributed frequencies and non-uniform couplings would bring it closer to the work of Sirovich, Knight, and Omurtag [2000], and help unify detailed (Hodgkin-Huxley type) neural models, simpler integrate-and-fire and phase models, and connectionist networks.

Acknowledgments: This work was partially supported by DoE: DE-FG02-95ER25238 and NIMH: MH62196 (Cognitive and Neural Mechanisms of Conflict and Control, Silvio M. Conte Center). Eric Brown was supported by a National Science Foundation Graduate Fellowship and a Burroughs-Wellcome Training Grant in Biological Dynamics: 1001782, and

Jeff Moehlis by a National Science Foundation Postdoctoral Research Fellowship. We thank Jason Ritt for assistance with incorporating the stimulus into the density approach as well as other helpful comments, Georgi Medvedev for useful suggestions and discussions, and Martin Golubitsky for an insightful and stimulating communication early in this project. Weinan E provided motivation for studying and insights about symmetric solutions in the $N \to \infty$ limit.

References

Arnold, V. [1973], *Ordinary Differential Equations.* MIT Press, Boston.

Ashwin, P. and J. Swift [1992], The dynamics of N weakly coupled identical oscillators, *J. Nonlin. Sci.* **2**, 69–108.

Aston-Jones, G., J. Rajkowski, and T. Alexinsky [1994], Locus coeruleus neurons in the monkey are selectively activated by attended stimuli in a vigilance task., *J. Neurosci.* **14**, 4467–4480.

Bressloff, P. and S. Coombes [1998], Spike train dynamics underlying pattern formation in integrate-and-fire oscillator networks, *Phys. Rev. Lett.* **81**, 2384–2387.

Brown, E., J. Moehlis, P. Holmes, G. Aston-Jones, and E. Clayton [2002], *The influence of spike rate on response in locus coeruleus*, Unpublished manuscript, Program in Applied and Computational Mathematics, Princeton University, 2002.

Chow, C. and N. Kopell [2000], Dynamics of spiking neurons with electrotonic coupling, *Neural Comp.* **12**, 1643–1678.

Crawford, J. [1995], Scaling and singularities in the entrainment of globally coupled oscillators, *Phys. Rev. Lett.* **74**, 4341–4344.

Crawford, J. and K. Davies [1999], Synchronization of globally coupled phase oscillators: singularities and scaling for general couplings, *Physica D* **125 (1-2)**, 1–46.

Daido, H. [1994], Generic scaling at the onset of macroscopic mutual entrainment in limit cycles with uniform all-to-all coupling, *Phys. Rev. Lett.* **73(5)**, 760–763.

E, W. [2001], *personal communication*, 2001.

Ermentrout, B. [1996], Type I membranes, phase resetting curves, and synchrony, *Neural Comp.* **8**, 979–1001.

Ermentrout, G. and N. Kopell [1990], Oscillator death in systems of coupled neural oscillators, *SIAM J. on Appl. Math.* **50**, 125–146.

Fenichel, N. [1971], Persistence and smoothness of invariant manifolds for flows, *Ind. Univ. Math. J.* **21**, 193–225.

Freidlin, M. and A. Wentzell [1998], *Random perturbations of dynamical systems.* Springer, New York.

Gerstner, W., L. van Hemmen, and J. Cowan [1996], What matters in neuronal locking?, *Neural Comp.* **8**, 1653–1676.

Golomb, D., D. Hansel, B. Shraiman, and H. Sompolinsky [1992], Clustering in globally coupled phase oscillators, *Phys. Rev. A* **45(6)**, 3516–3530.

Golubitsky, M., I. Stewart, and D. Schaeffer [1988], *Singularities and Groups in Bifurcation Theory, Vol. 2.* Springer, New York.

Grant, S., G. Aston-Jones, and D. Redmond [1988], Responses of primate locus coeruleus neurons to simple and complex sensory stimuli., *Brain Res. Bull.* **21 (3)**, 401–410.

Guckenheimer, J. and P. Holmes [1983], *Nonlinear Oscillations, Dynamical Systems and Bifurcations of Vector Fields.* Springer-Verlag, New York.

Hansel, D., G. Mato, and C. Meunier [1995], Synchrony in excitatory neural networks, *Neural Comp.* **7**, 307–337.

Hodgkin, A. and A. Huxley [1952], A quantitative description of membrane current and its application to conduction and excitation in nerve, *J. Physiol.* **117**, 500–544.

Honeycutt, R. [1992], Stochastic Runge-Kutta Algorithms I. White Noise, *Phys. Rev. A* **45**, 600–603.

Hoppensteadt, F. and E. Izhikevich [1997], *Weakly Connected Neural Networks.* Springer-Verlag, New York.

Izhikevich, E. [2000], Phase equations for relaxation oscillators, *SIAM J. on Appl. Math.* **60**, 1789–1804.

Johnston, D. and S. Wu [1997], *Foundations of Cellular Neurophysiology.* MIT Press, Cambridge, MA.

Keener, J. and J. Sneyd [1998], *Mathematical Physiology.* Springer, New York.

Kim, S. and S. Lee [2000], Phase dynamics in the biological neural networks, *Physica D* **288**, 380–396.

Kopell, N. and G. Ermentrout [1990], Phase transitions and other phenomena in chains of coupled oscillators, *SIAM J. on Appl. Math.* **50**, 1014–1052.

Kopell, N. and G. Ermentrout [1994], Inhibition-produced patterning in chains of coupled nonlinear oscillators, *SIAM J. Appl. Math.* **54**, 478–507.

Kopell, N., G. Ermentrout, and T. Williams [1991], On chains of osciallators forced at one end, *SIAM J. on Appl. Math.* **51**, 1397–1417.

Kuramoto, Y. [1984], *Chemical Oscillations, Waves, and Turbulence.* Springer, Berlin.

Kuramoto, Y. [1997], Phase- and center-manifold reductions for large populations of coupled oscillators with application to non-locally coupled systems, *Int. J. Bif. Chaos* **7**, 789–805.

Murray, J. [2001], *Mathematical Biology, 3rd. Ed.* Springer, New York.

Nichols, S. and K. Wiesenfeld [1992], Ubiquitous neutral stability of splay states, *Phys. Rev. A* **45(12)**, 8430–8435.

Okuda, K. [1993], Variety and generality of clustering in globally coupled oscillators, *Physica D* **63**, 424–436.

Omurtag, A., E. Kaplan, B. Knight, and L. Sirovich [2000a], A population approach to cortical dynamics with an application to orientation tuning, *Network* **11**, 247–260.

Omurtag, A., B. Knight, and L. Sirovich [2000b], On the simulation of large populations of neurons, *J. Comp. Neurosci.* **8**, 51–63.

Ritt, J. and N. Kopell [2002], *In preparation*, 2002.

Rumelhart, D. and J. McClelland [1986], *Parallel Distributed Processing: Explorations in the Microstructure of Cognition.* MIT Press, Cambridge, MA.

Servan-Schreiber, D., H. Printz, and J. Cohen [1990], A network model of catecholamine effects: Gain, signal-to-noise ratio, and behavior, *Science* **249**, 892–895.

Sirovich, L., B. Knight, and A. Omurtag [2000], Dynamics of neuronal populations: The equilibrium solution, *SIAM J. on Appl. Math.* **60**, 2009–2028.

Strogatz, S. [2000], From Kuramoto to Crawford: Exploring the onset of synchronization in populations of coupled oscillators, *Physica D* **143**, 1–20.

Taylor, D. and P. Holmes [1998], Simple models for excitable and oscillatory neural networks, *J. Math. Biol.* **37**, 419–446.

Tsang, K., R. Mirollo, and S. Strogatz [1991], Dynamics of a globally coupled oscillator array, *Physica D* **48**, 102–112.

Usher, M., J. Cohen, D. Servan-Schreiber, J. Rajkowski, and G. Aston-Jones [1999], The role of locus coeruleus in the regulation of cognitive performance, *Science* **283**, 549–554.

van Vreeswijk, C., L. Abbot, and B. Ermentrout [1994], When inhibition not excitation synchronizes neural firing, *J. Comp. Neurosci.* **1**, 313–321.

Watanabe, S. and S. Strogatz [1994], Constants of the motion for superconducting Josephson arrays, *Physica D* **74**, 195–253.

Watanabe, S. and J. Swift [1997], Stability of periodic solutions in series arrays of Josephson Junctions with internal capacitance, *J. Nonlin. Sci.* **7**, 503–536.

Wiesenfeld, K., P. Colet, and S. Strogatz [1998], Frequency locking in Josephson arrays: Connection with the Kuramoto model, *Phys. Rev. E* **57 (2)**, 1563–1569.

Wiggins, S. [1994], *Normally Hyperbolic Invariant Manifolds in Dynamical Systems.* Springer, New York.

Williams, J., R. North, A. Shefner, S. Nishi, and T. Egan [1984], Membrane properties of rat locus coeruleus neurons, *Neuroscience* **13**, 137–156.

Zhu, W. [1988], Stochastic averaging methods in random vibration, *Appl. Mech. Rev.* **41**, 189–199.

6

Recent Results in the Kinetic Theory of Granular Materials

Carlo Cercignani

To Larry Sirovich, on the occasion of his 70th birthday.

ABSTRACT A survey of the recent results concerning the Boltzmann equation for granular materials is given. In particular the model proposed by Bobylev et al. for Maxwell grains is discussed with particular attention to the possibility of obtaining closed form solutions. The fact that the kinetic equation for grains has different a priori estimates with respect to those for gas molecules and its consequences for existence proofs is also presented.

Contents

1 Introduction

Larry Sirovich made several contributions to the theory of the Boltzmann equation in the 1960's. I just recall here his work on kinetic modelling (Sirovich [1962]) and sound propagation in a rarefied gas (Sirovich and Thurber [1965a,b]). Since those years, the theory of the Boltzmann equation has made remarkable progress in both abstract theory and applications (Cercignani [1990, 1988]; Cercignani, Illner, and Pulvirenti [1994]; Cercignani [2000]). Regretfully for me, who had the pleasure of reading most of his work in this area, he moved to other fields and people working in kinetic theory lost the privilege of having him among their ranks. Since I have stubbornly continued to work on the Boltzmann equation, my contribution to this volume will be in this area.

Recently kinetic theory has been applied to the less microscopic dynamics of granular materials. The latter has undergone a notable development in the last few years, because of its growing importance in the applications (sands, powders, rock and snow avalanches, landslides, grains, fluidized beds). The problems related to the study of fast flows of grain materials, which arise, more and more frequently, in industrial processes and are of growing importance in the study of natural phenomena, have been the object of much attention and have been treated with various methods that differ in rigor and complexity. In the majority of these studies one adopts the assumption of one-dimensional flow and neglects the interaction between grains and air. The various methods applied to these simplified problems have also been used to model other important cases of collisional granular motion, including fluidized beds .

Some recent work in the field (Herrmann [1992]; Hutter and Rajagopal [1994]; Jaeger, Nagel, and Behringer [1996]) deals with the quasi-static flow regime, usually characterized by relatively large densities and prolonged contacts between the grains, as well as more than two-body interactions. Here we are interested in the rapid granular flow (Campbell [1990]; Lun [1991]). The methods used in this regime include: i) Development of physical and experimental models; ii) computer simulations; iii) kinetic theory.

Actually many recent studies are based on the assumption that, in certain conditions of motion, collisions between particles supply the main mechanism of momentum and energy exchange. This assumption spontaneously suggests an analogy with the kinetic theory of gases. In the latter theory the particles are of course molecules and there are thus essential differences between the two situations, that must be duly taken into account. In particular, the intermolecular collisions are frequently elastic, whereas this is not a reasonable assumption when dealing with particles of a granular material, which is accordingly modelled as a system of hard spheres undergoing dissipative collisions. This has two consequences: the medium we are describing does not conserve energy and entropy is not necessarily increasing when the system is isolated. We should emphasize that one is only looking at the motion of grains when discussing energy and entropy, but ignoring energy and entropy on the atomic level within individual grains.

Equations derived for rapid granular flows may be of some use in the quasi-static flow regime, since the dissipative nature of the particle interactions is a feature common to all regimes of granular flow. A direct analysis based on the Boltzmann equation is due to Goldshtein and Shapiro [1995]; a more systematic approach has been provided by Sela and Goldhirsch [1998].

The evolution equation for the system of the aforementioned particles undergoing interparticle inelastic collisions contains a Boltzmann-like collision term. Hence it is an integro-differential equation to be solved with suitable boundary conditions.

In the present paper, following previous work (Bobylev, Carrillo and Gamba [2000]; Carrillo, Cercignani and Gamba [2000]), we shall also consider pseudo-Maxwellian particles approximating dissipative hard spheres, with the aim of describing the shear flow of a granular material between two parallel plates.

The paper is organized as follows. In Section 2 we recall our basic equation, and give its model version for Maxwell grains. In Section 3 we obtain the form of this equation for homoenergetic flows of a granular material and briefly hint at the possibility of proving an existence theorem for this equation. In Section 4 we derive exact equations for the second order moments and show that these equations possess an exact steady solution. This produces a constitutive equation showing that the shear stress is proportional to the square of the strain rate, as is usually found in fluid approximations of the kinetic theory of granular materials; this is the only case, as far as we know, when this result is obtained from an exact solution. In Section 5 we compare the exact solution for Maxwell particles with an approximate solution for hard sphere grains, in order to examine some possible deficiencies of the Maxwell model. In Section 6 we discuss the main differences between the a priori inequalities required in the existence theories in the cases of gases and granular materials.

2 The Kinetic Equation for Granular Materials

Let us denote by $f(\mathbf{x}, \mathbf{v}, t)$ the distribution function of the grains ($\mathbf{x} \in \mathbf{R}^3$, $\mathbf{v} \in \mathbf{R}^3$, $t \in \mathbf{R}_+$ being the position and velocity of a grain).

For smooth spherical grains of diameter d, f satisfies the inelastic Boltzmann equation, which reads as follows:

$$\frac{\partial f}{\partial t} + \mathbf{v} \cdot \frac{\partial f}{\partial \mathbf{x}} = Q(f, f)$$

where

$$Q(f, f) = \mathsf{d}^2 \int_{\Re^3 \times \mathbf{S}^2} d\mathbf{w} d\mathbf{n} (\mathbf{V} \cdot \mathbf{n})_+ \left[\frac{1}{e} J f({}^{\backprime}\mathbf{v}) f({}^{\backprime}\mathbf{w}) - f(\mathbf{v}) f(\mathbf{w}) \right]$$

Here $\mathbf{V} = \mathbf{v} - \mathbf{w}$ is the relative velocity, and $\mathbf{v}'$ and $\mathbf{w}'$ are the *precollisional* velocities (note the position of the (inverted) prime) related to $\mathbf{v}$ and $\mathbf{w}$ by

$$\begin{aligned} {}^{\backprime}\mathbf{v} &= \tfrac{1}{2}(\mathbf{v} + \mathbf{w}) + \frac{1-e}{4}(\mathbf{v} - \mathbf{w}) + \frac{1+e}{4}|\mathbf{v} - \mathbf{w}|\mathbf{n} \\ {}^{\backprime}\mathbf{w} &= \tfrac{1}{2}(\mathbf{v} + \mathbf{w}) - \frac{1-e}{4}(\mathbf{v} - \mathbf{w}) - \frac{1+e}{4}|\mathbf{v} - \mathbf{w}|\mathbf{n} \end{aligned}$$

and J is the Jacobian of the transformation:

$$J = \frac{1}{e^2}\frac{|\mathbf{v}-\mathbf{w}|}{|\mathbf{v}_* - \mathbf{w}_*|} \tag{2.1}$$

Here $0 < e \leq 1$ is the so-called restitution coefficient ($e = 1$ for elastic collisions).

As suggested by Bobylev, Carrillo and Gamba [2000], to simplify the mathematics, we shall also consider a system of pseudo-Maxwellian particles, satisfying the following Boltzmann equation:

$$\frac{\partial f}{\partial t} + \mathbf{v}\cdot\frac{\partial f}{\partial \mathbf{x}} = B(\rho,\theta)Q(f,f)\,.$$

The explicit form of the collision term is given by the following formulas:

$$Q(f,f) = \frac{1}{4\pi}\int_{\mathbf{R}^3}\int_{\mathbf{S}^2}[f(t,{}^{\backprime}\mathbf{v})f(t,{}^{\backprime}\mathbf{w})J - f(t,\mathbf{v})f(t,\mathbf{w})]\,d\mathbf{n}\,d\mathbf{w}\,.$$

Here and in the following we use the notation:

$$\rho = \int_{\mathbf{R}^3} f(\mathbf{v},t)d\mathbf{v}\,,\quad \rho\mathbf{u} = \int_{\mathbf{R}^3}\mathbf{v}f(\mathbf{v},t)d\mathbf{v}\,,$$
$$3\rho\theta = \int_{\mathbf{R}^3}|\mathbf{v}-\mathbf{u}|^2 f(\mathbf{v},t)d\mathbf{v}\,,$$

where $\rho \in \mathbf{R}_+$, $\mathbf{u} \in \mathbf{R}^3$, and $\theta \in \mathbf{R}_+$ are the density, the bulk velocity and the temperature of the granular material. The coefficient $B(\rho,\theta)$ is given by

$$B(\rho,\theta) = B(\rho)\sqrt{\theta}\,,$$

where $B(\rho)$ is a given positive function of ρ. This function can be taken to be a constant if the proper volume of all the grains is negligible compared to the volume occupied by the granular material.

3 The Evolution Equation for Homoenergetic Flows

We now consider a granular material with an average density ρ_0, in motion between two parallel plates located at $x = 0$ and $x = L$. The upper plate moves with velocity V, the lower one is at rest.

We look for self-similar solutions: if we cut the slab at $x = L'$ and put a wall there, the solution between $x = 0$ and $x = L'$ remains the same, if the plate at $x = L'$ has now velocity $V' = VL'/L$. Then the basic parameter must be $K = V/L$.

We assume a bounce-back boundary condition in the reference frame of each plate:

$$f(t,0,y,\mathbf{v}) = f(t,0,y,-\mathbf{v}),$$
$$f(t,L,y,\mathbf{v}) = f(t,L,y,2V\mathbf{j}-\mathbf{v}).$$

Following a recent paper (Cercignani [2001]), we look for solutions such that $\mathbf{x}$ appears in f only through the bulk velocity $\mathbf{u}$

$$f = f(\mathbf{c}, t)$$

where $\mathbf{c} = \mathbf{v} - \mathbf{u}$ is the random velocity. These solutions turn out to be homoenergetic, *i.e.* the energy density is constant in space.

The book by Truesdell and Muncaster [1980] gives a unified discussion of homoenergetic affine flows for a general medium. The defining properties read as follows:

a) The body force (per unit mass) $\mathbf{X}$ acting on the particles is constant:

$$\mathbf{X} = \text{const.};$$

b) The central moments

$$p_{i_1 i_2 \dots i_n} = \int_{\mathbf{R}^3} c_{i_1} c_{i_2} \dots c_{i_n} f \, d\mathbf{c} \qquad (\text{for } i_k = 1,2,3)$$

are space-homogeneous;

c) The bulk velocity $\mathbf{u}$ is an affine function of position $\mathbf{x}$:

$$\mathbf{u} = \mathsf{K}(t)\mathbf{x} + \mathbf{u}_0(t).$$

An analysis of the momentum balance equation based on a), b) and c) immediately leads to the following restrictions on K and $\mathbf{u}_0$:

$$\begin{aligned} \dot{\mathsf{K}} + \mathsf{K}^2 &= 0, \\ \dot{\mathbf{u}}_0 + \mathsf{K}\mathbf{u}_0 &= \mathbf{X}. \end{aligned} \tag{3.1}$$

The general solution of this system is:

$$\begin{aligned} \mathsf{K}(t) &= [\mathsf{I} + t\mathsf{K}(0)]^{-1}\mathsf{K}(0) \\ \mathbf{u}_0(t) &= [\mathsf{I} + t\mathsf{K}(0)]^{-1}\left[\mathbf{u}_0(0) + t\mathbf{X} + \tfrac{1}{2}t^2\mathsf{K}(0)\mathbf{X}\right] \end{aligned} \tag{3.2}$$

where I is the 3×3 identity matrix. The solution of these equations exists globally for $t > 0$ if the eigenvalues of $\mathsf{K}(0)$ are nonnegative; otherwise the solution ceases to exist for $t = t_0$, where $-t_0^{-1}$ is the largest, in absolute value, among the negative eigenvalues of $\mathsf{K}(0)$.

In particular, if

$$[\mathsf{K}(0)]^2 = 0\,,$$

then $[\mathsf{I} + t\mathsf{K}(0)]^{-1} = \mathsf{I} - t\mathsf{K}(0)$ and therefore $\mathsf{K}(t)$ is independent of time. $\mathbf{u}$ is then steady if and only if

$$\mathsf{K}(0)\mathbf{X} = \mathbf{0}\,,$$

and $\mathbf{u}_0(0)$ is chosen in such a way that

$$\mathsf{K}(0)\mathbf{u}_0(0) = \mathbf{X}\,.$$

In particular, this is always possible if the body force vanishes, $(\mathbf{X} = \mathbf{0})$ as we have implicitly assumed.

The matrix K has a vanishing square if and only if a coordinate system exists for which the matrix representation of $\mathsf{K}(0)$ is given by

$$((K_{ij})) = \begin{pmatrix} 0 & 0 & 0 \\ K & 0 & 0 \\ 0 & 0 & 0 \end{pmatrix}\,.$$

For a proof see a previously quoted paper (Cercignani [1989]). When this applies one talks, for obvious reasons, of a homoenergetic shear flow. The density is not only space-homogeneous but also constant in time for this flow.

Solutions of the Boltzmann equation for homoenergetic affine flows exist, even for models more general than Maxwell's. In particular one can easily show that $f(\mathbf{c}, t)$ must satisfy the following equation

$$\frac{\partial f}{\partial t} - \frac{\partial f}{\partial \mathbf{c}} \cdot \mathsf{K}\mathbf{c} = Q(f, f)\,,$$

In this case one should prove that this equation admits an existence theorem. This was proved for energy conserving collisions (Cercignani [1989]), but the result should hold for the case of granular materials, albeit modifications in the proof are certainly required.

4 The Moment Equations and Their Form for Homoenergetic Flows

The moment equations do not form, in general, a closed system. To find a closed form solutions there are two ways out:

1. Use the pseudo-Maxwell grains (Cercignani [2001]).

2. Use a Gaussian approximation for f (Bobylev, Groppi, and Spiga [2001]).

It is well-known that the moments of distribution function are particularly useful if we assume (pseudo-)Maxwell particles, first introduced for granular materials by Bobylev, Carrillo and Gamba [2000]. This occurs because when we try to form the equations satisfied by moments, the contribution from the collision term contains a finite number of moments of order not higher than the order of the moment appearing in the time derivative term in the Boltzmann equation. This is a property discovered by Maxwell and characteristic of the particles named after him.

As shown by Galkin [1956] and Truesdell [1956], the second order moment equations for a Maxwell gas, associated with a homoenergetic affine flow, decouple from those of higher order and can be solved explicitly.

In order to obtain the moment equations, one multiplies by the appropriate monomial $c_{i_1}c_{i_2}\dots c_{i_n}$ and integrate over velocity space. The term with the derivative w.r.t. the velocity variables must be handled by partial integration; the collision term is complicated unless we use Maxwell particles.

The most interesting moment equations are those for $p_{22} = \int c_1^2 f d\mathbf{v}$, $p_{12} = \int c_1 c_2 f d\mathbf{v}$, $p = \frac{1}{3}\int |\mathbf{c}|^2 f d\mathbf{v} = \rho\theta$:

$$\begin{aligned} \dot{p} + \tfrac{2}{3}Kp_{12} + \epsilon(1-\epsilon)B(\rho)\sqrt{\theta}p &= 0\,, \\ \dot{p}_{12} + \tfrac{1}{2}(1-\epsilon^2)B(\rho)\sqrt{\theta}p_{12} + Kp_{11} &= 0\,, \\ \dot{p}_{11} + \tfrac{1}{2}(1-\epsilon^2)B(\rho)\sqrt{\theta}(p_{11}-p) &= 0\,. \end{aligned}$$

These equations form a system of three nonlinear first order differential equations that possesses a steady solution with $p_{11} = p$ and

$$\begin{aligned} p_{12} &= -\tfrac{3}{2}\frac{1}{K}\epsilon(1-\epsilon)B(\rho)\sqrt{\theta}p \\ \theta &= \tfrac{4}{3}\frac{1}{(1-\epsilon^2)\epsilon(1-\epsilon)[B(\rho)]^2}K^2 \end{aligned}$$

and hence

$$p_{12} = -\tfrac{3}{2}\frac{1}{K}\epsilon(1-\epsilon)B(\rho)\rho\theta^{\frac{3}{2}}.$$

Hence

$$p_{12} = -\eta(\epsilon)\rho[B(\rho)]^{-2}K|K|$$

where $\eta(\epsilon)$ is a constant which, of course, tends to infinity when ϵ vanishes. Also the "temperature" of the granular material θ reaches a well defined value proportional to K^2.

5 Gaussian Approximation for Hard Spheres

Here we recall the approximate method introduced by Bobylev, Groppi, and Spiga [2001]. One assumes that f is given by an anisotropic Gaussian

$$f(\mathbf{v}) = \frac{1}{\sqrt{(2\pi)^3 \det \mathsf{P}}} \exp\left(-\tfrac{1}{2}\mathbf{c} \cdot \mathsf{P}^{-1}\mathbf{c}\right),$$

where

$$\mathsf{P} = \int_{\Re^3} dv\, f(\mathbf{v})\mathbf{c} \otimes \mathbf{c}\,.$$

We remark that, for any $\mathbf{v} \neq \mathbf{0}$, we have:

$$\mathbf{c} \cdot \mathsf{P}\mathbf{c} > 0, \qquad \mathrm{Tr}\mathsf{P} = p_{11} + p_{22} + p_{33} = 3p\,.$$

We also have:

$$p_{3j} = p_{33}\delta_{3j} = p_{j3}\,.$$

The resulting moment equations must be solved numerically.

The solution indicates a remarkable degeneracy of the pseudo-Maxwell model, which yields:

$$p_{11} = p_{33}\,,$$

whereas for hard spheres one can fit numerical calculations with

$$\begin{aligned} p_{11} - p_{33} &= A(1-\theta) + B(1-\theta)^2\,, \\ \theta &= \frac{1-\beta}{1+\beta} = \frac{1+e}{3-e}\,, \\ A &= -0.1851 \qquad B = 0.1253\,. \end{aligned}$$

Thus the comparison between the exact solution of the pseudo-Maxwell model and the solution for hard sphere grains shows that the idea to model the grains as (pseudo)-Maxwellian particles, which is so attractive because of the resulting analytic simplicity, has a serious drawback. The stresses are different; it remains to be seen whether this is just a quantitative effect or can also have important inaccuracies from the qualitative viewpoint as well. Numerical simulations might shed light on this issue.

We end the discussion of the shear flow problem by a remark: the results in the paper are obtained using a constant velocity gradient. This has the consequence that all the central moments of the distribution function (including the density) are uniform.

On the other hand, most experimentally observed granular flows exhibit non-uniform densities (e.g. the clustering instability of a freely cooling granular material, which can be observed both experimentally and numerically) and nonlinear velocity gradients. A nonlinear velocity profile would yield qualitatively different results, but it does not seem to be possible to weaken the assumptions in the theory and still have an exact solution. An approximately uniformly dense flow with constant velocity gradients could be achieved if we could have adiabatic walls, *i.e.* walls which exchange momentum but do not exchange energy with the granular gas (this is the case for the bounce-back boundary conditions used before).

6 A Priori Inequalities and Existence Theory

The existence theory for general (space inhomogeneous) solutions of the standard Boltzmann equation has undergone a notable development after the paper by DiPerna and Lions [1989], who proved that there are weak solutions of the Cauchy problem for a modified form of the Boltzmann equation, which they called 'renormalized'. Renormalized solutions satisfy the Boltzmann equation divided by $1 + f$ (or an analogous function of f, capable of producing at most a linear growth for renormalized collision term); if we were looking for a classical (strong) solution of the Boltzmann equation, then this renormalization does not change the set of solutions; it may, however, when we look for weak solutions.

In order to prove existence, DiPerna and Lions [1989] constructed a sequence of smooth approximations to a solution and extracted a weakly converging sequence. Renormalization was required in order to show that the weak limit is actually a solution of the equation. Since the unit sphere of L^1 is not compact, the fact that the L^1 norm of the functions of the approximating sequence is uniformly bounded by its initial value is not enough to have a weakly converging sequence: one needs also energy conservation to control the high speed behavior and Boltzmann's H-theorem to avoid singularity concentrations capable of producing a measure rather than an L^1 function in the limit. In fact, this theorem leads to the uniform boundedness of $\int f \log f d\mathbf{x}d\mathbf{v}$.

When we pass to the case of granular materials, the lack of energy conservation is not a problem. It is actually an advantage; in fact, since energy cannot grow, we can still bound it by its initial value and in addition we have an extra a priori bound from the energy source:

$$\int_{\Re^3 \times \Re^3 \times \mathbf{S}^2} d\mathbf{v}\, d\mathbf{w}\, d\mathbf{n}\, (\mathbf{V} \cdot \mathbf{n})_+ |\mathbf{V}|^2 f(\mathbf{v})\, f(\mathbf{w}) < K\,,$$

where K only depends on initial data.

On the other hand, we no longer have the H-theorem. In fact, if we try to form an evolution equation for $\int f \log f d\mathbf{x}d\mathbf{x}$, we find that the source due to collisions (the so-called entropy source) is proportional to

$$D(f) = \int_{\Re^3\times\Re^3\times\mathbf{S}^2} d\mathbf{v}\, d\mathbf{w}\, d\mathbf{n}\, (\mathbf{V}\cdot\mathbf{n})_+ f(\mathbf{v})f(\mathbf{w}) \log\Big[\frac{f(\mathbf{v}')f(\mathbf{w}')}{f(\mathbf{v})f(\mathbf{w})}\Big],$$

where $\mathbf{v}'$ and $\mathbf{w}'$ are the *post-collisional* velocity, which, at variance with the case of a molecular gas, differ from the pre-collisional ones, $'\mathbf{v}$ and $'\mathbf{w}$. It is not possible to show that $D(f)$ is non-positive and thus that $\int f \log f d\mathbf{x} d\mathbf{x}$ is bounded by its own initial value. The best possible estimate seems to be (Panferov [2001]):

$$D(f) \le \frac{1-e^2}{e^2} \int_{\Re^3\times\Re^3\times\mathbf{S}^2} d\mathbf{v} d\mathbf{w} d\mathbf{n} (\mathbf{V}\cdot\mathbf{n})_+ f(\mathbf{v})f(\mathbf{w}),$$

which is obtained via the elementary inequality $\log x \le x - 1$ by letting $x = [f(\mathbf{v}')f(\mathbf{w}')]/[f(\mathbf{v})f(\mathbf{w})]$.

Now, in general we cannot prove that the integral in the right hand side of the last inequality is bounded, even for a finite time, in terms of the initial data and thus it is not clear how to carry out the program of extending the proof by DiPerna and Lions to the case of granular material, just because the first important step, extraction of a sequence weakly converging to a candidate solution, cannot be performed. A way out of this situation is to cutoff small values of the relative speed V (Cercignani, Gamba and Panferov [2002]); then the bound of the energy source can be used to bound the entropy source as well. In this case the strong bound now available can even be used to prove that the solution is actually a weak solution of the Boltzmann equation (in the usual sense) and not just a renormalized one (Cercignani, Gamba and Panferov [2002]).

7 Concluding Remarks

The kinetic theory of granular materials is reaching now a mature stage, where one can see new interesting problems arising. These are not just small modifications of problems in the kinetic theory of gases, because the lack of both conservation of energy and growth of entropy produce a major modification in the mathematical structure. Here we have given a few examples of problems which may be fruitfully studied: in particular, by using the model proposed by Bobylev, Carrillo and Gamba [2000] for Maxwell grains is discussed with particular attention to the possibility of obtaining closed form solutions; we have also seen how the different a priori estimates, due to the aforementioned change of structure, have important consequences for existence proofs.

Acknowledgments: The research described in the paper was performed in the frame of European TMR (contract n. ERBFMRXCT970157) and was also partially supported by MURST of Italy.

References

Bobylev, A. V., J. A. Carrillo, and I. A. Gamba [2000], On some kinetic properties and hydrodynamics equations for inelastic interaction, *J. Stat. Phys.* **98**, 743-773.

Bobylev, A.V., Groppi, M. and Spiga G. [2001], Approximate solutions to the problem of stationary shear flow of granular material, Submitted to *European Journal of Mechanics B/Fluids.*

Campbell, C. S. [1990], Rapid granular flows, *Ann. Rev. Fluid Mech.* **22**, 57–92.

Carrillo, J. A., C. Cercignani and I .A. Gamba [2000], Steady states of a Boltzmann equation for driven granular media, *Phys. Rev. E*, **62**, 7700-7707.

Cercignani, C. [1988], *The Boltzmann Equation and Its Applications*, Springer-Verlag, New York.

Cercignani, C. [1989], Existence of homoenergetic affine flows for the Boltzmann equation, *Archive for Rational Mechanics and Analysis* **105**, 377-387.

Cercignani, C. [1990], *Mathematical Methods in Kinetic Theory*, Plenum Press, New York (2nd edition).

Cercignani, C. [2000], *Rarefied gas dynamics. From basic concepts to actual calculations*, Cambridge University Press, Cambridge.

Cercignani, C. [2001], Shear flow of a granular material, *Jour. Stat. Phys.*, **102**, 1407-1415.

Cercignani, C., I .A. Gamba and V. Panferov [2002], to appear.

Cercignani, C., R. Illner and M. Pulvirenti [1994], *The Mathematical Theory of Dilute Gases*, Springer-Verlag, New York.

DiPerna, R. and P. L. Lions [1989], On the Cauchy problem for Boltzmann equations: Global existence and weak stability, *Ann. of Math.* **130**, 321–366.

Galkin, V. S. [1956], On a solution of the kinetic equation. PMM (in Russian) **20**, 445-446.

Goldshtein, A. and M. Shapiro [1995], Mechanics of collisional motion of granular materials. Part I. General hydrodynamic equations, *J. Fluid Mech.* **282**, 75–114.

Herrmann, H. J. [1992], Simulation of granular media, *Physica A* **191**, 263–276.

Hutter, K., and K. R. Rajagopal [1994], On flows of granular materials, *Continuum Mech. Thermodyn.* **6**, 81–139.

Jaeger, H. N., S. R. Nagel, and R. P. Behringer [1996], The physics of granular materials, *Physics Today* 32–38.

Lun, C. K. K. [1991], Kinetic theory for granular flow of dense, slightly inelastic, slightly rough spheres, *J. Fluid Mech.* **223**, 539–559.

Panferov, V. [2001], Private communication.

Sela, N., and I. Goldhirsch [1998], Hydrodynamic equations for rapid flows of smooth inelastic spheres, to Burnett order, *J. Fluid Mech.* **361**, 41–74.

Sirovich, L. [1962], Kinetic modeling of gas mixtures, Phys. Fluids **5**, 908–918.

Sirovich, L. and J. K. Thurber [1965a], Propagation of forced sound waves in rarefied gas-dynamics, *J. Acoust. Soc. Am.* **37**, 329-339.

Sirovich, L. and J. K. Thurber [1965b], Comparison of theory and experiment for forced sound-wave propagation in rarefied gasdynamics: reply to comments on 'Propagation of forced sound-waves in rarefied gasdynamics', *J. Acoust. Soc. Am.* **38**, 478-480.

Truesdell, C. [1956], On the pressures and the flux of energy in a gas according to Maxwell's kinetic theory, II, *Journal of Rational Mechanics and Analysis* **5**, 55-128.

Truesdell, C. and R. G. Muncaster [1980], *Fundamentals of Maxwell's Kinetic Theory of a Simple Monatomic Gas*, Academic Press, New York.

7

Variational Multisymplectic Formulations of Nonsmooth Continuum Mechanics

R. C. Fetecau
J. E. Marsden
M. West

To Larry Sirovich, on the occasion of his 70th birthday.

ABSTRACT This paper develops the foundations of the multisymplectic formulation of *nonsmooth* continuum mechanics. It may be regarded as a PDE generalization of previous techniques that developed a variational approach to collision problems. These methods have already proved of value in computational mechanics, particularly in the development of asynchronous integrators and efficient collision methods. The present formulation also includes solid–fluid interactions and material interfaces and, in addition, lays the groundwork for a treatment of shocks.

Contents

1 Introduction

There has been much interest lately in using variational methods for computational mechanics, such as Kane, Repetto, Ortiz, and Marsden [1999], Kane, Marsden, Ortiz, and West [2000] and Pandolfi, Kane, Marsden, and Ortiz [2002]. These variational methods have the attractive property that one can give the precise sense in which the algorithms used preserve the mechanical structure. This sense is a natural consequence of the variational structure of the algorithms and involves the symplectic and multisymplectic character of the algorithm. We refer the reader to the survey Marsden and West [2001], references therein, as well as the references in the following paragraphs for additional details about the general setting and properties of variational integrators.

There have been two developments in this area that bear directly on the present work. First of all, variational collision algorithms for finite dimensional problems (such as the collision of rigid bodies) have been developed that also share the conservation properties of smooth variational integrators; for example, the paper of Fetecau, Marsden, Ortiz, and West [2002] shows that the collision algorithms are symplectic, including the dynamics through the collision.

Second, in the area of variational algorithms for PDE's, a basic work was that of Marsden, Patrick, and Shkoller [1998] which laid the foundations of the method. The continuous part of these techniques were applied to the context of continuum mechanics in Marsden, Pekarsky, Shkoller, and West [2001]. This approach to continuum mechanics and discrete multisymplectic mechanics was developed further in Lew, Marsden, Ortiz, and West [2002], which introduced the notion of AVI's (asynchronous variational integrators). AVI's allow one to spatially and temporally adapt the algorithm and still retain its variational and multisymplectic character. This technique was also shown to be efficient computationally for two- and three-dimensional elasticity.

The purpose of the present paper is to combine the ideas from the continuous variational collision theory with the continuous part of multisymplectic theory of continuum mechanics. The result is a variational theory for PDE's of the sort arising in continuum mechanics that allow for material interfaces, elastic collisions, shocks and fluid–solid interactions. We shall not attempt to give a full account of all the literature in this area as it is extensive and complex, but we do mention the important works of Moreau [1982, 1986, 1988] that bear on the topics treated here.

In §5.2 we classify a collection of nonsmooth dynamic models that we will study. In this paper we do not address the discrete and algorithmic aspects of this theory. In fact, we plan to merge this work with variational algorithms and discrete mechanics (such as the work on AVI's) in future publications.

2 Multisymplectic Geometry

In this section we will review some aspects of basic covariant field theory in the framework of multisymplectic geometry. The multisymplectic framework is a PDE generalization of classical non-relativistic mechanics (or particle mechanics) and has diverse applications, including to electromagnetism, continuum mechanics, gravity, bosonic strings, etc.

The traditional approach to the multisymplectic geometric structure closely follows the derivation of the canonical symplectic structure in particle mechanics. The derivation first defines the field theoretic analogues of the tangent and cotangent bundles (called the first jet bundle and the dual jet bundle, respectively). It then introduces a canonical multisymplectic form on the dual jet bundle and pulls it back to the Lagrangian side using the covariant Legendre transform. As an alternative, Marsden, Patrick, and Shkoller [1998] gave a very elegant approach of deriving the multisymplectic structure by staying entirely on the Lagrangian side, which we will use here.

We start by reviewing the main concepts of the multisymplectic field-theoretic setting.

Let X be an oriented manifold, which in many examples is spacetime, and let $\pi_{XY} : Y \to X$ be a finite-dimensional fiber bundle called the ***covariant configuration bundle***. The physical fields will be sections of this bundle, which is the covariant analogue of the configuration space in classical mechanics.

The role of the tangent bundle is played by J^1Y (or $J^1(Y)$), the ***first jet bundle*** of Y. We identify J^1Y with the *affine* bundle over Y whose fiber over $y \in Y_x = \pi_{XY}^{-1}(x)$ consists of those linear maps $\gamma : T_xX \to T_yY$ satisfying

$$T\pi_{XY} \circ \gamma = \mathrm{Id}_{T_xX}\,.$$

We let $\dim X = n+1$ and the fiber dimension of Y be N. Coordinates on X are denoted $x^\mu, \mu = 1, 2, \ldots, n, 0$, and fiber coordinates on Y are denoted by $y^A, A = 1, \ldots, N$. These induce coordinates $v^A{}_\mu$ on the fibers of J^1Y. For a section $\varphi : X \to Y$, its tangent map at $x \in X$, denoted $T_x\varphi$, is an element of $J^1Y_{\varphi(x)}$. Thus, the map $x \mapsto T_x\varphi$ is a local section of J^1Y regarded as a bundle over X. This section is denoted $j^1(\varphi)$ or $j^1\varphi$ and is called the first jet of φ. In coordinates, $j^1(\varphi)$ is given by

$$x^\mu \mapsto (x^\mu, \varphi^A(x^\mu), \partial_\nu\varphi^A(x^\mu))\,, \tag{2.1}$$

where $\partial_\nu = \frac{\partial}{\partial x^\nu}$.

We will study Lagrangians defined on J^1Y and derive the Euler-Lagrange equations by a procedure similar to that used in Lagrangian mechanics on the tangent bundle of a configuration manifold (see Marsden and Ratiu [1999]). We thus consider theories for which Lagrangians depend at most on the fields and their *first* derivatives (first order field theories). For a geometric-variational approach to second-order field theories we refer the reader to Kouranbaeva and Shkoller [2000].

Higher order jet bundles of Y, J^mY, can be defined as $J^1(\cdots(J^1(Y)))$ and are used in the higher order field theories. In this paper we will use only J^1Y and a specific subbundle Y'' of J^2Y which we will define below.

Let $\gamma \in J^1Y$ so that $\pi_{X,J^1Y}(\gamma) = x$. Analogous to the tangent map of the projection π_{Y,J^1Y}, $T\pi_{Y,J^1Y} : TJ^1Y \to TY$, we may define the jet map of this projection which takes J^2Y onto J^1Y:

$$J\pi_{Y,J^1Y} : \mathrm{Aff}(T_xX, T_\gamma J^1Y) \to \mathrm{Aff}(T_xX, T\pi_{Y,J^1Y} \cdot T_\gamma J^1Y)\,.$$

We define the subbundle Y'' of J^2Y over X which consists of second-order jets so that on each fiber

$$Y''_x = \{s \in J^2Y_\gamma \mid J\pi_{Y,J^1Y}(s) = \gamma\}\,. \tag{2.2}$$

In coordinates, if $\gamma \in J^1Y$ is given by $(x^\mu, y^A, v^A{}_\mu)$, and $s \in J^2Y_\gamma$ is given by $(x^\mu, y^A, v^A{}_\mu, w^A{}_\mu, k^A{}_{\mu\nu})$, then s is a second order jet if $v^A{}_\mu = w^A{}_\mu$. Thus, the second jet of a section φ, $j^2(\varphi)$, given in coordinates by the map $x^\mu \mapsto (x^\mu, \varphi^A, \partial_\nu\varphi^A, \partial_\mu\partial_\nu\varphi^A)$, is an example of a second-order jet.

Next we introduce the field theoretic analogue of the cotangent bundle. We define the ***dual jet bundle*** J^1Y^* to be the *vector* bundle over Y whose fiber at $y \in Y_x$ is the set of affine maps from J^1Y_y to $\Lambda^{n+1}(X)_x$, the bundle of $(n+1)$-forms on X. A smooth section of J^1Y^* is therefore an affine bundle map of J^1Y to $\Lambda^{n+1}(X)$ covering π_{XY}.

Fiber coordinates on J^1Y^* are $(p, p_A{}^\mu)$, which correspond to the affine map given in coordinates by

$$v^A{}_\mu \mapsto (p + p_A{}^\mu v^A{}_\mu) d^{n+1}x\,, \tag{2.3}$$

where

$$d^{n+1}x = dx^1 \wedge \cdots \wedge dx^n \wedge dx^0\,.$$

Analogous to the canonical one- and two-forms on a cotangent bundle, there are canonical $(n+1)$- and $(n+2)$-forms on the dual jet bundle J^1Y^*. We will omit here the intrinsic definitions of these canonical forms (see Gotay, Isenberg, and Marsden [1997] for details). In coordinates, with $d^nx_\mu = \partial_\mu \,\lrcorner\, d^{n+1}x$, these forms are given by

$$\Theta = p_A{}^\mu dy^A \wedge d^nx_\mu + p d^{n+1}x \tag{2.4}$$

and

$$\Omega = dy^A \wedge dp_A{}^\mu \wedge d^nx_\mu - dp \wedge d^{n+1}x\,. \tag{2.5}$$

A ***Lagrangian density*** $\mathcal{L} : J^1Y \to \Lambda^{n+1}(X)$ is a smooth bundle map over X. In coordinates, we write

$$\mathcal{L}(\gamma) = L(x^\mu, y^A, v^A_\mu) d^{n+1}x\,. \tag{2.6}$$

The covariant Legendre transform for $\mathcal{L}$ is a fiber preserving map over Y, $\mathbb{F}\mathcal{L} : J^1Y \to J^1Y^*$, expressed intrinsically as the first order vertical Taylor approximation to $\mathcal{L}$:

$$\mathbb{F}\mathcal{L}(\gamma) \cdot \gamma' = \mathcal{L}(\gamma) + \left.\frac{d}{d\epsilon}\right|_{\epsilon=0} \mathcal{L}(\gamma + \epsilon(\gamma' - \gamma)), \tag{2.7}$$

where $\gamma, \gamma' \in J^1Y_y$.

The coordinate expression of $\mathbb{F}\mathcal{L}$ is given by

$$p_A^\mu = \frac{\partial L}{\partial v^A{}_\mu}, \quad \text{and} \quad p = L - \frac{\partial L}{\partial v^A{}_\mu} v^A{}_\mu \tag{2.8}$$

for the multimomenta p_A^μ and the covariant Hamiltonian p.

Now we can use the covariant Legendre transform to pull back to the Lagrangian side the multisymplectic canonical structure on the dual jet bundle. We define the ***Cartan form*** as the $(n+1)$-form $\Theta_\mathcal{L}$ on J^1Y given by

$$\Theta_\mathcal{L} = (\mathbb{F}\mathcal{L})^* \Theta \tag{2.9}$$

and the $(n+2)$-form $\Omega_\mathcal{L}$ by

$$\Omega_\mathcal{L} = -d\Theta_\mathcal{L} = (\mathbb{F}\mathcal{L})^* \Omega, \tag{2.10}$$

with local coordinate expressions

$$\Theta_\mathcal{L} = \frac{\partial L}{\partial v^A_\mu} dy^A \wedge d^n x_\mu + \left(L - \frac{\partial L}{\partial v^A_\mu} v^A_\mu \right) d^{n+1}x \tag{2.11}$$

and

$$\Omega_\mathcal{L} = dy^A \wedge d \left(\frac{\partial L}{\partial v^A_\mu} \right) \wedge d^n x_\mu - d \left(L - \frac{\partial L}{\partial v^A_\mu} v^A_\mu \right) \wedge d^{n+1}x. \tag{2.12}$$

To lay the groundwork for the following sections we introduce the concept of jet prolongations. We will show how automorphisms of Y lift naturally to automorphisms of J^1Y and we will construct the covariant analogue of the tangent map.

Let $\eta_Y : Y \to Y$ be a π_{XY}-bundle automorphism covering a diffeomorphism $\eta_X : X \to X$. If $\gamma : T_xX \to T_yY$ is an element of J^1Y, let $\eta_{J^1Y}(\gamma) : T_{\eta_X(x)}X \to T_{\eta_Y(y)}Y$ be defined by

$$\eta_{J^1Y}(\gamma) = T\eta_Y \circ \gamma \circ T\eta_X^{-1}. \tag{2.13}$$

The π_{Y,J^1Y}-bundle automorphism $j^1(\eta_Y)$, also denoted η_{J^1Y}, is called the ***first jet extension*** or ***prolongation*** of η_Y to J^1Y and has the coordinate expression

$$\eta_{J^1Y}(\gamma) = \left(\eta_X^\mu(x), \eta_Y^A(x,y), \left[\partial_\nu \eta_Y^A + (\partial_B \eta_Y^A) v^B{}_\nu \right] \partial_\mu \left(\eta_X^{-1} \right)^\nu \right), \tag{2.14}$$

where $\gamma = (x^\mu, y^A, v^A{}_\mu)$.

If V is a vector field on Y whose flow is η_λ, so that

$$V \circ \eta_\lambda = \frac{d\eta_\lambda}{d\lambda},$$

then its ***first jet extension*** or ***prolongation***, denoted $j^1(V)$ or V_{J^1Y}, is the vector field on J^1Y whose flow is $j^1(\eta_\lambda)$; that is

$$j^1(V) \circ j^1(\eta_\lambda) = \frac{d}{d\lambda} j^1(\eta_\lambda). \tag{2.15}$$

In coordinates, $j^1(V)$ has the expression

$$j^1(V) = \left(V^\mu, V^A, \frac{\partial V^A}{\partial x^\mu} + \frac{\partial V^A}{\partial y^B} v^B_\mu - v^A_\nu \frac{\partial V^\nu}{\partial x^\mu} \right). \tag{2.16}$$

We note that one can also view V as a section of the bundle $TY \mapsto Y$ and take its first jet in the sense of (2.1). Then one obtains a section of $J^1(TY) \mapsto Y$ which is not to be confused with $j^1(V)$ as defined by (2.15) and (2.16); they are two different objects.

This is the differential-geometric formulation of the multisymplectic structure. However, as we mentioned before, there is a very elegant and interesting way to construct $\Theta_\mathcal{L}$ directly from the variational principle, staying entirely on the Lagrangian side. It is this variational approach that we will use in the next sections to extend the multisymplectic formalism to the nonsmooth context.

3 Variational Multisymplectic Geometry in a Nonsmooth Setting

We now consider the variational approach to multisymplectic field theory of Marsden, Patrick, and Shkoller [1998] and formulate it a nonsmooth setting. A novelty of this variational approach is that it considers *arbitrary* and not only *vertical* variations of sections. The motivation for such a generalization is that, even though both the vertical and arbitrary variations result in the same Euler-Lagrange equations, the Cartan form obtained from the vertical variations is missing one term (corresponding to the $d^{n+1}x$ form). However, the horizontal variations account precisely for this extra term and make the Cartan form complete.

We reconsider the need for horizontal variations in the nonsmooth context and adapt the formalism developed in Marsden, Patrick, and Shkoller [1998] to give a rigorous derivation of the jump conditions when fields are allowed to be nonsmooth.

3.1 Nonsmooth Multisymplectic Geometry

Let U be a manifold with smooth closed boundary. In the smooth context, the configuration space is the infinite-dimensional manifold defined by an appropriate closure of the set of smooth maps

$$\mathcal{C}^\infty = \{\phi : U \to Y \mid \pi_{XY} \circ \phi : U \to X \text{ is an embedding}\}.$$

In the the nonsmooth setting, we must also introduce a codimension 1 submanifold $D \subset U$, called the ***singularity submanifolds*** across which the fields ϕ may have singularities. For example, the submanifold D may be the spacetime surface separating two regions of a continuous medium or, in the case of two elastic bodies colliding, D may be the spacetime contact set.

For the smooth case, the observation that the configuration space is a smooth manifold enables the use of differential calculus on the manifold of mappings as required by variational principles (see Marsden and Ratiu [1999],Marsden, Patrick, and Shkoller [1998]). In this subsection we will present various types of configuration spaces that one must consider in the nonsmooth context and discuss their manifold structure.

Configuration spaces. The applications that we present in the next sections require different configuration spaces, according to the type of singularities that we allow across the singularity submanifold D.

Case (a). Continuous but nonsmooth. For the first examples presented in this paper (such as rigid-body dynamics with impact and propagating singular surfaces within a continuum medium), the configuration space is the set of continuous maps

$$\begin{aligned}\mathcal{C}^a = \{\phi : U \to Y \mid\ & \pi_{XY} \circ \phi : U \to X \text{ is an embedding},\\ & \phi \text{ is } C^0 \text{ in } U \text{ and of class } C^2 \text{ in } U \setminus D\}\end{aligned} \tag{3.1}$$

For each $\phi \in \mathcal{C}^a$, we set $\phi_X = \pi_{XY} \circ \phi$ and $U_X = \pi_{XY} \circ \phi(U)$, $D_X = \pi_{XY} \circ \phi(D)$, so that $\phi_X : U \to U_X$ and its restriction to D are diffeomorphisms. We also denote the section $\phi \circ {\phi_X}^{-1}$ by φ, as in the diagram

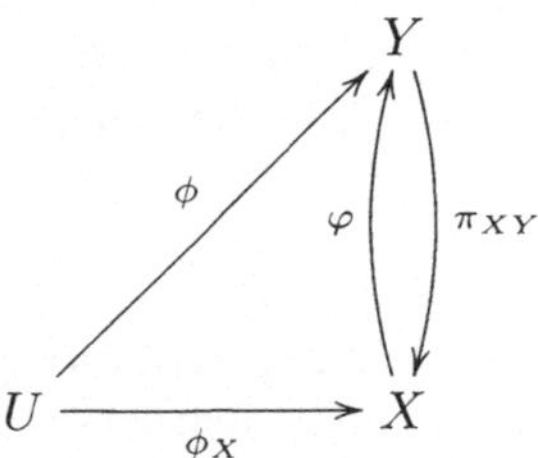

The submanifold D separates the interior of U into two disjoint open subsets U^+ and U^-, that is $\text{int}(U) = U^+ \cup U^- \cup (U \cap D)$ and we let

$U_X^+ = \phi_X(U^+)$ and $U_X^- = \phi_X(U^-)$ be their corresponding images in X. It follows that $\text{int}(U_X) = U_X^+ \cup U_X^- \cup (U_X \cap D_X)$.

Remark. For particle mechanics, the formalism reduces to the spacetime formulation that we developed in Fetecau, Marsden, Ortiz, and West [2002] to study nonsmooth rigid-body dynamics. We will discuss this example in detail in §4.

Case (b). Discontinuous without separation (slip). For problems such as propagation of free surfaces in fluids or interaction of an elastic body and a fluid, the configuration map ϕ is no longer continuous. We must therefore choose a new configuration space to include these cases. Observe that in such problems the fluid–fluid and the solid–fluid boundaries are material surfaces, in the sense that particles which are on the separating surface at a given time remain on the surface at later times.

Let U_S and U_F two open subsets of U such that $\partial U_S = \partial U_F = D$ is a codimension 1 submanifold in U. We adopt the subscripts S and F because one of the applications of this general setting will be solid–fluid interactions; U_S, U_F and D will be interpreted as the spacetime regions of the solid, the fluid and of the surface separating the two materials, respectively. For fluid–fluid boundaries, D is the spacetime free surface.

The requirement that there be no flow across the material interface is expressed by considering the configuration space

$$\begin{aligned} \mathcal{C}^b = \{\phi : U \to Y \mid \pi_{XY} \circ \phi : U \to X \text{ is an embedding,} \\ \phi \text{ is of class } C^2 \text{ in } U_S \cup U_F \text{ and } \overline{\phi_S}(D) = \overline{\phi_F}(D) = \phi(D)\}, \end{aligned} \tag{3.2}$$

where ϕ_S, ϕ_F are the restrictions of the map ϕ on U_S, U_F, respectively, and the notation $\overline{f}$ represents the continuous extension of the map f to the closure of its domain.

Remark. One may alternatively denote U_S and U_F by U^+ and U^-, respectively. This will be particularly useful in §3.2, where we retain only these notations for the domains where there are no singularities. As in case (a), we denote by U_X^+, U_X^- and D_X the images in X of U^+, U^- and D, under ϕ_X.

Case (c). Discontinuous with separation (collisions). Collisions of elastic bodies may exhibit both of the features of the two classes of configuration maps presented so far. The mechanical impact of two solids generates stress waves that propagate through their bodies, reflect on the boundaries and then return to the contact interface. At the instant when a returning wave first reaches the contact surface, the release begins and separation will eventually occur. Because of the complicated, non-linear structure of the governing equations, little of a general nature may be presented for impact problems. We refer to Graff [1991] for a detailed discussion on the longitudinal impact of two elastic rods and on the impact of an elastic sphere with a rod, to demonstrate some of the complexities encountered in such problems.

We will consider the frictionless impact with slipping of two elastic bodies. The analog of D from the previous paragraphs will be the spacetime contact set. However, to the contact set in the spatial configuration, there correspond two distinct surfaces in the reference configuration. In the multisymplectic formalism we consider two disjoint open sets U_1 and U_2 in U and $D_1 \subset \partial U_1$, $D_2 \subset \partial U_2$ two codimension 1 submanifolds. We consider the following set as the configuration space for collision problems

$$\begin{aligned} \mathcal{C}^c = \{\phi : U \to Y \mid \pi_{XY} \circ \phi : U \to X \text{ is an embedding}, \\ \phi \text{ is of class } C^2 \text{ in } U \text{ and } \overline{\phi}(D_1) = \overline{\phi}(D_2)\}\,, \end{aligned} \tag{3.3}$$

where the notation $\overline{f}$ represents, as before, the continuous extension of the map f.

Remarks.

1. The set $\overline{\phi}(D_1)$ (or equivalently, $\overline{\phi}(D_2)$) must be interpreted as a *subset* of the spacetime contact set. The subset does not contain points which belong to other types of discontinuity surfaces (such as, for example, the points on the interface at the very moment of impact, which also belong to $\mathcal{C}^a$ type waves that are generated by the mechanical impact and propagate through the bodies). Intersections of different types of discontinuity surfaces are extremely important and we intend to treat this subject in our future work on this topic. We will thus not discuss this point further here.

2. For a more unified presentation and to include all three cases a-c in the general result from §3.2, we will refer later to U_1 and U_2 as U^+ and U^-.

3. For purposes of connecting this work with PDE methods, one can consider the closure of either of $\mathcal{C}^{a-c}$ in the topology of a larger space such as $H^s(U, Y)$ or $\mathcal{C}^\infty(U^+, Y) \times \mathcal{C}^\infty(U^-, Y)$. This enables one to regard $\mathcal{C}^{a-c}$ as subsets in a manifold of mappings of the appropriate Sobolev class, as in Palais [1968] and Ebin and Marsden [1970].

4. In the remainder of the paper we will write $\mathcal{C}^{a-c}$ to indicate the appropriate configuration space with the manifold structure obtained

by the procedure explained above. However, whenever we state a general result which applies to all configuration manifolds a-c, we will write $\mathcal{C}$ to mean any of the three.

Variations and tangent spaces. We will account for general variations of maps $\phi \in \mathcal{C}$ induced by a family of maps ϕ^λ defined by the action of some Lie group. More precisely, let $\mathcal{G}$ be a Lie group of π_{XY}-bundle automorphism η_Y covering diffeomorphisms η_X, with Lie algebra $\mathfrak{g}$, acting on $\mathcal{C}$ by $\Phi : \mathcal{G} \times \mathcal{C} \to \mathcal{C}$, where

$$\Phi(\eta_Y, \phi) = \eta_Y \circ \phi. \tag{3.4}$$

Now let $\lambda \mapsto \eta_Y^\lambda$ be an arbitrary smooth path in $\mathcal{G}$ such that $\eta_Y^0 = \mathrm{Id}_Y$, and let $V \in T_\phi\mathcal{C}$ be given by

$$V = \left.\frac{d}{d\lambda}\right|_{\lambda=0} \Phi(\eta_Y^\lambda, \phi), \quad \text{and} \quad V_X = \left.\frac{d}{d\lambda}\right|_{\lambda=0} \eta_X^\lambda \circ \phi_X. \tag{3.5}$$

We define the vertical component V_yY of the tangent space at y to be

$$V_yY = \{\mathcal{V} \in T_yY \mid T\pi_{XY} \cdot \mathcal{V} = 0\}. \tag{3.6}$$

Using this, we can naturally split the tangent space at each point $y = \phi(u)$ in the image of ϕ into vertical and horizontal components, $T_yY = V_yY \oplus H_yY$, where

$$H_yY = T_u\phi \cdot T_uU.$$

This decomposition of T_yY induces a decomposition of $T_\phi\mathcal{C}$, so that any vector $V \in T_\phi\mathcal{C}$ may be decomposed as $V = V^h + V^v$, where

$$V^h = T(\phi \circ \phi_X^{-1}) \cdot V_X, \tag{3.7}$$

and by (3.5), $V_X = T\pi_{XY} \cdot V$.

Case (a). For $\phi \in \mathcal{C}^a$, it is easy to show that the tangent space $T_\phi\mathcal{C}^a$ is given by

$$\begin{aligned} T_\phi\mathcal{C}^a = \{V : U \to TY \mid & V \text{ is } C^0 \text{ in } U \text{ and of class } C^2 \text{ in } U \setminus D, \\ & \pi_{Y,TY} \circ V = \phi \text{ and } T_{\pi_{XY}} \circ V = V_X \text{ is a vector field on } X\}. \end{aligned} \tag{3.8}$$

Case (b). In the multisymplectic description of a continuum medium, the bundle Y over X is trivial. It consists of a fiber manifold M (also called the ambient space) attached to each point of the spacetime $X = B \times \mathbb{R}$ (B is called the reference configuration). The fiber components of the points $\phi(u)$ with $u \in D$ (the interface in the reference configuration) constitute the image of the interface in the spatial configuration.

By constructing variations of maps $\phi \in \mathcal{C}^b$ as in (3.4), we can prove the following lemma

3.1 Lemma. *Let N_A be the outward unit normal of the current configuration interface and $V = (V^\mu, V^A)$ be a tangent vector obtained by (3.5). Then, $[\![V^A(u)]\!]N_A = 0$, for all $u \in D$.*

Proof. Let us first explain the sense in which the jump condition of this lemma must be interpreted. A point $y \in D$ is mapped by ϕ to a point $\phi(y) \in Y$ and denote by y the fiber component of $\phi(y)$ (y is a point on the current interface). By definition (3.2), there exist two points $y_S, y_F \in D$ such that $\overline{\phi_S}(y_S) = \overline{\phi_F}(y_F) = \phi(y)$.

Consider now a variation of a map $\phi \in \mathcal{C}^b$ given by $\phi^\lambda = \eta_Y^\lambda \circ \phi$, where η_Y^λ are π_{XY}-bundle automorphisms covering diffeomorphisms η_X and $\eta_Y^0 = \mathrm{Id}_Y$. By $[\![V(x)]\!]$ we mean

$$[\![V(u)]\!] = \left.\frac{d}{d\lambda}\right|_{\lambda=0} \overline{\phi_S^\lambda}(u_S) - \left.\frac{d}{d\lambda}\right|_{\lambda=0} \overline{\phi_F^\lambda}(u_F)\,, \tag{3.9}$$

where ϕ_S^λ, ϕ_F^λ are the restrictions of the map ϕ^λ on U_S, U_F, respectively. The two terms in the right-hand-side of (3.9) are vectors in $T_{\phi(u)}Y$, so the addition operation makes sense.

Since $\phi^\lambda \in \mathcal{C}^b$, there exists a point $u_F^\lambda \in D$ such that

$$\overline{\phi_S^\lambda}(u_S) = \overline{\phi_F^\lambda}(u_F^\lambda)\,.$$

We note that $u_F^0 = u_F$. Then, we can derive

$$\left.\frac{d}{d\lambda}\right|_{\lambda=0} \overline{\phi_S^\lambda}(u_S) = \left.\frac{d}{d\lambda}\right|_{\lambda=0} \overline{\phi_F^\lambda}(u_F) + T\overline{\phi_F}(u_F)\cdot v\,, \tag{3.10}$$

where

$$v = \left.\frac{d}{d\lambda}\right|_{\lambda=0} u_F^\lambda \in T_{u_F}D\,. \tag{3.11}$$

By using (3.10), (3.9) becomes

$$[\![V(u)]\!] = T\overline{\phi_F}(u_F)\cdot v\,. \tag{3.12}$$

Now, (3.11) and (3.12) prove the lemma. ■

Using the same notation as before, we thus proved

$$\begin{aligned} T_\phi\mathcal{C}^b = \{V : U \to TY \mid\ & V \text{ is of class } C^2 \text{ in } U_S \cup U_F, \pi_{Y,TY} \circ V = \phi, \\ & T_{\pi_{XY}} \circ V = V_X \text{ is a vector field on } X, \text{ and } [\![V^A]\!]N_A = 0 \text{ on } D\}\,. \end{aligned} \tag{3.13}$$

Case (c). A result very similar to Lemma 3.1 will hold for case (c) as well. More precisely, consider two points $u_1 \in D_1$ and $u_2 \in D_2$ such that $\overline{\phi}(u_1) = \overline{\phi}(u_2) = y$. By an argument similar to the one used in the proof of Lemma 3.1, we can prove

$$V(u_2) - V(u_1) = T\overline{\phi}(u_1) \cdot v\,, \tag{3.14}$$

where $v \in T_{u_1}D_1$. Then, $V(u_2) - V(u_1) \in T_yY$ and, if we denote by N_A the components of the outward unit normal of the contact set in the current configuration, we abuse the notation and write

$$[\![V^A]\!]N_A = 0\,.$$

Hence, the tangent space $T_\phi \mathcal{C}^c$ is given by

$$\begin{aligned} T_\phi \mathcal{C}^c = \{V : U \to TY \mid V \text{ is of class } C^2 \text{ in } U_1 \cup U_2, \pi_{Y,TY} \circ V = \phi, \\ T\pi_{XY} \circ V = V_X \text{ is a vector field on } X \text{ and } [\![V^A]\!]N_A = 0 \text{ on } D\}\,, \end{aligned} \tag{3.15}$$

where the jump relation has the interpretation explained before.

3.2 Variational Approach

We will show next how to derive the equations of motion and the jump conditions directly from the variational principle, staying entirely on the Lagrangian side.

The ***action function*** $S : \mathcal{C} \to \mathbb{R}$ is defined by

$$S(\phi) = \int_{U_X} \mathcal{L}(j^1(\phi \circ \phi_X^{-1}))\,. \tag{3.16}$$

We say that $\phi \in \mathcal{C}$ is a ***stationary point*** or ***critical point*** of S if

$$\left.\frac{d}{d\lambda}\right|_{\lambda=0} S(\Phi(\eta_Y^\lambda, \phi)) = 0 \tag{3.17}$$

for all curves η_Y^λ with $\eta_Y^0 = \mathrm{Id}_Y$.

Using the infinitesimal generators defined in (3.5), we compute:

$$\begin{aligned} dS_\phi \cdot V &= \left.\frac{d}{d\lambda}\right|_{\lambda=0} S(\Phi(\eta_Y^\lambda, \phi)) \\ &= \left.\frac{d}{d\lambda}\right|_{\lambda=0} \int_{\eta_X^\lambda(U_X)} \mathcal{L}(j^1(\Phi(\eta_Y^\lambda, \phi))) \\ &= \int_{U_X} \left.\frac{d}{d\lambda}\right|_{\lambda=0} \mathcal{L}(j^1(\Phi(\eta_Y^\lambda, \phi))) + \int_{U_X} \mathfrak{L}_{V_X}\left[\mathcal{L}(j^1(\phi \circ \phi_X^{-1}))\right]\,. \end{aligned}$$

In Marsden, Patrick, and Shkoller [1998] the following lemma is proved.

3.2 Lemma. *For any $V \in T_\phi \mathcal{C}$,*

$$dS_\phi \cdot V^h = \int_{\partial U_X} V_X \lrcorner \left[\mathcal{L}(j^1(\phi \circ \phi_X^{-1})) \right], \tag{3.18}$$

and

$$dS_\phi \cdot V^v = \int_{U_X} \left. \frac{d}{d\lambda} \right|_{\lambda=0} \mathcal{L}\big(j^1\big(\Phi(\eta_Y^\lambda, \phi)\big)\big). \tag{3.19}$$

The previous lemma leads to the following fundamental theorem.

3.3 Theorem. *Given a Lagrangian density $\mathcal{L} : J^1Y \longrightarrow \Lambda^{n+1}(X)$, which is smooth away from the discontinuity, there exists a unique smooth section $D_{EL}\mathcal{L} \in C^\infty(Y'', \Lambda^{n+1}(X) \otimes T^*Y))$ and a unique differential form $\Theta_\mathcal{L} \in \Lambda^{n+1}(J^1Y)$ such that for any $V \in T_\phi \mathcal{C}$ which is compactly supported in U and any open subset U_X such that $\overline{U_X} \cap \partial X = \emptyset$,*

$$\begin{aligned} dS_\phi \cdot V = & \int_{U_X^+} D_{EL}\mathcal{L}(j^2(\phi \circ \phi_X^{-1})) \cdot V + \int_{U_X^-} D_{EL}\mathcal{L}(j^2(\phi \circ \phi_X^{-1})) \cdot V \\ & + \int_{U_X \cap D_X} [\![j^1(\phi \circ \phi_X^{-1})^* \left(j^1(V) \lrcorner \Theta_\mathcal{L} \right)]\!], \end{aligned} \tag{3.20}$$

where $[\![\cdot]\!]$ denotes the jump.

Furthermore,

$$D_{EL}\mathcal{L}(j^2(\phi \circ \phi_X^{-1})) \cdot V = j^1(\phi \circ \phi_X^{-1})^*[j^1(V) \lrcorner \Omega_\mathcal{L}]. \tag{3.21}$$

In coordinates, the action of the ***Euler-Lagrange derivative*** *$D_{EL}\mathcal{L}$ on Y'' is given by*

$$\begin{aligned} D_{EL}\mathcal{L}(j^2(\phi \circ \phi_X^{-1})) = & \left[\frac{\partial L}{\partial y^A}(j^1(\phi \circ \phi_X^{-1})) - \frac{\partial^2 L}{\partial x^\mu \partial v^A{}_\mu}(j^1(\phi \circ \phi_X^{-1})) \right. \\ & - \frac{\partial^2 L}{\partial y^B \partial v^A{}_\mu}(j^1(\phi \circ \phi_X^{-1})) \cdot (\phi \circ \phi_X^{-1})^B_{,\mu} \\ & \left. - \frac{\partial^2 L}{\partial v^B{}_\mu \partial v^A{}_\nu}(j^1(\phi \circ \phi_X^{-1})) \cdot (\phi \circ \phi_X^{-1})^B_{,\mu\nu} \right] d^{n+1}x \otimes dy^A, \end{aligned} \tag{3.22}$$

while the form $\Theta_\mathcal{L}$ matches the definition of the ***Cartan form*** *obtained via the Legendre transform and has the coordinate expression*

$$\Theta_\mathcal{L} = \frac{\partial L}{\partial v^A{}_\mu} dy^A \wedge d^n x_\mu + \left(L - \frac{\partial L}{\partial v^A{}_\mu} v^A{}_\mu \right) d^{n+1}x. \tag{3.23}$$

Proof. We choose $U_X = \phi_X(U)$ small enough so that it is contained in a coordinate chart O. In the coordinates on O, let $V = (V^\mu, V^A)$ so that along $\phi \circ \phi_X^{-1}$, the decomposition (3.7) can be written as

$$V_X = V^\mu \frac{\partial}{\partial x^\mu} \text{ and } V^v = (V^v)^A \frac{\partial}{\partial y^A} = \left(V^A - V^\mu \frac{\partial(\phi \circ \phi_X^{-1})^A}{\partial x^\mu} \right) \frac{\partial}{\partial y^A}.$$

Now, we use (3.19) to obtain

$$dS_\phi \cdot V^v = \int_{U_X} \left[\frac{\partial L}{\partial y^A}(j^1(\phi \circ \phi_X^{-1})) \cdot (V^v)^A \right. \\ \left. + \frac{\partial L}{\partial v^A{}_\mu}(j^1(\phi \circ \phi_X^{-1})) \cdot \frac{\partial (V^v)^A}{\partial x^\mu} \right] d^{n+1}x \,. \tag{3.24}$$

We split the integral $\int_{U_X}$ into $\int_{U_X^+} + \int_{U_X^-}$ and integrate by parts to obtain

$$dS_\phi \cdot V^v = \int_{U_X^+} \left[\frac{\partial L}{\partial y^A}(j^1(\phi \circ \phi_X^{-1})) - \frac{\partial}{\partial x^\mu} \frac{\partial L}{\partial v^A{}_\mu}(j^1(\phi \circ \phi_X^{-1})) \right] \\ \cdot (V^v)^A d^{n+1}x \\ + \int_{U_X^-} \left[\frac{\partial L}{\partial y^A}(j^1(\phi \circ \phi_X^{-1})) - \frac{\partial}{\partial x^\mu} \frac{\partial L}{\partial v^A{}_\mu}(j^1(\phi \circ \phi_X^{-1})) \right] \\ \cdot (V^v)^A d^{n+1}x \\ + \int_{U_X \cap D_X} \left[\!\left[\frac{\partial L}{\partial v^A{}_\mu}(j^1(\phi \circ \phi_X^{-1})) \cdot (V^v)^A \right]\!\right] d^n x_\mu \,. \tag{3.25}$$

The jump arises from the different orientations of D_X when we use Stokes theorem in U_X^+ and U_X^-. Additionally, from (3.18) we obtain the horizontal contribution

$$dS_\phi \cdot V^h = \int_{U_X \cap D_X} [\![V^\mu L]\!] d^{n+1}x \,. \tag{3.26}$$

We note that the terms corresponding to $\int_{\partial U_X}$ vanish in both (3.25) and (3.26) since V is compactly supported in U. Now, we can combine (3.25) and (3.26) to obtain

$$dS_\phi \cdot V = \int_{U_X^+ \cup U_X^-} \left\{ \left[\frac{\partial L}{\partial y^A} - \frac{\partial}{\partial x^\mu} \frac{\partial L}{\partial v^A_\mu} \right] (j^1(\phi \circ \phi_X^{-1})) \right\} d^{n+1}x \otimes dy^A \cdot V \\ + \int_{U_X \cap D_X} \left[\!\left[V \,\lrcorner\, \left\{ \frac{\partial L}{\partial v^A_\mu}(j^1(\phi \circ \phi_X^{-1})) dy^A \wedge d^n x_\mu \right. \right.\right. \\ \left.\left.\left. + \left[L - \frac{\partial L}{\partial v^A_\mu}(j^1(\phi \circ \phi_X^{-1})) \frac{\partial (\phi \circ \phi_X^{-1})^A}{\partial x^\mu} \right] d^{n+1}x \right\} \right]\!\right] . \tag{3.27}$$

Let α be the n-form in the jump brackets of the integrand of the boundary integral in (3.27). Then, $\int_{U_X \cap D_X} \alpha = \int_{j^1(\phi \circ \phi_X^{-1})(U_X \cap D_X)} \alpha$, since α is invariant under this lift. Moreover, the vector V in the second term of (3.27) (now written as an integral over $j^1(\phi \circ \phi_X^{-1})(U_X \cap D_X)$) may be replaced by $j^1(V)$ since $\pi_{Y,J^1(Y)}$-vertical vectors are in the kernel of the form that V is acting on. Now we can pull back the integrand with $j^1(\phi \circ \phi_X^{-1})^*$ to get an n-form on $U_X \cap D_X$. To summarize, we proved that the boundary integral in (3.27) can be written as

$$\int_{U_X \cap D_X} [\![j^1(\phi \circ \phi_X^{-1})^* \, (j^1(V) \lrcorner \Theta_{\mathcal{L}})]\!] \,,$$

where $\Theta_{\mathcal{L}} \in \Lambda^{n+1}(J^1(Y))$ has the coordinate expression given by (3.23).

The integrand of the first integral in (3.27) defines the coordinate expression of the Euler-Lagrange derivative $D_{EL}\mathcal{L}$. However, if we choose another coordinate chart O', the coordinate expressions of $D_{EL}\mathcal{L}$ and $\Theta_{\mathcal{L}}$ must agree on the overlap $O \cap O'$ since the left hand side of (3.20) is intrinsically defined. Thus, we have uniquely defined $D_{EL}\mathcal{L}$ and $\Theta_{\mathcal{L}}$.

Now, we can define intrinsically $\Omega_{\mathcal{L}} = -d\Theta_{\mathcal{L}}$ and check that (3.21) holds, as both sides have the same coordinate expressions. ■

We now use Hamilton's principle of critical action and look for those paths $\phi \in \mathcal{C}$ which are critical points of the action function. More precisely, we call a field $\phi \in \mathcal{C}$ a ***solution*** if

$$dS(\phi) \cdot V = 0 \,, \tag{3.28}$$

for all vector fields $V \in T_\phi \mathcal{C}$ which vanish on the boundary ∂U.

From Theorem 3.3 it follows that a field ϕ is a solution if and only if the Euler-Lagrange derivative (evaluated at $j^2(\phi \circ \phi_X^{-1})$) is zero on U_X^+ and U_X^- and the n-form $j^1(\phi \circ \phi_X^{-1})^* \left[j^1(V) \lrcorner \Theta_{\mathcal{L}} \right]$ has a zero jump across $U_X \cap D_X$.

We thus obtain the Euler-Lagrange equations in U_X^+ and U_X^-, away from the singularities. In coordinates, they read

$$\frac{\partial L}{\partial y^A}(j^1(\phi \circ \phi_X^{-1})) - \frac{\partial}{\partial x^\mu} \frac{\partial L}{\partial v^A{}_\mu}(j^1(\phi \circ \phi_X^{-1})) = 0 \quad \text{in} \quad U_X^+ \cup U_X^-. \tag{3.29}$$

Finally, the intrinsic jump condition

$$\int_{U_X \cap D_X} [\![j^1(\phi \circ \phi_X^{-1})^* \, (j^1(V) \lrcorner \Theta_{\mathcal{L}})]\!] = 0 \tag{3.30}$$

has the following coordinate expression

$$\begin{aligned} \int_{U_X \cap D_X} \Bigg(& \left[\!\!\left[\frac{\partial L}{\partial v^A{}_\mu}(j^1(\phi \circ \phi_X^{-1})) \cdot V^A \right]\!\!\right] \\ & + \left[\!\!\left[L V^\mu - \frac{\partial L}{\partial v^A{}_\mu}(j^1(\phi \circ \phi_X^{-1})) \frac{\partial (\phi \circ \phi_X^{-1})^A}{\partial x^\nu} V^\nu \right]\!\!\right] \Bigg) d^n x_\mu = 0 \,. \end{aligned} \tag{3.31}$$

In the next section we will write the jump conditions (3.31) for the particle and continuum mechanics multisymplectic models and give their physical interpretations. Here, we simply note that by taking vertical variations only ($V^\mu = 0$) we obtain a jump condition involving only momenta $p_A{}^\mu$; this will represent the jump in linear momentum condition. Horizontal variations will in turn give the correct energy jump and a kinematic compatibility condition.

4 Classical Mechanics

For a classical mechanical system (such as particles or rigid bodies) with configuration space Q, let $X = \mathbb{R}$ (parameter time) and $Y = \mathbb{R} \times Q$, with π_{XY} the projection onto the first factor. The first jet bundle J^1Y is the bundle whose holonomic sections are tangents of sections $\phi : X \to Y$, so we can identify $J^1Y = \mathbb{R} \times TQ$. Using coordinates (t, q^A) on $\mathbb{R} \times Q$, the induced coordinates on J^1Y are the usual tangent coordinates (t, q^A, v^A).

We will apply the multisymplectic formalism described in §2 to nonsmooth rigid-body dynamics. We are particularly interested in the problem of rigid-body collisions, for which the velocity, acceleration and forces are all nonsmooth or even discontinuous. The multisymplectic formalism will elegantly recover the spacetime formulation of nonsmooth Lagrangian mechanics of Fetecau, Marsden, Ortiz, and West [2002].

In Fetecau, Marsden, Ortiz, and West [2002] a mechanical system with configuration manifold Q is considered, but with the dynamics restricted to a submanifold with boundary $C \subset Q$, which represents the subset of admissible configurations. The boundary ∂C is called the contact set; for rigid body collision problems, the submanifold ∂C is obtained from the condition that interpenetration of matter cannot occur. The dynamics is specified by a regular Lagrangian $L : TQ \to \mathbb{R}$. We note that the multisymplectic framework allows us to consider time dependent Lagrangians as well (see the general definition of a Lagrangian density in §2), but we will restrict our discussion here to only autonomous systems.

To apply the multisymplectic formalism for such systems, we choose U to be the interval $[0,1]$ and D the set containing only one element $\tau_i \in [0,1]$. The set $\mathcal{C}^a$ from (3.1) becomes:

$$\begin{aligned} \mathcal{C}' = \{\phi : [0,1] \to \mathbb{R} \times Q \mid \phi \text{ is a } C^0, PW\ C^2 \text{ curve },\\ \phi(\tau) \text{ has only one singularity at } \tau_i\}\,. \end{aligned} \tag{4.1}$$

Now let $U_X = [t_0, t_1]$ be the image in $X = \mathbb{R}$ of the embedding $\phi_X = \pi_{XY} \circ \phi$. The section $\varphi = \phi \circ \phi_X^{-1} : [t_0, t_1] \to \mathbb{R} \times Q$ can be written in coordinates as $t \mapsto (t, q^A(t))$. Let $t_i = \phi_X(\tau_i)$ be the the moment of impact, so that $q(t_i) \in \partial C$. We note that, even though the singularity parameter time τ_i is fixed, it is allowed to vary in the t space according to $t_i = \phi_X(\tau_i)$ and thus, the setting is not restrictive in this sense.

Hence, the map $\phi_X : [0,1] \to [t_0, t_1]$ is just a time reparametrization. The need for a nonautonomous formulation of an autonomous mechanical system is explained in the following remarks.

Remarks.

1. In the smooth context, the dynamics of a mechanical system can be described by sections of smooth fields $\varphi : [t_0, t_1] \to Q$. As we

noted in the general setting, the key observation that the set of such smooth fields is a C^∞ infinite-dimensional manifold enables the use of differential calculus on the manifold of mappings (see Marsden and Ratiu [1999]). However, generalization to the nonsmooth setting is not straightforward and this is one of the main issues addressed in Fetecau, Marsden, Ortiz, and West [2002].

2. The approach used in Fetecau, Marsden, Ortiz, and West [2002] is to extend the problem to the nonautonomous case, so that both configuration variables and time are functions of a separate parameter τ. This allows the impact to be fixed in τ space while remaining variable in both configuration and time spaces, and it means that the relevant space of configurations will indeed be a smooth manifold, as proved in that reference. The nonsmooth multisymplectic formalism applied to this problem leads to essentially the same extended formulation.

The Cartan form (3.23) becomes the extended Lagrange 1-form of particle mechanics, with coordinate expression

$$\Theta_{\mathcal{L}} = \frac{\partial L}{\partial \dot{q}^A} dq^A + \left(L - \frac{\partial L}{\partial \dot{q}^A} \dot{q}^A \right) dt\,. \tag{4.2}$$

Define the ***energy*** $E : TQ \to \mathbb{R}$ by

$$E(q, \dot{q}) = \frac{\partial L}{\partial \dot{q}}(q, \dot{q}) \cdot \dot{q} - L(q, \dot{q})\,,$$

which allows us to write the Lagrangian 1-form in the compact notation

$$\Theta_{\mathcal{L}} = \frac{\partial L}{\partial \dot{q}} dq - E dt\,. \tag{4.3}$$

Let $V \in T_\phi \mathcal{C}$ be a tangent vector constructed as in (3.5) with coordinates $V = (V^0, V^A)$. As the fiber component of $\varphi(t_i)$ is varied in ∂C, we can write the jump condition (3.30) as

$$\Theta_{\mathcal{L}}|_{t_i^-} = \Theta_{\mathcal{L}}|_{t_i^+} \;\text{ on } \mathbb{R} \times TQ|\partial C\,. \tag{4.4}$$

In coordinates, the jump condition (4.4) reads

$$V^A \left. \frac{\partial L}{\partial \dot{q}^A} \right|_{t_i^-}^{t_i^+} + V^0 \left. \left(L - \frac{\partial L}{\partial \dot{q}^A} \dot{q}^A \right) \right|_{t_i^-}^{t_i^+} = 0\,. \tag{4.5}$$

Splitting this into the two components gives

$$\left. \frac{\partial L}{\partial \dot{q}} \right|_{t=t_i^-} \cdot \delta q = \left. \frac{\partial L}{\partial \dot{q}} \right|_{t=t_i^+} \cdot \delta q \tag{4.6}$$

for any $\delta q = V^A \frac{\partial}{\partial q^A} \in T_{q(t_i)}\partial C$ and

$$E\big(q(t_i^-), \dot{q}(t_i^-)\big) = E\big(q(t_i^+), \dot{q}(t_i^+)\big) . \tag{4.7}$$

These equations are the Weierstrass-Erdmann type conditions for impact. That is, equation (4.6) states that the linear momentum must be conserved in the tangent direction to ∂C, while equation (4.7) states that the energy must be conserved during an elastic impact.

Hence, horizontal variations (V^0) give conservation of energy and vertical variations (V^A) give conservation of the Lagrange 1-form on $T\partial C$.

5 Continuum Mechanics

5.1 Multisymplectic Formulation of Continuum Mechanics

Configuration Spaces in the Multisymplectic Formalism. We will use here the formalism constructed in Marsden, Pekarsky, Shkoller, and West [2001] to describe the configurations of a continuous medium. Let (B, G) be a smooth n-dimensional compact oriented Riemannian manifold with smooth boundary and let (M, g) be a smooth N-dimensional compact oriented Riemannian manifold. The space (B, G) will represent what is traditionally called the ***reference configuration***, while (M, g) will denote the ***ambient space***.

We choose $X = B \times \mathbb{R}$; the coordinates on X are $x^\mu = (x^i, x^0) = (x^i, t)$, with $\mu = 0, \ldots, n$, $i = 1, \ldots, n$. Let $Y = X \times M$ be a trivial bundle over X with M being a fiber at each point and let $\pi_{XY} : Y \to X$; $(x, t, y) \mapsto (x, t)$ be the projection on the first factor ($y \in M$ is the fiber coordinate). Let y^A, $A = 1, \ldots, N$ be fiber coordinates; they induce the coordinates on J^1Y denoted $\gamma = (x^\mu, y^A, v^A{}_\mu)$. We denote the fiber coordinates on J^1Y^* by $(\Pi, p_A{}^\mu)$; they correspond to the affine map given in coordinates by $v^A{}_\mu \mapsto (\Pi + p_A{}^\mu v^A{}_\mu) d^{n+1}x$.

A section $\varphi : X \to Y$ of π_{XY} has coordinate representation $\varphi(x) = (x^\mu, \varphi^A(x))$, while its first jet $j^1\varphi$ is given by

$$x^\mu \mapsto \big(x^\mu, \varphi^A(x), \partial_\mu \varphi^A(x)\big) , \tag{5.1}$$

where $\partial_0 = \frac{\partial}{\partial t}$ and $\partial_k = \frac{\partial}{\partial x^k}$.

We note that we introduced two different Riemannian structures on the spatial part of the base manifold X and on the fiber M. Thus, the formalism is general enough to apply for continuum models where the metric spaces (B, G) and (M, g) are essentially different (rods, shells models, fluids with free boundary). However, for classical 2 or 3 dimensional elasticity or for fluid dynamics in a domain with fixed boundaries, the two Riemannian structures may coincide.

Define the function $J : J^1Y \to \mathbb{R}$ with coordinate expression

$$J(x,t,y,v) = \det[v]\sqrt{\frac{\det[g(y)]}{\det[G(x)]}}\,. \tag{5.2}$$

For a section φ, $J(j^1(\varphi))$ represents the Jacobian of the linear transformation $D\varphi_t$. We note that, even in the cases where the metrics G and g coincide, there is no cancellation in (5.2), as the metric tensors are evaluated at different points ($g(y)$ is different from $G(x)$ unless $y = x$ or both tensors are constant).

Lagrangian dynamics. To describe the dynamics of a particular continuum medium in the variational multisymplectic framework, one needs to specify a Lagrangian density $\mathcal{L}$. The Lagrangian density $\mathcal{L} : J^1Y \to \Lambda^{n+1}X$ is defined as a smooth bundle map

$$\begin{aligned}\mathcal{L}(\gamma) = L(\gamma)d^{n+1}x = \mathbb{K} - \mathbb{P} &= \tfrac{1}{2}\sqrt{\det[G]}\rho(x)g_{AB}v^A{}_0\, v^B{}_0\, d^{n+1}x \\ &\quad - \sqrt{\det[G]}\rho(x)W(x, G(x), g(y), v^A{}_j)d^{n+1}x\,,\end{aligned} \tag{5.3}$$

where $\gamma \in J^1Y$, $\rho : B \to \mathbb{R}$ is the mass density and W is the ***stored energy function***.

The first term in (5.3), when restricted to first jet extensions, represents the kinetic energy, as $v^A{}_0$ becomes the time derivative $\partial_t\varphi^A$ of the section φ. The second term represents the potential energy and different choices of the function W specify particular models of continuous media. Typically, for elasticity, W depends on the field's partial derivatives through the Green deformation tensor C (see Marsden and Hughes [1983], for example), while for ideal fluid dynamics, W is only a function of the Jacobian J (5.2).

The Lagrangian density (5.3) determines the Legendre transformation $\mathbb{F}\mathcal{L} : J^1Y \to J^1Y^*$. The conjugate momenta are given by

$$p_A{}^0 = \frac{\partial L}{\partial v^A{}_0} = \rho g_{AB}v^B{}_0\sqrt{\det[G]}\,, \quad p_A{}^j = \frac{\partial L}{\partial v^A{}_j} = -\rho\frac{\partial W}{\partial v^A{}_j}\sqrt{\det[G]}\,,$$

and

$$\Pi = L - \frac{\partial L}{\partial v^A{}_\mu}v^A{}_\mu = \left[-\tfrac{1}{2}g_{AB}v^A{}_0 v^B{}_0 - W + \frac{\partial W}{\partial v^A{}_j}v^A{}_j\right]\rho\sqrt{\det[G]}\,.$$

We define the ***energy density*** e by

$$e = \frac{\partial L}{\partial v^A{}_0}v^A{}_0 - L \quad \text{or, equivalently} \quad e\,d^{n+1}x = \mathbb{K} + \mathbb{P}\,. \tag{5.4}$$

The Cartan form on J^1Y can be obtained either by using the Legendre transformation and pulling back the canonical $(n+1)$-form on the dual jet bundle as in (2.9), or by a variational route as in Theorem 3.3. The resulting coordinate expression is given by

$$\Theta_{\mathcal{L}} = \rho g_{AB} v^B{}_0 \sqrt{\det[G]} dy^A \wedge d^n x_0 - \rho \frac{\partial W}{\partial v^A{}_j} \sqrt{\det[G]} dy^A \wedge d^n x_j$$
$$+ \left[-\tfrac{1}{2} g_{AB} v^A{}_0 v^B{}_0 - W + \frac{\partial W}{\partial v^A{}_j} v^A{}_j \right] \rho \sqrt{\det[G]} d^{n+1} x \,. \tag{5.5}$$

Substituting the Lagrangian density (5.3) into equation (3.29) we obtain the Euler-Lagrange equations for a continuous medium

$$\rho g_{AB} \left(\frac{D_g \dot{\varphi}}{Dt} \right)^B - \frac{1}{\sqrt{\det[G]}} \frac{\partial}{\partial x^k} \left(\rho \frac{\partial W}{\partial v^A_k} (j^1 \varphi) \sqrt{\det[G]} \right)$$
$$= -\rho \frac{\partial W}{\partial g_{BC}} \frac{\partial g_{BC}}{\partial y^A} (j^1 \varphi) \,, \tag{5.6}$$

where

$$\left(\frac{D_g \dot{\varphi}}{Dt} \right)^A = \frac{\partial \dot{\varphi}^A}{\partial t} + \gamma^A_{BC} \dot{\varphi}^B \dot{\varphi}^C \tag{5.7}$$

is the covariant time derivative, and

$$\gamma^A_{BC} = \tfrac{1}{2} g^{AD} \left(\frac{\partial g_{BD}}{\partial y^C} + \frac{\partial g_{CD}}{\partial y^B} - \frac{\partial g_{BC}}{\partial y^D} \right)$$

are the Christoffel symbols associated with the metric g.

Given a potential energy W which specifies the material, equation (5.6) is a system of PDE's to be solved for a section $\varphi(x,t)$. We remark that all terms in this equation are functions of x and t and hence have the interpretation of material quantities. In particular, (5.7) corresponds to ***material acceleration.***

We define the multisymplectic analogue of the ***Cauchy stress tensor*** σ by

$$\sigma^{AB}(\varphi, x) = \frac{2\rho(x)}{J} \frac{\partial W}{\partial g_{AB}} \left(j^1 \varphi(x) \right) . \tag{5.8}$$

Equation (5.8) is known in the elasticity literature as the Doyle-Ericksen formula. We make the important remark that the balance of angular momentum

$$\sigma^T = \sigma$$

follows from the definition (5.8) and the symmetry of the metric tensor g.

In the case of Euclidean manifolds with constant metrics g and G, equation (5.6) simplifies to the familiar expression

$$\rho \frac{\partial^2 \varphi_A}{\partial t^2} = \frac{\partial}{\partial x^k} \left(\rho \frac{\partial W}{\partial v^A{}_k} (j^1 \varphi) \right) . \tag{5.9}$$

Next we will describe the multisymplectic formalism for the two main application of the theory we developed: elasticity and ideal fluid dynamics.

Elasticity. As we noted before, for the theory of elasticity the reference configuration (B, G) and the ambient space (M, g) are generally different. The spatial part B of the base manifold X has the interpretation of the reference configuration and the extra dimension of X corresponds to time. Later configurations of the elastic body are captured by a section φ of the bundle Y. For a fixed time t, the sections φ_t play the role of ***deformations***; they map the reference configuration B onto the spatial configuration, which is a subset of the ambient space M.

The fiber coordinates of the first jet $j^1\varphi$ of a section φ, as defined by (5.1), consist of the time derivative of the deformation $\dot{\varphi}^A$ and the ***deformation gradient*** F_i^A given by

$$F_i^A(x,t) = \frac{\partial \varphi^A}{\partial x^i}\,. \tag{5.10}$$

Hence, the first jet of a section φ has the local representation

$$j^1\varphi : (x,t) \mapsto \left((x,t), \varphi(x,t), \dot{\varphi}(x,t), F(x,t)\right).$$

For a given section φ, the ***first Piola-Kirchhoff stress tensor*** $\mathcal{P}_A{}^j$ is defined by

$$\mathcal{P}_A{}^j(\varphi, x) = \rho(x)\frac{\partial W}{\partial v^A{}_j}\left(j^1\varphi(x)\right). \tag{5.11}$$

We also define ***the Green deformation tensor*** (also called the right Cauchy-Green tensor) C by $C = \varphi_t^*(g)$; in coordinates we have

$$C_{ij}(x,t) = g_{AB}F_i^A F_j^B\,. \tag{5.12}$$

Using definitions (5.8) and (5.11), the Euler-Lagrange equations (5.6) become

$$\rho g_{AB}\left(\frac{D_g\dot{\varphi}}{Dt}\right)^B = \mathcal{P}_A{}^i{}_{|i} + \gamma^B_{AC}(\mathcal{P}^j_B F^C_j - J g_{BD}\sigma^{DC})\,, \tag{5.13}$$

where we have introduced the ***covariant divergence*** defined by

$$\mathcal{P}_A{}^i{}_{|i} = \mathrm{DIV}\mathcal{P} = \frac{\partial \mathcal{P}_A{}^i}{\partial x^i} + \mathcal{P}^j_A\Gamma^k_{jk} - \mathcal{P}^i_B\gamma^B_{AC}F^C_i\,. \tag{5.14}$$

Here, the Γ^i_{jk} are the Christoffel symbols corresponding to the base metric G.

We note that in (5.13) there is no a-priori relationship between the first Piola-Kirchhoff stress tensor and the Cauchy stress tensor, as W is assumed to have the most general form $W(x, G, g, v)$. However, such a relationship can be derived by imposing material frame indifference on the energy function. This assumption will imply that the energy function W depends on the deformation gradient F (equivalently, on v) and on the field metric g only through the Green deformation tensor given by (5.12) , that is $W = W(C(v,g))$. For this particular form of W, definitions (5.8) and (5.11) lead to

$$\mathcal{P}_A^{\ i} = J(\sigma F^{-1})_A^{\ i} . \tag{5.15}$$

Relation (5.15) is known as the Piola transformation law. Substituting it into (5.13), one obtains the Euler-Lagrange equations for the standard elasticity model

$$\rho g_{AB} \left(\frac{D_g \dot{\varphi}}{Dt} \right)^B = \mathcal{P}_{A\ |i}^{\ i} . \tag{5.16}$$

For elasticity in a Euclidean space, this equation simplifies to

$$\rho \frac{\partial^2 \varphi^A}{\partial t^2} = \frac{\partial \mathcal{P}^{Ai}}{\partial x^i} . \tag{5.17}$$

Barotropic Fluids. For ideal fluid dynamics we have the same multisymplectic bundle picture as that described for elasticity. For fluids moving in a fixed region we set $B = M$ and call it the ***reference fluid container***. However, for fluid dynamics with free boundary, the structures (B, G) and (M, g) are generally different. Configurations of the fluid are captured by a section φ of the bundle Y, which has the interpretation of the particle placement field. In coordinates, the spatial point $y \in Y_{(x,t)}$ corresponds to a position $y = \varphi(x,t)$ of the fluid particle x at time t.

For standard models of barotropic fluids, the potential energy of the fluid depends only on the Jacobian of the deformation, that is $W = W(J(g, G, v))$. The ***pressure*** function is defined to be

$$P(\varphi, x) = -\rho(x) \frac{\partial W}{\partial J} \left(j^1 \varphi(x) \right) . \tag{5.18}$$

For a given section φ, $P(\varphi) : X \to \mathbb{R}$ has the interpretation of the material pressure which is a function of the material density. Using (5.18), the Cauchy stress tensor (5.8) becomes

$$\sigma^{AB}(x) = \frac{2\rho}{J} \frac{\partial W}{\partial J} \frac{\partial J}{\partial g_{AB}} (j^1 \varphi) = -P(x) g^{AB}(y(x)) . \tag{5.19}$$

We refer to Marsden, Pekarsky, Shkoller, and West [2001] for a discussion on how the pressure function arises in both the compressible and incompressible models. We also remark here that one could also consider (5.19) as a defining equation for pressure, from which (5.18) would follow.

With these notations, the Euler-Lagrange equations (5.6) become

$$\rho g_{AB} \left(\frac{D_g \dot{\varphi}}{Dt} \right)^B = -\frac{\partial P}{\partial x^k} J \left(\left(\frac{\partial \varphi}{\partial x} \right)^{-1} \right)_A^k . \tag{5.20}$$

We introduce the ***spatial density*** $\rho_{\rm sp} = \rho / J$ and define the ***spatial pressure*** $p(y)$ by $p(y(x)) = P(x)$; then (5.20) can be re-written in the familiar form

$$\frac{D_g \dot{\varphi}}{Dt}(x,t) = -\frac{1}{\rho_{\rm sp}} \operatorname{grad} p \circ \varphi(x,t) . \tag{5.21}$$

5.2 Propagating Singular Surfaces Within an Elastic Body

In this subsection we apply the theory developed in §3.1 to investigate the motion of a singular surface of order 1 within a compressible elastic body. The ***order*** of a singular surface is given by the lowest order of the derivatives of the configuration map $\phi(x,t)$ that suffer a non-zero jump across the surface. For a singular surface of order 1, the configuration map $\phi(x,t)$ is continuous, but its first order derivatives (the velocity $\dot{\phi}$ and the deformation gradient F) may suffer jump discontinuities upon the surface. Thus, the configuration space for this problem belongs to class (a) of the classification considered in §3.1.

The multisymplectic formalism will lead to the derivation of the correct jumps in linear momentum and energy across the discontinuity surface. Moreover, spatial horizontal variations will lead to a kinematic condition known as the Maxwell compatibility condition.

We use the same notation as previously, so let U be diffeomorphic to an open subset of the spacetime X and let D be a codimension 1 submanifold in U representing a discontinuity surface in spacetime, moving within the elastic body. The configuration space $\mathcal{C}$ is given by (3.1) and all the results from §3 apply for this example.

We note first that Theorem 3.3 implies that the Euler-Lagrange equations (5.13) will be satisfied on either side of the discontinuity.

Now let $V \in T_\phi \mathcal{C}$ be a tangent vector with coordinates (V^μ, V^A) and consider initially only *vertical* variations ($V^\mu = 0$). From (3.31) we obtain the jump conditions

$$\int_{D_X} \left[\!\left[\frac{\partial L}{\partial {v^A}_\mu}(j^1(\phi \circ \phi_X^{-1})) \right]\!\right] \cdot V^A d^n x_\mu = 0\,, \tag{5.22}$$

where we used the continuity of the vector field V.

For simplicity, consider the Euclidean case, where $G_{\mu\nu} = \delta_{\mu\nu}$ and $g_{AB} = \delta_{AB}$. The jump relation (5.22) becomes

$$\int_{D_X} \left[\!\left[\rho(x) \frac{\partial \varphi^A}{\partial t} \right]\!\right] \cdot V^A d^n x_0 + \int_{D_X} [\![\mathcal{P}_A^j]\!] \cdot V^A d^n x_j = 0\,, \tag{5.23}$$

where, as before, φ denotes the section $\varphi = \phi \circ \phi_X^{-1}$.

The 1-forms dt and dx^j, $j = 1, \ldots, n$, on D_X are not independent. More precisely, if D_X is given locally by $f(t, x^1, \ldots, x^n) = 0$, then by differentiating we obtain

$$\partial_j f\, dx^j + \partial_t f\, dt = 0\,. \tag{5.24}$$

Define N_j by $N_j = \partial_j f / |\nabla_x f|$ and define the propagation speed U by

$$U = -\frac{\partial_t f}{|\nabla_x f|}\,, \tag{5.25}$$

where $|\cdot|$ represents the Euclidean norm. This speed is a measure of the rate at which the moving surface traverses the material; it also gives the excess of the normal speed of the surface over the normal speed of the particles comprising it. Then (5.23) becomes

$$\int_{D_X} \left(\left[\!\left[\rho U \frac{\partial \varphi^A}{\partial t} \right]\!\right] + [\![\mathcal{P}_A^j]\!] \, N_j \right) \cdot V^A d^n x_0 = 0 \,. \tag{5.26}$$

By a standard argument in the calculus of variations we can pass to the local form and recover the standard jump of linear momentum across a propagating singular surface of order 1, which is

$$\left[\!\left[\rho U \frac{\partial \varphi^A}{\partial t} \right]\!\right] + [\![\mathcal{P}_A^j]\!] \, N_j = 0 \,. \tag{5.27}$$

An alternative approach to derive the jump in linear momentum uses the balance of linear momentum for domains traversed by singular surfaces (see Truesdell and Toupin [1960], pg. 545 for example). For such derivations, the jump conditions are usually expressed in spatial coordinates, where the propagation speed U and the Piola–Kirchhoff stress tensor are replaced by a local propagation speed and the Cauchy stress tensor, respectively.

Remarks.

1. The conservation of mass implies the following jump relation, known as the *Stokes–Christoffel condition* (see Truesdell and Toupin [1960], pg. 522)
$$[\![\rho U]\!] = 0 \,. \tag{5.28}$$
In Courant and Friedrichs [1948], the continuous quantity ρU is denoted by m and is called the *mass flux* through the surface. Using (5.28), we can re-write (5.27) as
$$\rho U \left[\!\left[\frac{\partial \varphi^A}{\partial t} \right]\!\right] + [\![\mathcal{P}_A^j]\!] N_j = 0 \,. \tag{5.29}$$

2. In the terminology of Truesdell and Toupin [1960], the singular surfaces that have a non-zero propagation speed are called *propagating* singular surfaces or *waves*. We first note that this definition excludes material surfaces, for which $U = 0$. Moreover, Truesdell and Toupin [1960] classifies the singular surfaces with non-zero jump in velocity into two categories: surfaces with transversal discontinuities ($[\![U]\!] = 0$), called ***vortex sheets***, and surfaces with arbitrary discontinuities in velocity ($[\![U]\!] \neq 0$), called ***shock surfaces***. One can conclude that there are no material shock surfaces, i.e., shock surfaces are always waves. Also, there is a nonzero mass flux ($m \neq 0$) through a shock surface or through a vortex sheet which is not material.

3. In the context of gas dynamics, Courant and Friedrichs [1948] defines a ***shock front*** as a discontinuity surface across which there is a nonzero gas flow ($m \neq 0$). Then, a shock surface in the sense of Truesdell and Toupin [1960] is also a shock front.

The linear momentum jump was derived by taking vertical variations of the sections. Next we will focus on *horizontal* variations ($V^\mu \neq 0$) and derive the corresponding jump laws. Using (5.27), the jump conditions (3.31) become

$$\int_{D_X} \left[\!\left[LV^\mu - \frac{\partial L}{\partial v^A_\mu}(j^1(\phi \circ \phi_X^{-1}))\frac{\partial(\phi \circ \phi_X^{-1})^A}{\partial x^\nu} V^\nu \right]\!\right] d^n x_\mu = 0\,. \tag{5.30}$$

Consider first only time component variations ($V^0 \neq 0$, $V^j = 0$ for $j = 1, \ldots, n$); then (5.30) gives

$$\begin{aligned}\int_{D_X} &\left[\!\left[L - \frac{\partial L}{\partial v^A_0}\frac{\partial(\phi \circ \phi_X^{-1})^A}{\partial t} \right]\!\right] V^0 d^n x_0 \\ &- \int_{D_X} \left[\!\left[\frac{\partial L}{\partial v^A_j}\frac{\partial(\phi \circ \phi_X^{-1})^A}{\partial t} \right]\!\right] V^0 d^n x_j = 0\,.\end{aligned} \tag{5.31}$$

Using (5.11), (5.24) and (5.25), (5.31) becomes

$$\int_{D_X} \left([\![Ue]\!] + \left[\!\left[\mathcal{P}^j_A \frac{\partial \varphi^A}{\partial t} \right]\!\right] N_j \right) V^0 d^n x_0 = 0\,. \tag{5.32}$$

From (5.32) we recover the standard jump of energy (see Truesdell and Toupin [1960], pg. 610 for example),

$$[\![Ue]\!] + \left[\!\left[\mathcal{P}^j_A \frac{\partial \varphi^A}{\partial t} \right]\!\right] N_j = 0\,. \tag{5.33}$$

Finally we consider space component variations ($V^j \neq 0$) in (5.30) and use (5.32) to obtain

$$\begin{aligned}\int_{D_X} &\left[\!\left[\frac{\partial L}{\partial v^A_0}\frac{\partial(\phi \circ \phi_X^{-1})^A}{\partial x^j} V^j \right]\!\right] d^n x_0 \\ &- \int_{D_X} \left[\!\left[LV^j - \frac{\partial L}{\partial v^A_j}\frac{\partial(\phi \circ \phi_X^{-1})^A}{\partial x^k} V^k \right]\!\right] d^n x_j = 0\,.\end{aligned} \tag{5.34}$$

Then, by using (5.11), (5.24), (5.25) and (5.28), (5.34) becomes

$$\int_{D_X} \left(\rho U \left[\!\left[F^A{}_j \frac{\partial \varphi^A}{\partial t} \right]\!\right] + [\![L]\!] N_j + [\![F^A{}_j \mathcal{P}_A{}^k]\!] N_k \right) V^j d^n x_0 = 0\,. \tag{5.35}$$

Since the components V^j are arbitrary we conclude that

$$\rho U \left[\!\!\left[F_j^A \frac{\partial \varphi^A}{\partial t} \right]\!\!\right] + [\![L]\!] N_j + [\![F_j^A \mathcal{P}_A^k]\!] N_k = 0 \,. \tag{5.36}$$

Even though (5.36) does not resemble any standard conservation law, after some algebraic manipulations using (5.27) and (5.33), (5.36) can be rewritten, for continuous U, as

$$U \, [\![F_j^A]\!] + \left[\!\!\left[\frac{\partial \varphi^A}{\partial t} \right]\!\!\right] N_j = 0 \,, \tag{5.37}$$

which is the statement of the Maxwell compatibility condition (see Jaunzemis [1967], Chapter 2 or Truesdell and Toupin [1960], Chapter C.III. for the derivation of the kinematical conditions of compatibility from Hadamard's lemma).

To summarize, the *vertical* variations of the sections led us to derive the jump in linear momentum, while *horizontal* time and space variations accounted for the energy balance and the kinematic compatibility condition, respectively.

5.3 Free Surfaces in Fluids and Solid–Fluid Boundaries

Now, we investigate in the multisymplectic framework a different type of discontinuous motion that will illustrate the case (b) of the classification from §3.1. We consider two types of discontinuity surfaces, namely *free surfaces* in fluids and *solid–fluid boundaries*. A free surface or a free boundary is a surface separating two immiscible fluids or two regions of the same fluid in different states of motion (Karamcheti [1966]). The second type of discontinuity considers the interaction of a deformable elastic body with a surrounding barotropic fluid.

As we already noted in §3.1, these types of discontinuous surfaces have one feature in common, namely they are material surfaces (particles which are on the surface at a given time remain on the surface at later times). Equivalently, there is no flow across the discontinuity and the surface is stationary relative to the medium. Hence, in the reference configuration, the surface D_X is given locally by $f(x^1, \ldots, x^n) = 0$ (no dependence of the function f on t). Moreover, from (5.25) we have that the propagation speed U for such surfaces is zero. In the terminology of Truesdell and Toupin [1960], these surfaces are material vortex sheets of order 0.

Free Surfaces in Fluids. Theorem 3.3 implies that Euler-Lagrange equations of type (5.20) will be satisfied on either side of the surface separating the two fluid regions. Next, we will show that Theorem 3.3 gives the correct force balance on the separating surface and the other physical conditions that must be satisfied on such boundaries.

Let $V \in T_\phi \mathcal{C}$ be a tangent vector with coordinates (V^μ, V^A). We consider first only *vertical* variations ($V^\mu = 0$); from (3.31) we obtain the following jump conditions

$$\int_{D_X} \left[\!\left[\frac{\partial L}{\partial v^A_\mu} (j^1(\phi \circ \phi_X^{-1})) \cdot V^A \right]\!\right] d^n x_\mu = 0 . \tag{5.38}$$

For simplicity, we will consider Euclidean geometries, that is $G_{\mu\nu} = \delta_{\mu\nu}$ and $g_{AB} = \delta_{AB}$. We recall that for fluids, $W = W(J)$; this relation and the stationarity of the discontinuity surface ($U = 0$ on D_X) simplifies the jump relation (5.38) to

$$\int_{D_X} \left[\!\left[\rho \frac{\partial W}{\partial J} \frac{\partial J}{\partial v^A_j} N_j \cdot V^A \right]\!\right] d^n x_0 = 0 , \tag{5.39}$$

where $N_j = \frac{\partial_j f}{|\nabla_x f|}$ is the normal vector to D_X. From the definition of the Jacobian J (5.2), one can derive

$$\frac{\partial J}{\partial v^A_j} = J(v^{-1})^j_A .$$

We use this relation and the definition of the material pressure (5.18) to re-write (5.39) as

$$\int_{D_X} \left[\!\left[PJ \left(\left(\frac{\partial \varphi}{\partial x} \right)^{-1} \right)^j_A N_j V^A \right]\!\right] d^n x_0 = 0 . \tag{5.40}$$

We notice that in (5.40), the term $J \left(\left(\frac{\partial \varphi}{\partial x} \right)^{-1} \right)^j_A N_j d^n x_0$ represents the A-th component of the area element in the spatial configuration, as given by the formula of Nanson (see Truesdell and Toupin [1960], pg. 249 or Jaunzemis [1967], pg. 154, for example). Hence, substituting $y = \varphi_t(x)$ in (5.40) and then passing to the local form we can obtain the jump relation

$$[\![pV^A N_A]\!] = 0 , \tag{5.41}$$

where p is the spatial pressure defined by $p(y(x)) = P(x)$. Now, we combine (5.41) with the property that the vector field V has a zero normal jump (see (3.13)), to obtain

$$[\![p]\!] = 0 , \tag{5.42}$$

which is the standard pressure balance at a free surface.

We take now horizontal variations $V^\mu \neq 0$ such that $V^0 \neq 0$ and $V^j = 0$, for $j = 1, \ldots, n$. Then, (3.31) simplifies to

$$\int_{D_X} \left[\!\left[\frac{\partial L}{\partial v^A_j} \frac{\partial (\phi \circ \phi_X^{-1})^A}{\partial t} N_j V^0 \right]\!\right] d^n x_0 = 0 . \tag{5.43}$$

Furthermore, using the continuity of V^0 and the particular form of $W = W(J)$ in the Lagrangian (5.3), we can write (5.43) as

$$\int_{D_X} \left[\!\left[PJ \left(\left(\frac{\partial \varphi}{\partial x} \right)^{-1} \right)^j_A \frac{\partial \varphi^A}{\partial t} N_j \right]\!\right] V^0 d^n x_0 = 0 . \tag{5.44}$$

As before, we use the formula of Nanson, substitute $y = \varphi_t(x)$ in (5.44), and then pass to the local form to obtain

$$\left[\!\left[p \frac{\partial \varphi^A}{\partial t} N_A \right]\!\right] = 0 . \tag{5.45}$$

Using the pressure continuity (5.42), the jump condition (5.45) becomes

$$\left[\!\left[\frac{\partial \varphi^A}{\partial t} N_A \right]\!\right] = 0 . \tag{5.46}$$

We can also use the continuity of the normal vector to write (5.46) as

$$\left[\!\left[\frac{\partial \varphi^A}{\partial t} \right]\!\right] N_A = 0 . \tag{5.47}$$

The jump condition (5.46) is a kinematic condition which restricts the possible jumps of the fluid velocity only to tangential discontinuities (the normal component is continuous). In the literature, this condition may appear either as a boundary condition (see Karamcheti [1966]) or as a definition for vortex sheets (see Truesdell and Toupin [1960]). However, we recover it through a variational procedure, as a consequence of the general theorem of the §3.2, using the particular form of the space of configurations (3.2) and of its admissible variations (3.13).

Finally, let consider only space component horizontal variations ($V^0 = 0$ and $V^j \neq 0$, for $j = 1, \dots, n$). The vector field $V^j \frac{\partial}{\partial x^j}$ on X is lifted by $T(\phi \circ \phi_X^{-1})$ to a horizontal vector field on Y (see decomposition (3.7)) with coordinates

$$V^A = \frac{\partial (\phi \circ \phi_X^{-1})^A}{\partial x^j} V^j . \tag{5.48}$$

Then, by using the previous jump conditions (5.42), (3.31) simplifies to

$$\int_{D_X} \left[\!\left[\frac{\partial L}{\partial v^A_k} \frac{\partial (\phi \circ \phi_X^{-1})^A}{\partial x^j} N_k V^j \right]\!\right] d^n x_0 = 0 , \tag{5.49}$$

where we also used that $V^j N_j = 0$ for material surfaces. Using (5.48) and the definition of the material pressure (5.18), (5.49) becomes exactly (5.40), so it will provide the already known jump condition (5.42).

Solid–Fluid Boundaries. We again apply Theorem 3.3 to find that the Euler-Lagrange equations (5.13) will be satisfied in the domain occupied by the elastic body, while the fluid dynamics in the outer region will be described by (5.20). As for free surfaces, the boundary terms in Theorem 3.3 will give the correct pressure-traction balance on the boundary of the elastic body, as well as restrictions on the jumps in velocity.

For vertical variations only, the jump conditions are those given by (5.38). For Euclidean geometries, these conditions become

$$\int_{D_X} \left[\mathcal{P}^j_A N_j (V^A)^+ - PJ\left(\left(\frac{\partial\varphi}{\partial x}\right)^{-1}\right)^j_A N_j (V^A)^- \right] d^n x_0 = 0\,, \tag{5.50}$$

where we adopt the usual notation with superscript + and − for the limit values of a discontinuous function at a point on the singular surface by approaching the point from each side of the discontinuity.

Using the Piola transformation (5.15) and the formula of Nanson we can make the substitution $y = \varphi_t(x)$ in (5.50) and then pass to the local form; we obtain

$$\sigma^{AB} N_B (V^A)^+ - pN_A (V^A)^- = 0\,. \tag{5.51}$$

We use the property of the vector field V from (3.13),

$$(V^A)^+ N_A - (V^A)^- N_A = 0\,,$$

to write (5.51) as

$$(\sigma^{AB} N_B - pN_A)\cdot (V^A)^+ = 0\,. \tag{5.52}$$

As there are no restrictions on $(V^A)^+$, we have

$$\sigma^{AB} N_B - pN_A = 0\,. \tag{5.53}$$

Moreover, by denoting by $t^A = \sigma^{AB} N_B$ the stress vector, we obtain

$$t^A N_A - p = 0\,, \tag{5.54}$$

which is the pressure-traction balance on the boundary of the elastic body.

We now consider horizontal variations $V^\mu \neq 0$ such that $V^0 \neq 0$ and $V^j = 0$, for $j = 1, \ldots, n$. Using the previous result (5.54), the general jump conditions (3.31) reduce to

$$\int_{D_X} \left[\!\left[\frac{\partial L}{\partial v^A_j} \frac{\partial (\phi \circ \phi_X^{-1})^A}{\partial t} N_j V^0 \right]\!\right] d^n x_0 = 0\,. \tag{5.55}$$

For solid–fluid interactions, the jump conditions (5.55) become

$$\int_{D_X} \left[\mathcal{P}^j_A N_j \left(\frac{\partial \varphi^A}{\partial t}\right)^+ - PJ\left(\left(\frac{\partial\varphi}{\partial x}\right)^{-1}\right)^j_A N_j \left(\frac{\partial \varphi^A}{\partial t}\right)^- \right] V^0 d^n x_0 = 0\,, \tag{5.56}$$

By using the Piola transformation (5.15) and Nanson's formula, we make the substitution $y = \varphi_t(x)$ in (5.56) and then pass to the local form to get

$$\sigma^{AB} N_B \left(\frac{\partial \varphi^A}{\partial t}\right)^+ - pN_A \left(\frac{\partial \varphi^A}{\partial t}\right)^- = 0 . \tag{5.57}$$

Now, from (5.53) and (5.57) we can derive

$$\left[\!\left[\frac{\partial \varphi^A}{\partial t}\right]\!\right] N_A = 0 , \tag{5.58}$$

which implies the continuity of the normal component of the velocity. Thus, only tangential discontinuities in the velocity are possible. We emphasize again that we obtain this restriction as a consequence of the choice of the configuration space (see (3.2) and (3.13)) and not by prescribing it as a boundary condition.

By an argument similar to the one used for fluid–fluid interfaces, we can show that the space component horizontal variations do not provide new jump conditions; they will lead in fact to the jump condition (5.51), from which the pressure-traction balance (5.54) can be derived.

5.4 Collisions of Elastic Bodies

We now illustrate the last category of the classification of configuration spaces from §3.1. We will apply the general formalism to investigate the collision of two elastic bodies, where the configuration manifold is given by $\mathcal{C}^c$ defined in (3.3) and the analog of the singular surfaces from the previous subsections is the codimension 1 spacetime contact surface. The interface is a material surface, so it has a zero propagation speed $U = 0$. By the choice of the configuration space $\mathcal{C}^c$ we allow the elastic bodies to slip on each other during the collision, but they do so without friction.

If we consider only vertical variations, the jump conditions will be given by (5.38). In the Euclidean case these conditions become

$$\int_{D_X} [\![\mathcal{P}_A^{\,j} N_j V^A]\!] d^n x_0 = 0 . \tag{5.59}$$

By making the change of variables $y = \phi_t(x)$ in (5.59) and using the Piola transformation (5.15), we can write the integral in the spatial configuration and then pass to the local form to obtain

$$[\![\sigma^{AB} N_B V^A]\!] = 0 , \tag{5.60}$$

where N_A are the components of the outward unit normal to the contact set in the current configuration.

Let $t^A = \sigma^{AB} N_B$ denote the stress vector, as before. By using the jump restriction on V from (3.15) we can derive

$$[\![t^A N_A]\!] = 0\,, \tag{5.61}$$

which represents the balance of the normal tractions on the contact set during a collision. From the derivation of (5.61) we also obtain that the tangential tractions are zero on the contact surface.

Let us consider now time component horizontal variations ($V^0 \neq 0$ and $V^j = 0$, for $j = 1, \ldots, n$); the general jump conditions (3.31) reduce to (5.55), which in turn become

$$\int_{D_X} \left[\!\!\left[\mathcal{P}_A^j N_j \left(\frac{\partial \varphi^A}{\partial t} \right) \right]\!\!\right] V^0 d^n x_0 = 0\,. \tag{5.62}$$

By the same procedure used before, we can pass to the local form in the spatial configuration and obtain

$$\left[\!\!\left[t^A \frac{\partial \varphi^A}{\partial t} \right]\!\!\right] = 0\,. \tag{5.63}$$

From (5.61) and (5.63) we can derive

$$\left[\!\!\left[\frac{\partial \varphi^A}{\partial t} N_A \right]\!\!\right] = 0\,, \tag{5.64}$$

which gives the continuity of the normal components of the velocities, once the contact is established. However, the tangential discontinuity in velocities, due to slipping, may be arbitrary.

The space component horizontal variations will not provide new jump conditions; we can show this by the same procedure used in §5.3.

6 Concluding Remarks and Future Directions

There are several directions to pursue in the future to complete the foundations laid in this paper. Perhaps the most important task is to develop algorithms and a discrete mechanics for nonsmooth multisymplectic variational mechanics and to take advantage of the current algorithms (such as that of Pandolfi, Kane, Marsden, and Ortiz [2002]) that are already developing in this direction.

Another task is to further develop the theory of shock waves by combining the geometric approach here with more analytical techniques, such as those used in hyperbolic systems of conservation laws, as well as incorporating appropriate thermodynamic notions.

We also need to extend the basic theory to incorporate constraints, similar to the way that Marsden, Pekarsky, Shkoller, and West [2001] deal with incompressibility constraints.

For some systems, there will be surface tension and other boundary effects; for some of these systems a Hamiltonian structure is already understood (see Lewis, Marsden, Montgomery, and Ratiu [1986] and references therein) but not a multisymplectic structure. Developing the multisymplectic formalism for fluid–solid interactions would also be of interest.

Here we have only considered isolated discontinuities, but there may be degeneracies caused by the intersections of different types and dimensions of discontinuity surfaces that require further attention.

Finally, as in Kane, Marsden, Ortiz, and West [2000] and Pandolfi, Kane, Marsden, and Ortiz [2002], friction (or other dissipative phenomena) and forcing need to be included in the formalism.

Acknowledgments: We are grateful for helpful remarks from Michael Ortiz and Steve Shkoller. This work was partially supported by AFOSR contract F49620-02-1-0176.

References

Abraham, R., J. E. Marsden, and T. S. Ratiu [1988], *Manifolds, Tensor Analysis and Applications*, second edition, *Applied Mathematical Sciences*, vol. **75**, Springer-Verlag, New York.

Benjamin, T. B. [1984], Impulse, flow force and variational principles, *IMA J. Appl. Math.* **32**, 3–68.

Courant, R. and K. O. Friedrichs [1948], *Supersonic Flow and Shock Waves*, volume 1 of *Pure and applied mathematics*. Interscience Publishers, Inc., New York.

Ebin, D. G. and J. E. Marsden [1970], Groups of diffeomorphisms and the motion of an incompressible fluid, *Ann. of Math.* **92**, 102–163.

Fetecau, R., J. E. Marsden, M. Ortiz, and M. West [2002], Nonsmooth Lagrangian mechanics and variational collision algorithms, *(preprint)*.

Gotay, M., J. Isenberg, and J. E. Marsden [1997], Momentum Maps and the Hamiltonian Structure of Classical Relativistic Field Theories I, *available at* `http://www.cds.caltech.edu/~marsden/`.

Graff, K. F. [1991], *Wave Motion in Elastic Solids.* Dover, New York, reprinted edition.

Jaunzemis, W. [1967], *Continuum Mechanics.* The Macmillan Company, New York.

Kane, C., E. A. Repetto, M. Ortiz, and J. E. Marsden. [1999], Finite element analysis of nonsmooth contact, *Computer Meth. in Appl. Mech. and Eng.* **180**, 1–26.

Kane, C., J. E. Marsden, M. Ortiz, and M. West [2000], Variational integrators and the Newmark algorithm for conservative and dissipative mechanical systems, *Int. J. Num. Math. Eng.* **49**, 1295–1325.

Karamcheti, K. [1966], *Principles of Ideal-Fluid Aerodynamics.* Robert E. Krieger Publishing Company, Inc., Florida, corrected reprint edition, 1980.

Kouranbaeva, S. and S. Shkoller [2000], A variational approach to second-order mulitysmplectic field theory, *J. of Geom. and Phys.* **35**, 333–366.

Lew, A., J. E. Marsden, M. Ortiz, and M. West [2002], Asynchronous variational integrators, *Arch. Rat. Mech. Anal. (to appear).*

Lewis, D., J. E. Marsden, R. Montgomery, and T. S. Ratiu [1986], The Hamiltonian structure for dynamic free boundary problems, *Physica D* **18**, 391–404.

Marsden, J. E. and T. J. R. Hughes [1983], *Mathematical Foundations of Elasticity.* Prentice Hall. Reprinted by Dover Publications, NY, 1994.

Marsden, J. E., G. W. Patrick, and S. Shkoller [1998], Multisymplectic geometry, variational integrators and nonlinear PDEs, *Comm. Math. Phys.* **199**, 351–395.

Marsden, J. E., S. Pekarsky, S. Shkoller, and M. West [2001], Variational methods, multisymplectic geometry and continuum mechanics, *J. Geom. and Physics* **38**, 253–284.

Marsden, J. E. and T. S. Ratiu [1999], *Introduction to Mechanics and Symmetry*, volume 17 of *Texts in Applied Mathematics*, vol. **17**; 1994, Second Edition, 1999. Springer-Verlag.

Marsden, J. E. and M. West [2001], Discrete mechanics and variational integrators, *Acta Numerica* **10**, 357–514.

Moreau, J.-J. [1982], Fluid dynamics and the calculus of horizontal variations, *Internat. J. Engrg. Sci.* **20**, 389–411.

Moreau, J. [1986], Une formulation du contact à frottement sec; application au calcul numérique, *C. R. Acad. Sci. Paris Sér. II* **302**, 799–801.

Moreau, J.-J. [1988], Free boundaries and nonsmooth solutions to some field equations: variational characterization through the transport method. In *Boundary control and boundary variations (Nice, 1986)*, volume 100 of *Lecture Notes in Comput. Sci.*, pages 235–264. Springer, New York.

Palais, R. S. [1968], *Foundations of Global Non-Linear Analysis.* Benjamin/Cummings Publishing Co., Reading, MA.

Pandolfi, A., C. Kane, J. E. Marsden, and M. Ortiz [2002], Time-discretized variational formulation of nonsmooth frictional contact, *Int. J. Num. Methods in Engineering* **53**, 1801–1829.

Truesdell, C. A. and R. Toupin [1960], The Classical Field Theories. In Flügge, S., editor, *Encyclopedia of Physics*, volume III/1. Springer-Verlag OHG, Berlin.

8

Geometric Analysis for the Characterization of Nonstationary Time Series

Michael Kirby
Charles Anderson

To Larry Sirovich, on the occasion of his 70th birthday.

ABSTRACT Subspace methodologies, such as the Karhunen-Loève (KL) transform, are powerful geometric tools for the characterization of high-dimensional data sets. The KL transform, or the related singular value decomposition (SVD), maximizes the mean-square projection of the data ensemble on subspaces of reduced rank. Other interesting subspace approaches solve modified optimization problems and have received comparably less attention in the literature. Here we present two such methodologies: 1) signal fraction analysis (SFA), a method that optimizes the amount of signal retained when signals are superposed and 2) canonical correlation analysis (CCA), a method for contructing transformations that allow the comparison of two data sets. We compare these methods to the more widely employed SVD in the context of real data. We address the important and practical problem of whether two time-series are generated by the same process. As a specific example, the classification of noisy multivariate electroencephalogram (EEG) time-series data is considered.

Contents

1 Empirical Geometric Data Analysis

Innovations in computer technology have now fundamentally altered the manner in which information is collected and processed. Of central importance is our significantly enhanced ability to obtain a detailed empirical view of the world. The impact of *high-resolution* representations of information on the process of knowledge discovery is only beginning to be realized. Current applications are wide ranging and include, for example, airplane design, medical imaging, space exploration and weather prediction. Indeed, the general problem of extracting useful information from large data sets is now common across disciplines (Case [1999]). Terabyte hard drives combined with gigabyte memories and gigahertz CPUs are driving the development of new algorithms to process, characterize and understand distinguishing features buried in data.

One of the most revealing trends has been the discovery of structures in data. For example, it is now well known that apparently turbulent and chaotic fluid flows possess coherent structure, see, e.g., (Sirovich [1987a]) and that understanding this structure is paramount to understanding the underlying phenomenon. Perhaps surprisingly, it appears that many processes, both natural and artificial, have a tendency towards self-organization and as a result produce patterns and structure that can be directly investigated (Camazine, Deneubourg, Franks, Sneyd, Theraulaz, Bonabeau [2001]; Haken [1983]).

The purpose of this paper is to present several basic methods for extracting structure from data and to explore their application on a hard problem in spatio-temporal pattern analysis, i.e., automated interpretation of electroencephalogram (EEG) waves. The long-term goal of this research program is ambitious, i.e., to develop an efficient procedure to process spontaneous EEG signals for the purposes of developing a practical and noninvasive machine-human interface. The state-of-the-art of this line of research is available in the proceedings of the first NIH-sponsored workshop on "Brain-Computer Interfaces" published in a special issue of *IEEE Transactions on Rehabilitation Engineering*, summarized by Wolpaw, Birbaumer, Heetderks, McFarland, Peckham, Schalk, Donchin, Quatrano, Robinson and Vaughan [2000].

The outline of the paper is as follows: In Section 2 the optimization problem for the standard KL procedure is introduced. In Section 3 we consider an alternative optimization problem for comparing two data sets known as canonical correlation analysis. In Section 4 we present a geometric approach for splitting signal and noise based on yet another optimization criterion. In Section 5 we apply the suite of methodologies presented to the problem of extracting structure in data and in Section 6 we examine the relative utility of these representations for classifying multivariate time series. In summary, we will argue that there is, based on the techniques presented here, significant opportunity to quantify and exploit discriminatory structure in EEG data.

2 Karhunen-Loève Analysis (KLA)

A now fundamental technique (as well as geometric prototype) for the low-dimensional characterization of large data sets data sets is based on the Karhunen-Loève (KL) expansion (Karhunen [1946]; Loeve [1946]).[1] A major advance in this direction was made in a suite of papers by Larry Sirovich that outlined the theoretical and computational issues of its application to the field of fluid dynamics (see Sirovich [1987a,b]; Sirovich, [1987c]).

Applications to image processing (Sirovich and Kirby [1987]), and the management of large scientific data bases (Everson and Sirovich [1992]) were made possible by the introduction of the *snapshot method*. This innovation permitted the analysis of families of patterns, such as faces, residing in high-dimensional vector spaces. Prior work employing the KL transform to digital images revolved around a technique known as *block coding* (Rao and Yip [1990]). A key aspect to this innovation proposed by Sirovich was the geometric viewpoint: now patterns were to be viewed as points in high-dimensional vector spaces.

2.1 A Method Before Its Time?

The application of the KL transform to the characterization of large data sets has been referred to by Sirovich as *data hungry*. Indeed, in many ap-

[1]The discrete Karhunen-Loève transform is also referred to as principal component analysis (Jolliffe [1986]; Preisendorfer [1988]), the method of empirical orthogonal eigenfunctions (Lorenz [1956]) as well as the singular value decomposition (SVD), (see Kirby [2001] for a general overview). We use these terms interchangeably.

plications in the literature the available "large" data sets are too small to adequately characterize the patterns under investigation. The mathematical methodology is rarely the limiting factor in such analyses but rather the limitations associated with data generation and collection. As such, the KL technique has been awaiting the massive data sets that are only recently becoming available. Thus, unlike many other approaches, the utility of this powerful mathematical methodology is actually increasing as the computer technology improves. Furthermore, as experimental techniques in areas such as fluid mechanics continue to improve and ultimately produce highly-resolved, three-dimensional vector fields, there will be even more need for data hungry subspace methods such as the KL transform for addressing their analysis.

The computational limitations of the KL transform, i.e., those aside from technological barriers such as data storage and retrieval, stem primarily from the rank of the data matrix under investigation. The numerical estimation of the best basis requires the solution of a symmetric eigenvector problem of the rank of the data matrix. It appears that there has been relatively little research that addresses the problem of applying the KL procedure to data matrices that generate very high rank covariance matrices; a notable exception is Sirovich's *method of partitions* (Sirovich [1987a]). Adaptive methods motivated by computation in biological systems have also been proposed; (see Diamantaras and Kung [1996] for a review).

2.2 The KL Optimization Problem and SVD

In this section we briefly review the framework of the KL procedure in the vector space setting. This development will provide a context for the comparison of methods employing alternative optimization criteria presented in later sections. For further details and additional references on the KL procedure and applications see, e.g., Jolliffe [1986]; Kirby [2001]; Preisendorfer [1988].

Given an ensemble of P points $\{x^{(\mu)} \in \mathbb{R}^N\}$, where the pattern index $\mu = 1, \ldots, P$, the KL procedure produces a *best basis* $\{u^{(i)}\}$ in the sense that the D-term expansion

$$x_D = \sum_{i=1}^{D} a_i^{(\mu)} u^{(i)}$$

minimizes the mean-square truncation error, i.e.,

$$\text{mse} = \langle \|x - x_D\|^2 \rangle \tag{2.1}$$

for any D; the angle brackets denote ensemble average and the norm is taken as the usual Euclidean norm. If we write the set of best basis vectors that span this best D-dimensional subspace as the columns of a $N \times D$

matrix, i.e., $U_D = [u^{(1)}|\cdots|u^{(D)}]$, then the KL change of basis may be expressed as

$$a = U_D^T x \tag{2.2}$$

Now, the ensemble $\{a^{(\mu)} \in \mathbb{R}^D\}$ is a D-dimensional representation of the originally N-dimensional data. The assumption in general is that $D \ll N$; of course, whether this is in fact true or not is an empirical fact and depends on the data set under investigation.

Note that the pattern index μ may simply be a label, as for the Rogues Gallery problem (Sirovich and Kirby [1987]; Kirby and Sirovich [1990]), or, alternatively, it may denote time as for the processing of a sequence of digital images (Kirby, Weisser and Dangelmayr [1991]; Kirby, Weisser, and Dangelmayr [1993]). Computationally, the best basis vectors (with non-zero eigenvalues) $\{u^{(i)}\}$ may be found efficiently and accurately by computing the *thin* singular value decomposition of the data matrix, i.e.,

$$X = U\Sigma V^T. \tag{2.3}$$

Note that this approach avoids forming the normal equations XX^T or X^TX and the associated loss of numerical precision in the computation of the eigenvalues and eigenvectors (Golub and Van Loan [1996]; Trefethen and Bau [1997]).

3 Canonical Correlation Analysis (CCA)

Here we present canonical correlation analysis—another geometric subspace tool for analyzing large data sets initially proposed by Hotelling [1936]. To motivate this presentation, we first discuss some of the interesting questions that can be addressed more appropriately with CCA than with the SVD (or KL transform). This discussion is followed by a derivation of the method; the results of the application of this problem to time-series analysis are presented in Section 5.2; see also Section 5.4 for an example of how the methodology extracts highly correlated structure between two data sets.

3.1 Why CCA?

Often the investigation of a physical system or systems involves the comparison of two or more data sets. Questions arise such as

- Are two data sets generated by the same process?
- Are there relationships between variables in a single realization of a process that reveal latent structure in the data?
- Can certain variables in a process be identified as providing superior predictive skill?

For the third question we are motivated in part by the application of CCA to the time-series prediction problem investigated in the context of weather forecasting (Barnett and Preisendorfer [1987]). The following brief description of CCA follows, in part, that found in Mardia, Kent and Bibby [1979].

3.2 The CCA Optimization Problem

Given two data matrices $X \in \mathbb{R}^{M\times P}$ and $Y \in \mathbb{R}^{N\times P}$, it is possible to determine a relationship between them by applying an appropriate change of coordinates to each. To determine vectors $a \in \mathbb{R}^M$ and $b \in \mathbb{R}^N$ that extract the similarity in the two data sets, it is natural to attempt to maximize the correlation between the scalar values $c_x = a^T x$ and $c_y = b^T y$. Of course this should be done ensemble-wise leading to the optimization problem

$$\max_{a,b} \quad a^T C_{xy} b$$

subject to the constraints

$$a^T C_{xx} a = 1$$

and

$$b^T C_{yy} b = 1$$

where

$$C_{xy} = XY^T,$$
$$C_{xx} = XX^T$$

and

$$C_{yy} = YY^T.$$

This maximization problem leads to the generalized eigenvalue problem

$$\begin{pmatrix} 0 & C_{xy} \\ C_{yx} & 0 \end{pmatrix} \begin{pmatrix} a \\ b \end{pmatrix} = \mu \begin{pmatrix} C_{xx} & 0 \\ 0 & C_{yy} \end{pmatrix} \begin{pmatrix} a \\ b \end{pmatrix}$$

Substituting the transformations $\alpha = C_{xx}^{\frac{1}{2}} a$ and $\beta = C_{yy}^{\frac{1}{2}} b$ into the generalized eigenvalue problem produces the equations

$$C_{xx}^{-\frac{1}{2}} C_{xy} C_{yy}^{-\frac{1}{2}} b = \mu a$$

$$C_{yy}^{-\frac{1}{2}} C_{yx} C_{xx}^{-\frac{1}{2}} a = \mu b$$

Introducing

$$Z = C_{xx}^{-\frac{1}{2}} C_{xy} C_{yy}^{-\frac{1}{2}}$$

these equations may be rewritten as the system

$$\begin{pmatrix} 0 & Z \\ Z^T & 0 \end{pmatrix} \begin{pmatrix} a \\ b \end{pmatrix} = \mu \begin{pmatrix} a \\ b \end{pmatrix} \tag{3.1}$$

Thus, we recognize that a and b are the left and right singular vectors of the matrix Z, respectively. There are high precision routines for calculating the solutions $\{a^{i)}, b^{(i)}, \mu^{(i)}\}$. In addition, one may alternatively compute the eigenvectors of the $M \times M$ problem

$$ZZ^T a = \lambda a$$

or the $N \times N$ problem

$$Z^T Z b = \lambda b$$

where

$$ZZ^T = C_{xx}^{-\frac{1}{2}} C_{xy} C_{yy}^{-1} C_{yx} C_{xx}^{-\frac{1}{2}}$$

and

$$Z^T Z = C_{yy}^{-\frac{1}{2}} C_{yx} C_{xx}^{-1} C_{xy} C_{yy}^{-\frac{1}{2}}$$

The result of this computation is the singular value decomposition

$$Z = U \Sigma V^T \tag{3.2}$$

where the diagonal matrix $\Sigma_{ii} = \sqrt{\lambda_i}$. The solution vectors are collected in matrices as $A = [a^{(1)}| \cdots |a^{(L)}]$ and $B = [b^{(1)}| \cdots |b^{(L)}]$ where $L = \min\{M, N\}$.

4 Signal Fraction Analysis (SFA)

The KL procedure produces an optimal set of data dependent eigenvectors. Given a data matrix X, the eigenvectors $\{u^{(i)}\}$may be expressed

$$u^{(i)} = X\beta^{(i)}$$

where the coefficient vectors $\beta^{(i)}$ reflect the weighting of the data required to make an optimal set of eigenvectors according to the minimum mean square error criterion. As a result of this data dependency expressed above, it appears clear that if the data itself consists of signal plus noise, then so will the optimal eigenvectors. Signal fraction analysis is a method that acts to filter out the noise component of a data set from the optimal eigenvectors.

4.1 Why SFA?

Each element of a family of patterns $\{x^{(\mu)}\}$ may be composed of a signal component $p^{(\mu)}$ and a mixture of other extraneous signals that may be viewed collectively as a *noise* component $q^{(\mu)}$, i.e., $x^{(\mu)} = p^{(\mu)} + q^{(\mu)}$, or, in terms of data matrices,

$$X = P + Q.$$

In the general situation, neither the de-noised signal $p^{(\mu)}$ nor the noise component $q^{(\mu)}$ is observable; only the noisy signal $x^{(\mu)}$ is available.

As alluded to above, an optimal representation of such noisy data in terms of the KL transform generally results in both the signal and the noise combining to construct the data dependent eigenvectors–even those associated with large variance; (see Anderle and Kirby [2002] for an example in the context of filtering noisy time series). A notable exception occurs for the special case where the *noise* covariance matrix has the form

$$QQ^T = \alpha^2 I,$$

i.e., *white* noise which is assumed to have zero mean and be uncorrelated with the signal; here α^2 is the variance of the noise and I is the $N \times N$ identity matrix. In this instance, the covariance matrix of the signal may be decomposed as

$$\begin{aligned} C_x &= \langle (p+q)(p+q)^T \rangle \\ &= \langle pp^T \rangle + \langle qq^T \rangle \\ &= C_p + \alpha^2 I, \end{aligned}$$

where $C_p = \langle pp^T \rangle$. As the observed (composite signal) covariance matrix C_x and the covariance matrix of the noise C_q are related by a shift of the identity, it follows that they share the same eigenvectors. The associated eigenvalues are all shifted upwards by the variance of the noise α^2, leaving the differences of the eigenvalues as well as their ordering preserved (Fukunaga and Olsen [1971]); the preceding discussion follows (Kirby [2001]).

In situations where the noise is not white, an alternative to the KL procedure may be desirable. For example, EEG time series will be seen to possess colored noise in the application considered in this paper. Thus, here we consider an alternative subspace optimization criterion that will lead us to the maximum signal fraction method originally proposed by Green, Berman, Switzer and Craig [1988] in the context of the analysis of multi-spectral satellite data; (see also Switzer and Green [1984]). This approach provides an alternative to the KL procedure for treating data with additive non-white noise. It is also useful as a subspace approach for preprocessing data.

4.2 The SFA Optimization Problem

One approach to separating P and Q, given X, is based on determining the subspace that best describes the data in P and another subspace that best describes the data in Q. Depending on the manner in which P and Q reside in their respective spaces this procedure may result in a third subspace that describes the overlapping region common to both P and Q.

The subspace that best represents the data matrix P may be found by expressing the data dependence of an eigenvector as

$$\begin{aligned} \phi &= X\psi && (4.1)\\ &= P\psi + Q\psi. && (4.2) \end{aligned}$$

In other words, each basis vector may be decomposed as

$$\phi = \phi_P + \phi_Q$$

where the component $\phi_P = P\psi$ represents the P data dependency and $\phi_Q = Q\psi$ represents the Q data dependency.

Now the *Q-fraction* of a basis vector ϕ is defined as

$$D(\phi) = \frac{\phi_Q^T \phi_Q}{\phi^T \phi}.$$

Maximizing the function $D(\phi)$ maximizes the representation of Q by the basis vector ϕ. This may now be rewritten as

$$D(\phi(\psi)) = \frac{\psi^T Q^T Q \psi}{\psi^T X^T X \psi}.$$

The maximization problem leads to a *symmetric definite generalized eigenproblem*

$$\alpha X^T X \psi = \beta Q^T Q \psi. \qquad (4.3)$$

Thus, given a data matrix X and the solution matrix $\Psi_m = [\psi^{(1)}|\dots|\psi^{(m)}]$ of the generalized singular vector problem, given in Equation (4.3), the orthonormal basis for $\mathbb{R}^m$ is given by $\Phi_m = [\phi^{(1)}|\dots|\phi^{(m)}]$ where $\Phi_m = X\Psi_m$. The basis vectors are ordered by increasing noise fraction, so a truncation of the basis corresponds to noise filtering. If we include the complete basis consisting of m vectors, we may express the data without loss as $X = \Phi_m \Phi_m^T X$. Truncating columns of Φ_m filters the noise, i.e., the data may be decomposed as

$$X_D = \Phi_D B_D,$$

where the smaller matrix

$$B_D = \Phi_D^T X$$

consists of reconstruction coefficients and $D < m$.

4.3 Estimating the Covariance Matrix of the Noise

Note that to compute the Ψ matrix (and ultimately the maximum signal basis vectors) it is necessary to estimate $Q^T Q$.[2] As suggested by Switzer

[2]Alternatively, one could choose to estimate the matrix $P^T P$.

and Green [1984], the covariance matrix of the noise may be estimated by shifting the data and computing differences. One can show in this case, under the appropriate assumptions[3], that the covariance matrix of the differences Σ is approximately twice $Q^T Q$. Note that this method requires that the data be smooth, i.e., $x_t \approx x_{t+1}$. It should be noted that the procedure for estimating the covariance matrix of the noise need not be absolutely perfect. The penalty of error is a modest rotation of the subspace spanning the maximum noise fraction eigenvectors. For further details concerning the estimation of the covariance matrix see Green, Berman, Switzer and Craig [1988] as well as Anderle [2001].

5 Representations of EEG Time Series

In this section we compare and contrast the utility of the SVD, CCA and SFA techniques to produce low-dimensional representations of the EEG data. In particular, we are interested in characterizations of the data that extract coherent structures that may be exploited for the classification of mental tasks. Thus, in each case, the raw data is converted to an alternative representation associated with the appropriate optimality condition. These representations are selected as follows:

- SVD: right singular vectors in V obtained in Equation (2.3)
- CCA: correlation vectors a, b obtained in Equation (3.1)
- SFA: the generalized singular vectors ψ obtained in Equation (4.3)

Further discussion on the selection and form of these representations in the context of time lagging is discussed below.

We used data obtained previously by Keirn and Aunon (Keirn and Aunon [1990]; Keirn [1988]) who used the following procedure. The subjects were seated in an Industrial Acoustics Company sound controlled booth with dim lighting and noiseless ventilation fans. An Electro-Cap elastic electrode cap was used to record from positions C3, C4, O1, O2, P3, and P4, shown in Figure 5.1 and defined by the 10–20 system of electrode placement (Jasper [1958]). These six channels were referenced to electrically linked mastoids at A1 and A2. The impedance of all electrodes was kept below five Kohms. Data were recorded at a sampling rate of 250 Hz with a Lab Master 12 bit A/D converter mounted in an IBM-AT computer. Before each recording session, the system was calibrated with a known voltage. The electrodes were connected through a bank of Grass 7P511 amplifiers with analog bandpass filters from 0.1–100 Hz. Eye blinks were detected by

[3]Note that these assumptions are only required as a means to estimate the covariance matrix of the noise and are not intrinsic to the SFA approach itself

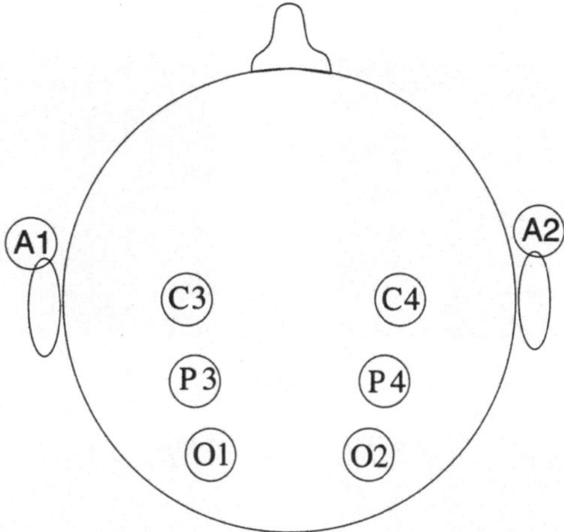

FIGURE 5.1. Location of the electrodes on the head. Note that electrodes A1 and A2 are used for calibration and are not used in the data analysis.

means of a separate channel of data recorded from two electrodes placed above and below the subject's left eye. In addition, subjects were asked to remain as motionless as possible as extraneous muscle movement appears as artifacts in the EEG data.

In this application, we focus on distinguishing between two distinct mental tasks involving one subject and recorded under controlled conditions as multivariate time-series. The first data class corresponds to EEG data recorded while the subject performed mental arithmetic, i.e., the Math Task. During the second task the subject was asked to imagine writing a letter to someone, i.e., the Letter Task.

Data from two trials of each task were used. Each trial lasted 10 seconds and data were recorded at 250 Hz resulting in a total of 2500 samples for each trial. In addition, the data in each trial was segmented into sliding windows of length 64. This 1/4 second window length was selected to permit a reasonably fast response time for the human-machine interface problem. Note also that studies suggest that EEG data is stationary over windows of up to one second. In the absence of data lagging, a data point consists of a vector of concatenated 64 6-tuples of electrode values. For a time lag of k points this becomes a $64 \times 6 \cdot (k+1)$ dimensional point. All four trials were chosen as ones in which the subject did not blink their eyes.

To assess the utility of each method for the classification problem one trial, labelled Trial 1 here, from each task is employed as the training data while a different trial, i.e., Trial 2, is employed as the test data for assessing the discriminating information present in the structures extracted from each methodology. Once the set of parameters is computed for each training window all the windows in both test sets are classified using the Fisher's linear discriminant as well as k-nearest neighbors (described in

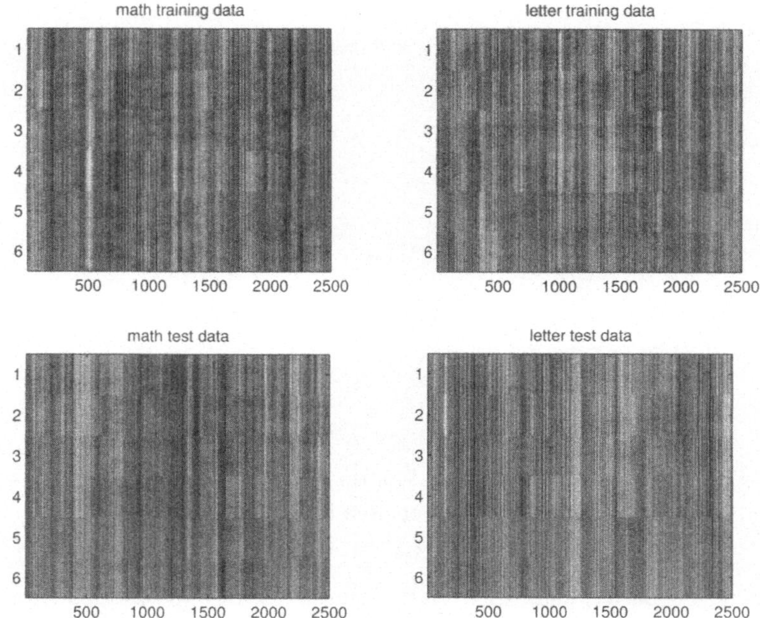

FIGURE 5.2. Top left: training data for math task. Bottom left: test data for math task. Top right: training data for letter task. Bottom right: test data for letter task. EEG signals recorded from each of the six electrodes at positions, C3, C4, P3, P4, O1 and O2 are displayed in rows 1–6, respectively. The signals (in microvolts) are mean subtracted and scaled to display coordinates.

Section 6.1.2). The classification rates are determined by the ratio of the number of windows correctly classified for each task and the total number of windows in both test sets. The classification rates are computed for all the methods with lags ranging from $k = 0$ to $k = 4$.

The raw training and test data for both the math and letter tasks is shown in Figure 5.2. Here the data has been mean-subtracted and scaled into display coordinates. Each of the six time series associated with the electrodes is displayed as a row of length 2500, corresponding to ten seconds of recording.

Note that the classification of the raw data, i.e., without employing any of the information packing representations discussed, is very poor, i.e., slightly better than chance — see Figure 5.3.

5.1 Representation Via SVD

The SVD was calculated for each $64 \times 6 \cdot (k+1)$ data matrix X_i associated with the ith window where $k = 0, \ldots, 4$ is the data time lag. Writing the factorization

$$X_i = U_i \Sigma_i V_i^T,$$

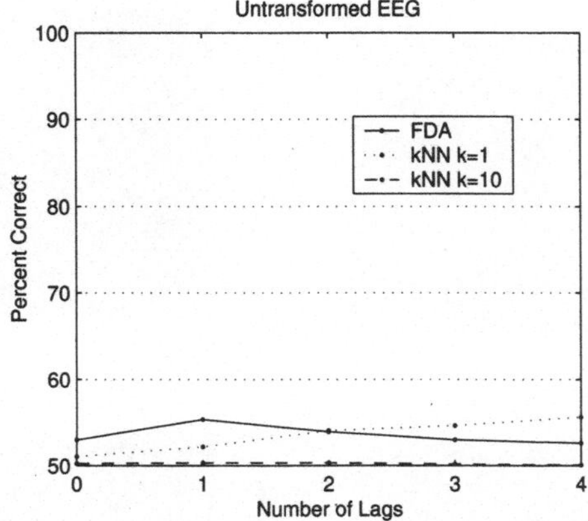

FIGURE 5.3. Percent of test samples correctly classified by Fisher and kNN classifiers for data with zero to four lags represented as untransformed data; see Section 6 for a discussion of these classification techniques. The best accuracy is approximately 55%.

we selected the right singular vectors as candidates for classification of the time series. For example, if $k = 2$ each right singular vector is an 18-tuple vector, a reduction in dimension for each window from $64 \cdot 12 = 728$ values to 18 numbers.

An example of the representation of the data in terms of SVD (for the training and testing samples for the math and letter tasks) is displayed in Figure 5.4. As a result of its superior classification performance (see Section 6) only the seventh mode is shown for the case of time lag 2. This mode is computed for each window and is evolving in time. Note that although the raw data appears to have high correlation along a column (fixed point in time) the right singular vector shown does not. Indeed, some of the components are seemingly stationary along the rows while others have significant variance. In particular, the variance along the rows 13-18 of the letter task appears much smaller than the math task along the same rows. These rows correspond to the twice-lagged values from all electrodes.

Thus, it appears visually that this mode has information that should permit a discrimination between the two mental tasks. We see later that this is indeed the case.

5.2 Representation Via CCA Transform

As discussed above, CCA is designed to compare two data sets. There are many ways to form such pairs with the EEG data, such as using data from two different trials or partitioning one trial into two sets of electrodes. In this investigation we present the results for the latter experiment, motivated by the possible discovery of time-dependent interactions between the

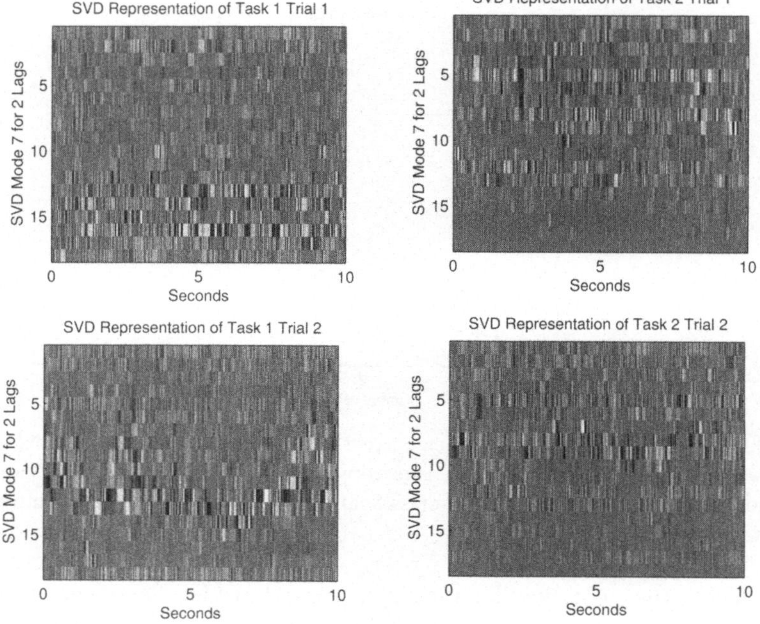

FIGURE 5.4. Seventh SVD modes for windowed data with two lags. This representation is one for which classification performance was over 90%. (See also Figure 6.2.) The two displays on the left are the two Math trials and on the right are the two Letter trials. The horizontal axis is time; each 64-sample window (1/4 second) is represented by the seventh SVD mode displayed as a column of gray strips. Dark strips are most negative and bright strips are most positive. Each SVD mode (column of image) consists of 18 values, six corresponding to data samples from each of the six channels for the current time step, C3, C4, P3, P4, O1, and O2, and six for the one lag step, C3', C4', P3', P4', O1', and O2' and six more for the two lag step C3", C4", P3", P4", O1", and O2".

electrodes for different modes of thought.

Thus, for the following experiments, two data sets were formed from the single trials (shown in Figure 5.2) by partitioning the data into samples from electrodes on the left side of the head versus the right side—samples from C3, P3, and O1 comprised one data set and samples from C4, P4, and O2 comprised the second data set. Asymmetry in EEG across the two hemispheres of the brain are often observed during mental tasks.

Although the data is now partitioned into two subsets as described above, the windowing and lagging procedures are exactly as describe in the previous section. Thus, here we are investigating a *local* CCA applied to each window of each trial independently. Applying CCA to samples with k time lags results in the time dependent matrices of correlation vectors A_i and B_i being of dimension $(k+1) \cdot 3 \times (k+1) \cdot 3$.

Using the A_i and B_i matrices as a representation for the once-lagged data in a window results in a dimensionality reduction of $64 \cdot 12 = 768$ down to $2 \cdot 6 \cdot 6 = 72$ (two matrices A and B each of size 6×6).

This can be reduced even further if we consider just the first few columns of A and B as the linear combinations of signals from each channel that produce the strongest correlation. For example, if just the first columns of A and B are used, only 12 parameters represent each window of data.

Figure 5.5 shows the CCA representation of once-lagged data consisting of the single highest-correlation canonical vectors of A and B as a column vector of 12 components for each window.[4] The first three components are the components of the first canonical vector from A that correspond to the C3 sample at time i, at time $i-1$ and at time $i-2$. Lagged components are indicated by electrode names followed by quotation marks.

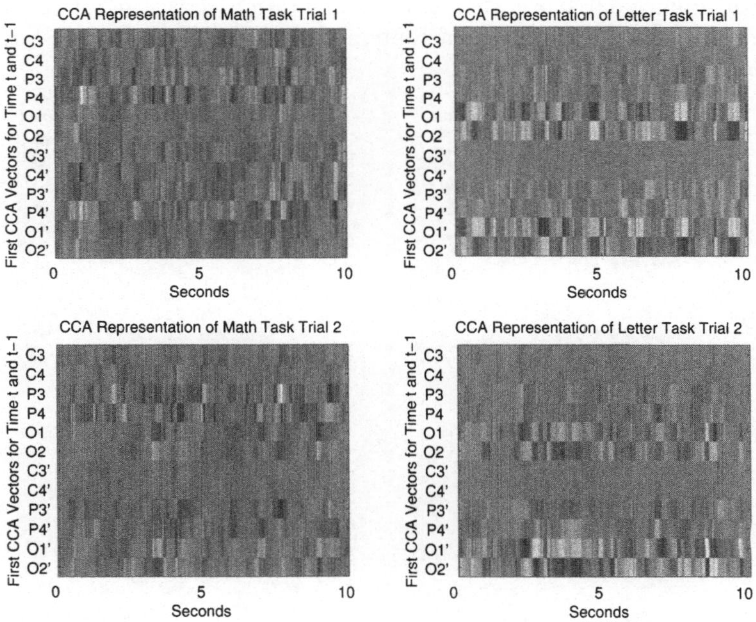

FIGURE 5.5. First CCA mode for matrices A and B for windowed data with one lag. The components for first mode of A appear in the odd numbered rows (C3, P3, O1, C3', P3', O1') and the components of B appear in the even numbered rows (C4, P4, O2, C4', P4', O2'). This figure suggests that the O1 and O2 electrodes possess discriminatory information for the letter task. (See caption for Figure 5.4 for additional explanation.)

[4] As the matrices A_i and B_i for each window CCA transform are only computed up to a sign, for consistency the dot product between A_i and A_{i-1} was compared to the dot product between B_i and B_{i-1}. The sign of the dot product of greater magnitude was used as the factor to multiply both matrices A_i and B_i.

5.3 Representation Via Signal Fraction Analysis

Recall that the SFA eigenvectors of a data matrix X may be written $\Phi = X\Psi$. In analogy with our treatment via the CCA, we select the coding of the data in the matrix Ψ as our representation for classification.

To classify segments of EEG data windowed as in the previous sections, a local version of the SFA transform was employed. The SFA transform of for each window produces a 12 x 12 Ψ matrix. Using each Ψ as a representation for the once-lagged data in a window results in a dimensionality reduction of $64 \cdot 12 = 768$ down to $12 \cdot 12 = 144$. This can be reduced even further if we consider the last few columns of Ψ as the linear combinations of signals from each channel that produce the cleanest, least noisy signals. For example, if just the last column of Ψ is used, only 12 parameters represent each window of data. Figure 5.6 shows the SFA representation consisting of the single strongest-signal mode of Ψ as a column vector of 12 components for each window.

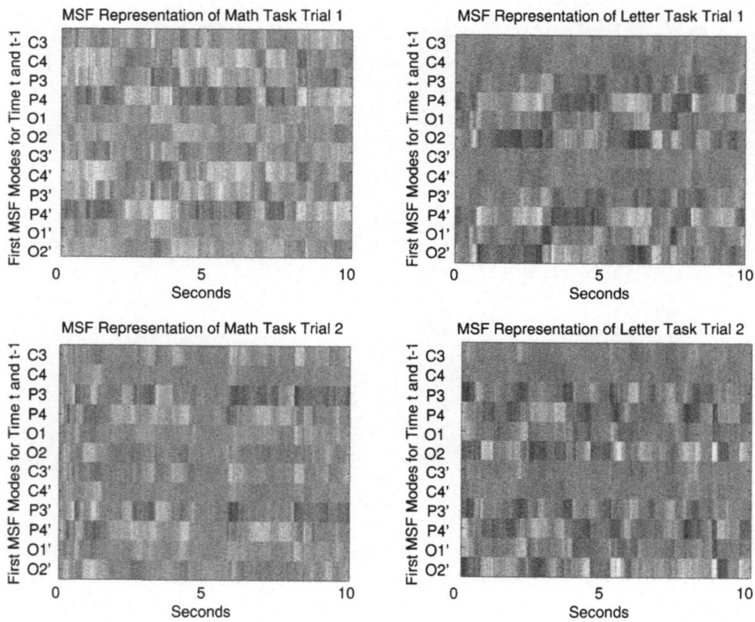

FIGURE 5.6. First SFA modes for windowed data with one lag. Again, as with CCA (see Figure 5.5), the O1 and O2 electrodes seem to possess discriminatory information for the letter task with the SFA representation. Compare Figure 5.4.

A number of differences between tasks are apparent. The most obvious differences are the C3, C4 components, which have relatively large values for math task trials, and the O1, O2 components, which have relatively large values for the letter tasks trials. Both tasks show significant values in the P3, P4 components.

5.4 CCA Extracts 60 Hz Noise

In this section we give an example of how CCA was able to identify structure in the data, i.e., the electrical hum of a 60Hz power source recorded by the EEG electrodes (see Section 5 for a detailed discussion of the data and how it was collected). Note that special precautions were taken to eliminate such experimental artifacts so the discovery of the 60Hz signal in a subset of the data came as a surprise.[5]

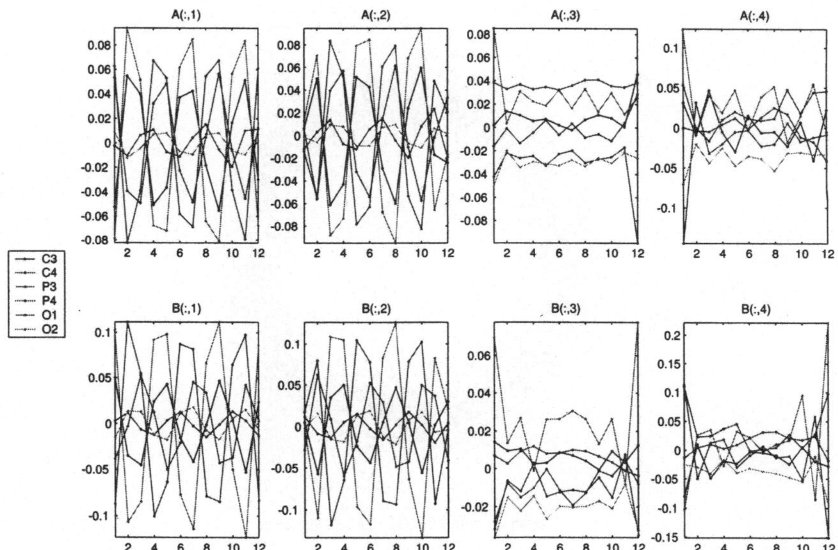

FIGURE 5.7. The first two canonical vectors for both data sets reveal a periodicity associated with 60Hz electrical artifact. See text for further explanation.

The data in this experiment has a 12 order lag (see Section for a description of the lagging procedure A). Figure 5.7 plots the first four columns of matrices A and B. Each column is plotted as separate curves for each of the 12 (6 in A and 6 in B) components corresponding to the 12 lagged values from each channel. The first two columns show curves with a period of approximately four. Since the sampling rate is 250 Hz, this corresponds to a period of about 1/60th of a second, indicating a strong 60 Hz noise. This is present in the C3, C4, P3, and P4 channels, with opposite phase between the two hemispheres of the head (C3 vs. C4 and P3 vs P4). The 60 Hz signal is also present in O1 and O2, but it is not opposite phase in the two hemispheres. The 60 Hz signal appears to be mostly captured by the first two canonical sequences represented by the first two columns of

[5]Of course one can identify structure such as this using Fourier analysis. In this instance we actually discovered the 60Hz hum first using CCA and then verified its presence by looking at the power spectrum of the data.

the matrices A and B. It does not appear in the third and fourth.

Figure 5.8 shows a plot of the projection of the EEG data windows onto the first two columns of A, the first two canonical sequences. (Actually, data from the first task is projected onto the first two A columns, and data from the second task is projected onto the first two B columns.) The two dimensional projections approximately lie on a circle as would be expected if the first two canonical sequences are truly capturing the 60 Hz noise in the signal.

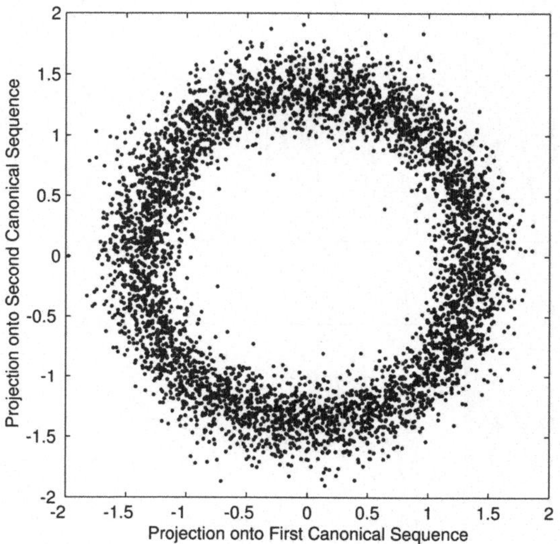

FIGURE 5.8. A two-dimensional representation of the data in terms of the most highly correlated coordinates revealing another view of the periodicity evidenced in Figure 5.7.

We remark that the signal fraction analysis, developed in Section 4, provides an excellent subspace approach for filtering out the 60Hz noise. Indeed, the two modes are clearly pure signal and noise free. Note that filtering this artifact via Fourier methods can be less clean due to leakage of the Fourier representation.

6 Classification Results

In this section we compare the results of applying Fisher's discriminant analysis (FDA) and k-nearest neighbor classification (kNN) algorithms to the representations of the EEG data computed in Section 5. The classification rates as a function of the number of modes retained for the classification task are shown in Figure 6.1. The predictive skill of individual modes for all representations is also shown in Figure 6.2. The number of data lags and its effect on the classification rates is also discussed.

6.1 Classification Methodologies

6.1.1 Fisher's Discriminant Analysis (FDA)

To keep this paper as self-contained is possible, we present a short description of a classical approach for classification known as Fisher's linear discriminant analysis. This discussion roughly follows (Duda and Hart [1973]).

Fisher's method addresses the question of how an optimal separating hyperplane may be constructed for the classification of two sets of data. It is assumed that samples of each class are available and that the class membership of these classes is known. The goal is to now classify (correctly), in terms of this hyperplane, previously unseen examples of both classes. If the data sits in $\mathbb{R}^N$, then the hyperplane is completely specified by its normal.

Fisher proposed an objective function that considered the projection of the data onto a space of one dimension (the hyperplane normal) that would maximize the *separation* of the data. Denote this projection by

$$y = w^T x$$

where it is assumed w has unit length. Fisher quantified separation as the ratio of two terms. The numerator of this cost function is proportional to the distance between classes as measured by the distance between the projected means, i.e.,

$$N(w) = |\langle y\rangle_{y\in C_1} - \langle y\rangle_{y\in C_2}|^2$$

where the mean of class i is given by

$$\langle y\rangle_{y\in C_i} = \frac{1}{|C_i|}\sum_{x\in C_i} w^T x$$

The denominator of the cost function is a measure of how scattered the data in each class is about its mean. The scatter for each class is measured by

$$S_i = \sum_{y\in C_i} (y - \langle y\rangle_{y\in C_i})^2$$

producing a denominator of the total scatter

$$D(w) = S_1 + S_2.$$

In summary, Fisher proposed that w be found such that the cost function

$$J(w) = \frac{N(w)}{D(w)}$$

is maximized. This leads to the following generalized singular value problem

$$S_B w = \lambda S_w w$$

for the normal w where the *between class scatter* matrix is defined as

$$S_B = (\langle x\rangle_{x\in C_1} - \langle x\rangle_{x\in C_2})(\langle x\rangle_{x\in C_1} - \langle x\rangle_{x\in C_2})^T$$

and the *within class scatter* matrix is defined as

$$S_w = \sum_{x\in C_1}(x - \langle x\rangle)(x - \langle x\rangle)^T + \sum_{x\in C_2}(x - \langle x\rangle)(x - \langle x\rangle)^T$$

Because of the rank deficiency of S_B this problem has the simple solution

$$w = S_W^{-1}(\langle x\rangle_{x\in C_1} - \langle x\rangle_{x\in C_2})$$

(See Duda and Hart [1973] for details of this derivation.)

6.1.2 k-Nearest Neighbors (kNN)

An alternative to the separating hyperplane approach for classification described above is k nearest neighbors (kNN). In kNN, all of the pairwise distances of the training data are computed. When a new point is presented to the system, the k closest points, or nearest neighbors, in the training set are found. The class of the new point is assigned to be that of the most frequent class in the k nearest neighbors. In this investigation we employed $k = 1$ and $k = 10$ classification with comparable results. For $k = 1$ the new point is assigned the class of its nearest neighbor while for $k = 10$ the new point is assigned that class occuring 6 or more times (for the two class problem). The selection of the metric to establish the distance between points is a parameter of the procedure. Here we have restricted our attention to the usual Euclidean distance.

6.2 Multi-Mode Classification

Here we discuss the classification results as a function of the number of modes in the representations of the data via SVD, CCA and SFA.

As can be seen in the first column of Figure 6.1, the classification of the EEG data using the right singular vectors of the SVD depends significantly on the number of lags used. No time lagging results in very poor classification rates for any number of modes; superior results are found empirically for lag two data. Five modes are required to obtain classification rates over 90%. Note, however, that SVD does not order modes according to their capacity for discrimination; predictive skill of individual modes is considered below. Interestingly, lagging the data beyond two data points actually degrades classification performance.

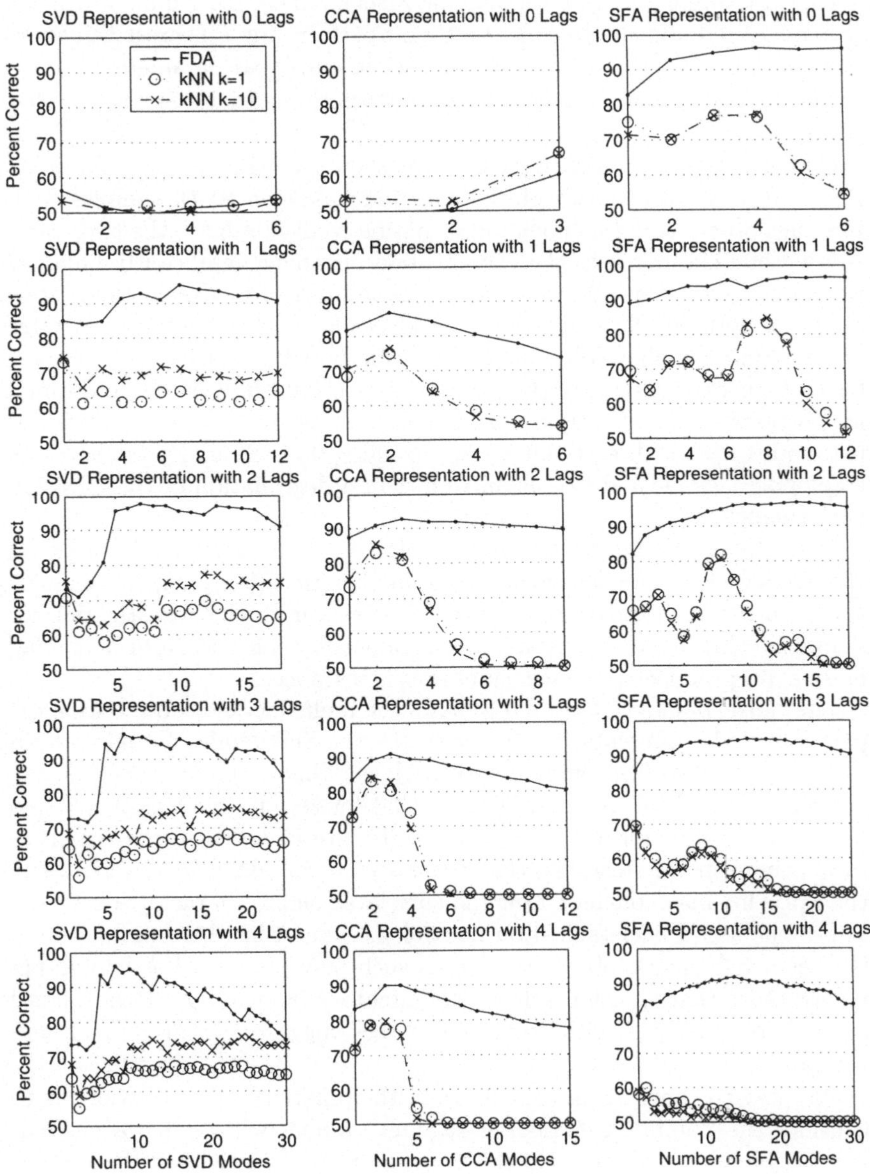

FIGURE 6.1. Percent of test samples correctly classified by FDA and kNN classifiers for data with zero to four lags represented as SVD, CCA, and SFA transforms. Each row of graphs is for a different number of lags, starting with zero in the top row to four in the bottom row. The first column of graphs is for the SVD representation, the second column is for the CCA representation, and the third column is for the SFA representation. The horizontal axis in each graph is the number of modes used to perform the classification.

It is also interesting to note that the Fisher's linear discriminant classification method appears to consistently outperform the k nearest neighbors method. This suggests that only a subset of employed parameters are actually performing the discrimination, an hypothesis that warrants further study.

Classification results using CCA are shown in the second column of Figure 6.1. As in the case of classification based on SVD, samples with zero lags do not contain enough information to discriminate the two tasks. Adding one lag increases the classification accuracy significantly but now we notice a significant dependence on the number of modes employed–two modes is optimal. Note also that the kNN and Fisher classification rates are now more comparable, but Fisher's method still consistently outperforms k-nearest neighbors. As with SVD, two lags appears to be optimal for purposes of classification; in addition, the first mode is seen to discriminate quite well with an accuracy approaching 90%. Similarly to the SVD, employing lags beyond two again results in a degradation of classification performance.

It is also appealing to attempt to classify data using CCA based on the windowed canonical correlations Σ in Equation (3.2), i.e., the singular values rather than the canonical vectors. However, in this investigation we found the classification accuracy was significantly inferior. Note that a lag of four did produce the best results in this instance.

The third column of Figure 6.1 shows the multi-mode discrimination capacity of SFA representation. Surprisingly, the SFA modes actually afford a reasonable classification rate with no lags. This is the only method examined that had this feature. Observe that for zero to two lags there is no degradation in performance as more modes are included in the representation; note in this instance that the *noise* is contained in the modes in the tail. One may conclude that the correlated signal characterized by the tail of the series is classification neutral, i.e., knowing this information is neither helpful nor problematic and as such may be viewed as containing no structure that translates into discriminatory information. Performance of kNN does degrade with additional modes suggesting that this method is more sensitive to noise.

One way to compare the results from the different representations is to consider the minimum number of lags and modes with which a particular classification accuracy is achieved. Figure 6.1 shows that the highest accuracy achieved by CCA is about 92%. Table 6.1 shows for each representation the minimum number of lags, the minimum number of included modes, and the resulting dimensionality of the representation of each 64-sample window for which a 92% accuracy is achieved. The table shows that the SFA representation captures the discriminatory information in considerably fewer dimensions than either SVD or CCA.

Another way to compare the results for the different representations is to consider the maximum accuracy achieved for each. Table 6.2 lists the

	Lags	Modes	Dim.
SVD	1	5	60
CCA	2	3	54
SFA	0	2	12

TABLE 6.1. Minimum configurations of each representation for which 92% classification accuracy is achieved. "Dim." is the dimensionality of each 64-sample window of data for the corresponding representation configuration, calculated as $6(l+1)m$ where l is the number of lags and m is the number of modes.

	Accuracy	Lags	Modes	Dim.
SVD	98%	2	7	126
CCA	92%	2	3	54
SFA	98%	2	14	252

TABLE 6.2. Maximum classification accuracies and configurations for each representation.

maximum accuracy and the number of lags, modes, and resulting dimensionality. This table summarizes a few observations. SVD and SFA representations resulted in higher accuracies than did CCA. Considering all configurations, SVD and SFA reach the same high accuracy level of 98%. All three representations achieve the best results for 2 lags.

6.3 Single-Mode Classification

The classification experiments were repeated for all methodologies using single mode representations, see the graphs in Figure 6.2 for the results. This study was motivated by the fact that the modes in the multi-mode representations considered above are not ordered according to their discriminatory capacity. Therefore it was deemed of interest to identify modes with maximum discriminatory capacity across the methods and to attempt to identify relevant relationships and structure in the variables that are associated with discriminatory capacity.

Most obvious, and perhaps surprising, is the superior single mode classification of the right singular vectors. In particular, mode seven with lag two data appears to be the most discriminating single mode from all the methods with a classification rate of just over 90% (see Figure 5.4 for the *coherent structure* revealed by this mode). There are also other single modes with high classification rates, particularly for the SFA approach. It is noteworthy that the variance in the classification rates seems to be highest for the SVD (modes are either really good or really bad) and smallest for the SFA with the lowest order modes, i.e., those containing the maximum signal, being the best while the highest order modes associated with the noise have little discriminatory information, as would be expected.

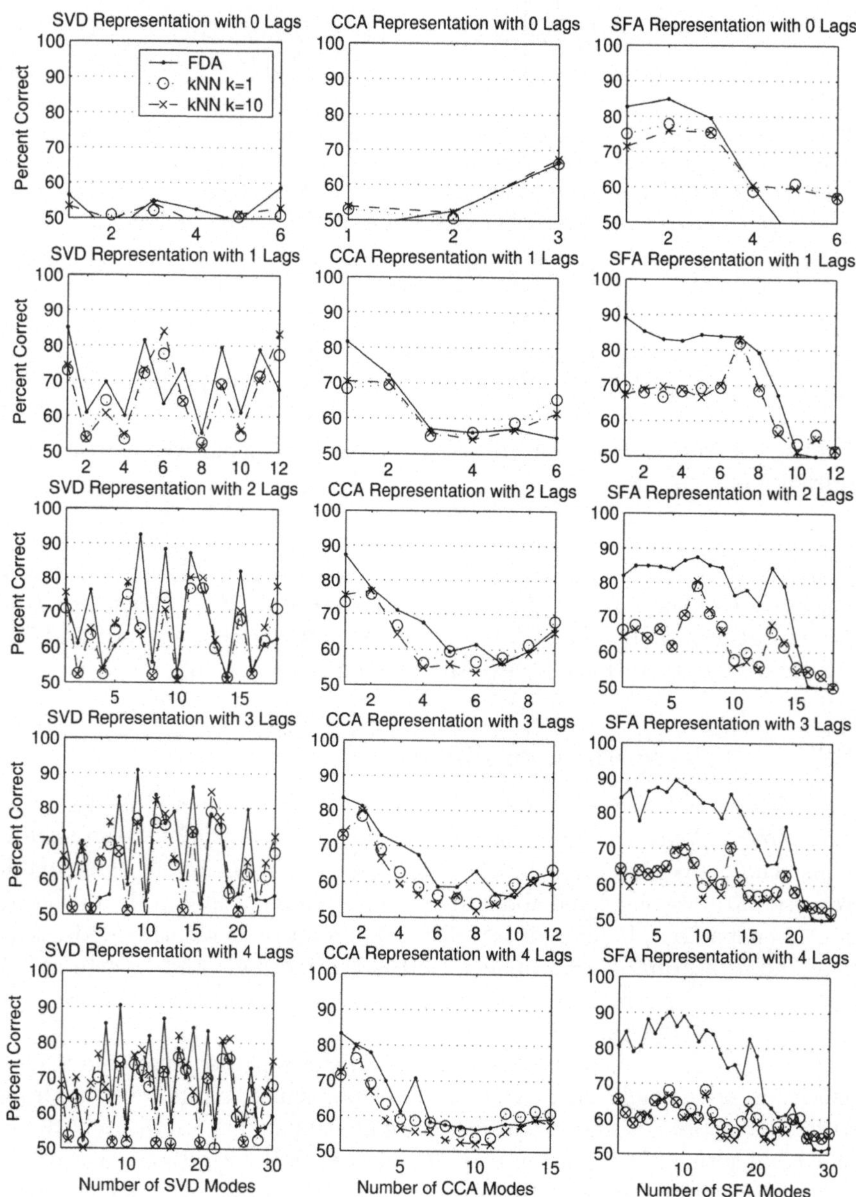

FIGURE 6.2. Percent of test samples correctly classified. Like Figure 6.1, except that the horizontal axis in each graph is which single mode is used to perform the classification.

	Accuracy	Lags	Dim.
SVD	92%	2	18
CCA	87%	2	18
SFA	89%	1	12

TABLE 6.3. Maximum classification accuracies and configurations for each representation for the single-mode results shown in Figure 6.2.

The single-mode results are summarized in Table 6.3 where the maximum accuracies are listed with the corresponding representation configurations. When single modes are used, accuracy with the SVD representation is a little higher than the accuracy with other representations. As mentioned earlier, the variation in accuracy is high from one SVD mode to the next.

7 Conclusions

In this paper we have compared the SVD, CCA and SFA methodologies by investigating the two class classification problem where each class is initially represented as a noisy multivariate time series. We address two general questions within the context of the EEG problem. First, given sample time series of two processes, is it possible to identify accurately by which process a new data set is generated? Second, given representations that solve the first problem, what can be learned, or deduced, about the differences between the processes?

Each method under consideration may be viewed as an approach for empirically constructing dimensionality reducing transformations. The distinct optimization problems that characterize each method make them attractive for such a comparative study. In particular, we observe that each method extracts *different* coherent structures—each associated with its governing optimization criterion. In each case these structures reveal an organization that may be exploited for the classification problem. Indeed, given the very high classification rates obtained by these methods for very messy data, it may be concluded that these approaches have effectively extracted coherent structures with significant discriminatory information.[6] Furthermore, we view this structure as geometric in nature in the sense each method provides alternative subspaces in which to characterize the data.

For comparison, in prior work, we investigated EEG signal representations based on autoregressive models (Anderson and Peterson [2001]). This statistical approach for time-series analysis concluded that information in

[6] Recall that classification attempts on raw data were essentially no better than guessing.

the O1 and O2 channels was most discriminatory, while information in C3 and C4 was also helpful. Visual inspection of the transformed EEG signals presented here also suggests that O1, O2, C3, and C4 are most discriminatory.

The high classification accuracies and variations with the representations considered here lead to many interesting questions regarding how each representation is capturing disciminatory information and what this information says about how the brain is performing each task. The following discussion is an example of the kind of analyses these representations afford. Consider the SFA modes shown in Figure 5.6. The C3 and C4 components are more extreme and variable for the Math Task, while the O1 and O2 are more extreme and variable for the Letter Task. Since this is the first SFA mode, we conclude that the SFA transformation is showing that there is more signal in C3, C4 when the Math Task is performed, and there is more signal in O1, O2 when the Letter Task is performed. It would be very useful to relate this finding to the cognitive psychology literature on our current understanding of how the human brain processes each kind of problem. Another observation can be made regarding the covariance of components. For both tasks the components and their lagged values vary in very similar patterns. This suggests that the projection space resulting in the strongest signal is one in which each electrode's voltage does not vary much from one sample to the next. This is hardly surprising, since the sampling rate is 256 Hz; what is surprising is that an additional lag or two does increase the classification accuracy. The fact that the C3, C4, C3', and C4' components are weakest (close to zero) for the Letter Task could mean that over a 64-sample, or 1/4 second, window the voltages in these electrodes are not as synchronized as in other electrodes. Since desynchronization is believed to indicate involvement in mental activity Pilla and Lopez [2000], this representation might be showing a strong involvement in the C3, C4 areas of the brain, which are over parts of the cortex involved in sensory-motor activity. Yet another observation is that the P4 and O2 components are negatively correlated for the Letter Task, but not for the Math Task. P3 and O1 show this also, but to a much lesser extent. This indicates a bias in activity between the right hemisphere of the brain, where P4 and O2 are located, and the left hemisphere, where P3 and O1 are located.

Finally, we conclude that subspace methods for studies in EEG classification provide a new perspective on the information content of EEG signals and may lead to new ways of interpreting the activity of the brain. The work presented here has stimulated a number of questions that will be investigated further. In particular, given more electrodes it may be possible to isolate areas of cortical activity in real time and shed light on the interactions between parts of the brain for various types of mental tasks. In particular, based on these preliminary results, front to back interactions as well as left/right brain correlations in electrical activity may provide key discriminatory information for the human-machine interface problem.

A Appendix: Dynamic Data and the Method of Delays

In this investigation we employ the method of delays to incorporate additional dynamic, or time-structure, present in the data.[7] As illustrated in Section 6, the size of the delay greatly impacts the efficacy of the data representation. Thus, constructing delays has a pragmatic motivation: it has the potential to significantly improve the data representation for classification. In addition to this, constructing delay vectors is rooted in the more general setting of dynamical systems theory. Here we briefly describe the method of delays (Packard, Crutchfield, Farmer and Shaw [1980]) and an argument for why it works based on Takens' theorem (Takens [1981]).

Modeling the brain using tools from dynamical systems theory has been an active area of research (see, e.g., Fuchs and Kelso [1993]). In our context, the multivariate time-series being recorded by the six EEG electrodes shown in Figure 5.1 may be viewed as a coarse measure of a complex dynamical system representing the state space of cortical electrical activity in the brain. Can such a limited view of the state space be the basis of a meaningful model?

Denoting the voltage measurement at the ith electrode $x_i(t_n)$ observed at discrete time t_n a *delay vector* may be formed by appending lagged values as

$$y_i(t_n) = \big(x_i(t_n), x_i(t_{n+1}), \dots, x_i(t_{n+d})\big) .$$

Since there are six electrodes, i.e., $i = 1, \dots, 6$, each column c_n of the lagged data matrix now has the form

$$c_n = [y_1(t_n)| \cdots |y_6(t_n)]^T .$$

Taken's theorem suggests that one may obtain a reconstruction of a dynamical system using this simple lag trick. In particular, if d is large enough [if the dynamical system resides on a manifold of dimensions m the large enough means $d > 2m+1$ (Takens [1981])], then the trajectory traced out by the lagged vector forms a set that is related to the actual dynamics by a change of coordinates. It is possible that the variation in classification results obtained for different lag vectors may be related to the effectiveness with which the underlying dynamical system is being reconstructed.

In Figure A.1 we see the effect of applying the method of delays on the canonical correlation coefficients. These coefficients clearly depend on the lag value as well as the nature of the structure in the data.

---

[7]We note that in the literature the method of delays has been employed in the context of PCA and referred to as dynamic PCA. We are not aware of instances where the method of delays has been employed with either CCA or SFA.

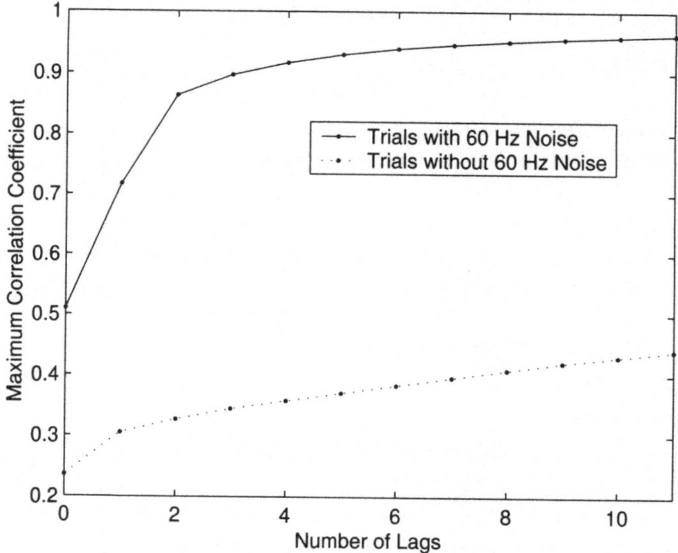

FIGURE A.1. Maximum CCA correlation coefficients for varying number of lags. The maximum coefficient increases with the number of lags. Trials for which data includes a strong 60 Hz component result in much higher maximum correlation coefficients.

Acknowledgments: M.K. thanks Larry Sirovich for directing him along a path that has proven to be more fun than one should be allowed to have while working. Research was partially supported by the NSF-MPS-Mathematical & Physical Sciences award DMS-9973303 and by DOD-USAF-Office of Scientific Research under contract #: F49620-99-1-0034 P00003. C.A. would like to acknowledge NSF grant IRI-9202100.

References

Anderle, M. and M. Kirby [2002], *An Application of the Maximum Noise Fraction Method to Filtering Noisy Time-Series*, in Mathematics in Signal Processing V, Editors: J. G. McWhirter and I. K. Proudler, Oxford University Press, June 2002.

Anderle, M. [2001], *Modeling Geometric Structure in Noisy Data*, Ph.D. Dissertation, Department of Mathematics, Colorado State University

Anderson, Charles W. and David A. Peterson [2001], Recent Advances in EEG Signal Analysis and Classification, *Clinical Applications of Artificial Neural Network*, Editors R. Dybowski and V. Gant, Cambridge University Press, 175–191.

Barnett, T. P. and R. Preisendorfer [1987], Origins and Levels of Monthly and Seasonal Forecast Skill for United States Surface Air Temperatures Determined by Canonical Correlation Analysis, *Monthly Weather Review* **115**, 1825–1850.

Brockwell, P. J. and R. A. Davis [1988], *Time Series: Theory and Methods.* Springer Verlag, New York.

Camazine, S., J. L. Deneubourg, N. R. Franks, J. Sneyd, G. Theraulaz, and E. Bonabeau [2001], *Self-Organization in Biological Systems*, Princeton Studies in Complexity.

Case, James [1999], Data Mining Emerges as A New Discipline in a World of Increasingly Massive Data Sets, *SIAM News* **32**(10).

Diamantaras, K. and S. Y. Kung [1996], *Principal Component Neural Networks: Theory and Applications*, Wiley and Sons.

Duda, Richard O. and Peter E. Hart [1971], *Pattern Classification and Scene Analysis*, Wiley and Sons.

Everson, R. M. and L. Sirovich [1992], Analysis and management of large scientific databases, *International Journal of Supercomputing Applications* **6**(1), 50–68.

Fuchs, A. and J. A. S. Kelso [1993], Self-organization in brain and behavior: Critical instabilities and dynamics of spatial modes, *Nonlinear Dynamical Analysis of the EEG*, B. H. Jansen, M. E. Brandt, eds., World Scientific, Singapore (1993)

Fukunaga, K. and D. R. Olsen [1971], An algorithm for finding intrinsic dimensionality of data. *IEEE Transactions on Computers*, C-20(2):176–183.

Golub, Gene H. and Charles F. Van Loan [1996], *Matrix Computations.* Johns Hopkins U.P., Baltimore, third edition.

Green, Andrew A., Mark Berman, Paul Switzer, and Maurice D. Craig. [1988], A transformation for ordering multispectral data in terms of image quality with implications for noise removal. *IEEE Transactions on Geoscience and Remote Sensing*, 26(1):65–74, January 1988.

Haken, H. [1983], *Synergetics: An Introduction: Nonequilibrium Phase Transitions and Self-Organization in Physics, Chemistry, and Biology*, Springer-Verlag, New York.

Hotelling, H. [1936], Relations Between Two Sets of Variables, *Biometrika*, **28**, 321–377.

Jolliffe, I. T. [1986], *Principal Component Analysis.* Springer, New York, 1986.

Karhunen, K. [1946], Über lineare Methoden in der Wahrscheinlichkeitsrechnung, *Ann. Acad. Sci. Fennicae Ser A1, Math. Phys.* **37**.

Keirn, Z. A., and J. I. Aunon [1990], A New Mode of Communication Between Man and His Surroundings, *IEEE Transactions on Biomedical Engineering*, **37** (12), 1209–1214.

Keirn, Z. A. [1988], Alternative Modes of Communication Between Man and Machine, *Masters Thesis*, Purdue University, Lafayette, IN.

Jasper, H. [1958], The ten twenty electrode system of the international federation, *Electroencephalography and Clinical Neurophysiology*, **10**, 371–375.

Kirby, M. [2001], *Geometric Data Analysis.* Wiley, New York, 2001.

Kirby, M. and L. Sirovich [1990], Application of the Karhunen-Loève procedure for the characterization of human faces. *IEEE Trans. Pattern Anal. Mach. Intell.*, 12(1):103–108.

Kirby, M., F. Weisser and G. Dangelmayr [1991], A problem in facial animation: Analysis and synthesis of lip motion. In *Proc. 7th Scandinavian Conf. on Image Analysis*, pages 529–536, Aalborg, Denmark.

Kirby, M., F. Weisser and G. Dangelmayr [1993], Speaking with images: A model problem in the representation of still and moving images. *Pattern Recognition*, 26(1):63–73.

Loève, K. [1955], *Probability Theory*, Von Nostrand.

Lorenz, E. [1956], Empirical orthogonal eigenfunctions and statistical weather prediction. Science Report No. 1, Statistical Forecasting Project 1, M.I.T., Cambridge, MA, 1956.

Mardia, K. V. J. T. Kent and J. M. Bibby [1979], *Multivariate Analysis.* Academic Press, New York, 1979.

Packard, N. H., J. P. Crutchfield, J. D. Farmer and R. S. Shaw [1980], Geometry from a time series, *Phys. Rev. Lett.*, 45(9), 712–716.

Packard, N. H., J. P. Crutchfield, J. D. Farmer and R. S. Shaw [1980], Detection of movement-related desynchronization of the EEG using neural networks , *Proceeding of the 22nd annual internations conference of the IEEE Engineering in medicine and biology society*, 1372–1376.

Preisendorfer, Rudolph W. [1988], *Principal Component Analysis in Meteorology and Oceanography.* Developments in Atmospheric Science. Elsevier, New York, 1988.

Rao, K. R. and R. Yip [1990], *Discrete Cosine Transform.* Academic Press, New York, 1990.

Sirovich, L. and M. Kirby [1987], A low-dimensional procedure for the characterization of human faces, *J. of the Optical Society of America A* **4**, 524–529.

Sirovich, L. [1987a], Turbulence and the dynamics of coherent structures, Part I: Coherent structures. *Quart. of Appl. Math.*, XLV(3):561–571.

Sirovich, L. [1987b], Turbulence and the dynamics of coherent structures, Part II: Symmetries and transformations. *Quart. of Appl. Math.*, XLV(3):573–582.

Sirovich, L. [1987c], Turbulence and the dynamics of coherent structures, Part III: Dynamics and scaling. *Quart. of Appl. Math.*, XLV(3):583–590.

Switzer, P. and A. Green.[1984], Min/max autocorrelation factors for multivariate spatial imagery. Technical Report 6, Stanford University, Department of Statistics.

Takens, F. [1981], Detecting Strange Attractors in Turbulence, in *Detecting Strange Attractors in Turbulence*, edited by D. A. Rand and L. S. Young, Lecture notes in mathematics, **898**, 336–381, Warwick, 1980.

Trefethen, Lloyd N. and David Bau [1997], *Numerical Linear Algebra.* SIAM, Philadelphia, 1997.

Wolpaw, J. R., N. Birbaumer, W. J. Heetderks, D. J. McFarland, P. H. Peckham, G. Schalk, E. Donchin, L. A. Quatrano, C. J. Robinson, T. M. Vaughan [2000], Brain-Computer Interface Technology: A Review of the First International Meeting, *IEEE Transactions on Rehabilitation Engineering* **8**(2), 164–173.

9

High Conductance Dynamics of the Primary Visual Cortex

David McLaughlin
Robert Shapley
Michael Shelley
Jim Jin

It is our pleasure to dedicate this article to Larry Sirovich on the occasion of his 70th *birthday. Given Larry's many contributions to scientific modeling using asymptotic analysis, we believe that the concept of an "emergent separation of time-scales" in cortical processing will be particularly appealing to him.*

ABSTRACT This article discusses large synaptic conductance increases, and their consequent separation of time-scales, which can result from cortical activity, together with some implications of these large conductances for cortical processing. The specific setting is the primary visual cortex (Area V1) of the Macaque monkey, as described by a large-scale neuronal network model (McLaughlin, Shapley, Shelley, and Wielaard [2000]), and the comparison of this model's performance (Shelley, McLaughlin, Shapley, and Wielaard [2002]) with observations from laboratory experiments (Borg-Graham, Monier, and Fregnac [1998]; Hirsch, J.M.Alonso, Reid, and Martinez [1998]; Anderson, Carandini, and Ferster [2000]). In the model, the source of large conductances is traced to inhibitory cortico-cortical synapses, which set a high conductance operating point in which the model cortical layer achieves orientation selectivity, neuronal dynamics and response magnitude, and the linear dependence of simple cells on visual stimuli (e.g. Wielaard, Shelley, Shapley, and McLaughlin [2001]) consistent with experimental observations. In the absence of visual stimulation, the high conductance operating point sets relaxation time-scales at the synaptic time scale ($\simeq 4 - -5\,\text{ms}$). High contrast visual stimulation produces cortical activity which further reduces this relaxation time scale by a factor of 2, so that it becomes even shorter than synaptic time-scales. While only a factor of 2, this is shown to be sufficient to produce the dynamical consequences of time scale separation, including near instantaneous (with respect to the synaptic time scale) responses of neurons to temporal changes in synaptic drive, with each intracellular membrane potential tracking closely an *effective reversal potential* composed of the instantaneous synaptic inputs. This time scale separation, which emerges from cortical activity, permits the use of methods from asymptotic analysis to express the spiking activity of a cell in simplified manner. These asymptotic results suggest how accurate and smoothly graded responses are achieved in the model network. Further, since neurons in this high-conductance state respond quickly, they are also good candidates as coincidence detectors and as burst transmitters.

Contents

1 Introduction

Neurons communicate with each other by detecting and responding to voltage spikes in the membrane potentials of other neurons. The dynamics of a neuron's membrane potential is described by the voltage difference across the cell's membrane, and is governed by the conductance of the membrane (the inverse of its electrical resistance). (See, for example Koch [1999].) The value of this conductance is not a static membrane property, but changes with time through synaptic processes which are initiated by impinging voltage spikes from other neurons. Thus, the value of the conductance is influenced by the activity of the neuronal network.

For fixed capacitance C (a cellular property which is static), the value of the conductance g sets a time-scale $\tau_g \equiv C/g$ upon which the neuron responds to synaptic input. Until recently, this time-scale was treated as well estimated by conductance values measured in cells within cortical slices, of the order of 20 ms for cortical neurons. There was even some experimental evidence that cortical activity stimulated by high contrast visual signals did not alter significantly the mean value of the conductance of cortical neurons. However, in the early 1990s, theoretical considerations of Koch, Segev and others emphasized the possibility that the membrane time-scale could depend upon cortical activity, as well as possible functional implications of this dependence (see Koch, Rapp, and Segev [1996]; Bernander, Douglas, Martin, and Koch [1991]; Bernander, Koch, and Usher [1994]).

In vivo conductances are extremely difficult to measure accurately. Hence, recent experiments have focused on the demonstration of the presence of large conductance *changes* under visual stimulation. In the cat visual cortex, Borg-Graham, Monier, and Fregnac [1998] have studied the intracellular responses of simple cells in visual cortex to flashed bars. Using *in vivo* whole-cell voltage clamping, they measured the dynamics of conductance during visual stimulation and found two- to three-fold conductance increases over unstimulated background. They associate these large conductances with the activation of strong cortico-cortical inhibition. Hirsch, J.M.Alonso, Reid, and Martinez [1998] have also performed *in vivo* whole-cell patch recordings of simple cells in cat cortex. Their results likewise suggest strong cortico-cortical inhibition arising in response to visually-driven, excitatory geniculate input, with this inhibition driving large increases in total membrane conductance. Most recently, Anderson, Carandini, and Ferster [2000] confirmed the presence of these large conductance changes induced by high contrast visual stimulation for cat.

Our large-scale computational model (McLaughlin, Shapley, Shelley, and Wielaard [2000]) for primary visual cortex of Macaque operates in a high-conductance state (Shelley, McLaughlin, Shapley, and Wielaard [2002]). This state depends upon cortical activity, with conductance changes in qualitative agreement with the recent experimental observations just described. We emphasize that this high conductance state is not an individual cellular property, and that we are not free to place the model network arbitrarily into a high-conductance regime. Rather, in our model the total conductance g must be large in order to obtain background firing rates which agree in magnitude with experimental observations (see Mechler [1997]). High contrast visual stimulation further increases the conductance, providing enough inhibition to produce neuronal responses in agreement with experiment. Background activity increases the conductance by a factor of 4–5 when compared with slice values, and high contrast visual stimulation further increases it by another factor of 2. Together, this combined cortical activity increases the total conductance in model neurons by a factor of 8–10 over base-line cellular conductances. Thus, our simulations suggest that background cortical activity is sufficient to make the relaxation time-scale τ_g comparable with synaptic time-scales, of the order 4 ms, which is much shorter than the order of 20 ms as measured in cortical slices. Moreover, high contrast visual stimulation further decreases the relaxation time-scale to the order of 2 ms, thereby converting the relaxation time-scale τ_g to a value even shorter than that of synaptic time-scales.

If the emergent conductance is high enough, it is intuitively clear that the membrane potential will respond with near instantaneity to changes in synaptic input. But how high is high enough? The answer is a matter of the competing time-scales. Within the model discussed below, a neuron's membrane potential $v(t)$ satisfies a dynamical equation of the form

$$\frac{1}{g}\frac{dv}{dt} = -\left[v - V_S(t)\right], \tag{1.1}$$

First, we note that the capacitance C does not appear in this equation, as all conductances have been defined as rates, with units of sec^{-1}, by dividing through by C. Throughout this article we will use conductances normalized as rates in order to emphasize the time-scales which they represent. For example, the leakage conductance $g_L = 50\,\text{s}^{-1}$ produces a leakage time-scale of $\tau_L = g_L^{-1} = 1/(50\,\text{s}^{-1}) = 20\,\text{ms}$. True conductances are obtained by multiplication by $C = 10^{-6}F\,\text{cm}^{-2}$; for example, $g_L = 50\,\text{s}^{-1} \times 10^{-6}F\,\text{cm}^{-2} = 50\,\text{nS}$. Also, note that the total conductance g and the *effective reversal potential* $V_S(t)$ depend entirely upon synaptic inputs, and thus vary only on the synaptic time-scale τ_{syn}. If $\tau_g \equiv g^{-1} \ll \tau_{\text{syn}}$, i.e. there is an asymptotic separation between the relaxation and synaptic time-scales, then we can proceed to simplify mathematically the dynamical description. However, in our large-scale model the relaxation time-scale $[\tau_g = O(2\,\text{ms})]$ is a only factor of two smaller than the synaptic time-scale $[\tau_{\text{syn}} = O(4\,\text{ms})]$; therefore, within our large-scale model we have verified this is nonetheless sufficient to realize the consequences of asymptotic scale separation (see Shelley, McLaughlin, Shapley, and Wielaard [2002], and below). With this scale separation, cortical neurons respond to synaptic input almost instantaneously (with respect to the synaptic time-scale τ_{syn}), and we are led to the novel concept of an "emergent time-scale separation"; that is, a scale separation which emerges from cortical activity.

Neurons whose membrane potential responds nearly instantaneously with respect to the synaptic time-scale, rather than integrating over the synaptic time-scale, could act effectively as coincidence detectors even in the presence of synaptic noise. This rapid response could permit the neurons to encode and decode information in detailed temporal spiking patterns, which would be difficult with longer time constants in the presence of synaptic noise. However, at the same time that they are acting as coincidence detectors, neurons in our network model produce an average firing rate that is approximately proportional to synaptic input.

Section II offers background material about our large-scale model of primary visual cortex of Macaque. In Section III, we establish that the model operates in a high conductance state, while in Section IV we discuss the mechanistic properties of this state, i.e., the properties which follow from time-scale separation. Finally, in Section V, we discuss possible functional consequences. Much of this material also appears in Shelley, McLaughlin, Shapley, and Wielaard [2002], although with somewhat different emphasis.

2 Background about the Model

The primary visual cortex (V1) is a layered structure. Our model, described in McLaughlin, Shapley, Shelley, and Wielaard [2000], focuses upon the dynamical responses of neurons, such as their orientation selectivity (their quality as detectors of the orientation of edges within a visual scene), within a single sublayer. In particular, it is a large-scale, detailed cortical model of a small local patch ($1\,\text{mm}^2$) of input layer $4C\alpha$ for V1 of the adult Macaque monkey. This input layer receives visual input from the retina through the lateral geniculate nucleus (LGN). Our model consists of a two-dimensional lattice of 128^2 coupled integrate-and-fire point neurons, of which 75% are excitatory and 25% are inhibitory. It belongs to a class of models with *recurrent network filtering* (see Shapley, McLaughlin, and Shelley [2002]) which is intermediate between "feedforward filtering" (Hubel and Wiesel [1962]) (a direct transformation of the visual signal), and strong excitatory nonlinear cortical feedback (Ben-Yishai, Bar-Or, and Sompolinsky [1995]; Hansel and Sompolinsky [1996, 1998]). Distinctive features of our model include:

- A regular map of orientation preference modeled after the beautiful "pinwheel" patterns which are revealed by optical imaging experiments, e.g. Bonhoeffer and Grinvald [1991]; Blasdel [1992a,b]; Maldonado, Godecke, Gray, and Bonhoeffer [1997]; Everson, Prashanth, Gabbay, Knight, Sirovich, and Kaplan [1998], as tiling the superficial layers of the cortex. We assume that these patterns originate in $4C\alpha$ and extend to the upper layers (based upon classical orientation columns in V1 cortex (Hubel and Wiesel [1962])). This regular orientation map is assumed to be hard-wired into the cortex through the orientation preference of each group of LGN cells that converge onto cortical cells (Reid and Alonso [1995]; Alonso, Usrey, and Reid [2001]). This local patch ($1\,\text{mm}^2$) contains four orientation hypercolumns with pinwheel centers.

- An irregular random map of spatial phase preference, with a broad distribution (DeAngelis, Ghose, Ohzawa, and Freeman [1999]). In the model, this is also conferred by the convergent LGN input.

- Local excitatory and inhibitory cortical synaptic interactions that (i) are isotropic, (ii) with excitatory length-scales ($200\,\mu\text{m}$) longer than inhibitory length-scales ($100\,\mu\text{m}$), and (iii) with cortico-cortical inhibition dominant.

We emphasize that both the construction and performance of the model are consistent with experimental observations. Imaging, anatomical and physiological measurements constrain the representation of the model's neuronal components as well as their coupling architecture; moreover, the model's performance with respect to dynamics and orientation selectivity is consistent qualitatively with laboratory observations.

2.1 The Equations of the Model

The model itself consists in equations for the membrane potentials of excitatory (and inhibitory) neurons, denoted by v_E^j (and v_I^j):

$$\frac{dv_P^j}{dt} = -g_L\, v_P^j - g_{PE}^j(t)\, [v_P^j - V_E] - g_{PI}^j(t)\, [v_P^j - V_I], \quad P = E, I \quad (2.1)$$

together with the definition of the mth *spike-times* t_m^j and reset,

$$v_P^j(t_m^j) = 1; \quad v_P^j(t_m^j + 0^+) = 0. \quad (2.2)$$

Here the superscript $j = (j_1, j_2)$ indexes the spatial location of the neuron within the cortical layer. In these equations, we have normalized the potentials, making them dimensionless quantities. This normalization sets the spiking threshold to unity, the reset potential to zero, $V_E = 14/3$, and $V_I = -2/3$. Within this normalization, the potentials range over $-2/3 \leq v_E^j, v_I^j \leq 1$.

2.2 Conductances

The time-dependent conductances arise from the input forcing (through the LGN) and from noise to the layer, as well as from the cortical network activity of the excitatory and inhibitory populations. They have the form:

$$g_{EE}^j(t) = F_{EE}(t) + S_{EE} \sum_k a_{j-k} \sum_l G_E(t - t_l^k),$$

$$g_{EI}^j(t) = f_{EI}^0(t) + S_{EI} \sum_k b_{j-k} \sum_l G_I(t - T_l^k),$$

with similar expressions for g_{IE}^j and g_{II}^j, where

$$F_{PE}(t) = g_{\text{lgn}}^j(t) + f_{PE}^0(t), \quad P = E, I.$$

Here t_l^k (T_l^k) denotes the time of the lth spike of the kth excitatory (inhibitory) neuron. Note that g_{EE} (g_{EI}) is the conductance driven by excitatory (inhibitory) network activity, and that the first "E" labels the postsynaptic target as an excitatory cell. The conductance $g_{\text{lgn}}^j(t)$ denotes the drive from the retina by which the visual signal (standing, drifting, or random gratings) is relayed through convergent LGN neurons to the jth cortical neuron. The tiling of the cortex by pinwheel patterns of orientation preference, and the random map of spatial phase preference, are injected into the model through this LGN drive. The conductances $f^0_{PP'}(t)$ are stochastic and represent activity from other areas of the brain.

The kernels $(a, b, \dots)$ represent the spatial coupling between neurons, and reflect our understanding of cortical anatomy. This local (within one hypercolumn) coupling is spatially isotropic, with the spatial length-scale of excitation (radius $200\,\mu$m) exceeding that of inhibition (radius 100μm). The temporal kernels $G_\sigma(t)$ model the time course of synaptic conductance changes in response to arriving spikes from the other neurons, which introduces the time-scale $\tau_{\text{syn}} = 4 - -6\,\text{ms}$.

The computational model's behavior depends on the choice of the cortico-cortical synaptic coupling coefficients: $S_{EE}, S_{EI}, S_{IE}, S_{II}$. All cortical kernels have been normalized to unit area. Hence, these coupling coefficients represent the strength of interaction, and are treated as adjustable parameters in the model. In the numerical experiments reported here, the strength matrix $(S_{EE}, S_{EI}, S_{IE}, S_{II})$ was set to be (0.8, 9.4, 1.5, 9.4). This matrix means cortical inhibition dominates in that inhibitory neurons have much stronger coupling to all other cortical neurons than do excitatory neurons. The matrix given here generates simple cells that have the orientation selectivity, and the magnitude and dynamics of response, seen in physiological experiments (McLaughlin, Shapley, Shelley, and Wielaard [2000]).

3 The High Conductance Operating Point of the Model

First, we show that this large-scale cortical network develops large conductances when under high contrast visual stimulation, and that the predicted conductance changes are consistent with recent experimental and theoretical observations. Figures 3.1a and c show the time-dependent total conductances of a typical, orientation selective, excitatory neuron within our large-scale cortical model. This neuron is stimulated at its preferred, and orthogonal-to-preferred, orientations by an $8Hz$ drifting grating (also at optimal spatial frequency and high contrast), with the visual stimulation switched on at $t = 0.5\,\text{sec}$. Shown is the neuron's *total conductance* (called g in Eq. (1.1)),

$$g_T(t) \equiv g_L + \ g_{EE}(t) + \ g_{EI}(t)\,, \tag{3.1}$$

where $g_{EE}(t)$ and $g_{EI}(t)$ are the excitatory and inhibitory conductances at time t. The conductance g_T reflects several time-scales inherent in the operation of computational cortex under visual stimulation: (i) the time-scale of the visual stimulus, $\tau_{\text{lgn}} = O(10^2\,\text{ms})$; (ii) the base cellular time-

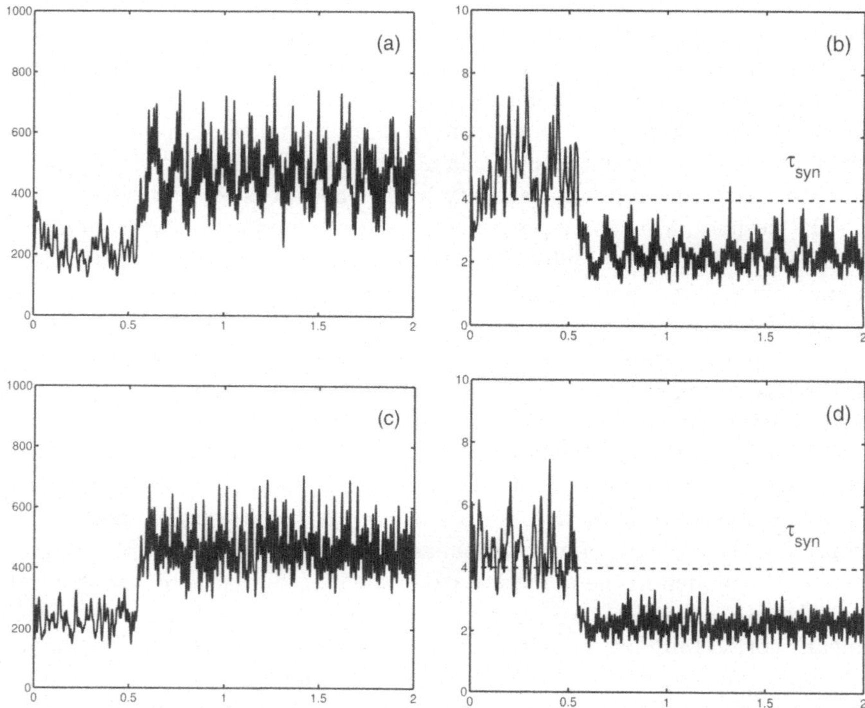

FIGURE 3.1. (a) and (c): Total conductance $g_T(t)$ ($\sec^{-1}$) vs t (sec) for an orientation selective excitatory neuron. A drifting grating stimulus, switched on at $t = 0.5$ sec is at (a) preferred and (c) orthogonal-to-preferred orientation (drifting at 8 Hz, at optimal spatial and temporal frequency, 100% contrast). Panels (b) and (d) show the associated relaxation time-scale g_T^{-1} (msec), where the dashed line marks the level of τ_{syn}.

scale of leakage, $\tau_L = 20\,\text{ms}$; (iii) the synaptically mediated cortico-cortical interactions which govern the temporal fluctuations in $g_{EE}(t)$ and $g_{EI}(t)$, $\tau_{\text{syn}} = O(4\,\text{ms})$; and (iv) the conductance time-scale, $\tau_g = [g_T^{-1}]$ (where $[\cdot]$ denotes some characteristic size).

Note that even when unstimulated ($t < 0.5\,\text{s}$), the total conductance is always much higher than the leakage conductance of individual cells: In unstimulated background, $g_T = 230 \pm 40\,\text{s}^{-1}$ (mean $\pm$ standard deviation) versus $g_L = 50\,\text{s}^{-1}$, or a time-scale of $\tau_g \simeq 4\,\text{ms}$ versus $\tau_L = 20\,\text{ms}$. In this case of unstimulated background, τ_g and τ_{syn} are roughly in balance, and are the smallest time-scales. However, with the onset of visual stimulation at the preferred orientation, Fig. 3.1a shows that the total conductance of the model neuron undergoes a two-fold increase, producing $g_T = 550 \pm 100\,\text{s}^{-1}$, with the associated conductance time-scale decreasing to $\tau_g \simeq 2\,\text{ms}$. When stimulated at the orthogonal-to-preferred orientation, $g_T = 400 \pm 80\,\text{s}^{-1}$ (Fig. 3.1c), which while considerably less than at preferred, is still nearly twice its unstimulated value.

In all cases, drifting grating stimuli at either preferred or orthogonal-to-preferred orientation or using randomly flashed grating stimuli (Shelley, McLaughlin, Shapley, and Wielaard [2002]), the large conductances induced by visual stimulation cause the membrane time-scale τ_g to drop below $\tau_{\rm syn}$, the synaptic time-scale, and hence become the smallest time-scale in the system. To emphasize this point we show g_T^{-1}, the dynamical relaxation time-scale, in Figs. 3.1b and d. Again, before the onset of visual stimulation, τ_g and $\tau_{\rm syn}$ are roughly in balance. With the onset of stimulation, a temporal scale separation develops as τ_g drops below $\tau_{\rm syn}$.

Our model requires strong inhibition to operate realistically when compared with experimental observations. The network's consequent cortical activity produces conductance values that are very large: Background activity provides a factor of 4-5 over cellular leakage values (as suggested by Bernander, Douglas, Martin, and Koch [1991]), and high contrast visual stimulation provides an additional factor of 2. These numerical observations of conductance changes within our model of more than 200%, caused by visual stimulation, are qualitatively consistent with recent intracellular measurements for cat (see Borg-Graham, Monier, and Fregnac [1998]; Hirsch, J.M.Alonso, Reid, and Martinez [1998]; Anderson, Carandini, and Ferster [2000]) , as described above in the Introduction.

3.1 Components of the Conductances

In addition to high conductance, the parameter regime at which the model cortex operates is one of near balance of excitatory and inhibitory currents,

$$g_E V_E \simeq g_I |V_I|.$$

This balance, together with the relative sizes of the two reversal potentials ($V_E >> |V_I|$), implies relatively larger inhibitory conductances ($g_I \simeq [V_E/V_I \; g_E] > g_E$). The dominance of inhibitory conductances in our large-scale model is seen in Figs. 3.2.

The total conductance g_T arises from several sources: the LGN, excitatory and inhibitory cortico-cortical activity, and from noise. Fig. 3.2a shows the individual excitatory components—from LGN (blue), other excitatory cortical cells (red), and noise (green)—that make up g_T for the neuron of Figs. 3.1a,c (under drifting grating stimulation). Fig. 3.2b shows the two constituent inhibitory conductances, from inhibitory cortical cells (red), and noise (green). Note that cortico-cortical inhibition dominates all of the component conductances.

3.2 Spatial Distribution of Conductances

As with orientation selectivity (McLaughlin, Shapley, Shelley, and Wielaard [2000]), the value of total conductance for a model neuron depends on its

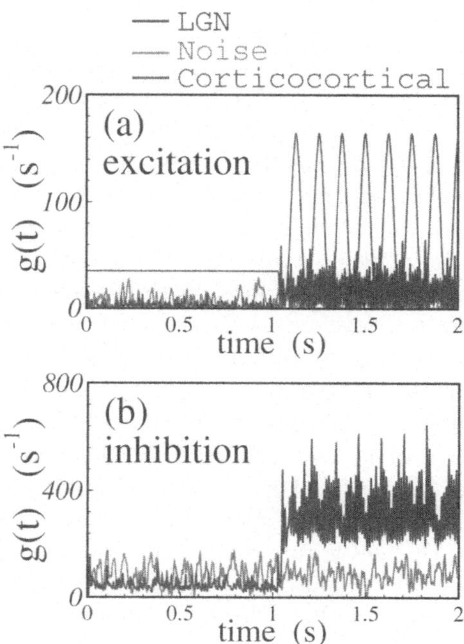

FIGURE 3.2. The five components of the total conductance for a typical excitatory neuron, without visual stimulation and for a drifting grating stimulus (at 8 Hz, switched on after one second). (a) Excitatory: LGN (blue), cortical interaction from layer $4C\alpha$ (red) and noise (green). (b) Inhibitory: cortical interaction from layer $4C\alpha$ (red) and noise (green). Notice the difference in the vertical scales for (a) and (b).

location within the cortical layer, relative to pinwheel centers. With large-scale computational models, one can readily examine the spatial distribution of the total conductance, as well as its temporal fluctuations, across the entire network. For drifting grating stimuli, the spatial distribution of the temporal average of the conductances across the local patch of model input layer is shown in Fig. 3.3a, with the spatial distribution of standard deviation of the temporal fluctuations shown in Fig. 3.3b. For the average total conductance (Fig. 3.3a), there is rather uniform behavior in regions far from the pinwheel centers. For example, there is uniformly high conductance in the "purple" region of preferred orientation (centered at (1,1) in Fig. 3.3a), and uniformly lower conductance in the "yellow" region of orthogonal-to-preferred orientation (centered at (0,0)). Note, however, that even regions of non-preferred orientation have conductances well above background values ($400\,s^{-1}$ as compared with background values of $230\,s^{-1}$). Near the pinwheel centers, the average conductance changes more rapidly in space, over distances of 100 microns, a scale set by the axonal arbors of inhibitory neurons. The fluctuations (Fig. 3.3b) are significantly

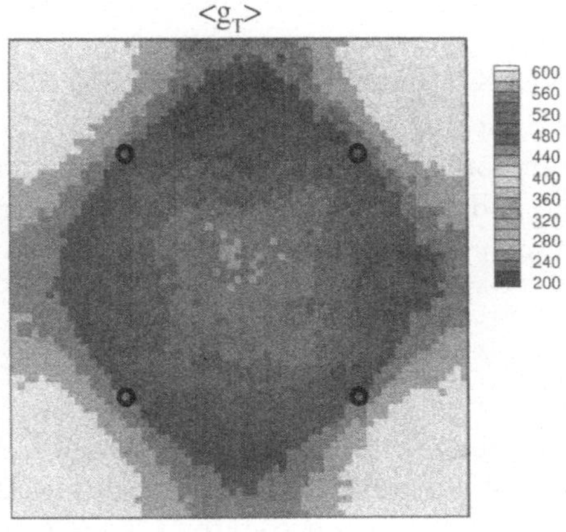

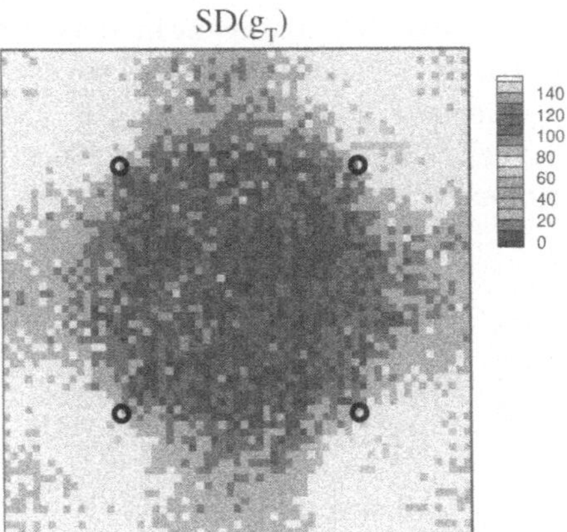

FIGURE 3.3. The spatial distribution of the neurons' local patch of the input layer: Upper: temporal averages; Lower: standard deviations of the temporal fluctuations.

larger far from pinwheels than near.

As emphasized in McLaughlin, Shapley, Shelley, and Wielaard [2000], these near-far distinctions can be traced to a transfer of the "circle of influence upon a given neuron", from spatial interaction lengths within the cortical layer to interaction lengths in terms of the angle of orientation preference. Because of the spatial extent of the cortico-cortical interactions, and their relationship to the tiling of the cortex by pinwheel patterns of orientation preference, neurons near (or at) a pinwheel center experience cortico-cortical inhibition which is effectively *global in orientation preference.* This is in contrast to those far from pinwheel centers which experience only cortico-cortical inhibition which is *local in orientation preference*, where these far neurons interact monosynaptically only with neurons of similar orientation preference.

Other stimuli, such as randomly flashed gratings, create very different and distinct conductance maps on the model cortex. In contrast to drifting gratings, the temporal mean and standard deviation of the temporal fluctuations of the conductance for randomly flashed grating stimuli are distributed relatively uniformly throughout the entire layer (not shown).

3.3 Summary

These spatial distributions define dynamical, spatially dependent "operating points" or states of cortical activity. They emphasize that our model cortex operates in a high-conductance state—a state caused by cortical activity. We emphasize that this high conductance state is not an individual cellular property, and that we are not free to place the network arbitrarily into a high-conductance regime. Rather, in our model the total conductance g_T must be large in order to obtain orientation selectivity (McLaughlin, Shapley, Shelley, and Wielaard [2000]) in agreement with experiment and the linearity of simple cells (Shelley, McLaughlin, Shapley, and Wielaard [2002]) also observed in experiment.

4 Asymptotic Analysis: Effective Reversal Potential, Slaving, and Spiking

In our model cortex, large conductances induced by visual stimulation cause τ_g to be the shortest time-scale present. If τ_g is so short that it is well separated from synaptic time-scales in the cortex, one can use asymptotic analysis to derive useful input-output (I/O) relations that express the cell's membrane potential and the interspike times directly in terms of its synaptic input conductances. (Within the model, the synaptic time-scale τ_{syn} is the next shortest time-scale, and we will return later in this Section to show that τ_g is sufficiently shorter than τ_{syn} to achieve the consequences of the time-scale separation used in this asymptotic analysis.)

4.1 Asymptotic Analysis

For this analysis, we turn to Eq. (2.1), which governs a single cell's response to conductance changes, written in the form:

$$\frac{dv}{dt} = -g_T(t)v(t) + I_D(t) = -g_T(t)\big[v(t) - V_S(t)\big] \tag{4.1}$$

where, as in Eq. (3.1), g_T is the total conductance, and

$$I_D(t) = I_D\left[g_E(t)\,, g_I(t)\right] \equiv g_E(t)V_E - g_I(t)\,\left|V_I\right|\,, \tag{4.2}$$

$$V_S(t) = V_S\left[g_E(t), g_I(t)\right] \equiv I_D(t)/g_T(t)\,. \tag{4.3}$$

We call I_D the *difference current* (as it is a difference of currents arising from excitation and inhibition), and V_S the *effective reversal potential.* Note that (i) $g_T(t)$ and $V_S(t)$ are explicit functions of the instantaneous synaptic input conductances $g_E(t)$ & $g_I(t)$, and (ii) a *necessary condition* for $v(t)$ to cross the spiking threshold is that $V_S > 1$, i.e., V_S itself is above threshold. (This follows directly from Eq. (4.1), and the requirement that $v(t_s) = 1$ while $dv(t)/dt\big|_{t=t_s} > 1$ at a spike time t_s.)

It is clear intuitively from Eq. (4.1) that if g_T is sufficiently large, a neuron in this high conductance state will possess two characteristic response properties:

1. When *subthreshold*, the membrane potential is well approximated by the effective reversal potential $V_S(t)$, which is expressed explicitly in terms of the instantaneous input conductances to the cell,

$$v(t) \simeq V_S(t) = V_S\left[g_E(t)\,, g_I(t)\right]\,. \tag{4.4}$$

2. When forced away from $V_S(t)$ (as when $v(t)$ spikes), the membrane potential returns quickly to being very close to $V_S(t)$.

We now define precisely the notion of time-scale separation, and use it to develop I/O relationships through asymptotic arguments. Let τ_g be an average value of g_T^{-1}, and let $\tau_{\rm syn}$ be a time-scale governing the fluctuations of synaptic conductances. We then express V_S and g_T as $V_S(t/\tau_{\rm syn})$ and $g_T(t/\tau_{\rm syn})$, as both are defined in terms of these conductances and thus fluctuate only on the synaptic time-scale. Then rescaling time and total conductance as $\tilde{t} \equiv t/\tau_{\rm syn}$ and $\tilde{g}_T \equiv \tau_g\, g_T$, and defining the ratio of time-scales, $\epsilon \equiv \tau_g/\tau_{\rm syn}$, Eq. (4.1) becomes:

$$\epsilon\frac{dv}{d\tilde{t}} = -\tilde{g}_T(\tilde{t})\big(v - V_S(\tilde{t})\big)\,. \tag{4.5}$$

If $\epsilon = O(1)$ or larger, then the cell smoothes the synaptic input over an $O(1)$ (synaptic) time-scale. That is, the cell acts as an integrator. If instead $\epsilon \ll 1$, the cell will respond almost instantaneously to its synaptic input.

By introducing an "intrinsic" and dimensionless time $u \equiv \int_0^{\tilde{t}} g(\tilde{t}')d\tilde{t}'$, and expressing v and V_S in terms of it, Eq. (4.5) takes the form

$$\epsilon\frac{dv}{du} = -\big(v - V_S(u)\big)\,.$$

The equation can be immediately integrated to yield

$$v(u) = v_{in}\;\exp\left\{-\frac{u-u_{in}}{\epsilon}\right\} \;+\; \frac{1}{\epsilon}\int_{u_{in}}^{u} \exp\left\{-\frac{u-u'}{\epsilon}\right\}\, V_S(u')\,du',$$

which can be successively integrated by parts to develop the following asymptotic expansion in small ϵ:

$$v(u) = \left[V_S(u) - \epsilon \frac{dV_S}{du}(u) + \cdots\right]$$
$$+\exp\left\{-\frac{u-u_{in}}{\epsilon}\right\}\left[v_{in} - V_S(u_{in}) + \epsilon \frac{dV_S}{du}(u_{in}) + \cdots\right]. \tag{4.6}$$

Note that the first term (in brackets) is independent of initial data, and is V_S to leading order, while the second term shows the rapid loss of dependence of the solution upon its initial data. Indeed, Formula (4.6) shows the two characteristic features of neurons within a high conductance state:

1. When *subthreshold* and $u - u_{in} >> \epsilon$, Formula (4.6) shows that the membrane potential is indeed slaved to the effective reversal potential V_S:

$$v(\tilde{t}) = V_S(\tilde{t}) - \frac{\epsilon}{\tilde{g}_T(\tilde{t})}\frac{dV_S}{d\tilde{t}} + O(\epsilon^2)\,, \tag{4.7}$$

where V_S is, again, defined in terms of the instantaneous synaptic conductances by Eq. (4.3). Moreover, this analysis also shows that membrane potential lags in time behind V_S, as the expansion in Eq. (4.7) can be reordered and expressed as

$$v(\tilde{t}) = V_S\left(\tilde{t} - \frac{\epsilon}{\tilde{g}_T(\tilde{t})}\right) + O(\epsilon^2)\,, \tag{4.8}$$

That is, the first-order correction can be reinterpreted as a backwards time shift of $V_S(t)$, with the time shift given by the scaled total conductance itself, and with amplitude corrections to v only appearing only at next order, $O(\epsilon^2)$.

2. The second feature described by Eq. (4.3) is the very rapid response of neurons in the high conductance regime: After $v(\tilde{t})$ is forced away from $V_S(\tilde{t})$ (as when the neuron spikes at $\tilde{t}_k$), the membrane potential very rapidly attempts to return to $V_S(\tilde{t})$:

$$v(\tilde{t}) - V_S(\tilde{t}) = O(\epsilon) + O(e^{-u/\epsilon})\,, \quad \text{where} \quad u \equiv \int_{\tilde{t}_k}^{\tilde{t}} g(\tilde{t}')d\tilde{t}', \quad \tilde{t} > \tilde{t}_k\,. \tag{4.9}$$

That is, within a basically $O(\epsilon)$ time from v being displaced from V_S, say at a spike time $\tilde{t}_k$, v returns to V_S. If V_S happens to be above threshold, this rapid return will likely lead to another spike.

Comments

(i) The I/O relation (4.7) and its derivation clarify the assumptions made by experimentalists in their estimates of conductances that rely on measuring the membrane potential for different holding currents and using linear regression (e.g. Borg-Graham, Monier, and Fregnac [1998]; Hirsch, J.M.Alonso, Reid, and Martinez [1998]; Anderson, Carandini, and Ferster [2000]). If large conductances are indeed present *in vivo*, this analysis shows that, when the neuron is not spiking, one can treat the *in vivo* membrane, approximately, as a passive resistance to current flow, and calculate the membrane potential directly from the effective reversal potential, as in Eq. (4.3).

(ii) We have also used this analysis to understand the generation of the linear-like responses of simple cells in our nonlinear cortical model (see Wielaard, Shelley, Shapley, and McLaughlin [2001]). A sensitive test of simple cell behavior is the absence of higher harmonic response under visual stimulation by standing gratings undergoing contrast reversal. This absence is seen both in extracellular spiking (De Valois, Albrecht, and Thorell [1982]) and in the intracellular potential (Ferster, Chung, and Wheat [1996]; Jagadeesh, Wheat, Kontsevich, Tyler, and Ferster [1997]). I/O formula (4.7), $v \simeq V_S$, helps explain how geniculate and cortico-cortical conductances combine to produce little higher harmonic content.

4.2 Slaving in the Model Cortex

Next, we show that neurons in this computational cortex are described with surprising accuracy by this asymptotic analysis, even though τ_g lies below the synaptic time-scale τ_{syn} by *only a factor of two* (see Fig. 3.1.) This claim is well illustrated by Fig. 4.1a: It shows the membrane potential $v(t)$ (green curve) for an excitatory neuron, where the model cortex is being driven by the randomly flashed grating stimulus, and its effective reversal potential $V_S(t)$ (red curve), calculated from Eq. (4.3) as the ratio of the difference current to total conductance (time is again dimensional). The total conductance g_T is also shown (dashed blue, in units of $\sec^{-1}$. Note that the conductance sizes are about those for the drifting grating stimulation used in Fig. 3.1.

While V_S is subthreshold, the two are nearly identical, with $v(t)$ tracking $V_S(t)$ closely. First, recall that Eq. (4.8) predicts that $v(t)$ is nearly a leftward shifted copy of V_S, with the size of the shift inversely proportional to g_T. Figs. 4.1a,c confirm this feature of the potential. Note too that the rises in V_S above threshold, and thus the onset of spiking activity, are associated with dips in the total conductance. These decreases in g_T are tied to decreases in the inhibitory conductance, and thus the spiking activity shown here arises from a release from inhibition.

There is some experimental indication that neurons in the cortex *in vivo* actually respond in the slaved fashion of the neurons in the model. Specifically, Figure 1 of Borg-Graham, Monier, and Fregnac [1998] can be inter-

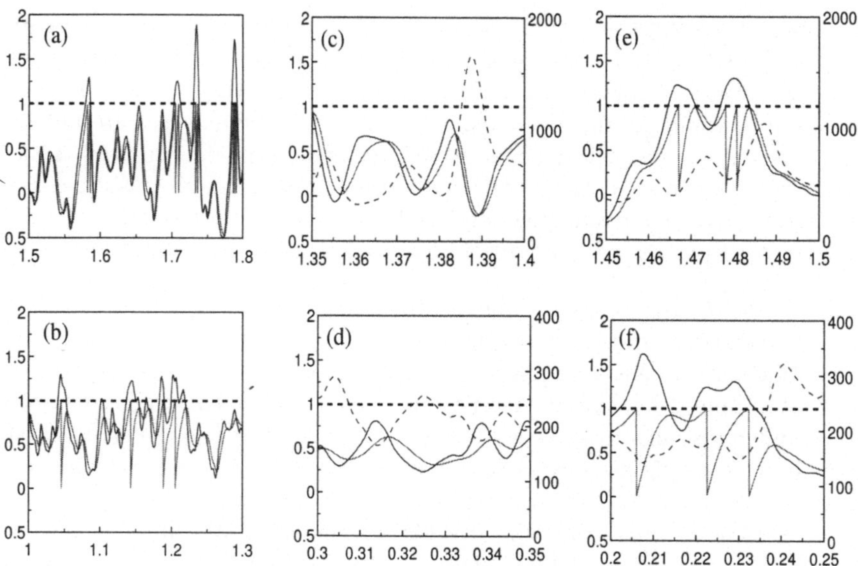

FIGURE 4.1. The membrane potential $v(t)$ (green), calculated from the large-scale simulation of the full network, Eq. (2.1); and the effective reversal potential $V_S(t)$ (red), calculated from the ratio of intracellular current and conductance as in Eq. (4.3), for an orientation selective excitatory neuron. The neuron (a) is being stimulated by randomly flashed gratings (at optimal spatial and temporal frequency and 100% contrast); and (b) without stimulation. The temporal scale is enlarged in (c, d, e, f), with the neuron subthreshold in (c) and (d), with (c) and without (d) stimulation; while in (e) and (f) the neuron is suprathreshold (and spiking), with (e) and without (f) stimulation. In (c,d,e,f) the total conductance is shown as dashed blue, with scale in inverse seconds indicated on the right axis. Intracellular potentials have been normalized to set the threshold to firing at unity (the dashed horizontal line) and reset potential to zero.

preted as showing that, immediately after the stimulus is switched on, the voltage is "slaved" to an effective reversal potential, during which time the conductance is very large.

For comparison, Fig. 4.1b shows $v(t)$ and $V_S(t)$ for an unstimulated neuron. Again the scale of g_T is set by leakage, the stochastic conductances, and a now *weaker* network activity. Thus, the membrane time-scale τ_g is now larger than that under stimulation, and comparable to that of synaptically mediated fluctuations in V_S. Nonetheless, Fig. 4.1b shows that $v(t)$ still tracks $V_S(t)$, though more loosely since ϵ is larger than under stimulation. Figs. 4.1c, d, e & f show the dynamics over a smaller (50 ms) time interval. In Figs. 4.1c & d, the neuron is subthreshold, with and without stimulation; while in Figs. 4.1e & f, the neuron is spiking, with and without stimulation. A significant difference between the stimulated and unstimulated cases, as seen in Fig. 4.1, is the recovery time after spike reset, which sets a (local in time) frequency for spiking. During stimulation by randomly flashed gratings, this recovery time is one-half of its background value, which is in register with the changes in the conductance time-scales: $\tau_g^{\text{back}} \simeq 4\,\text{ms}$ versus $\tau_g^{stim} \simeq 2\,\text{ms}$.

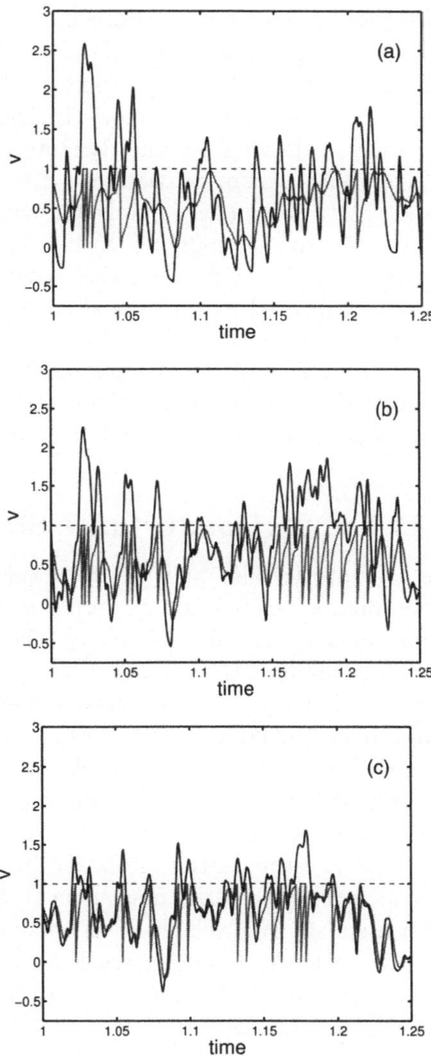

FIGURE 4.2. The membrane potential $v(t)$ (green) of a single model neuron impinged upon presynaptically by inhibitory and excitatory spike trains (with Poisson distributed spike-times), and the effective reversal potential $V_S(t)$ (red). Here the Poisson rates are chosen so that $\langle g_T \rangle = 125\,\text{s}^{-1}$ (a), $250\,\text{s}^{-1}$ (b), and $500\,\text{s}^{-1}$ (c), while simultaneously requiring that $\langle I_D \rangle / \langle g_T \rangle = 0.75$. The horizontal dashed line marks the threshold to firing.

We see that even in background, with the two time-scales τ_g and τ_{syn} in approximate balance and not separated, v still tracks V_S rather well. To investigate whether this balance might constitute the transition from low to high conductance behavior, we consider the dynamics of a single model neuron that is impinged upon presynaptically by excitatory and inhibitory spike trains (here, spike times are Poisson processes). The synaptic strengths are fixed, and the presynaptic spike rates are adjusted to achieve a given mean total conductance $\langle g_T \rangle = \tau_g^{-1}$ (here $\langle \cdot \rangle$ denotes expectation), while keeping $\langle I_D \rangle / \langle g_T \rangle$ fixed at 0.75 (below threshold). As in Fig. 4.1, Figs. 4.2a,b, & c shows a detail from the dynamics of v and V_S in response to these spike trains, where respectively, $\tau_g = 2 \cdot \tau_{\mathrm{syn}}$ (Fig. 4.2a), τ_{syn} (Fig. 4.2b), and $\tau_{\mathrm{syn}}/2$ (Fig. 4.2c). The latter two correspond roughly in conductance time-scales to the model network in background and under stimulation, respectively, while the first is a case where τ_{syn} is the shortest time-scale—a state not captured by our model cortex in its normal operating regime. Fig. 4.2a shows that with a conductance time-scale larger (but not much larger) than that of the synaptic fluctuations, $\tau_g = O(8\,\mathrm{ms})$ versus $\tau_{\mathrm{syn}} = O(4\,\mathrm{ms})$, the cell's response is an integrator over V_S, not a tracker of its fluctuations, as it is for $\tau_g = O(4\,\mathrm{ms})$ and $\tau_g = O(2\,\mathrm{ms})$ cases.

4.3 Spiking in a High Conductance State

High conductance asymptotics also allow us to develop simple analytical expressions for the spike rate of a cell in terms of its synaptic inputs. When the effective reversal potential $V_S(t)$ crosses the spiking threshold (a necessary condition for spiking), the slaving of the intracellular potential v to V_S means that $v(t)$ is usually also dragged across threshold to spike. Upon reset to zero, the large conductance causes $v(t)$ to relax rapidly back towards $V_S(t)$ on the short relaxation time-scale, $O(\tau_g)$. If $V_S(t)$ still lies above threshold, then v will cross threshold again to produce another spike. Using Eq. (4.6), we can estimate this time to recover from reset to spike threshold, or equivalently, estimate the interspike interval τ_{ISI}. First, let $u_{in} = u_s$ in Eq. (4.6), where u_s is the first spike time, and so set $v(u = u_{\mathrm{in}}) = 0$ (reset after a spike). We seek the next spike time $u_s^+ = u_s + \Delta u$, where $\Delta u = O(\epsilon)$. Over this short time interval, V_S is only slowly varying and can be considered frozen. Evaluating Eq. (4.6) at $u = u_s^+$, for which $v(u_s^+) = 1$, we find to leading order

$$\Delta u = -\epsilon \log(1 - V_S^{-1}), \quad V_S > 1,$$

or, returning to dimensional time, and using that g_T also varies slowly over the interspike interval,

$$\tau_{ISI} \simeq -\frac{\log\left(1 - V_S^{-1}\right)}{g_T}, \quad V_S > 1. \tag{4.10}$$

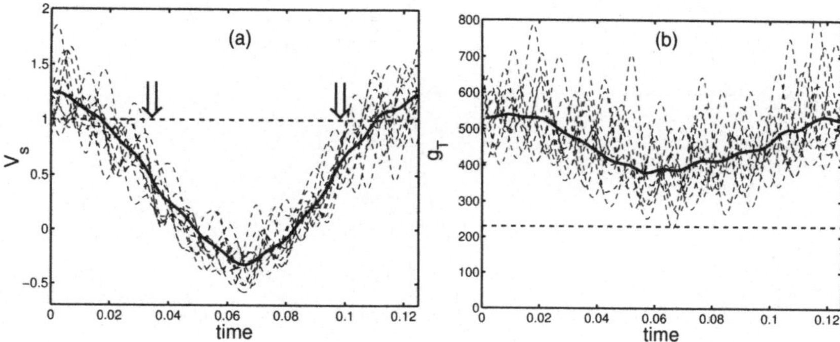

FIGURE 4.3. Panel (a): The dynamics of the effective reversal potential V_S for a neuron of the cortical network model. The network is being stimulated by an $8Hz$, full contrast drifting grating, set at the preferred orientation of the neuron. The dashed curves are V_S for 10 successive cycles of the stimulus. The solid curve is the cycle average of V_S, computed from 72 stimulus cycles. The horizontal dashed line marks the threshold to firing, and the arrows mark the times of offset and onset of firing during the stimulus cycle, as seen in Fig. 4.4a. Panel (b): The corresponding dynamics of the total conductance g_T. The horizontal dashed line marks the mean conductance level in the absence of visual stimulation ($230\,\mathrm{s}^{-1}$).

We see that when $V_S(t)$ is above the spiking threshold, and $v(t)$ is spiking, it is the total conductance g_T that sets the scale for the interspike interval size. When the conductance is doubled τ_{ISI} is halved, as is demonstrated in Figs. 4.1e,f.

Equation (4.10) can then be used to estimate the spiking rate, $M(T;\Delta)$, of a cell during the time interval $(T, T+\Delta)$ as:

$$\frac{\#\,\text{of spikes in time}\ \Delta}{\Delta} \simeq \frac{1}{\Delta}\int_T^{T+\Delta} \left.\frac{-g_T(t)}{\log(1 - V_S^{-1}(t))}\right|_{V_S>1}. \tag{4.11}$$

That is, given the synaptic inputs to a cell, the evaluation of an integral over the suprathreshold time course of V_S estimates the spike rate of that cell.

To study the utility and accuracy of Eq. (4.11) in predicting spike rates, first consider Fig. 4.3, which shows V_S and g_T for a neuron being stimulated by an 8 Hz, full contrast drifting grating (set at preferred orientation). V_S and g_T are shown (the dashed curves) for 10 successive cycles of the stimulus. Also shown are their cycle averages $\bar{V}_S$ and $\bar{g}_T$ (solid curves, computed from 72 cycles). Clearly, both V_S and g_T have substantial temporal fluctuations on the τ_{syn} time-scale, relative to their cycle averages, and these fluctuations will also have a substantial effect on the average modulation and onset of firing. For example, the cycle average $\bar{V}_S$ being at spiking threshold will not predict the onset of cell firing, as the fluctuations of V_S above threshold can persist even when $\bar{V}_S$ is well below. Even though these

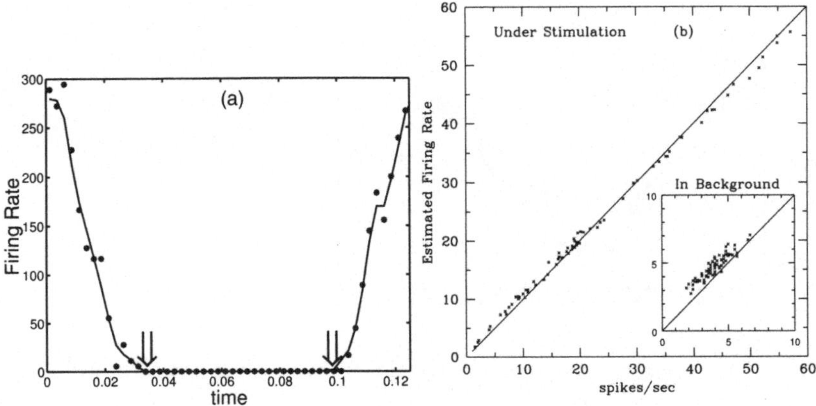

FIGURE 4.4. Panel (a): Comparison of firing rates calculated from spiking in the large-scale simulation with asymptotic estimates. Panel (a) shows cycle-averaged firing rates for the neuron of Fig. 4.3. The points are calculated by counting spikes in the large-scale simulation of the full network, Eqs. (2.1). The solid curve is calculated using the asymptotic formula Eq. (4.11). All cycle averages are found from 72 cycles of the stimulus. The arrows mark the times of offset and onset of firing during the stimulus cycle; *cf.* Fig. 4.3a. While this particular neuron is among those with the highest firing rates in our simulations, lower rate neurons also give similar results. Panel (b) shows a further comparison of asymptotic estimates of firing rates with results from simulation: A "scatter plot", over a subpopulation of 70 neurons, of the time-averaged firing rates measured in the large-scale simulations, against those predicted by Eq. (4.11). The inset shows these results for the same set of neurons, but now in the absence of visual stimulation.

fluctuations occur on the relatively short synaptic time-scale, in the high conductance state they are resolved, and their contributions captured by Eq. (4.11).

Figure 4.4a compares the cycle averaged spike rate, $\bar{M}$, over the course of the stimulus cycle, found by counting spikes in the cortical network simulation (points), with the estimate to $\bar{M}$ given by Eq. (4.11) (solid curve). For the former, the spike count was binned into 50 subintervals of the stimulus period (of size $\Delta = 0.125/50$), with the cycle average taken over 72 cycles of the stimulus. In the same way, the integral in Eq. (4.11) was found by numerical quadrature over each subinterval of the stimulus cycle (again of size Δ, with $\Delta >> \delta t$, the temporal resolution of the simulation), and also cycle averaged over the 72 cycles of the stimulus. The agreement is very good, with Eq. (4.11) capturing the magnitude, onset, and time-course of firing over the stimulus cycle. Further, for a group of 70 neurons from the simulation, Fig. 4.4b shows a "scatter plot" comparing firing rates over the stimulus period as predicted by Eq. (4.11), against those measured by counting spikes in the full network simulations. Again, the agreement between the full simulations and the asymptotic estimates is excellent.

5 Possible Functions of a High Conductance State

Although the functional consequences of this high conductance regime are far from certain, the near instantaneous response of the membrane potential to synaptic changes in input conductance could enable neurons to act as coincidence detectors; and it could permit transferal of detailed incoming temporal spiking patterns to the next level along the cortical pathway. Related ideas have been discussed in the literature. (See the recent reviews of Koch, Rapp, and Segev [1996], Usrey and Reid [1999], and references therein.)

5.1 Coincidence Detection

Results from our computational model are consistent with the interpretation that cortical activity provides neurons with the ability to respond "nearly instantaneously" (with respect to synaptic time-scales) to changes in their synaptic drive. Thus, two spikes arriving at nearly identical times could act together and cause $V_S(t)$ to cross threshold and fire a postsynaptic spike. On the other hand, at a larger time separation Δ which exceeds the synaptic time-scale of $G_E(t)$ ($\Delta > \tau_{syn}$), the two would act independently, and if neither individually produced voltage changes of sufficient size to cross threshold, the postsynaptic neuron would not fire. It follows that neurons in a high conductance regime, with $g_T^{-1} \ll \tau_{\text{syn}}$, could act as coincidence detectors with a temporal resolution of τ_{syn}. This temporal resolution is supported by recent measurements (Roy and Alloway [2000]) on cat somatosensory cortex, which find that thalmacortical spikes separated in time by less than 6-8 ms enhance cortical response, while little enhancement is found when the temporal separation of the incoming spikes exceeds 6-8 ms. While the synaptic time-scale τ_{syn} sets the temporal window for two spikes to be considered as "coincident", we emphasize that the near instantaneous response guaranteed by the separation of time-scales $\tau_g \ll \tau_{\text{syn}}$ enables cortical neurons to act effectively as coincidence detectors—especially in the presence of noise on the synaptic time-scale. This effectiveness is enhanced by cortical activity; that is, *cortical activity converts a neuronal integrator into a coincidence detector.*

5.2 "Burst" Generation

Next, consider an overdamped cortical neuron acting as a "burst generator": Assume that, because of incident spikes, its effective reversal potential crosses the firing threshold from below at time t_1, and remains above threshold until time $t_2 = t_1 + \Delta$. If g_T is sufficiently large, the neuron will fire a sequence of spikes in time interval $t_2 - t_1 = \Delta$, where the number of spikes (N) in the sequence is estimated as

$$N \propto \frac{\Delta}{\tau_g} = \Delta\ g_T$$

(i.e. the interspike times within the sequence scale as $O[\tau_g = g_T^{-1}]$). An incoming sequence of spikes could then cause the postsynaptic neuron to likewise fire a sequence of spikes, and therefore through this mechanism the cortical neuron acts as a "burst transmitter".

Thus, I/O analysis identifies two distinct time-scales for an overdamped neuron when acting as a coincidence detector or as a burst transmitter. For detecting coincidence in the arrival of two spikes, the temporal resolution of the input is set by τ_{syn}, the synaptic time-scale. For burst generation, the conductance size itself sets the interspike intervals within an output sequence as g_T^{-1}. Moreover, high contrast visual stimulation, which induces conductance increases of more than 200%, reduces (by more than a factor of two) the temporal separations of the outgoing spikes in the pattern transmitted to the next level of cortical processing.

5.3 Graded Response through Strong Temporal Fluctuations

It is clear from Figs. 4.3a and 4.4a that there is a strong correlation between the long temporal scale modulation of the effective reversal potential V_S (on the τ_{lgn} time-scale) and the modulation in the firing rate. We consider this relation in Fig. 5.1 by plotting the cycle averaged firing rate $\bar{M}$ (and its estimate) from Fig. 4.4a, against the cycle averaged potential $\bar{V}_S$, for 8 Hz drifting grating stimulus. (During the the stimulus cycle, the firing rate both rises and falls, transiting through nearly equal values of $\bar{V}_S$. This gives the two, nearly overlying curves—one for rising, one for falling—seen in the figure.) This relation shows a monotonic and smooth (i.e. graded) increase in firing rate with increase in potential $\bar{V}_S$, with the onset of firing occurring well below the threshold to firing (the long-dashed vertical line) set at the level of the intracellular potential v in the integrate-and-fire dynamics.

This spanning of the firing threshold illustrates how fluctuations in the synaptic inputs contribute to creating a network response, even when the mean potential lies below threshold. To emphasize this further, the dashed curve in Fig. 5.1 shows the simple firing rate estimate

$$\text{Firing Rate}\left[\bar{V}_S\right] \sim -\frac{\langle \bar{g}_T \rangle}{\log(1-\bar{V}_S^{-1})}, \quad \text{for } \bar{V}_S > 1, \tag{5.1}$$

where $\langle \bar{g}_T \rangle$ is simply the mean of the relatively unmodulated g_T. This estimate of firing rate *vs* $\bar{V}_S$ does capture the magnitude and rise of firing at high rates, but does not capture the overall curvature of the data, nor its spanning the threshold to firing, as both arise from the substantial synaptic fluctuations relative to average behavior (Figs. 4.3a & b).

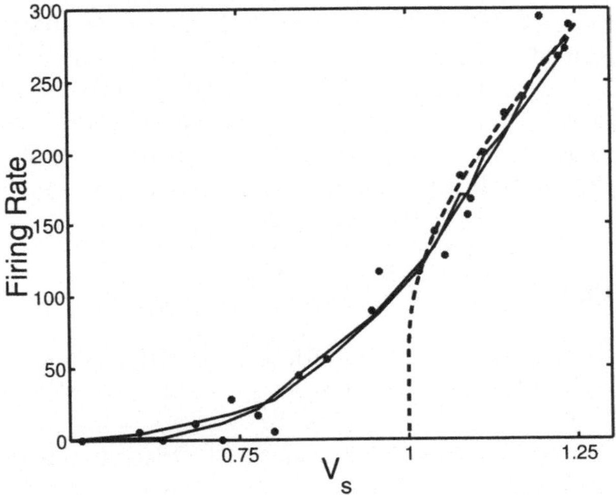

FIGURE 5.1. The cycle averaged firing rates (points and solid curves, as in Fig. 4.4) plotted against the cycle averaged effective reversal potential $\bar{V}_S$. During the the stimulus cycle, the firing rate rises and falls, and in so doing transits through nearly equal values of $\bar{V}_S$. This gives the two, nearly overlying curves, and sets of data points, seen in the figure. The dashed curve shows the simplified estimate of firing rate versus $\bar{V}_S$ given in Eq. (5.1). Note that this estimate obtains only for $\bar{V}_S > 1$.

5.4 Conclusion

Our network produces the nearly linear responses of Simple cells. This is accomplished through a nonlinear network whose conductance is high, and dominated by inhibition. In this high conductance state, Eq. (4.4) indicates that $V_S \simeq V_I\left(1 + K \cdot (g_E/g_I)\right)$, or that the modulation of the effective reversal potential of a neuron will be approximately proportional to its ratio of excitatory to inhibitory conductances. This is important because in our simple cell network, the excitatory conductance g_E is dominated by, and so follows closely, its geniculate component. Following our understanding, and instantiation, of the cortical architecture (see McLaughlin, Shapley, Shelley, and Wielaard [2000]; Wielaard, Shelley, Shapley, and McLaughlin [2001]), the inhibitory conductance, that divides g_E, is only cortico-cortical, being a sum over the activity of many nearby inhibitory neurons. Thus in V_S we see the geniculate input conductance being normalized by a population response.

Again, as Fig. 5.1 shows, in this case of steady-state tuning the firing rate is a smoothly increasing function of V_S. The overall scale of firing, or the gain, is set by the total conductance, which in our model is dominated by cortico-cortical inhibition. This may be the most significant of the given, possible functions of the cortex operating in high conductance: The firing rate is the result of an accurate and smoothly graded computation of the ratio of excitatory to inhibitory conductances. This may be a necessary task that underlies the computations of Simple cells whose computations are believed to be crucial for many tasks of visual perception (see Wielaard, Shelley, Shapley, and McLaughlin [2001] and the references therein).

References

Alonso, J., W. Usrey, and R. Reid [2001], Rules of connectivity between geniculate cells and simple cells in cat primary visual cortex, *J. Neuroscience* **21**, 4002–4015.

Anderson, J., M. Carandini, and D. Ferster [2000], Orientation tuning of input conductance, excitation, and inhibition in cat primary visual cortex., *J. Neurophysiol.* **84**, 909–926.

Ben-Yishai, R., R. Bar-Or, and H. Sompolinsky [1995], Theory of orientation tuning in the visual cortex, *Proc Nat Acad Sci USA* **92**, 3844–3848.

Bernander, O., R. Douglas, K. Martin, and C. Koch [1991], Synaptic background activity influences spatiotemporal integration in single phyrmidal cells, *Proc. Nat. Acad. Sci. USA* **88**, 11569–11573.

Bernander, O., C. Koch, and M. Usher [1994], The effect of synchronized inputs at the single neuron level, *Neural Computation* **6**, 622–641.

Blasdel, G. [1992a], Differential imaging of ocular dominance and orientation selectivity in monkey striate cortex, *Journal of Neuroscience* **12**, 3115–3138.

Blasdel, G. [1992b], Orientation selectivity, preference, and continuity in the monkey striate cortex, *Journal of Neuroscience* **12**, 3139–3161.

Bonhoeffer, T. and A. Grinvald [1991], Iso-orientation domains in cat visual cortex are arranged in pinwheel like patterns, *Nature* **353**, 429–431.

Borg-Graham, L., C. Monier, and Y. Fregnac [1998], Visual input evokes transient and strong shunting inhibition in visual cortical neurons, *Nature* **393**, 369–373.

De Valois, R., D. Albrecht, and L. Thorell [1982], Spatial frequency selectivity of cells in macaque visual cortex, *Vision Res* **22**, 545–559.

DeAngelis, G., R. Ghose, I. Ohzawa, and R. Freeman [1999], Functional Micro-Organization of Primary Visual Cortex: Receptive Field Analysis of Nearby Neurons, *Journal of Neuroscience* **19**, 4046–4064.

Everson, R., A. Prashanth, M. Gabbay, B. Knight, L. Sirovich, and E. Kaplan [1998], Representation of spatial frequency and orientation in the visual cortex, *PNAS* **95**, 8334–8338.

Ferster, D., S. Chung, and H. Wheat [1996], Orientation selectivity of thalamic input to simple cells of cat visual cortex, *Nature* **380**, 249–252.

Hansel, D. and H. Sompolinsky [1996], Chaos and synchrony in a model of a hypercolumn in visual cortex, *J Comp Neuroscience* **3**, 7–34.

Hansel, D. and H. Sompolinsky [1998], *Modeling feature selectivity in local cortical circuits*, pages 499–567. MIT Press, Boston.

Hirsch, J., J.M.Alonso, R. Reid, and L. Martinez [1998], Synaptic integration in striate cortical simple cells, *J. Neuroscience* **15**, 9517–9528.

Hubel, D. and T. Wiesel [1962], Receptive fields, binocular interaction and functional architecture of the cat's visual cortex, *J Physiol (Lond)* **160**, 106–154.

Jagadeesh, B., H. Wheat, L. Kontsevich, C. Tyler, and D. Ferster [1997], Direction selectivity of synaptic potentials in simple cells of the cat visual cortex, *J. Neurophysiol.* **78**, 2772–2789.

Koch, C. [1999], *Biophysics of Computation.* Oxford University Press, Oxford.

Koch, C., M. Rapp, and I. Segev [1996], A brief history of time constants, *Cerebral Cortex* **6**, 93–101.

Maldonado, P., I. Godecke, C. Gray, and T. Bonhoeffer [1997], Orientation selectivity in pinwheel centers in cat striate cortex, *Science* **276**, 1551–1555.

McLaughlin, D., R. Shapley, M. Shelley, and J. Wielaard [2000], A Neuronal Network Model of Macaque Primary Visual Cortex (V1): Orientation Selectivity and Dynamics in the Input Layer $4C\alpha$, *Proc. Natl. Acad. Sci. USA* **97**, 8087–8092.

Mechler, F. [1997], *Neuronal response variability in the primary visual cortex*, PhD thesis, New York University.

Reid, R. and J. Alonso [1995], Specificity of monosynaptic connections from thalamus to visual cortex, *Nature* **378**, 281–284.

Roy, S. and K. Alloway [2000], Coincidence Detection or Temporal Integration? What the Neurons in Somatosensory Cortex are Doing, *Journal of Neuroscience* **21**, 2462–2473.

Shapley, R., D. McLaughlin, and M. Shelley [2002], *Orientation selectivity: models and neural mechanisms.* MIT Press, Boston.

Shelley, M., D. McLaughlin, R. Shapley, and J. Wielaard [2002], States of High Conductance State in a Large-Scale Model of the Visual Cortex, *to appear, J Comp Neuroscience.*

Usrey, W. M. and R. C. Reid [1999], Synchronous activity in the visual system, *Annu. Rev. Physiol.* **61**, 435–456.

Wielaard, J., M. Shelley, R. Shapley, and D. McLaughlin [2001], How Simple cells are made in a nonlinear network model of the visual cortex, *J. Neuroscience* **21**, 5203–5211.

10

Power Law Asymptotics for Nonlinear Eigenvalue Problems

Paul K. Newton
Vassilis G. Papanicolaou

To Larry Sirovich, on the occasion of his 70th birthday.

ABSTRACT Nonlinear eigenvalue problems arise in a wide variety of physical settings where oscillation amplitudes are too large to justify linearization. These problems have amplitude dependent frequencies, and one can ask whether scaling laws of the kind that arise in classical Sturm-Liouville theory still pertain. We prove in this paper that in the context of analyzing radially symmetric solutions to a class of nonlinear dispersive wave models, they do. We re-cast the equations as Hamiltonian systems with 'dissipation' where the radial variable 'r' plays the role of time. We treat the case of a symmetric double-well potential in detail and show that the appropriate nonlinear eigenfunctions are trajectories that start with energies above the center hump and ultimately decay to the peak as $r \to \infty$. The number of crossings over the center line in the double-well (related to the number of times the trajectory 'bounces' off the sides) corresponds to the number of zeroes of the nonlinear eigenfunction and the initial energy levels corresponding to these eigenfunctions form a discrete set of values which can be related to the eigenvalues. If $u_n(0) \equiv \gamma_n$ denotes the value for which the problem has an 'eigenfunction' $u_n(r)$ with exactly n zeroes for $r \in (0, \infty)$, we prove that the spacings $\gamma_{n+1} - \gamma_n$ follow power law scaling

$$\gamma_{n+1} - \gamma_n \sim \frac{F_p}{n^{1-(1/p)}}$$

as $n \to \infty$ where $2p+1$ is the power of the nonlinearity and F_p is a constant which depends on p but is otherwise independent of n. This in turn gives asymptotic information on the spacing of the initial energy levels

$$E_n = \frac{\gamma_n^{2p+2}}{2p+2} - \frac{\gamma_n^2}{2}$$

which we prove follows power law form

$$E_{n+1} - E_n \sim G_p \cdot n^{1+(2/p)}.$$

Although only symmetric double-well potentials are treated in this paper (both the direct and inverse problem), it is clear that much more general situations can and should be analyzed in the future, including scaling laws for asymmetric potentials and potentials with more than two local minima.

Contents

1 Introduction

Asymptotic scaling laws for linear eigenvalue problems are abundant in the classical literature, for ordinary differential equations (Coddington and Levinson [1995]), partial differential equations (Courant and Hilbert [1953]), and for linear integral equations (Sirovich and Knight [1985, 1986]). For example, if one solves the Dirichlet problem for Laplace's equation on domains of length l, area A, or volume V, we know the asymptotic distribution of the eigenvalues follows power law form

$$\lambda_j \sim \left(\frac{\pi}{l}\right)^2 j^2$$

$$\lambda_j \sim \left(\frac{4\pi}{A}\right) j$$

$$\lambda_j \sim \left(\frac{6\pi^2}{V}\right)^{2/3} j^{2/3}$$

More generally, for elliptic eigenvalue problems

$$\mathcal{L}[u] + \lambda\rho u = 0; \qquad u = 0 \in \partial D$$

where

$$\mathcal{L}[u] \equiv (pu_x)_x + (pu_y)_y - qu$$

with $p(x,y) > 0, \rho(x,y) > 0$, we know (Courant and Hilbert [1953]) that the number $N(\lambda)$ of eigenvalues less than a fixed λ is given by

$$\lim_{\lambda\to\infty} \frac{N(\lambda)}{\lambda} = \frac{1}{4\pi} \iint_A \frac{\rho}{p} dA,$$

while in 3D, the scaling follows the form

$$\lim_{\lambda\to\infty}\frac{N(\lambda)}{\lambda^{3/2}}=\frac{1}{6\pi^2}\iiint_V\left(\frac{\rho}{p}\right)^{3/2}dV.$$

One of the reasons one seeks spectral information for linear problems is to use the information in an inversion process to reconstruct classical or quantum potentials (Courant and Hilbert [1953]). For nonlinear problems, however, most of the classical techniques for both the direct and inverse problem fail. Nonetheless, these problems arise in practice and there is mounting evidence that interesting asymptotic scaling laws still pertain (Newton and O'Connor [1996]; Newton and Watanabe [1993]) for the direct problem, and that this information may be used in some instances to reconstruct the 'potential', which in this case is the nonlinearity.

To be concrete, suppose one looks for radially symmetric standing waves solutions to the nonlinear Schrödinger equation

$$i\psi_t+\Delta\psi+|\psi|^{2p}\psi=0,$$

which are of the form

$$\psi=\exp(i\lambda t)w(r),\qquad \text{where}\quad \lambda>0.$$

This leads to the nonlinear eigenvalue problem

$$w''(r)+\frac{d-1}{r}w'(r)+w^{2p+1}-\lambda w=0 \tag{1.1}$$

along with the boundary conditions $w(0)=H>0$, $w'(0)=0$, $w\to 0$, as $r\to\infty$ (notice also that we always have $w'\to 0$, as $r\to\infty$), where d is the underlying spatial dimension. Upon introducing the new scaled variables $t=\sqrt{\lambda}r$; $u=\lambda^{-1/2p}w$ in (1.1), we arrive at the simpler nonautonomous problem

$$u''(t)+\frac{\alpha}{t}u'(t)+u(t)^{2p+1}-u(t)=0,\qquad t>0, \tag{1.2}$$

$$u(0)=\gamma,\qquad u'(0)=0, \tag{1.3}$$

where $\gamma=H/\lambda^{1/2p}$, $\alpha=d-1$, and with the extra condition that $u\to 0$ as $t\to\infty$. One then looks for parameter values $\gamma=\gamma_j$ that give rise to solutions $u_j(t)$ having j zeroes. For this nonlinear problem, the eigenvalues are linearly ordered, namely $\gamma_0<\cdots<\gamma_j<\gamma_{j+1}<\cdots$, and there is numerical evidence that the eigenvalues follow the power law form

$$\gamma_j\sim const.\cdot j^{\mu}$$
$$\mu=\frac{\alpha}{[2-p(\alpha-1)]}.$$

More generally, one can look for θ-dependent solutions. For these 'defect' type solutions

$$\psi = \exp[i(\lambda t + m\theta + \theta_0)]w(r),$$

with $w \sim \beta r^m$ as $r \to 0$, m integer valued, we have

$$\frac{\beta_j}{\lambda^{(1-mp)/2p}} \sim const. \cdot j^{\mu}$$

with

$$\mu = \frac{\alpha}{2 - p(\alpha - 1)} + \frac{mp}{2}.$$

See Newton and Watanabe [1993] for the case $m = 0$, and Newton and O'Connor [1996] for $m \neq 0$. Although there are fundamental differences between the linear and the nonlinear case, to our knowledge there are no rigorous estimates for the kinds of scaling behavior that have been discovered (Newton and O'Connor [1996]; Newton and Watanabe [1993]) for the nonlinear case. As has been recognized since the papers of Finkelstein, LeLevier, and Ruderman [1951], Anderson and Derrick [1970a, 1971b], phase space techniques are quite useful in understanding these types of questions, which is the approach followed in Newton and O'Connor [1996]; Newton and Watanabe [1993]. The work of Nehari [1963] and Ryder [1968] were among the first to extend Sturm-Liouville oscillation theory to nonlinear problems, where they ask whether solutions could be ordered according to the number of zeroes they contain. Jones [1986]–Jones and Küpper [1986] studied these questions more recently, also from a dynamical systems point of view, while Kwong [1989] proved uniqueness for the positive solutions and Berestucki and Lions [1983] proved existence of infinitely many solutions. Grillakis [1990], on the other hand, uses PDE methods (as opposed to dynamical systems methods) to prove existence of solutions with fixed numbers of zeroes. Crucial to the existence of eigenvalues in (1.2) is the dissipative nature ($\alpha > 0$) of the equation, however what makes the problem tricky is the nonautonomous nature of the dissipative term. Hence, our goal in this paper is to prove rigorous scaling laws for an autonomous model based on the form of (1.2).

We hope it will be clear that the kinds of scaling laws described in this paper, while specific to problems with symmetric double-well potentials, in principle could be seen in more general settings — asymmetric potentials, for example, and potentials having more than two local minima. From this point of view, the topic is ripe for further investigation.

2 The Autonomous Model

We start by considering the simpler autononmous nonlinear eigenvalue problem

$$u''(t) + \alpha u'(t) + u(t)^{2p+1} - u(t) = 0, \qquad t > 0, \tag{2.1}$$

$$u(0) = \gamma, \qquad u'(0) = 0, \tag{2.2}$$

where $p \geq 1$ is an integer, $\alpha > 0$, and $\gamma > 0$ is a parameter. This problem has a unique solution for all $t \geq 0$. The global existence follows from the dissipative nature of the equation. Indeed, if we multiply (2.1) by $u'(t)$ and then integrate from 0 to t, we get

$$u'(t)^2 + 2\alpha \int_0^t u'(s)^2 ds + \frac{u(t)^{2p+2}}{p+1} - u(t)^2 = \frac{\gamma^{2p+2}}{p+1} - \gamma^2, \tag{2.3}$$

which implies that $u(t)$ stays bounded as $t \to \infty$. Adding the extra condition that

$$u \to 0, \qquad \text{as} \qquad t \to \infty, \tag{2.4}$$

one obtains a nonlinear eigenvalue-like problem where γ plays the role of the eigenvalue parameter. Then one can look for parameter values $\gamma = \gamma_n$ that give rise to solutions $u_n(t)$ having n zeroes.

To understand the underlying issues and to make contact with some of the other relevant literature, it is useful to look at the phase space structure and potential for the system (2.1)–(2.2) (notice that the system (1.2)–(1.3) has qualitatively similar phase space structure). For definiteness, we take the case $\alpha = 1$, $p = 1$ (corresponding to $d = 2$ with a cubic nonlinearity). If we plot the potential

$$V(u) = \frac{u^{2p+2}}{2p+2} - \frac{u^2}{2} = \frac{u^4}{4} - \frac{u^2}{2}$$

we see it has double well structure, as shown in Figure 2.1. With dissipation, any initial condition starting above the hump will decay and eventually settle into either the right minimum (i.e. $u \to 1$, as $t \to \infty$), the left minimum (i.e. $u \to -1$, as $t \to \infty$), or the hump maximum. The later correspond to the solutions with the extra condition $u \to 0$, as $t \to \infty$, imposed. The number of times the trajectory crosses the centerline corresponds to the number of zeroes of the solution. Since $u(0) = \gamma$ and $u'(0) = 0$, the initial *energy level* is $\gamma^4/4 - \gamma^2/2$. Shown in Figure 2.1 are the initial energy levels E_0, E_2 corresponding to the "eigenfunctions" having $0, 2$ zeroes respectively, and decaying to zero asymptotically. One by-product of this paper is the scaling law for the asymptotic distribution of the energy level spacings $E_{n+1} - E_n$ as $n \to \infty$. If we let

$$N(\gamma) = \text{the number of zeroes of the solution } u(t) \text{ of (2.1)–(2.2) on } (0, \infty), \tag{2.5}$$

it is not hard to see that $N(\gamma) \to \infty$ (monotonically), as $\gamma \to \infty$. One of the results of this paper is the derivation of the asymptotics of $N(\gamma)$, which in turn determine the asymptotics of γ_n. Our analysis will also give some insight for the more challenging case (1.2)–(1.3).

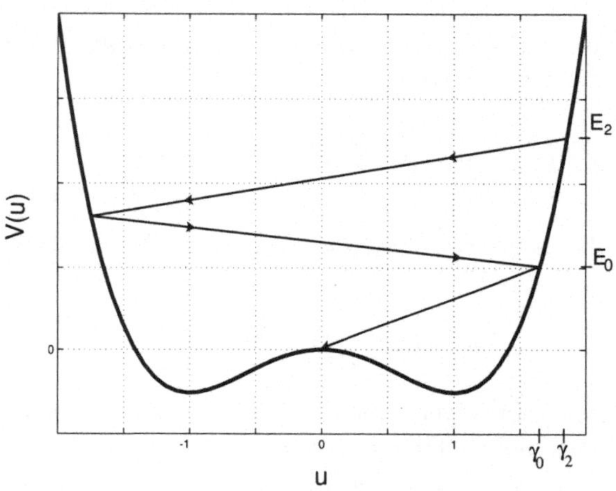

FIGURE 2.1. Double-well potential, energy level spacings E_0 and E_2, and corresponding eigenvalues γ_0 and γ_2 for the autonomous system (2.1).

2.1 The Hamiltonian Case

If we take $\alpha = 0$ in (2.1), (corresponding to the one-dimensional case (1.2)), our problem becomes Hamiltonian (i.e. nondissipative)

$$u''(t) + u(t)^{2p+1} - u(t) = 0, \qquad t > 0,$$

$$u(0) = \gamma, \qquad\qquad u'(0) = 0.$$

Here, of course, $N(\gamma) \equiv \infty$, where $N(\gamma)$, as in (2.5), is the number of zeroes of the solution $u(t)$ of (2.1)–(2.2) on $(0, \infty)$. In fact, in this case we can write down an explicit formula for $u(t)$ in terms of hyperelliptic functions, since (2.3) yields

$$u'(t) = \pm\sqrt{\frac{\gamma^{2p+2}}{p+1} - \gamma^2 - \frac{u(t)^{2p+2}}{p+1} + u(t)^2}$$

or

$$\int_{\gamma}^{u(t)} \frac{du}{\pm\sqrt{(p+1)^{-1}\gamma^{2p+2} - \gamma^2 - (p+1)^{-1}u^{2p+2} + u^2}} = t.$$

In particular, the above equation implies that, for any fixed γ, $u(t)$ is periodic with period T, where

$$T = T(\gamma) = 4\int_0^{\gamma} \frac{du}{\sqrt{(p+1)^{-1}\gamma^{2p+2} - \gamma^2 - (p+1)^{-1}u^{2p+2} + u^2}}.$$

If we substitute $u = \gamma x$ in the above integral, we obtain

$$T = \frac{4}{\gamma^p}\int_0^1 \frac{dx}{\sqrt{(p+1)^{-1} - (p+1)^{-1}x^{2p+2} - \gamma^{-2p}(1-x^2)}}.$$

It follows that, as $\gamma \to \infty$,

$$T = \frac{c_p}{\gamma^p}\left[1 + O\left(\gamma^{-2p}\right)\right], \tag{2.6}$$

where

$$c_p = 4\sqrt{p+1}\int_0^1 \frac{dx}{\sqrt{1-x^{2p+2}}} = \frac{2\sqrt{\pi}}{\sqrt{p+1}} \cdot \frac{\Gamma\left(\frac{1}{2p+2}\right)}{\Gamma\left(\frac{p+2}{2p+2}\right)} \tag{2.7}$$

(here $\Gamma(\cdot)$ is the gamma function).

2.2 A Formula for u''

Consider the inhomogeneous problem

$$u''(t) + k_1 u'(t) = f(t), \qquad t > 0,$$

$$u(0) = u_0, \qquad u'(0) = 0,$$

with $k_1 > 0$. The solution is

$$u(t) = -\frac{e^{-k_1 t}}{k_1}\int_0^t f(s)e^{k_1 s}ds + \frac{1}{k_1}\int_0^t f(s)ds + u_0$$

hence

$$u'(t) = e^{-k_1 t}\int_0^t f(s)e^{k_1 s}ds. \tag{2.8}$$

This formula is useful for us when considering the nonlinear problem

$$u''(t) + k_1 u'(t) + k_2 u(t)^{2p+1} - k_3 u(t) = 0, \qquad t > 0, \tag{2.9}$$

$$u(0) = u_0, \qquad u'(0) = 0, \tag{2.10}$$

where $p \geq 1$ is an integer and k_1, k_2, $k_3 > 0$. Its dissipative nature implies that $u(t)$ is bounded on $[0, \infty)$. If we set

$$f(t) = k_3 u(t) - k_2 u(t)^{2p+1}$$

in (2.8), we obtain that the solution $u(t)$ of (2.9)–(2.10) as

$$u'(t) = e^{-k_1 t} \int_0^t \left[k_3 u(s) - k_2 u(s)^{2p+1}\right] e^{k_1 s} ds$$

and hence

$$u''(t) = k_3 u(t) - k_2 u(t)^{2p+1} - k_1 e^{-k_1 t} \int_0^t \left[k_3 u(s) - k_2 u(s)^{2p+1}\right] e^{k_1 s} ds. \tag{2.11}$$

In particular, the last formula implies that $u \in C^2[0, \infty)$ (and, therefore, $u \in C^\infty[0, \infty)$ by induction).

3 Some Preliminary Scaling Lemmas

In the equation (2.1) we introduce the new independent variable

$$\tau = \frac{\gamma^p}{c_p} t,$$

where c_p is as in (2.7). This scaling is inspired by (2.6). If we use the notation

$$v(\tau) = \frac{1}{\gamma} u(t) = \frac{1}{\gamma} u\left(\frac{c_p}{\gamma^p}\tau\right), \tag{3.1}$$

then the problem (2.1)–(2.2) becomes

$$v''(\tau) + \alpha c_p \gamma^{-p} v'(\tau) + c_p^2 v(\tau)^{2p+1} - c_p^2 \gamma^{-2p} v(\tau) = 0, \qquad \tau > 0, \tag{3.2}$$

$$v(0) = 1, \qquad v'(0) = 0. \tag{3.3}$$

For large γ the problem (3.2)–(3.3) can be seen as a perturbation of the problem

$$v_0''(\tau) + c_p^2 v_0(\tau)^{2p+1} = 0, \qquad \tau > 0, \tag{3.4}$$

$$v_0(0) = 1, \qquad v_0'(0) = 0. \tag{3.5}$$

This is another Hamiltonian problem that can be solved explicitly, since (3.4)–(3.5) yield

$$v_0'(\tau)^2 + \frac{c_p^2}{p+1} v_0(\tau)^{2p+2} = \frac{c_p^2}{p+1}, \tag{3.6}$$

Observe that $v_0(\tau)$ makes sense for all $\tau \in \mathbf{R}$. The constant c_p is given by (2.7) and we have

$$v_0(\tau + 1) = v_0(\tau).$$

We also notice that (3.4)–(3.5)–(3.6) imply that $v_0(\tau)$ is an even function (hence $v_0'(\tau)$ is odd) such that

$$v_0\left(\frac{2n-1}{4}\right) = 0 \qquad \text{and} \qquad v_0'\left(\frac{n}{2}\right) = 0, \qquad \text{for every} \quad n \in \mathbf{Z},$$

and these are the only zeroes of $v_0(\tau)$ and $v_0'(\tau)$. All these zeroes are simple (if, for instance, $v_0'(\tau^*) = v_0''(\tau^*) = 0$, then (3.4) gives $v_0(\tau^*) = 0$, a contradiction).

3.1 Lemma. *Let $b > 0$ be a fixed number. If $v(\tau)$ and $v_0(\tau)$ satisfy (3.2)–(3.3) and (3.4)–(3.5) respectively, then there is a constant $A > 0$ (that depends only on b) such that*

$$|v(\tau) - v_0(\tau)| < A\gamma^{-p}, \qquad \textit{for all} \quad \tau \in [0, b],$$

in other words $v(\tau) = v_0(\tau) + O(\gamma^{-p})$, as $\gamma \to \infty$, uniformly on $[0, b]$.

Proof. Just notice that (3.2)–(3.3) is a regular perturbation of (3.4)–(3.5). ■

3.2 Corollary. *With the same assumptions as in the lemma we also have that there is an $A > 0$ such that*

$$|v''(\tau) - v_0''(\tau)| < A\gamma^{-p}, \qquad \textit{for all} \quad \tau \in [0, b].$$

Proof. Formula (2.11) applied to $v(\tau)$ gives

$$v''(\tau) = k_3 v(\tau) - k_2 v(\tau)^{2p+1} - k_1 e^{-k_1\tau} \int_0^\tau \left[k_3 v(\sigma) - k_2 v(\sigma)^{2p+1}\right] e^{k_1\sigma} d\sigma \tag{3.7}$$

where

$$k_1 = \alpha c_p \gamma^{-p}, \qquad k_2 = c_p^2, \qquad \text{and} \qquad k_3 = c_p^2 \gamma^{-2p}.$$

Now, the dissipative nature of (3.2) implies that there is a constant M, not depending on γ (as long as, say, $\gamma \geq 1$) such that

$$|v(\tau)| < M, \qquad \text{for all} \quad \tau \geq 0.$$

Thus (3.7) implies that

$$v''(\tau) = -c_p^2 v(\tau)^{2p+1} + \gamma^{-p} E(\tau),$$

where $E(\tau)$ is bounded on $[0, b]$ (by an upper bound that does not depend on γ). Since

$$v_0''(\tau) = -c_p^2 v_0(\tau)^{2p+1},$$

the proof is finished by using the estimate of Lemma 3.1. ∎

Remark. Since $v'(0) = v_0'(0) = 0$, the above corollary implies immediately that

$$|v'(\tau) - v_0'(\tau)| < A\gamma^{-p}, \qquad \text{for all} \quad \tau \in [0, b]. \tag{3.8}$$

To continue, we first observe that all zeroes of $v(\tau)$ and $v'(\tau)$, $\tau \in (0, \infty)$ are simple (otherwise (3.2) and the uniqueness theorem would imply that $v(\tau) \equiv 0$). Also (3.7) implies that, at least for $\gamma > 1$, $\tau = 0$ is a simple zero of $v'(\tau)$.

3.3 Lemma. *Let b, $v(\tau)$, and $v_0(\tau)$ be as in Lemma 3.1. We choose b so that $v_0(b)v_0'(b) \neq 0$ (i.e. $b \neq$ integer). If $v_0(z_0) = 0$, for some $z_0 \in (0, b)$, then, for all γ sufficiently large we have that there is a unique $z \in (0, b)$ such that $v(z) = 0$ and*

$$|z - z_0| < A\gamma^{-p},$$

where A is a positive constant. Likewise, if $v_0'(z_0') = 0$, for some $z_0' \in (0, b)$, then, for all γ sufficiently large we have that there is a unique $z' \in (0, b)$ such that $v(z') = 0$ and

$$|z' - z_0'| < A\gamma^{-p}.$$

Proof. We prove the second assertion only, since the proof of the first is similar. By (3.8)

$$|v'(z_0')| < A\gamma^{-p}.$$

Also, by (3.4) and (3.6) $v_0''(\tau)$ is away from 0 when τ is near z_0'. Therefore, the proof is finished by using the corollary of Lemma 3.1. ∎

Remark. Lemma 3.3 implies that, if

$$0 < \tau_1 < \tau_2 < \cdots < \tau_k \qquad \text{and} \qquad 0 < \tau_1' < \tau_2' < \cdots < \tau_k',$$

are the first k strictly positive zeroes of $v(\tau)$ and $v'(\tau)$ respectively (k is fixed), then

$$\left|\tau_j - \frac{2j-1}{4}\right| < A\gamma^{-p} \quad \text{and} \quad \left|\tau_j' - \frac{j}{2}\right| < A\gamma^{-p}, \quad j = 1, 2, ..., k. \tag{3.9}$$

We are now ready to prove the following important lemma:

3.4 Lemma. *Let $v(\tau)$ and $v_0(\tau)$ be given by (3.2)–(3.3) and (3.4)–(3.5) respectively. If τ_2' is the second strictly positive zero of $v'(\tau)$, then*

$$v(\tau_2') = 1 - \frac{\alpha}{c_p}\left[\int_0^1 v_0'(\tau)^2 d\tau\right]\gamma^{-p} + O\left(\gamma^{-2p}\right), \qquad as \quad \gamma \to \infty, \quad (3.10)$$

where the constant c_p is given by (2.7).

Proof. From (3.2)–(3.3) we obtain

$$v'(\tau)^2 + 2\alpha c_p \gamma^{-p} \int_0^\tau v'(\sigma)^2 d\sigma + \frac{c_p^2}{p+1} v(\tau)^{2p+2} - c_p^2 \gamma^{-2p} v(\tau)^2$$
$$= \frac{c_p^2}{p+1} - c_p^2 \gamma^{-2p}.$$

Now $v'(\tau_2') = 0$, thus, for $\tau = \tau_2'$ the above equation becomes

$$2\alpha c_p \gamma^{-p} \int_0^{\tau_2'} v'(\tau)^2 d\tau + \frac{c_p^2}{p+1} v(\tau_2')^{2p+2} - c_p^2 \gamma^{-2p} v(\tau_2')^2 = \frac{c_p^2}{p+1} - c_p^2 \gamma^{-2p}.$$

Hence

$$v(\tau_2')^{2p+2} = 1 - \frac{2(p+1)\alpha}{c_p} \gamma^{-p} \int_0^{\tau_2'} v'(\tau)^2 d\tau + O\left(\gamma^{-2p}\right).$$

Next we use (3.9) and (3.8) to conclude that

$$v(\tau_2')^{2p+2} = 1 - \frac{2(p+1)\alpha}{c_p} \gamma^{-p} \int_0^1 v_0'(\tau)^2 d\tau + O\left(\gamma^{-2p}\right).$$

The proof is completed by taking $(2p+2)$-th roots of both sides. ∎

Remark. In exactly the same way we can show that

$$|v(\tau_1')| = 1 - \frac{\alpha}{c_p}\left[\int_0^{1/2} v_0'(\tau)^2 d\tau\right]\gamma^{-p} + O\left(\gamma^{-2p}\right), \qquad \text{as} \quad \gamma \to \infty,$$

and since it is easy to see from (3.4)–(3.5) that

$$\int_0^{1/2} v_0'(\tau)^2 d\tau = \frac{1}{2}\int_0^1 v_0'(\tau)^2 d\tau$$

and $v(\tau_1')$ is (by Lemmas 3.1 and 3.3) close to $v_0(1/2) = -1$, we eventually get

$$v(\tau_1') = -1 + \frac{\alpha}{2c_p}\left[\int_0^1 v_0'(\tau)^2 d\tau\right]\gamma^{-p} + O\left(\gamma^{-2p}\right), \qquad \text{as} \quad \gamma \to \infty. \quad (3.11)$$

4 Rigorous Scaling Laws

Let $u(t)$ be the solution of (2.1)–(2.2) and $t_2' = c_p\gamma^{-p}\tau_2'$ be the second strictly positive zero of $u'(t)$. Then, we can use (3.1) to write (3.10) in terms of $u(t)$:

$$u(t_2') = \gamma\left[1 - \frac{\alpha B}{c_p}\gamma^{-p} + O\left(\gamma^{-2p}\right)\right], \qquad \text{as} \quad \gamma \to \infty, \tag{4.1}$$

where

$$B = \int_0^1 v_0'(\tau)^2 d\tau \tag{4.2}$$

($v_0(\tau)$ is, of course, given by (3.4)–(3.5)). Likewise (3.11) becomes

$$u(t_1') = -\gamma\left[1 - \frac{\alpha B}{2c_p}\gamma^{-p} + O\left(\gamma^{-2p}\right)\right], \qquad \text{as} \quad \gamma \to \infty, \tag{4.3}$$

Now, if t_{2n}' is the $(2n)$-th strictly positive zero of $u'(t)$, then we have

$$u(t_{2n}') \to \infty, \qquad \text{as} \qquad \gamma \to \infty.$$

Also, for any fixed γ, $\lim_n u(t_{2n}') = 1$ or -1 or 0 (this limit always exists). Hence (4.1) gives that, for any fixed positive integer n,

$$u(t_{2n+2}') = u(t_{2n}')\left[1 - \frac{\alpha B}{c_p}u(t_{2n}')^{-p} + O\left(u(t_{2n}')^{-2p}\right)\right], \qquad \text{as} \quad \gamma \to \infty, \tag{4.4}$$

where $t_0' = 0$ (as usual $u(0) = \gamma$) and t_k', $k = 1, 2, ...$, is the k-th strictly positive zero of $u'(t)$. It follows from (3.4)–(3.5)–(3.6) and (4.2) that

$$\begin{aligned} B &= 4\int_0^{1/4} v_0'(\tau)^2 d\tau = \frac{4c_p}{\sqrt{p+1}}\int_0^{1/4}\left(-\sqrt{1 - v_0(\tau)^{2p+2}}\right)v_0'(\tau)d\tau \\ &= \frac{4c_p}{\sqrt{p+1}}\int_0^1 \sqrt{1 - x^{2p+2}}\, dx, \end{aligned}$$

thus

$$\frac{B}{c_p} = \frac{4}{\sqrt{p+1}}\int_0^1 \sqrt{1 - x^{2p+2}}\, dx = \frac{\sqrt{\pi}}{(p+1)^{3/2}} \cdot \frac{\Gamma\left(\frac{1}{2p+2}\right)}{\Gamma\left(\frac{3p+4}{2p+2}\right)} \stackrel{\text{def}}{=} A_p, \tag{4.5}$$

and (4.4) becomes

$$u(t_{2n+2}') = u(t_{2n}')\left[1 - \alpha A_p u(t_{2n}')^{-p} + O\left(u(t_{2n}')^{-2p}\right)\right], \qquad \text{as} \quad \gamma \to \infty. \tag{4.6}$$

To continue, we need a technical lemma:

4.1 Lemma. *Consider a sequence of real numbers $\{b_n\}$ that satisfies the following:*

(i) $b_0 = \omega > 0$ (we think of ω as a large positive number);

(ii) there are constants K, $C > 0$ such that

$$|b_{n+1} - b_n + K| \leq \frac{C}{b_n}, \qquad \text{whenever} \quad b_n \geq 2.$$

Let $M \geq 2$ be fixed so that

$$K - \varepsilon > 0, \qquad \text{where} \quad \varepsilon = \frac{C}{M}.$$

Then, for all ω sufficiently large, we have the following:

(a) if k is an integer such that $0 \leq k \leq (\omega - M)/(K+\varepsilon)$, then $b_k \geq M$;

(b) there is a k such that: $0 \leq k \leq \lfloor(\omega - M)/(K-\varepsilon)\rfloor + 1$, and $b_k < M$ (here $\lfloor x \rfloor$ denotes the greatest integer which is less or equal to x).

Proof. Assume that there are k's such that $0 \leq k \leq (\omega - M)/(K+\varepsilon)$ and $b_k < M$, and let m be the smallest such k (for ω sufficiently large we will have $m > 0$). For $0 \leq k < m$ our assumptions for $\{b_n\}$ imply

$$|b_{k+1} - b_k + K| \leq \frac{C}{b_k} \leq \frac{C}{M} = \varepsilon$$

and hence

$$|b_m - \omega + Km| = \left|\sum_{k=0}^{m-1} (b_{k+1} - b_k + K)\right| \leq \sum_{k=0}^{m-1} \varepsilon = m\varepsilon. \tag{4.7}$$

Thus

$$-m\varepsilon \leq b_m - \omega + Km, \qquad \text{i.e.} \qquad b_m \geq \omega - Km - m\varepsilon.$$

But we have supposed that $b_m < M$; hence the above inequality implies

$$M > \omega - Km - m\varepsilon, \qquad \text{i.e.} \qquad m > \frac{\omega - M}{K+\varepsilon},$$

contradicting the assumption for m. This establishes (a).

Next assume $b_k \geq M$ for all k such that $0 \leq k \leq \lfloor(\omega - M)/(K-\varepsilon)\rfloor + 1$. Then again our assumptions for $\{b_n\}$ imply that for all these k's we have

$$|b_{k+1} - b_k + K| \leq \frac{C}{b_k} \leq \frac{C}{M} = \varepsilon$$

and hence (as long as m is in the above range) (4.7) is still valid. Thus

$$b_m - \omega + Km \leq m\varepsilon, \qquad \text{i.e.} \qquad b_m \leq \omega - Km + m\varepsilon. \tag{4.8}$$

Let us take

$$m = \left\lfloor \frac{\gamma - M}{K - \varepsilon} \right\rfloor + 1.$$

We have supposed that $b_m \geq M$. Thus (4.8) gives

$$M \leq \omega - Km + m\varepsilon, \qquad \text{i.e.} \qquad m \leq \frac{\omega - M}{K - \varepsilon},$$

contradiction. Thus (b) is established. ∎

We are now ready to present the following result

4.2 Theorem. *Let $N(\gamma)$ be the number of zeroes, on $[0, \infty)$, of the solution $u(t)$ of*

$$u''(t) + \alpha u'(t) + u(t)^{2p+1} - u(t) = 0, \qquad t > 0, \tag{4.9}$$

$$u(0) = \gamma, \qquad u'(0) = 0, \tag{4.10}$$

$$u \to 0 \quad as \quad t \to \infty \tag{4.11}$$

where $\alpha, \gamma > 0$ (these are equations (2.1)–(2.2)). Then (as usual, the notation $f(\gamma) \sim g(\gamma)$, as $\gamma \to \infty$, means that $\lim_{\gamma \to \infty} f(\gamma)/g(\gamma) = 1$),

$$N(\gamma) \sim \left(\frac{2}{p\alpha A_p} \right) \gamma^p, \qquad as \quad \gamma \to \infty,$$

where A_p is the constant of (4.5).

Proof. For typographical convenience we set

$$a_n = u(t'_{2n})$$

Thus (4.6) can be written as

$$a_0 = \gamma, \qquad a_{n+1} = a_n \left[1 - \alpha A_p a_n^{-p} + O\left(a_n^{-2p} \right) \right], \qquad \text{as} \quad \gamma \to \infty.$$

By raising to the p-th power we obtain

$$a_{n+1}^p = a_n^p \left[1 - p\alpha A_p a_n^{-p} + O\left(a_n^{-2p} \right) \right], \qquad \text{as} \quad \gamma \to \infty,$$

or, if we set $b_n = a_n^p$,

$$b_0 = \gamma^p, \qquad b_{n+1} = b_n \left[1 - p\alpha A_p b_n^{-1} + O\left(b_n^{-2} \right) \right], \qquad \text{as} \quad \gamma \to \infty.$$

Another way to express the above asymptotic formula is the following: As long as

$$b_n \geq 2 \quad \text{(say)},$$

there is a constant $C > 0$ such that

$$|b_{n+1} - b_n + p\alpha A_p| \leq \frac{C}{b_n}.$$

This means that $\{b_n\}$ satisfies the assumptions of Lemma 4 with $K = p\alpha A_p$ and $\omega = \gamma^p$.

Let us fix some $M \geq 2$ so that

$$p\alpha A_p - \varepsilon > 0, \qquad \text{where we have set} \quad \varepsilon = \frac{C}{M}.$$

Then, by (a) of Lemma 4.1 we have that

$$\text{if} \quad 0 \leq k \leq \frac{\gamma^p - M}{p\alpha A_p + \varepsilon}, \qquad \text{then} \quad b_k \geq M, \tag{4.12}$$

while, by (b) of the same lemma

$$\text{there is a } k \text{ such that: } 0 \leq k \leq \left\lfloor \frac{\gamma^p - M}{p\alpha A_p - \varepsilon} \right\rfloor + 1, \quad \text{and} \quad b_k < M. \tag{4.13}$$

A consequence of (4.12) is that

$$N(\gamma) \geq 2 \left[\frac{\gamma^p - M}{p\alpha A_p + \varepsilon} \right],$$

while (4.13) implies that

$$N(\gamma) \leq 2 \left[\frac{\gamma^p - M}{p\alpha A_p - \varepsilon} \right] + F(M),$$

where $F(M)$ is a number that depends only on M. Therefore

$$\frac{2}{p\alpha A_p + \varepsilon} \leq \liminf_{\gamma \to \infty} \frac{N(\gamma)}{\gamma^p} \leq \limsup_{\gamma \to \infty} \frac{N(\gamma)}{\gamma^p} \leq \frac{2}{p\alpha A_p - \varepsilon}$$

and the proof is finished since M (and hence ε) is arbitrary. ■

Recall that γ_n is the n-th eigenvalue of (2.1)–(2.2)–(2.4) so that the corresponding eigenfunction $u_n(t)$ has n zeroes in $(0, \infty)$. Since $N(\gamma_n) = n$, an immediate consequence of Theorem 4.2 is that the eigenvalue asymptotics of (2.1)–(2.2)–(2.4) are given by

$$\gamma_n \sim \left(\frac{p\alpha A_p}{2} \right)^{1/p} n^{1/p}, \qquad \text{as} \quad n \to \infty, \tag{4.14}$$

where the constant A_p is given by (4.5). But (4.14) can be strengthened considerably:

4.3 Theorem. *Let γ_n be the n-th eigenvalue of (2.1)–(2.2)–(2.4) (so that the corresponding eigenfunction $u_n(t)$ has n zeroes in $(0, \infty)$). Then*

$$\gamma_{n+1} - \gamma_n = \frac{1}{p}\left(\frac{p\alpha A_p}{2}\right)^{1/p} \frac{1}{n^{1-(1/p)}} + O\left(\frac{1}{n^{2-(1/p)}}\right), \qquad \textit{as} \quad n \to \infty. \tag{4.15}$$

Proof. First notice that, since (2.1) is autonomous, we have $u_{n+1}(t_1') = -u_n(0) = -\gamma_n$. Thus (4.3) applied to $u_{n+1}(t)$ implies

$$\gamma_n = \gamma_{n+1}\left[1 - \frac{\alpha B}{2c_p}\gamma_{n+1}^{-p} + O\left(\gamma_{n+1}^{-2p}\right)\right], \qquad \text{as} \quad n \to \infty,$$

or

$$\gamma_{n+1} - \gamma_n = \frac{\alpha A_p}{2}\gamma_{n+1}^{1-p} + O\left(\gamma_{n+1}^{1-2p}\right), \qquad \text{as} \quad n \to \infty.$$

By using (4.14) in the above equation, we obtain (4.15). ■

Theorem 4.3 has an immediate corollary:

4.4 Corollary. *Let E_n be the initial energy level corresponding to γ_n, namely*

$$E_n = \frac{\gamma_n^{2p+2}}{2p+2} - \frac{\gamma_n^2}{2}.$$

Then

$$E_{n+1} - E_n = \frac{1}{p}\left(\frac{p\alpha A_p}{2}\right)^{2+(2/p)} n^{1+(2/p)} + O\left(n^{2/p}\right), \qquad \textit{as} \quad n \to \infty. \tag{4.16}$$

5 A Nonlinear Inverse Problem

We now consider the inverse problem of how to reconstruct the nonlinearity from information about the eigenvalues. In particular, let $g(x)$ be a real-valued function analytic near $x = 0$. Without loss of generality, we assume that

$$g(x) = x^d \sum_{k=0}^{\infty} a_k x^k, \qquad \text{when} \quad |x| \leq 1, \tag{5.1}$$

where $d \geq 1$ is an integer and $a_0 > 0$, and also that

$$g'(x) > 0, \qquad \text{for all} \quad x \in (0, 1). \tag{5.2}$$

Consider the initial value problem

$$u''(t) + \lambda g'\left(u^2\right) u(t) = 0, \tag{5.3}$$

$$u(0) = 0, \qquad u'(0) = 1, \tag{5.4}$$

which is a Hamiltonian system with Hamiltonian

$$\mathcal{H}(u, u') = \frac{1}{2}\lambda g(u^2) + \frac{1}{2}(u')^2$$

where λ is a real parameter. If we multiply (5.3) by u', then integrate and use (5.4) and the fact that $g(0) = 0$ (which follows from our assumption (5.1)), we obtain

$$(u')^2 + \lambda g\left(u^2\right) = 1. \tag{5.5}$$

It is a straightforward exercise to show that, if $\lambda > 1/g\,(1)$, then (5.2) implies that $u(t)$ is periodic with period

$$T = 4\int_0^{\theta} \frac{du}{\sqrt{1 - \lambda g\left(u^2\right)}}, \tag{5.6}$$

where

$$\theta = \sqrt{g^{-1}\left(1/\lambda\right)}. \tag{5.7}$$

Here g^{-1} denotes the inverse function of g which takes positive values on $(0, g(1))$. Furthermore,

$$u(t) = 0 \qquad \text{if and only if} \qquad t = \frac{nT}{2}, \quad n \in \mathcal{Z}. \tag{5.8}$$

Our object of study is the nonlinear Dirichlet-type eigenvalue problem

$$u''(t) + \lambda g'\left(u^2\right) u(t) = 0, \qquad 0 < t < b, \tag{5.9}$$

$$u(0) = 0, \qquad u'(0) = 1, \qquad u(b) = 0, \tag{5.10}$$

where b is a fixed positive number and λ the eigenvalue parameter. Since, for any fixed λ, the problem (5.9)–(5.10) has at most one solution, we can say that all its eigenvalues are simple. We are interested in the eigenvalues λ such that

$$\lambda > \frac{1}{g\,(1)}. \tag{5.11}$$

The boundary condition $u(b) = 0$ and (5.8) imply that λ is an eigenvalue if and only if there is an integer $n > 0$ such that

$$T = \frac{2b}{n}. \tag{5.12}$$

Let us call λ_n the value of λ for which $T = 2b/n$. Then (5.6) and (5.7) imply

$$\frac{2b}{n} = 4\int_0^{\theta_n} \frac{du}{\sqrt{1-\lambda_n g\left(u^2\right)}}, \qquad \text{where} \quad \theta_n = \sqrt{g^{-1}\left(1/\lambda_n\right)}, \tag{5.13}$$

and, furthermore, the corresponding eigenfunction $u_n(t)$ has exactly $n-1$ zeros in the interval $(0, b)$.

The main result, to be proved in the next section, is that the eigenvalues λ_n of (5.9)–(5.10) determine the function g.

6 Reconstructing the Nonlinearity

Let g be a function satisfying (5.1) and (5.2). We introduce the auxiliary function

$$f(x) = \sqrt{g^{-1}(x)}, \tag{6.1}$$

where again g^{-1} is the inverse function of g, such that $g^{-1}(x) > 0$, for $x \in (0, g(1))$. By (5.1) (see, e.g. Lang [1999]) there is a $\delta > 0$ such that, if $x \in [0, \delta)$, $g^{-1}(x)$ can be expanded as

$$g^{-1}(x) = \sum_{k=1}^{\infty} b_k x^{k/d}, \qquad \text{with} \quad b_1 = \frac{1}{a_0^{1/d}} > 0.$$

Thus there is an $\varepsilon > 0$ such that

$$f(x) = \frac{x^{1/2d}}{a_0^{1/2d}}\left(1 + \sum_{k=1}^{\infty} c_k x^{k/d}\right), \qquad 0 \le x < \varepsilon. \tag{6.2}$$

We will now show how the sequence of eigenvalues may be used to determine the values of c_k. Assuming

$$\lambda > \frac{1}{g\left(1\right)},$$

we set

$$u = f\left(\frac{v}{\lambda}\right)$$

in the integral of (5.6) and thus obtain

$$T = T\left(\lambda\right) = \frac{4}{\lambda}\int_0^1 \frac{f'\left(v/\lambda\right)}{\sqrt{1-v}} dv. \tag{6.3}$$

Now by (6.2)

$$f'(x) = \frac{1}{2da_0^{1/2d}} \cdot \frac{1}{x^{(2d-1)/2d}} \left[1 + \sum_{k=1}^{\infty} (2k+1) c_k x^{k/d} \right], \qquad 0 < x < \varepsilon,$$

hence (6.3) becomes

$$T(\lambda) = \frac{2}{da_0^{1/2d}} \cdot \frac{1}{\lambda^{1/2d}} \int_0^1 \frac{1}{\sqrt{(1-v)}} \cdot \frac{1}{v^{(2d-1)/2d}} \left[1 + \sum_{k=1}^{\infty} (2k+1) c_k \left(\frac{v}{\lambda} \right)^{k/d} \right] dv,$$

or

$$T(\lambda) = \frac{2}{da_0^{1/2d}} \cdot \frac{1}{\lambda^{1/2d}} \left[I_0 + \sum_{k=1}^{\infty} \frac{(2k+1) I_k c_k}{\lambda^{k/d}} \right], \tag{6.4}$$

where

$$I_k = \int_0^1 \frac{v^{k/d}}{v^{(2d-1)/2d}} \cdot \frac{dv}{\sqrt{(1-v)}} = \frac{\Gamma\left(\frac{2k+1}{2d}\right)\sqrt{\pi}}{\Gamma\left(\frac{2k+1}{2d} + \frac{1}{2}\right)}, \qquad k = 0, 1, 2, \ldots, \tag{6.5}$$

where $\Gamma(\cdot)$ is the gamma function. In particular, if $d = 1$, then

$$I_k = \frac{(2k)!\pi}{2^{2k}(k!)^2}, \qquad k = 0, 1, 2, \ldots.$$

Observe that one important consequence of (6.4) is that there is a number Λ such that

$$T(\lambda) \text{ is decreasing for } \lambda \in [\Lambda, \infty), \qquad T(\infty) = 0. \tag{6.6}$$

Let $\lambda_n > 1/g(1)$ be an eigenvalue of (5.9)–(5.10). Since

$$T(\lambda_n) = \frac{2b}{n}, \tag{6.7}$$

we can conclude from (6.6) that there is an integer $n_0 \geq 1$ such that λ_n exists for every $n \geq n_0$ and that

$$\lambda_n \quad \text{is eventually increasing,} \quad \lim_n \lambda_n = \infty. \tag{6.8}$$

Furthermore, (6.4) and (6.7) give the asymptotic behavior of λ_n up to any desired order, e.g.

$$\lambda_n = \frac{I_0^{2d}}{d^{2d} a_0 b^{2d}} n^{2d} + O(n^{2d-2}),$$

thus the leading behavior of λ_n determines d and a_0.

$$\text{If} \quad d = 1, \qquad \lambda_n = \frac{\pi^2}{a_0 b^2} n^2 + O(1).$$

We are now ready for the main theorem concerning the reconstruction process.

6.1 Theorem. *Consider the nonlinear Dirichlet-type eigenvalue problem (5.9)–(5.10), where g is a function satisfying the assumptions (5.1) and (5.2). Then the eigenvalues $\{\lambda_n\}_{n=n_0}^{\infty}$ of (5.9)–(5.10) determine $g(x)$, for $|x| \leq 1$.*

Proof. In (6.4) we set $z = 1/\lambda^{1/d}$ and get

$$F(z) \stackrel{\text{def}}{=} \frac{1}{\sqrt{z}} T\left(1/z^d\right) = \frac{2}{d a_0^{1/2d}} \left[I_0 + \sum_{k=1}^{\infty} (2k+1)\, I_k c_k z^k \right].$$

Notice that, since Stirling's formula applied to (6.5) implies

$$I_k \sim \frac{\sqrt{d\pi}}{\sqrt{k}},$$

we have that $F(z)$ is analytic in the disk $|z| < \varepsilon$. Using (6.7) we obtain

$$F(z_n) = \frac{b}{n\sqrt{z_n}}, \qquad \text{where we have set} \quad z_n = \frac{1}{\lambda_n^{1/d}}$$

(thus $z_n \to 0$). It follows that the coefficients of the (convergent) power series

$$F(z) = \frac{2I_0}{d a_0^{1/2d}} + \frac{2}{d a_0^{1/2d}} \sum_{k=1}^{\infty} (2k+1)\, I_k c_k z^k, \qquad |z| < \varepsilon,$$

can be determined from $\{z_n\}_{n=n_0}^{\infty}$, and since I_k are given by (6.5), we can conclude that $\{\lambda_n\}_{n=n_0}^{\infty}$ determines $\{c_k\}_{k=1}^{\infty}$. Therefore we can determine f and thus (by (6.1)) g. In particular, we have $F(z_n) = w_n$, where $w_n = b/n\sqrt{z_n}$ and F is analytic in $|z| < \varepsilon$. Then, given $\{z_n\}_{n=n_0}^{\infty}$, there is an explicit way to obtain the Taylor coefficients of F:

$$F(0) = \lim w_n\,, \quad F'(0) = \lim \frac{w_n - F(0)}{z_n},$$

$$\frac{F''(0)}{2!} = \lim \frac{w_n - F(0) - F'(0) z_n}{z_n^2}, \qquad \text{etc.}$$

Notice that, in order to determine F (and thus g), it suffices to know a subsequence of $\{\lambda_n\}_{n=n_0}^{\infty}$. In practice, g is usually a polynomial, thus we need only a finite number of λ_n's. ■

Remark. (a) The assumption (5.1) regarding g can be relaxed. For example one can assume that

$$g(x) = x^{d/m} \sum_{k=0}^{\infty} a_k x^{k/m}, \qquad \text{when} \quad 0 < x < \delta,$$

but, it seems we need some analytic nature for the following reason: Equation (6.3) can be written as

$$T(\lambda) = 4 \int_0^{1/\lambda} \frac{f'(x)}{\sqrt{1 - \lambda x}} dx.$$

Thus by (6.7)

$$\int_0^{1/\lambda_n} \frac{f'(x)}{\sqrt{1 - \lambda_n x}} dx = \frac{b}{2n}.$$

From this equation it is clear that, to determine f (and thus g) from $\{\lambda_n\}_{n=n_0}^{\infty}$, we need some kind of analyticity.

(b) As an example of how delicate the determination of the nonlinearity from the eigenvalues is and how badly things can go wrong, consider the Neumann-type problem

$$u''(t) + \lambda g'\left(u^2\right) u(t) = 0, \qquad 0 < t < b,$$

$$u(0) = 1, \qquad u'(0) = 0, \qquad u'(b) = 0.$$

Then

$$\left(u'\right)^2 + \lambda g\left(u^2\right) = \lambda g(1)$$

and the period of $u(t)$ is

$$T = \frac{4}{\sqrt{\lambda}} \int_0^1 \frac{du}{\sqrt{g(1) - g\left(u^2\right)}}.$$

It follows that ν_n is a Neumann eigenvalue if and only if

$$\sqrt{\nu_n} = \frac{2n}{b} \int_0^1 \frac{du}{\sqrt{g(1) - g\left(u^2\right)}}.$$

Therefore, if both g and $\widetilde{g}$ satisfy (5.1)–(5.2) and

$$\int_0^1 \frac{du}{\sqrt{g(1) - g\left(u^2\right)}} = \int_0^1 \frac{du}{\sqrt{\widetilde{g}(1) - \widetilde{g}\left(u^2\right)}},$$

then their corresponding Neumann-type eigenvalue problems share the same eigenvalues.

Acknowledgments: We would like to acknowledge the now classical work of Sirovich and Knight [1981, 1982a,b, 1985, 1986] on eigenvalue analysis for integral operators, in which the authors, setting out to apply these techniques to vision analysis, developed, in the process, systematic treatments for Wigner transforms and other slowly varying integral operators. We dedicate this paper to Larry Sirovich. Always driven by questions of scientific significance, he has made wide-ranging contributions to applied mathematics, characterized by a "no holds barred" approach which consistently offers fresh looks at classical subjects and brings to bear classical techniques to emerging fields. Springer-Verlag's Applied Mathematical Sciences series, with its first 150 volumes produced under his steady and longstanding leadership, conceived in Fritz John's (undoubtedly) smoked filled New Rochelle living room in 1970, continues to set the scientific standard for technical quality in applied mathematics. The elegance of his lectures, delivered with polish, style, and humor, have inspired and influenced his many students (including the first author!) and post-docs throughout a distinguished career in the Division of Applied Mathematics at Brown University.

References

Anderson, D. L. T. and G. H. Derrick[1970], Stability of time dependent particle like solutions in nonlinear field theories I, *J. Math. Phys.* **11**, 1336.

Anderson, D. L. T. and G. H. Derrick[1971], Stability of time dependent particle like solutions in nonlinear field theories II, *J. Math. Phys.* **12**, 945.

Berestycki, H. and P.L. Lions[1983], Nonlinear scalar field equations II. Existence of infinitely many solutions, *Arch. Rat. Mech. Anal.*, 82, 347-376.

Coddington, E. A. and N. Levinson[1955], **Theory of Ordinary Differential Equations**, McGraw-Hill, New York.

Courant, R. and D. Hilbert[1953], **Methods of Mathematical Physics, Vol. I**, (p429) Interscience.

Finkelstein, R., R. LeLevier, and M. Ruderman [1951], Nonlinear spinor fields, *Phys. Rev.*, 83, 326.

Grillakis, M. [1990], Existence of nodal solutions of semilinear equations in R^n, *J. Diff. Eqns*, 85, 367-400.

Jones, C. K. R. T. [1986], On the infinitely many standing waves of some nonlinear Schrödinger equations, in **Nonlinear Systems of Partial Differential Equations in Applied Mathematics**, eds. B. Nicolaenko, D.D. Holm, J.M. Hyman, Lectures in Applied Mathematics, Vol. 23, AMS.

Jones, C. K. R. T., and T. Küpper [1986], On the infinitely many solutions of a semilinear elliptic equation, *SIAM J. Math. Anal.*, 17, 803-835.

Kwong, M. K. [1989], Uniqueness of positive solutions of $\Delta u - u + u^p = 0$ in R^n, *Arch. Rat. Mech. Anal.*, 105, 243.

Lang, S. [1999], **Complex Analysis**, 4th ed., Springer-Verlag, New York.

Nehari, Z. [1963], On a nonlinear differential equation arising in nuclear physics, *Proc. Roy. Irish Acad.*, 62, 117-135.

Newton, P. K. and M. O'Connor [April 1996], Scaling laws at nonlinear Schrödinger defect sites, *Phys. Rev. E*, Vol. 53, No. 4, 3442-3447.

Newton, P. K., and S. Watanabe [1993], The geometry of nonlinear Schrödinger standing waves: pure power nonlinearities, *Physica D*, 67, 19-44.

Ryder, G. H. [1968], Boundary value problems for a class of nonlinear differential equations, *Pac. Jour. Math.*, 22, 477-503.

Sirovich, L., and B. W. Knight [1981], On the eigentheory of operators which exhibit a slow variation, *Quarterly of Appl. Math.*, 38, 469-488.

Sirovich, L., and B.W. Knight [1982a] Contributions to the eigenvalue problem for slowly varying operators, *SIAM J. Appl. Math.*, 42, 356-377.

Sirovich, L., and B.W. Knight [1982b], The Wigner transform and some exact properties of linear operators, *SIAM J. Appl. Math.*, 42, 378-389.

Sirovich, L., and B.W. Knight [1985], The eigenfuntion problem in higher dimensions I. Asymptotic theory, *Proc. Nat. Acad. Sci.*, 82, 8274.

Sirovich, L., and B.W. Knight [1986], The eigenfuntion problem in higher dimensions II. Exact results, *Proc. Nat. Acad. Sci.*, 86, 527.

Weinberger, H. [1974], **Variational Methods for Eigenvalue Approximation**, SIAM, Philadelphia.

11

A KdV Model for Multi-Modal Internal Wave Propagation in Confined Basins

Larry G. Redekopp

To Larry Sirovich, on the occasion of his 70th birthday.

ABSTRACT The derivation of a reduced dynamics describing multi-modal evolution of long internal waves in a confined basin is presented. The model is rationally extracted from the primitive equations of motion and preserves the essential physics of bi-directional propagation of non-hydrostatic, weakly nonlinear wave motions. Even though evolution in closed, one-dimensional domains necessarily involves fields possessing both left- and right-running characteristics for each mode, it is shown that a computationally-efficient reduction of the fundamental system to a set of KdV equations, without imposing any prejudice regarding the direction of propagation, can be accomplished. The specific case for two modes leading to just a pair of KdV equations, and which captures the leading-order effects of both co- and counter-propagating wave components for both self- and cross-modal interactions, is described. Both cases of free response from prescribed initial conditions and forced response via excitation by wind stresses at the exposed upper surface are equally accessible by the model. It is argued that the resulting model comprises an efficient, rapid-simulation tool that faithfully captures essential physics relating to the energy-containing scales of motion in lakes and reservoirs, a tool with considerable practical relevance toward the goal of providing a physics-based guide for resource management of lakes and reservoirs.

Contents

1 Introduction

Among the most significant advances in applied mathematics in recent decades are the discovery of the soliton by Zabusky and Kruskal [1965] and the consequent development of the inverse scattering transform solution of the initial value problem for the Korteweg-deVries (KdV) equation on the infinite line by Gardner, Greene, Kruskal, and Muira [1967]; Gardner, Greene, Kruskal, and Muira [1974]. These historic works opened the way for truly dramatic leaps in our physical understanding of many wave-related natural phenomena, and especially in our ability to make quantitative predictions concerning a host of wave processes. This is particularly true for uni-modal systems developing in an unbounded domain and, to a somewhat lesser extent, for periodic domains (e.g., Osborne [1998]). By comparison, our ability to model and predict nonlinear wave dynamics in confined systems where bi-directional propagation is an essential element of the relevant physics is far less developed, and especially so for evolution in multi-modal systems. The present work is directed toward bridging this gap, describing the development of a rationally-based evolution model where the uni-directional dynamics associated with KdV theory can be effectively exploited to make quantitative predictions pertaining to bi-directional propagation in a closed domain.

The physical motivation for this work arises from increasing pressure to develop efficient simulation models which can be used to make predictions of the internal (i.e., subsurface) hydrodynamics in stratified lakes and reservoirs with reasonable fidelity. Such capability is crucial in the long term to devising effective strategies for the management of many fragile water resources, and to the gaining of an intuition concerning the mechanisms and controlling processes for downscale energy transfer in general. This need for quantitative predictions of sub-surface hydrodynamics in lakes and reservoirs stems from the fact that the ecology of a basin is inextricably linked, in large measure, to benthic, sidewall, and internal mixing stimulated by wind-generated internal waves (Mortimer and Horn [1982]; Imberger [1994]). It has been evident for some time (e.g., Mortimer [1952]) that the onset of wind blowing over a confined, stratified body initiates a predominantly horizontal transport of lighter surface waters toward the leeward shore, causing a slight upward tilt of the free surface and a much more pronounced downward tilt of the internal isopycnal surfaces in the down-wind direction. Since volume must be conserved in a closed domain, a concomitant deep return flow toward the windward end exists, forcing a tendency for the isopycnal surfaces at that end of the basin to upwell. Thus, a basin-scale, horizontal pressure gradient is established via a surface wind stress, an internal pressure gradient which is unbalanced when the wind relaxes. This non-equilibrium is relieved by the formation of a front which subsequently evolves into a propagating packet of long internal waves.

It has been shown by Horn, Imberger, and Ivey [2001] that the wind-generated non-equilibrium in a large class of natural lakes is such that soliton formation through nonlinear steepening is a frequent mechanism whereby a basin-scale initial condition degenerates into shorter-scale, propagating components of the typical spectrum. They show, in fact, that the duration of wind events and the scale of the basin in a great preponderance of practical cases is such that the resulting non-equilibrium degenerates into at least one solitary wave before the 'positive well' of the deformed thermoclinic structure undergoes a first reflection with an end wall. Of course, the fact that the basin is closed ensures that a 'negative well' also exists, one which would give rise to the formation of a purely dispersive packet in an unbounded domain. Admittedly, the delineation of the evolving wave field in terms of solitons and a dispersive packet is not precise for a closed domain, nevertheless the separation of the disturbance field into these two classes is quite useful for descriptive purposes. Employing this point of view, the dispersive packet necessarily coexists with and moves through any soliton groups, and the combined action of these interacting, and often counter-propagating, wave groups define the spectrum of shorter disturbances that evolve from a basin-scale non-equilibrium.

The transfer of energy to shorter scales within the evolutionary field, and the mixing processes associated with the evolutionary field, are crucial hydrodynamic processes that are not well understood, let alone quantified. The thesis underlying development of the evolutionary model described here is that the existence of a rapid-simulation tool that captures essential physics of these processes would be most valuable for building an intuitive understanding, and providing at least semi-quantitative inputs, that could guide an informed management strategy of stressed water resources. The emphasis on a need for a rapid-simulation capability is quickly reinforced when one considers the vast parameter space related to seasonal changes in density structure, the wide-ranging variations in basin shapes and sizes, the varying strength and duration of common wind events, etc. For these reasons and more, the extraction of reduced dynamical models from the primitive equations of motion, especially those that lead to efficient and faithful simulation of controlling physics, is much more than an abstract exercise. Also, it is motivation and analysis of this same genre, plied with impressive vision and compelling skill, that characterizes many contributions by Larry Sirovich, contributions that have impacted a host of diverse applications. Hence, it is a privilege to honor his scientific accomplishment and his disciplinary leadership by presenting this work as a moderate example of the success of his style.

2 Preliminary Foundations

An environmental model having considerable relevance for describing the wave climate in a lake consists of a closed basin possessing a vertical density structure consisting of three contiguous layers of homogeneous fluid arranged in a statically-stable configuration as shown in Figure 2.1. The metalimnion in this model has a finite thickness h_2 and any consequent evolutionary system for the internal wave field is expected to capture a reasonable facsimile of the two lowest baroclinic modes, and their interactions, in a real continuously stratified environment.

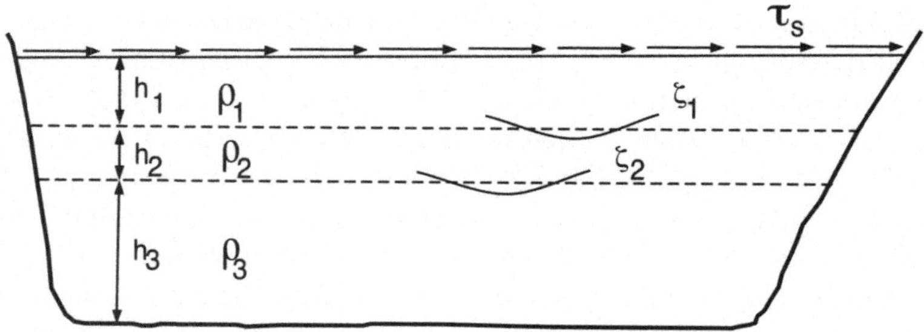

FIGURE 2.1. Schematic of a wind-driven, three layer lake model.

Since natural excitations (e.g., via surface wind stresses) typically deposit energy into the long wave portion of the internal wave spectrum, a useful analytical framework for describing the resultant wave field is provided by Boussinesq theory. The term 'Boussinesq theory' is employed here to specifically denote long wave evolution wherein no bias is imposed regarding the directionality of wave propagation. This is in sharp contradistinction to KdV theory where propagation is explicitly restricted along a single characteristic. Clearly, the presence of endwall boundaries in a closed basin, in general, requires the consideration of wave propagation along two families of characteristics (i.e., right and left). The other pivotal element in Boussinesq theory is the existence of a small parameter, the ratio of the 'trapping dimension' of the wave guide (herein, either the epilimnion depth h_1 or the depth of the basin) to a typical wave length. The existence of this asymptotic scale parameter permits the derivation of a reduced dynamics, one that is considerably more advantageous for both analysis and simulation of the controlling scales of motion than the primitive equations of motion. A second parameter, the ratio of the peak wave amplitude to the trapping dimension, can be introduced and its magnitude can be exploited to yield either a strongly or a weakly nonlinear model. Although this latter distinction is not fundamental to Boussinesq theory in its general form, the analytical description herein is limited exclusively to the weakly nonlinear case.

We restrict our attention to basins whose width is small compared to the basin length, but yet asymptotically large compared to the basin depth. In this case a lateral averaging operation can be applied to derive a one-space-dimensional dynamics cast in terms of the laterally-averaged vertical displacements $\zeta_1(x,t)$ and $\zeta_2(x,t)$ of the two interior density interfaces (see Figure 2.1). Also, the effect of variable basin width, sloping side walls and non-uniform basin depth will be neglected in what follows for the purpose of minimizing algebraic complexity and not obscuring the primary goal of this presentation — the development of a KdV system describing bi-directional propagation of coupled modes in a closed basin. The generalizations needed to account for these effects can be accomplished with minimal conceptual difficulty, but at the expense of considerable algebraic detail.

With the forgoing preliminaries in view, the following rational model, which forms the starting point for our consideration here, can be obtained from the underlying equations of motion:

$$\begin{aligned}\mathcal{N}_{tt} - \hat{c}_{12}^2\mathcal{N}_{xx} - \alpha_{21}\hat{c}_{23}^2\mathcal{Z}_{xx} &= \hat{\beta}_{11}\mathcal{N}_{xxxx} + \hat{\beta}_{13}\mathcal{Z}_{xxxx} + 2\hat{\gamma}_{111}\mathcal{N}_t\mathcal{N}_{xt} \\ &+ 2\hat{\gamma}_{133}\mathcal{Z}_t\mathcal{Z}_{xt} + 2\hat{\gamma}_{113}\mathcal{N}_t\mathcal{Z}_{xt} + 2\hat{\gamma}_{131}\mathcal{Z}_t\mathcal{N}_{xt} \\ &+ \hat{\mu}_{111}\mathcal{N}_x\mathcal{N}_{xx} + \hat{\mu}_{133}\mathcal{Z}_x\mathcal{Z}_{xx} + \hat{\mu}_{113}\mathcal{N}_x\mathcal{Z}_{xx} \\ &+ \hat{\mu}_{131}\mathcal{Z}_x\mathcal{N}_{xx}\,, \end{aligned} \tag{2.1a}$$

$$\begin{aligned}\mathcal{Z}_{tt} - \hat{c}_{23}^2\mathcal{Z}_{xx} - \alpha_{23}\hat{c}_{12}^2\mathcal{N}_{xx} &= \hat{\beta}_{21}\mathcal{N}_{xxxx} + \hat{\beta}_{23}\mathcal{Z}_{xxxx} + 2\hat{\gamma}_{211}\mathcal{N}_t\mathcal{N}_{xt} \\ &+ 2\hat{\gamma}_{233}\mathcal{Z}_t\mathcal{Z}_{xt} + 2\hat{\gamma}_{213}\mathcal{N}_t\mathcal{Z}_{xt} + 2\hat{\gamma}_{231}\mathcal{Z}_t\mathcal{N}_{xt} \\ &+ \hat{\mu}_{211}\mathcal{N}_x\mathcal{N}_{xx} + \hat{\mu}_{233}\mathcal{Z}_x\mathcal{Z}_{xx} + \hat{\mu}_{213}\mathcal{N}_x\mathcal{Z}_{xx} \\ &+ \hat{\mu}_{231}\mathcal{Z}_x\mathcal{N}_{xx}\,. \end{aligned} \tag{2.1b}$$

The dependent variables $\mathcal{N}$ and $\mathcal{Z}$ are related to the interface displacements ζ_1 and ζ_2, respectively, by the expressions

$$\zeta_1(x,t) = \mathcal{N}_x\,, \quad \zeta_2(x,t) = \mathcal{Z}_x\,. \tag{2.2}$$

Details concerning the derivation of the these equations plus presentation of the analytic form of the numerous coefficients are given in Redekopp [2002].

Equations (2.1a,2.1b) are written with terms defining linear, non-dispersive evolution on the left and the leading-order dispersive and nonlinear terms appearing on the right. We note that no constraints on the direction of propagation have been imposed, either in the linear operator on the left or in the nonlinear terms on the right, and that both linear and noninear coupling terms appear. The linear coupling terms on the left-hand side arise since the interface displacements are entirely independent and do not separately define a modal amplitude. Amplitudes of the linear wave modes for the wave guide consist of linear combinations of the two interface displacements of the form, say, $\mathcal{N} + s_i\mathcal{Z}$, $i = 1,2$, where the s_i are modal separation parameters that allow the left-hand sides of (2.1a,2.1b) to be placed in the form

$$\begin{bmatrix} \partial_{tt}^2 - u^2\partial_{xx}^2 & 0 \\ 0 & \partial_{tt}^2 - v^2\partial_{xx}^2 \end{bmatrix} \begin{bmatrix} U \\ V \end{bmatrix} = 0. \tag{2.3}$$

The modal phase speeds (u,v) are functions of the parameters (c_{ij}, α_{ij}) appearing in (2.1a,2.1b).

To construct the desired pair of modal Boussinesq equations corresponding to (2.1a,2.1b), the appropriate separation functions for the dispersive and nonlinear terms must be 'discovered'. The required form for the lowest baroclinic mode is

$$U(x,t) = \mathcal{N} + s_1\mathcal{Z} + s_1^{(d)}\mathcal{Z}_{xx} + n_{11}\mathcal{M}^{(1)} + n_{12}\mathcal{M}^{(2)}, \tag{2.4}$$

where $\mathcal{M}^{(1)}$ and $\mathcal{M}^{(2)}$ are the nonlinear functionals

$$\begin{aligned} \mathcal{M}^{(1)} &= \tfrac{1}{2}\alpha_{23}\int^x (\mathcal{N}_t^2 + c_{12}^2\mathcal{N}_x^2)dx + \tfrac{1}{2}\alpha_{21}\int^x (\mathcal{Z}_t^2 + c_{23}^2\mathcal{Z}_x^2)dx \\ &\quad - \alpha_{21}\alpha_{23}\int^x \mathcal{N}_t\mathcal{Z}_t dx\,, \end{aligned} \tag{2.5a}$$

$$\begin{aligned} \mathcal{M}^{(2)} &= \alpha_{23}c_{12}^4\int^t \mathcal{N}_t\mathcal{N}_x dt + \alpha_{21}c_{23}^4\int^t \mathcal{Z}_t\mathcal{Z}_x dt \\ &\quad + \alpha_{21}\alpha_{23}c_{12}^2c_{23}^2\int^t (\mathcal{N}_t\mathcal{Z}_x + \mathcal{Z}_t\mathcal{N}_x)dt\,. \end{aligned} \tag{2.5b}$$

These two nonlinear functionals are required to accomplish the modal separation for the two classes of nonlinear terms in (2.1a,2.1b). The second baroclinic mode $V(x,t)$ is defined in identical fashion as (2.4), albeit with a different set of separation parameters $(s_2, s_2^{(d)}, n_{21}, n_{22})$.

With the requisite modal separation relations in hand, the weakly nonlinear Boussinesq system for the given three-layer environmental model shown in Figure 2.1 becomes

$$\begin{aligned} U_{tt} - u^2U_{xx} &= \delta_1U_{xxxx} + \kappa_{11}U_tU_{xt} + \kappa_{12}U_tV_{xt} + \kappa_{21}V_tU_{xt} + \lambda_{11}U_xU_{xx} \\ &\quad + \lambda_{12}U_xV_{xx} + \lambda_{21}V_xU_{xx}\,, \end{aligned} \tag{2.6a}$$

$$\begin{aligned} V_{tt} - v^2V_{xx} &= \delta_2V_{xxxx} + \nu_{12}U_tV_{xt} + \nu_{21}V_tU_{xt} + \nu_{22}V_tV_{xt} \\ &\quad + \sigma_{12}U_xV_{xx} + \sigma_{21}V_xU_{xx} + \sigma_{22}V_xV_{xx}\,. \end{aligned} \tag{2.6b}$$

The linear, dispersive evolution for each mode is now separated into uncoupled forms, as required, and only self-mode and cross-mode nonlinear interaction terms appear, also as required by a consistent modal separation. This coupled pair of equations for the modal amplitude functions comprise an essential first step toward the objective of extracting a KdV model.

3 The KdV Model

3.1 The Multi-Modal Initial Value Problem

Equations (2.6a,2.6b) describe weakly-dispersive, weakly-nonlinear evolution of modally-coupled longwaves without any prejudice regarding the direction of propagation for either mode. As such, these equations are expected to capture much of the essential physics of the energy-containing scales of motion in a confined basin. They could be adopted directly, as could the original system (2.1a,2.1b), as a model for simulating the free modes of motion in a closed domain. However, the particular goal here is to seek a further simplication wherein the fourth-order-in-time system (2.6a,2.6b) is reduced to a second-order-in-time system (i.e., a pair of coupled KdV equations (3.11a,3.11b)). The motivation, as noted earlier, is to construct a truly rapid-simulation tool that retains the essential features of bi-directional propagation and the important physical effects of dispersion and nonlinearity.

Toward this goal we note that the d'Alembert solutions of the linear wave equations appearing on the left-hand sides of (2.6a,2.6b) are comprised of waves moving along the two families of left-going and right-going characteristics. Also, in a confined system with $x \in [0, L]$ say, the disturbance field propagating along a given characteristic is transformed to one propagating along its oppositely-directed counter part via either a complete or partial reflection process at endwalls. Physical considerations reveal that the leading-order horizontal velocity in any layer is proportional to a linear combination of $\mathcal{N}_t$ and $\mathcal{Z}_t$. Hence, for example, the case of perfect reflection requires the vanishing of this velocity component at endwalls, and this can be realized quite simply by constructing an even extension of the physical domain from $[0, L]$ to $[0, 2L]$, and subsequent use of periodic boundary conditions on the extended domain.

The folded extension of the domain about $x = L$ applies equally to any uneven bottom topography varying along the longitudinal (i.e., propagation) direction of the basin. This extended fundamental period for the physical basin is shown schematically in Figure 3.1, and is pivotal to the construction of the KdV model.

If the endwall reflection process is incomplete, as is almost always true in applications, the use of the same extended-folded domain is still relevant. The presence of a partial reflection process can be accommodated by inserting on the right-hand sides of (2.6a, 2.6b) a spatially-varying dissipative-dispersive term which is non-zero only in the vicinity of the physical endwall regions. Such a "model" term would dissipate wave energy near an endwell so that the reflected wave amplitude would be reduced and, perhaps, delayed and deformed. A few experimental studies of shoaling-reflecting, long interfacial internal waves (e.g, Michallet and Ivey [1999]) are available and provide quantitative results for the energy loss when endwall regions are

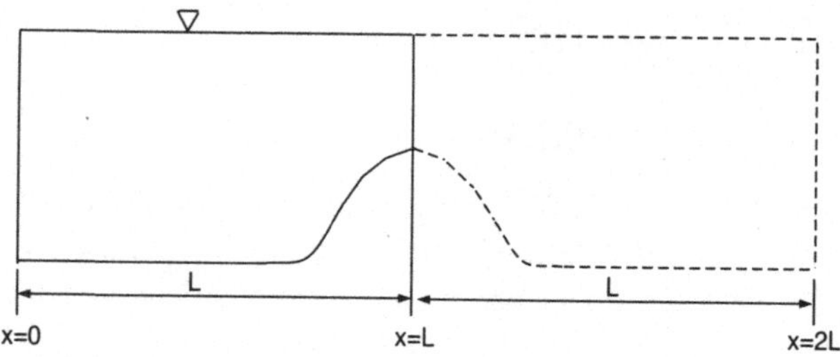

FIGURE 3.1. Sketch of the reflected-extended domain.

moderately steep. Such results could be employed to "design" a dissipative term spatally-localized around $x = L$, say, but the modeling of a term mimicking this process will not be pursued here. The point is, however, that there are procedures whereby the general development of the reduced dynamical system described here can be adapted to accomodate important aspects of physically-relevant boundary conditions.

The derivation of the intended KdV system is facilitated by the introduction of slow space-time scales $(X,T) = (\mu x, \mu t)$, where μ is the (small) long-wave parameter referred to earlier. Also, the amplitude parameter ϵ is introduced explicitly so that an asymptotic expansion of the modal amplitude functions having the following form can be constructed:

$$U(x,t) = \frac{\epsilon}{\mu}\left(U^{(1)}(X,T) + \epsilon U^{(2)}(X,T) + \dots\right). \tag{3.1}$$

A companion expression exists for $V(x,t)$. Reference to (2.2) reveals straightaway that the interface displacements are order ϵ. Furthermore, we adopt the principle of minimal simplication and take $\epsilon = \mu^2$, a scaling equivalent to stating that the Ursell number is of unit order and the leading effects of dispersion and nonlinearity appear at the same order.

Substitution of the proposed expansions into (2.6a,2.6b) reveals immediately that $U^{(1)}$ and $V^{(1)}$ satisfy homogeneous, linear wave equations which possess the general (d'Alembert) solutions

$$U^{(1)}(X,T) = F(r,\tau) + G(s,\tau)\,, \tag{3.2a}$$

$$V^{(1)}(X,T) = P(\xi,\tau) + Q(\eta,\tau)\,. \tag{3.2b}$$

In writing these solutions we have defined a slow time scale $\tau = \mu^3 t = \epsilon T$ and introduced the characteristic coordinates

$$r = X - uT\,, \qquad s = X + uT\,, \tag{3.3a}$$

$$\xi = X - vT\,, \qquad \eta = X + vT\,. \tag{3.3b}$$

To this order, the modal amplitudes are propagated separately along their respective families of characteristics and one can consider the variables (r, s)

and (ξ, η) as independent. However, it is clear that modal interactions will occur at higher order involving functions defined in terms of both modal families of left-going and right-going characteristics. Hence, at the next order in the limit process one needs to recognize that these two pairs of characteristic coordinates are not independent and write, for example, that

$$\xi = \tfrac{1}{2}\left(1 + \frac{v}{u}\right) r + \tfrac{1}{2}\left(1 - \frac{v}{u}\right) s\,, \tag{3.4a}$$

$$\eta = \tfrac{1}{2}\left(1 - \frac{v}{u}\right) r + \tfrac{1}{2}\left(1 + \frac{v}{u}\right) s\,, \tag{3.4b}$$

so that $P(\xi, \tau) = P(r, s, \tau)$, etc.

Proceeding to the next order in the expansion (3.1) one obtains an equation for $U^{(2)}(X, T)$ which can be placed in the following form:

$$\begin{aligned}
-4u^2 \frac{\partial^2 U^{(2)}}{\partial r \partial s} &= \Big\{ 2uF_{r\tau} + \delta_1 F_{rrrr} + (u^2\kappa_{11} + \lambda_{11})F_r F_{rr} \\
&+ \tfrac{1}{4}\left(1 + \frac{u}{v}\right)^2 (uv\kappa_{12} + \lambda_{12})F_r P_{rr} + \tfrac{1}{2}\left(1 + \frac{u}{v}\right)(uv\kappa_{21} + \lambda_{21})F_{rr}P_r \\
&- \tfrac{1}{4}\left(1 - \frac{u}{v}\right)^2 (uv\kappa_{12} - \lambda_{12})F_r Q_{rr} - \tfrac{1}{2}\left(1 - \frac{u}{v}\right)(uv\kappa_{21} - \lambda_{21})F_{rr}Q_r \Big\} \\
&+ \Big\{ -2uG_{s\tau} + \delta_1 G_{ssss} + (u^2\kappa_{11} + \lambda_{11})G_s G_{ss} \\
&+ \tfrac{1}{4}\left(1 + \frac{u}{v}\right)^2 (uv\kappa_{12} + \lambda_{12})G_s Q_{ss} + \tfrac{1}{2}\left(1 + \frac{u}{v}\right)(uv\kappa_{21} + \lambda_{21})G_{ss}Q_s \\
&- \tfrac{1}{4}\left(1 - \frac{u}{v}\right)^2 (uv\kappa_{12} - \lambda_{12})G_s P_{ss} - \tfrac{1}{2}\left(1 - \frac{u}{v}\right)(uv\kappa_{21} - \lambda_{21})G_{ss}P_s \Big\} \\
&+ (\text{non-secular interaction terms}).
\end{aligned} \tag{3.5}$$

The terms in each of the bracketed expressions are purposely organized in three groups. The first three terms in each expression are clearly secular, and by themselves correspond to KdV equations describing the long-time evolution along the respective characteristic. The remaining two groups of two terms each describe modal interactions between the field associated with a primary characteristic of the $U^{(1)}$ mode with the field associated with both characteristics of the $V^{(1)}$ mode. These latter terms may be described as only 'mildly secular', especially the latter two terms in each bracket as they describe the interaction, for example, of the field associated with a right-going characteristic of the U-mode and the left-going characteristic of the V-mode. The strength of these latter terms is clearly dependent on the relative magnitudes of the phase speeds (u, v) of the two modes. If these modes have quite similar phase speeds, these particular interaction terms are weak. Alternatively, the interactions may be quite strong when the speeds are disparate.

We note further that the two pairs of mildly secular terms in each bracket can be included with the leading secular terms in application of the secularity condition without causing the appearance of any non-local interaction terms in the resulting evolution equations. By contrast, the non-secular interaction terms not shown explicitly in (3.5) do contain non-local contributions to $U^{(2)}$.

After generating the inhomogeneous equation corresponding to (3.5) for the field $V^{(2)}(X,T)$, and after choosing the dependence of the four functions defined in (3.2a,3.2b) on the slow time scale τ so as to eliminate the appearance of secular terms, four coupled equations are obtained. Rewriting these equations in terms of laboratory coordinates (x,t) and defining the amplitude functions

$$(f,g,p,q)^T = \epsilon\frac{\partial}{\partial x}(F,G,P,Q)^T, \tag{3.6}$$

the following set of extended KdV equations emerge:

$$\begin{aligned} f_t + uf_x = &-\frac{\delta_1}{2u}f_{xxx} - \frac{1}{2u}(u^2\kappa_{11}+\lambda_{11})ff_x \\ &- \frac{1}{8u}\left(1+\frac{u}{v}\right)^2(uv\kappa_{12}+\lambda_{12})fp_x - \frac{1}{4u}\left(1+\frac{u}{v}\right)(uv\kappa_{21}+\lambda_{21})pf_x \\ &+ \frac{1}{8u}\left(1-\frac{u}{v}\right)^2(uv\kappa_{12}-\lambda_{12})fq_x + \frac{1}{4u}\left(1-\frac{u}{v}\right)(uv\kappa_{21}-\lambda_{21})qf_x; \end{aligned} \tag{3.7a}$$

$$\begin{aligned} g_t - ug_x = &\frac{\delta_1}{2u}g_{xxx} + \frac{1}{2u}(u^2\kappa_{11}+\lambda_{11})gg_x \\ &+ \frac{1}{8u}\left(1+\frac{u}{v}\right)^2(uv\kappa_{12}+\lambda_{12})gq_x + \frac{1}{4u}\left(1+\frac{u}{v}\right)(uv\kappa_{21}+\lambda_{21})qg_x \\ &- \frac{1}{8u}\left(1-\frac{u}{v}\right)^2(uv\kappa_{12}-\lambda_{12})gp_x - \frac{1}{4u}\left(1-\frac{u}{v}\right)(uv\kappa_{21}-\lambda_{21})pg_x\,; \end{aligned} \tag{3.7b}$$

$$\begin{aligned} p_t + vp_x = &-\frac{\delta_2}{2v}p_{xxx} - \frac{1}{2v}(v^2\mu_{22}+\sigma_{22})pp_x \\ &- \frac{1}{8v}\left(1+\frac{v}{u}\right)^2(uv\mu_{21}+\sigma_{21})pf_x - \frac{1}{4v}\left(1+\frac{v}{u}\right)(uv\mu_{12}+\sigma_{12})fp_x \\ &+ \frac{1}{8v}\left(1-\frac{v}{u}\right)^2(uv\mu_{21}-\sigma_{21})pg_x + \frac{1}{4v}\left(1-\frac{v}{u}\right)(uv\mu_{12}-\sigma_{12})gp_x\,; \end{aligned} \tag{3.8a}$$

$$\begin{aligned} q_t - vq_x = &\frac{\delta_2}{2v}q_{xxx} + \frac{1}{2v}(v^2\mu_{22}+\sigma_{22})qq_x \\ &+ \frac{1}{8v}\left(1+\frac{v}{u}\right)^2(uv\mu_{21}+\sigma_{21})qg_x + \frac{1}{4v}\left(1+\frac{v}{u}\right)(uv\mu_{12}+\sigma_{12})gq_x \\ &- \frac{1}{8v}\left(1-\frac{v}{u}\right)^2(uv\mu_{21}-\sigma_{21})qf_x - \frac{1}{4v}\left(1-\frac{v}{u}\right)(uv\mu_{12}-\sigma_{12})fq_x\,. \end{aligned} \tag{3.8b}$$

The modally-coupled wave dynamics developing from some prescribed initial condition in a basin of length L can be explored by simulating this set of four KdV equations using the folded-extended domain $[0, 2L]$ as discussed above. However, if we consider the free evolution arising from an initial displacement field (as might exist following cessation of a wind stress field applied to the surface of the basin), the foregoing model for the internal dynamics can be reduced to just two KdV equations, consisting of slightly-modified versions of only (3.7a) and (3.8a).

The d'Alembert solutions of the leading-order, linear wave equations partition the initial displacement field equally to their amplitude functions associated with their left-running and right-running characteristics. Recognition of this fact, together with the use of an even extension of the domain from $[0, L]$ to $[0, 2L]$, plus the symmetry properties of the two systems (3.7a,3.7b) and (3.8a,3.8b), reveals that

$$g(x,t) = f(2L - x, t)\,, \quad q(x,t) = p(2L - x, t)\,. \tag{3.9}$$

These relationships are pivotal to the development of the reduced model. They show that knowledge of the right-going fields $f(x,t)$ and $p(x,t)$ in the extended portion $[L, 2L]$ of the computational domain is equivalent to knowledge of the left-going fields $g(x,t)$ and $q(x,t)$ in the physical domain $[0, L]$. As a consequence, solving only for the fields associated with the right-running characteristics r and ξ in $[0, 2L]$, subject to the initial condition comprised of one-half the initial displacement field, is sufficient to obtain the fields associated with the complete family of characteristics throughout the physical domain $[0, L]$. The latter objective is realized by folding (alt., reflecting) the solution for the fields $f(x,t)$ and $p(x,t)$ in the domain $[L, 2L]$ about $x = L$ onto the physical domain $[0, L]$. This yields the solution for the complete displacement field, correct to the order of approximation implicit in (3.7a,3.7b) and (3.8a,3.8b), corresponding to the true initial condition in the physical domain. The procedure described here is illustrated schematically in Figure 3.2.

The procedure just outlined becomes quite simple when the environmental model has only a single internal density interface so the spectrum consists merely of one wave mode. In such a case, the solution for the wave field in the confined basin $[0, L]$ with the initial condition $f(x, 0) = f_0(x)$ requires simulation of the single KdV equation

$$\begin{aligned} f_t + u f_x &= -\frac{\delta_1}{2u} f_{xxx} - \frac{1}{2u}(u^2\kappa_{11} + \lambda_{11}) f f_x\,, \\ f(x,0) &= \tfrac{1}{2} f_0(x)\,, \end{aligned} \tag{3.10}$$

on the periodic interval $[0, 2L]$. Simulations based on this equation have been performed recently and compared with data from laboratory experiments involving a two-layer density structure in a long rectangular tank

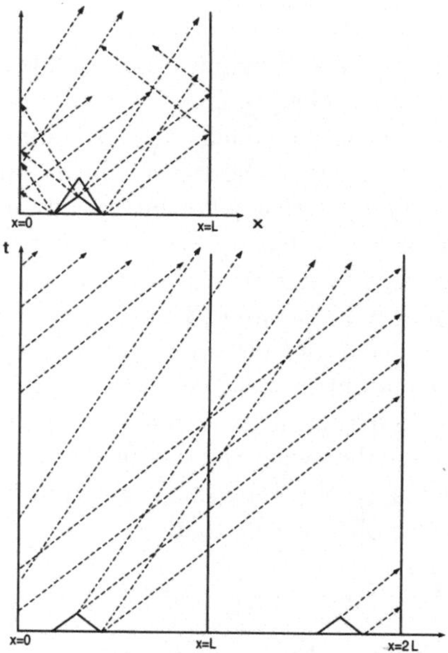

FIGURE 3.2. Schematic of the families of characteristics in the reflected-extended computational domain (lower sketch) and the folded-added solution in the physical domain (upper sketch).

(Horn, Imberger, Ivey, and Redekopp [2002]). The comparisons are quite favorable up to moderately large initial conditions, althrough there is evidence of underprediction of peak amplitudes and slight phase errors as the energy in the initial state increases. We conjecture that these erosions in the fidelity of the simulation could be overcome by inclusion of cubic nonlinear terms in (3.10).

The resultant pair of KdV equations for the two-mode system under consideration here are sightly more complex than (3.10) since interactions between modes having different characteristics must be included. Using the symmetry relations (3.9), the pair of equations (3.7a) and (3.8a) are cast in the following forms:

$$
\begin{aligned}
f_t + u f_x = &-\frac{\delta_1}{2u} f_{xxx} - \frac{1}{2u}(u^2\kappa_{11} + \lambda_{11})\, f f_x \\
&- \frac{1}{8u}\left(1 + \frac{u}{v}\right)^2 (uv\kappa_{12} + \lambda_{12})\, f p_x - \frac{1}{4u}\left(1 + \frac{u}{v}\right)(uv\kappa_{21} + \lambda_{21})\, p f_x \\
&- \frac{1}{8u}\left(1 - \frac{u}{v}\right)^2 (uv\kappa_{12} - \lambda_{12})\, f \left(\frac{\partial p}{\partial x}\Big|_{2L-x}\right) \\
&+ \frac{1}{4u}\left(1 - \frac{u}{v}\right)(uv\kappa_{21} - \lambda_{21})\left(p\big|_{2L-x}\right) f_x\,; \qquad (3.11a)
\end{aligned}
$$

$$
p_t + v p_x = -\frac{\delta_2}{2v} p_{xxx} - \frac{1}{2v}(\mu_{22} + \sigma_{22})\, p p_x
$$

$$
\begin{aligned}
&-\frac{1}{8v}\left(1+\frac{v}{u}\right)^2(uv\mu_{21}+\sigma_{21})pf_x-\frac{1}{4v}\left(1+\frac{v}{u}\right)(uv\mu_{12}+\sigma_{12})fp_x \\
&-\frac{1}{8v}\left(1-\frac{v}{u}\right)^2(uv\mu_{21}-\sigma_{21})p\Big(\frac{\partial f}{\partial x}\Big|_{2L-x}\Big) \\
&+\frac{1}{4v}\left(1-\frac{v}{u}\right)(uv\mu_{12}-\sigma_{12})\left(f\big|_{2L-x}\right)p_x\,. \qquad (3.11\text{b})
\end{aligned}
$$

The initial conditions for this system are related to the initial displacements of the independent interfaces through the relations

$$
\begin{aligned}
f(x,0) &= \tfrac{1}{2}\left[\zeta_1(x,0)+s_1\zeta_2(x,0)\right], \\
p(x,0) &= \tfrac{1}{2}\left[\zeta_1(x,0)+s_2\zeta_2(x,0)\right]. \qquad (3.12)
\end{aligned}
$$

Use of these expressions over the physical domain $[0,L]$, together with their even extension (reflection) over $[L,2L]$, provides the initial condition for the full periodic computational domain $[0,2L]$.

3.2 The Driven-Damped Problem

In the natural contexts that motivate this study, energy input to the large-scale internal wave field occurs principally through a wind stress applied at the free surface of the basin. This stress typically varies on a temporal scale that is commensurate with a significant fraction of the basin-scale pendulum (internal seiche) period, and with a spatial distribution that generally varies with fetch from the windward to the leeward shore. At the same time bottom boundary currents induced by the large-scale surface stress, and by waves evolving from the consequent internal non-equilibrium, exert a retarding frictional stress across benthic layers. A semi-quantitative lake model must, therefore, include a forcing term that mimics the spatio-temporal distortion of the equilibrium isopycnal structure generated by a surface wind stress plus a dissipative term that captures the damping of long internal waves by benthic boundary layers.

The shallow water theory underlying the set (2.1a,2.1b) can be generalized to describe a forced-dissipative system capturing the important physical effects just described. Including the action of imposed stresses acting at basin boundaries in the analysis gives rise to inhomogeneous terms on the right-hand sides of (2.1a,2.1b). Since it is beyond the present scope to describe all steps in the physical modeling, we focus exclusively on the extraction of a driven-damped KdV model for wind-generated, long internal waves in confined basins starting with an extended version of the weakly-nonlinear Boussinesq system. Furthermore, we choose to simplify the presentation by considering only the uni-modal case that emerges from an environmental model consisting of a single interior density interface (i.e., a metalimnion of vanishing thickness).

Assuming a turbulent model for benthic friction expressed in terms of a friction coefficient C_f, the driven-damped version of equation (2.1a) is

$$Z_{tt} - c^2 Z_{xx} = \beta Z_{xxxx} + \alpha(2Z_t Z_{xt} + c^2 Z_x Z_{xx}) - k_1 C_f Z_t |Z_t| + \mathcal{F}(x,t)\,. \tag{3.13}$$

The distributed wind stress on the basin surface (proportional to the square of a friction velocity) is represented by the function $\mathcal{F}(x,t)$. It is common to assume that this stress varies linearly with depth across the upper (epilimnion) layer (e.g., see Monismith [1985]), but this assumption is not essential. What is important for our purposes here is how these additional terms are to be represented in the reflected-extended portion of the computational domain. Employing an expansion of the form given in (3.1), and using the characteristic variables and related d'Alembert functions given in (3.3a) and (3.2a), respectively, one obtains the following counterpart to equation (3.5):

$$\begin{aligned} -4c^2 \frac{\partial^2 Z^{(2)}}{\partial r \partial s} &= \left\{ 2cF_{r\tau} + \beta F_{rrrr} + 3\alpha c^2 F_r F_{rr} + \frac{k_1}{\mu} C_f c^2 F_r \left| F_r - G_s \right| \right\} \\ &+ \left\{ -2cG_{s\tau} + \beta G_{rrrr} + 3\alpha c^2 G_s G_s s - \frac{k_1}{\mu} C_f c^2 G_s \left| F_r - G_s \right| \right\} \\ &+ \frac{\mathcal{F}}{\mu \epsilon^2} - \alpha c^2 (F_r G_{ss} + F_{rr} G_s)\,. \end{aligned} \tag{3.14}$$

The fact that the benthic friction and wind stress terms appear asymptotically large in this relation involving scaled coordinates is a consequence of not exposing explicitly the magnitude of multiplicative factors. Based on representative length and velocity scales for the internal waveguide, they are actually of order one or smaller. Alternatively, if the wave amplitude is scaled by balancing the internal pressure gradient with the wind stress, then $\mathcal{F} = O(\mu\epsilon^2)$.

The issue now is the splitting of the wind stress term $\mathcal{F}$ between the two characteristics and representation of its form in the extended $[L, 2L]$ portion of the domain. The term defining the frictional stress on the bottom is already apportioned to the two secular terms. It is argued that the wind stress should be partitioned equally between the two secular terms on the right-hand side. Then, after noting the symmetry relationship (3.9), the separate KdV expressions contained in the individual secular terms in (20) will be equivalent provided the wind stress function $\mathcal{F}$ is reflected with odd symmetry over the extended domain. Under these conditions, the KdV equation describing evolution of the field associated with the left-going characteristic in $[0, L]$ is equivalent to the KdV equation describing evolution of the field associated with the right-going characteristic in $[L, 2L]$.

With this construction, the driven-damped wave field in $[0, L]$ can be obtained by computing only the equation relevant to long-time evolution along the right-going characteristic over the extended-reflected domain $[0, 2L]$.

Recasting this equation in terms of laboratory coordinates leads to the reduced model

$$f_t + cf_x + \tfrac{3}{2}\alpha c f f_x + \frac{\beta}{2c} f_{xxx} = -\frac{ck_1}{2} C_f f \left| f(x,t) - f(2L-x,t) \right| - \frac{\mathcal{F}(x,t)}{4c}. \tag{3.15}$$

As described previously, the true displacement field can now be obtained at any time t in the physical domain $[0, L]$ by folding the computed solution for $f(x,t)$ in $[L, 2L]$ about the line $x = L$ and adding it to the existing solution $f(x,t)$ in $[0, L]$. The sum of these fields yields the leading order estimate for the dynamics driven by a prescribed wind stress subject to benthic dissipation.

It is generally accepted that the presence of a (nearly) stationary surface wind stress on a stratified basin is balanced primarily by the horizontal pressure gradient associated with tilted isopycnal surfaces (Heaps and Ramsbottom [1966]; Wu [1973]; Spigel and Imberger [1980]; Monismith [1985]). This particular balance is expressed in (3.15) between the term cf_x on the left-hand side and the wind stress term on the right-hand side. However, there is no *a priori* reason that this balance must dominate in the driven evolution. Clearly, the general balance is dynamic and must vary depending on the rate and distribution of energy input. The characteristic parameter affecting this balance is the Wedderburn number, (see, for example, Spigel and Imberger [1980] and Horn, Imberger, and Ivey [2001]). Scaling the interface displacement f by the upper layer depth h_1, the propagation path coordinate x by L, and using c_0 and u_{0*}^2 as characteristic measures for the phase speed c and the wind stress function $\mathcal{F}$, it is readily seen that

$$\frac{\mathcal{F}/c}{cf_x} \sim \frac{u_{0*}^2 L}{c_0^2 h_1} = \frac{1}{W} \tag{3.16}$$

where W is the Wedderburn number. Different interpretations regarding the meaning of this parameter exist, but one perspective is that it provides a gross measure of the ratio of the horizontal pressure gradient to the gradient of the horizontal stress across the mixed layer. However, the existence of free modes in the system and the intrinsic effect of nonlinear steepening can certainly influence the onset time for the appearance of nonstationary structures, even before the wind stress relaxes significantly. Hence, there are likely to be regimes of driven dynamics that do not scale exclusively in terms of the Wedderburn number. Issues of this nature, as well as the possibility of internal resonance under repetitive (daily) wind events, can be explored quite readily with the rapid-simulation tool developed here. Some preliminary simulations have been conducted to date, and the classification of different regimes of dynamic response is in progress.

4 Concluding Remarks

The analytical development described here seeks to provide a useful tool for simulating the internal dynamic response of stratified lakes and reservoirs to the application of a spatio-temporal wind stress at the free surface. The fact that these basins are closed argues immediately for the need to capture bi-directionally propagating waves. Also, a quick examination of typical non-equilibrium states generated by wind events in many basins reveals that nonlinear steepening plays a pivotal role in the degeneration of this non-equilibrium. As a consequence, it is imperative that any realistic model also contains the leading non-hydrostatic effects so that any frontal evolution is not artificially smeared by numerical dissipation and the potential for propagating nonlinear wave features is captured. These considerations suggest that the computational challenge for developing a quantitative tool for predicting the internal hydrodynamics in a closed basin can be quite severe. It is the author's contention that the model put forward here represents a vast reduction in this computational challenge.

A reduced evolutionary model has been described that holds considerable promise for efficient and rapid simulation of the energy-containing scales of internal motion in stratified lakes and reservoirs. The model emerges from a long wave assumption in which the small scales are filtered out, yielding a consistent basis for examining exclusively the energy containing scales. This filtering of the very broad range of accessible scales of motion translates into a significant contraction of a computationally intensive task. Furthermore, the numerical requirement implicit in the model proposed here is quite modest in the sense that the dynamics is described in terms of a single driven-damped KdV equation for each eigenmode of the vertical density structure in the basin. Hence, the vertical structure of the dynamics is related directly to the computed horizontal fields through analytical relations, and an important reduction in the dimension of the problem has been achieved. Other advantages include the fact that the KdV equation is first-order in time, and can be solved on a periodic domain via spectral methods. This represents a significant advantage over, say, even direct use of the underlying Boussinesq system. Lastly, the fact that non-hydrostatic effects are implicit in the Boussinesq and KdV models allows the capturing of important physics that are entirely precluded in the extant hydrostatic models. For at least these reasons, the model described here is viewed as a significant step toward providing a capability for assessing the impact of different forcing scenarios on the internal weather in a lake, and to provide the essential hydrodynamic input needed to evaluate potential threats to the ecology in such basins.

Acknowledgments: This work was supported in part by the Office of Naval Research under Grant N00014-95-0041. The author also acknowledges the assistance of the Gledden Trust in providing support via a Gledden Senior Visiting Fellowship at the University of Western Australia, where the initial ideas for this work were formed. Discussions with David Horn, Jorg Imberger and Greg Ivey at the Centre for Environmental Fluid Dynamics at the University of Western Australia were quite influential to this work and are gratefully acknowledged.

References

Gardner, C. S., J. M. Greene, M. D. Kruskal, and R. M. Muira [1967], Method for solving the Korteweg-deVries equation. *Phys. Rev. Lett.*, **19**, 1095–1097.

Gardner, C. S., J. M. Greene, M. D. Kruskal, and R. M. Muira [1974], Korteweg–deVries equation and generalizations. VI. Method for exact solution. *Comm. Pure Appl. Math.*, **27**, 97–133.

Heaps, N. S. and A. E. Ramsbottom [1966], Wind effects on water in a narrow two-layered lake, *Phil. Trans. R. Soc. London, A* **259**, 391–430.

Horn, D. A., J. Imberger, and G. N. Ivey [2001], The degeneration of large-scale interfacial gravity waves in lakes. *J. Fluid Mech.*, **434**, 181–207.

Horn, D. A., J. Imberger, G. N. Ivey, and L. G. Redekopp [2002], A weakly nonlinear model of long internal waves in closed basins. *J. Fluid Mech.*, to appear.

Imberger, J. [1994], Transport processes in lakes: A review. *In Limnology Now: A Paradigm of Planetary Problems (ed. R. Margalef).* Elsevier.

Michallet, H. and G. N. Ivey [1999], Experiments on mixing due to internal solitary waves breaking on uniform slopes. *J. Geophys. Res.*, **104**, 13,467–13,478.

Monismith, S. G. [1985], Wind-forced motions in stratified lakes and their effect on mixed-layer shear. *Limnol. Oceanogr.*, **30**, 771–783.

Mortimer, C. H. [1952], Water movements in lakes during summer stratification; evidence from the distribution of temperature in Windermere. *Mitt. Intl. Verein. Limnol.*, **20**, 124–197.

Mortimer, C. H. and W. Horn [1982], Internal wave dynamics and their implications for plankton biology in the Lake of Zurich. *Vierteljahresschr. Naturforsch. Ges. Zurich*, **127(4)**, 299–318.

Osborne, A. R. [1998], Solitons, cnoidal waves and nonlinear interactions in shallow-water ocean surface waves. *Physica D*, **123**, 64–81.

Redekopp, L. G. [2002], Evolution equations for bi-directional long internal wave propagation in confined basins. *J. Fluid Mech.*, submitted for publication.

Spigel, R. H. and J. Imberger [1980], The classification of mixed-layer dynamics in lakes of small to medium size. *J. Phys. Oceanogr.*, **10**, 1104–1121.

Wu, J. [1973], Wind induced entrainment across a stable density surface, *J. Fluid Mech.* **61**, 257–287.

Zabusky, N. and M. D. Kruskal [1965], Interaction of solitons in a collisionless plasma and the recurrence of initial states. *Phys. Rev. Lett.*, **15**, 240–243.

12

A Memory Model for Seasonal Variations of Temperature in Mid-Latitudes

K. R. Sreenivasan and D. D. Joseph

To Larry Sirovich, on the occasion of his 70th birthday.

ABSTRACT The Earth receives, on the average, the largest amount of radiation from the Sun on summer solstice (June 21), which is the longest day of the year. However, the warmest day occurs usually later; the time lag is about a month for 50 deg latitude, and decreases with increasing latitude. There is comparable time lag between the shortest day of the year (winter solstice, December 21) and the coldest day of the year. We model these and related observations by a linear Maxwell-type viscoelastic model. By comparing predictions of the model with observations, we extract, as functions of the latitude, two free parameters representing the memory and the effective viscosity coefficient. Some interpretation of the results is provided.

Contents

1 Introduction

The yearly seasons are related to Earth's position with respect to the Sun, so it is traditional and seemingly logical to assume that the longest day of the year, which is nominally when the Earth receives the largest amount of radiation from the Sun, is also the warmest. The following quote from Smart [1956] exemplifies this thinking: "...the days increase in warmth from March

21 to June 21 corresponding to the increasing [length of day]...; from the latter date, a decrease ensues." However, the reality is different. A survey of some fifty years of the temperature data for New Haven, CT, shows that the warmest day occurs on July 24 (with a standard deviation of about 10 days), which is 33 days after summer solstice (June 21). The time lag between the coldest day and winter solstice (December 21) is similar. The situation is akin to the commonly known fact that, often, the warmest temperature on a winter day is reached not at noon but a few hours later (see Figure 1.1). It is also of interest to be able to explain the fact that the temperature changes occur most rapidly around late April and October, a month or so after the occurrence of equinoxes (the two days of the year, March 21 and September 21, on which day and night have equal duration).

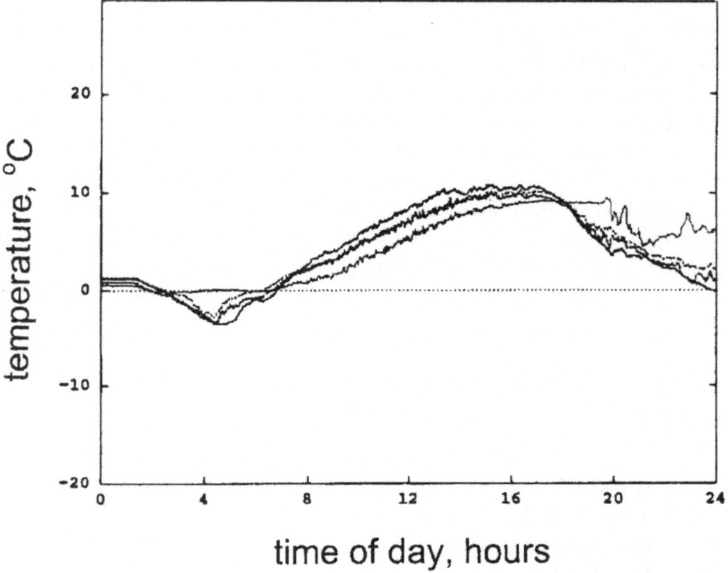

FIGURE 1.1. The variation of temperature as a function of time during a typical winter day. Different curves represent measurements at different places in the same neighbourhood.

A proper explanation of these observations should incorporate detailed modelling of solar radiation received by the Earth, the differential storage and rerelease of this radiation by oceans and the land, cloud coverage, albedo effect, ocean-land interactions, global circulation, and so forth. That would be a major task. Our limited goal here is to model these behaviors in a simple way by subsuming the details by two free parameters in a Maxwell memory model. One of the parameters is a relaxation time (or memory coefficient) and the other an (effective) viscosity coefficient; see, for example, Joseph [1990]. If the model has some value, one should be able to extract these parameters from comparison of its predictions with empirical observations, and to interpret them usefully. We show that it is possible to do so.

Section 2 describes the model briefly while Section 3 summarizes the data analysis. Section 4 is a disscussion of the model, with Section 5 providing a few concluding remarks.

2 The model

The simplest version of the model assumes that one can associate, with a given latitude, a mean temperature for the same day of the year. This is the value of the temperature averaged over many years, all of which correspond to a given day; further, the temperature is averaged for all positions on a given latitude so that the longitude-variations are ignored for this first look. Let us denote this mean temperature by $\theta(\ell, t)$, where ℓ is the latitude and t denotes, within a year, the number of days elapsed since the occurrence of one of two solstices. Let $\langle\theta\rangle$ be the time average of θ, taken over all days in a year, and T denote the difference $\theta - \langle\theta\rangle$. At any latitude, we are interested in the variation of T through the year, and wish to relate it to $L(t) = L^*(t) - \langle L^*\rangle$, where L^* is the length of day at the chosen latitude and $\langle L^*\rangle$, the yearly average of L^*, is approximately 12 hours for all latitudes. The proposed model is

$$\lambda\frac{\partial T}{\partial t} + T = \eta L(t)\,, \tag{2.1}$$

where λ is a relaxation (memory) parameter and η is a viscosity coefficient. The inevitable nonlinearities in the problem are hidden in the two parameters λ and η.

Models bearing some resemblance to equation (2.1) have indeed been proposed for seasonal variations of temperature, but they are overtly nonlinear and more complex; for example, Crowley and North [1991] incorporate ice-albedo feedback and turbulent eddy diffusivity—each of which is the subject of extensive and unfinished study in its own right. We know of no effort identical to ours, carrying out data analysis to the same degree.

3 Data analysis

Data on length of day are available in tabulated form in List [1951] and, in forms more suitable for computer manipulations, in Meeus [1988] and Montenbruck and Pflaeger [1994]. Figure 3.1 shows a plot of $L(t)$, taken from List [1951], for 50 deg latitude in the northern hemisphere over a

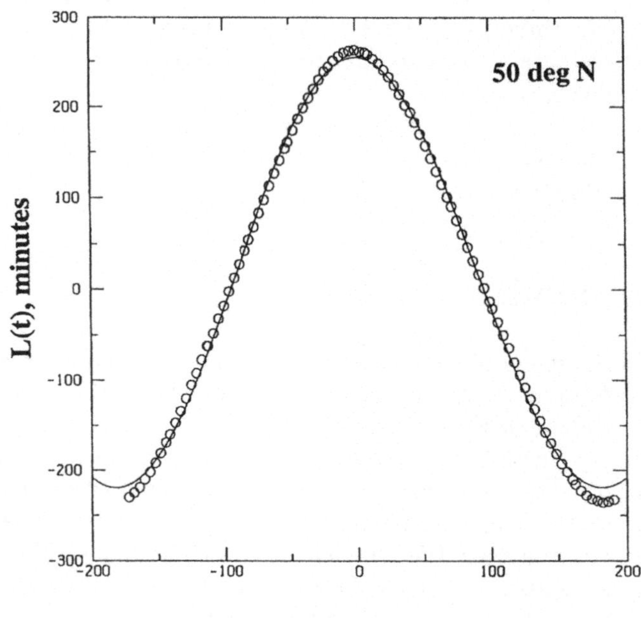

FIGURE 3.1. The variation of the length of day, $L(t)$, at 50 deg latitude, as a function of the number of days counted from the summer solstice. The data are from List [1951]. The cosine term shown by the full line fits the data well. The deviations from the fit can be accommodated by including higher harmonics, but the effort was deemed unnecessary for present purposes.

one-year period, as a function of the number of days from the winter solstice. For simplicity of further analysis, we fit the data in Figure 3.1 by $A\cos(\omega t)$, where A is the amplitude of the daylight variation and the circular frequency $\omega = 2\pi/365$ days^{-1}. This fit is adequate for our purposes; small discrepancies that exist between the fit and the data are not worth emphasizing, given the gross features we wish to understand. Figure 3.2 shows that the amplitude A, as determined from such fits, varies smoothly with the latitude and can be fitted by a simple polynomial.

If $L(t)$ can be fitted by a cosine term, the solution of equation (2.1) is simply

$$T(t) = a\cos\omega t + b\sin\omega t\,, \tag{3.1}$$

where

$$\lambda = \frac{b}{a\omega}\,, \quad \eta = (a + b\lambda\omega)\,. \tag{3.2}$$

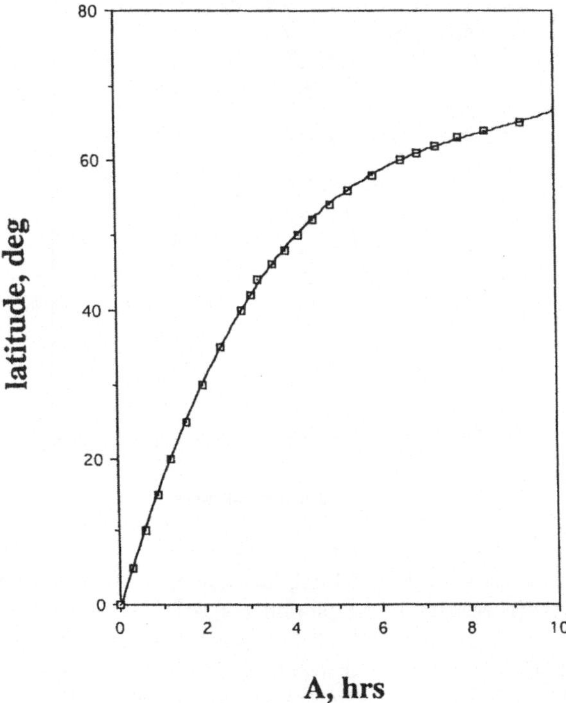

FIGURE 3.2. The variation of the coefficient A as a function of the latitude. The fit to the data (full line) is $l = 19.47A - 2.10A^2 + 0.082A^3$, where the latitude ℓ is expressed in degrees and the amplitude of daylight variations A in hours.

Equivalently, we have

$$T(t) = \frac{A\eta}{\sqrt{1+\lambda^2\omega^2}} \cos\, \omega(t-\phi)\,, \tag{3.3}$$

where

$$\phi = \omega^{-1} \tan^{-1}(\lambda\omega)\,. \tag{3.4}$$

We shall verify if this solution adequately expresses the mean temperature variation for a fixed latitude through the year, and if so, obtain a and b (and thus extract the parameters λ and η or ϕ). This will be done below.

The latitude-averaged temperature data have been compiled in Oort and Rasmuusen [1971] for a five-year period between 1955 and 1960. These data are compiled from observations from a number of weather stations (of the order of a few hundred) in the northern hemisphere. There are more weather stations over land than on water, so the data may be biased in some way. Figure 3.3 shows a comparison of these temperature data with the solution

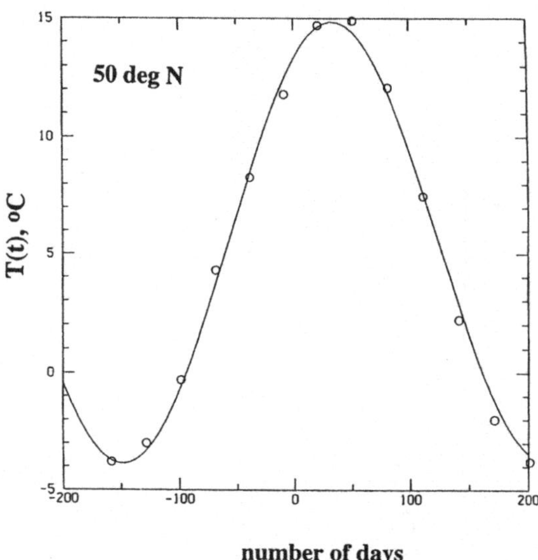

FIGURE 3.3. The variation of T at 50 deg latitude as a function of the number of days, counted with respect to the summer solstice, showing that it can be fitted quite well by equation (3.1). The observational data, taken from Oort and Rasmuusen [1971], were five-year averages. The authors remark that scanty evidence available over a period of some twenty years are substantially the same as the five-year averages, thus implying that reasonable stationarity has been reached. The weather stations are not distributed on a uniform grid on the globe (see, for example, figure 1a and 1b of Oort and Rasmuusen [1971]). This may introduce some unknown bias.

of the model equation. The agreement is good and one has

$$\lambda = 38 \text{ days}, \quad \text{and} \ \eta = 2.69°/\text{hr}.$$

This gives a phase lag ϕ of about 30 days, consistent with expectations raised earlier. Considering that η represents the increase in temperature per hour of heating by the Sun, a plausible interpretation is that, on the average, an hour of solar heating raises the temperature at 50 deg latitude by about 2.7 K.

In analyzing temperature data for other latitudes, two restrictions should be noted. First, data on length of day become increasingly uncertain for latitudes above 65 deg because small changes in atmospheric reflectivity can cause relatively large changes in daylight. This difficulty restricts useful consideration to lower latitudes (even though the model seems to work for latitudes at least as high as 75 deg). For latitudes below about 25 deg, the relatively small seasonal variations are influenced by a variety of minor effects, none of which—including the length of day—appears to have a particularly dominant influence. Possibilities for improving the model for low altitudes will be mentioned briefly later, but, for now, we restrict attention to the latitude range between 25 deg and 65 deg—which is what we mean by mid-latitudes.

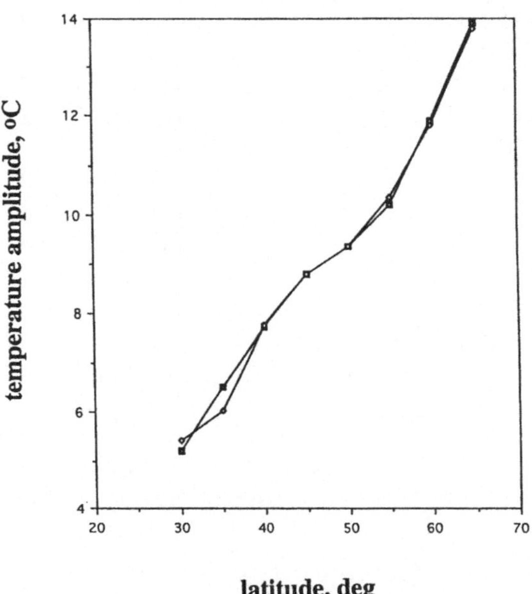

FIGURE 3.4. A comparison between the amplitude of the harmonic fit to the temperature data and the observed amplitude of temperature variation. Their good agreement is a measure of the goodness of the fit to the temperature variations in mid-latitudes.

Our experience is that temperature variations at all mid-latitudes can be fitted well by equation (3.1). A quick feel for the goodness of fit can be had from Figure 3.4, which compares half the difference amplitude of T at various mid-latitudes with the amplitude $(a^2+b^2)^{\frac{1}{2}}$ obtained by fitting equation (3.1) to the data. The agreement is good, and the data can be approximated crudely by a straight line which intercepts the latitude axis at a finite value of 13 deg (to which no particular attention need be paid because of the limitation of the model for tropical latitudes). One can combine this approximate and empirical formula with equations (3.1) and (3.2) to obtain

$$\frac{\eta A}{\sqrt{1+\lambda^2\omega^2}} \text{ (in } ^\circ C) \approx 0.23\ell \text{ (in deg)} - 3.06 \text{ (in } ^\circ C) . \tag{3.5}$$

The interpretation of the numbers on the right-hand-side of the equation is unclear. Equation (3.5) probably possesses no greater significance than the

city	latitude, deg.*	λ, days
Fairbanks, AL	64	20
Anchorage, AL	60	27
Minneapolis, MN	45	30
Burlington, VA	44	27
Boston, MA	42	38
Chicago, IL	42	34
New Haven, CT	41	39
Boulder, CO	40	36
Columbus, OH	40	34
Little Rock, AR	35	47
Jacksonville, FL	30	42
Miami, FL**	26	57

*Latitude rounded off to the nearest degree

** At this latitude, temperature variations are not fitted well by equation (3.1).

TABLE 3.1. The coefficient λ for a few cities in the United States. Despite the scatter, the trend towards larger λ for lower latitudes is unmistakable.

fact that it relates the three parameters λ, η and A. It can be simplified even further by noting that λ/η is approximately constant to within 15%.

A summary of the principal results is given in Figure 3.5 which plots the relaxation parameter λ (or, more usefully, the phase lag ϕ) and viscosity η as functions of the latitude. (Although the data come from the northern hemisphere, the model should hold equally for the southern hemisphere.) The increasing value of the time lag was not expected at the outset. To assuage the reader's skepticism, we plot in Figure 3.6 the temperature variations for latitudes of 30 and 75 deg. It is clear, without any help from the model, that the time lag is smaller for higher latitudes than for lower latitudes. The feature will be explained subsequently.

The data examined so far are averages for a given latitude, but the model works for local areas as well. Figure 3.7 shows the temperature variation through the year for the city of New Haven. Examination of these data for several other American cities shows that equations (3.1)–(3.3) adequately describe the observed temperature variations—although, not unexpectedly, λ varies somewhat between two cities that lie on the same latitude (see Table 3.1). The general trend, however, is quite similar to that for longitude-averaged data.

We now note three other aspects briefly:

a. *Temperature variations at different heights:* We have examined the temperature data at several altitudes from the ground. There are only minor variations with respect to height, at least until the tropopause is reached.

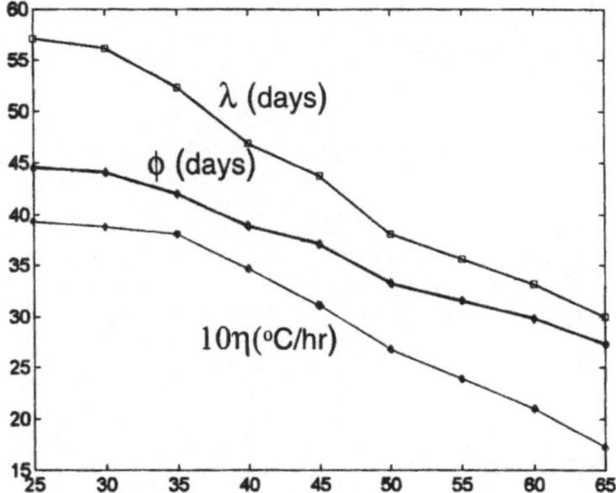

FIGURE 3.5. The parameters λ and η as functions of the latitude. Note the scale change between the two parameters. All data correspond to a pressure altitude of 1000 mb.

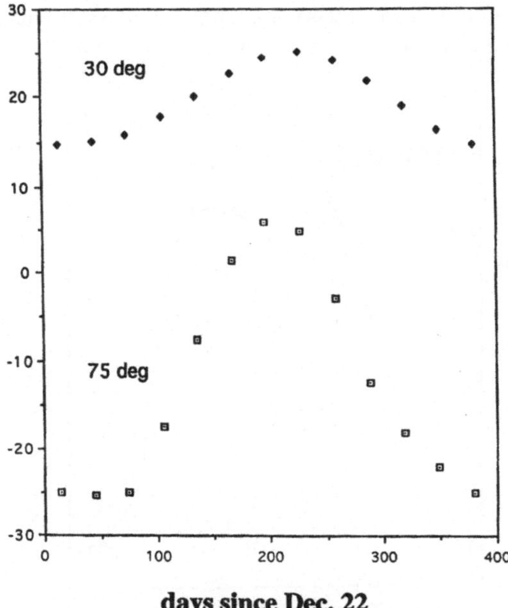

FIGURE 3.6. A comparison of the temperature variations at two substantially different latitudes. The data show without ambiguity that the warmest day occurs later in the year at lower latitudes than at higher latitudes.

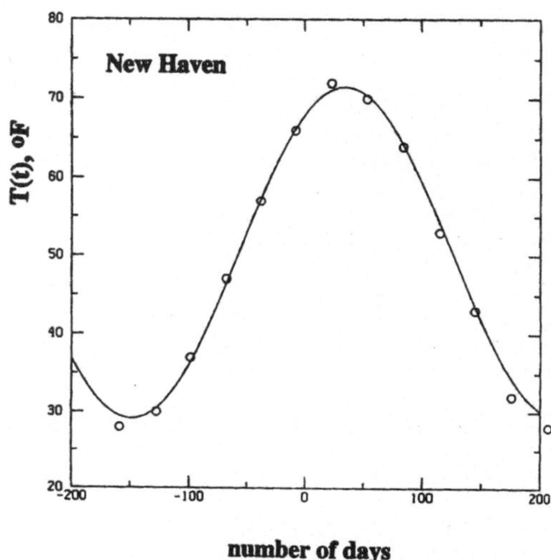

FIGURE 3.7. The variation of $T(t)$ for the city of New Haven as a function of the number of days, counted with respect to summer solstice, showing that it can be fitted well by equation (3.1). The observational data are from Miller and Thompson [1979] obtained over several (but unknown number of) years. A twelve-year average has a standard deviation of the order of 4°C.

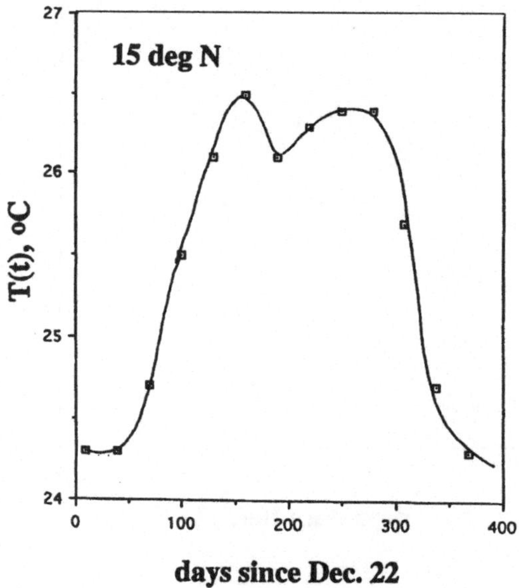

FIGURE 3.8. Temperature variations at 15 deg latitude, showing the limitations of the model for low latitudes.

Thereafter, temperature variations cannot be fitted by equation (3.1); and, if they can be fitted at all, the value of λ seems to be significantly smaller than at lower altitudes.

b. *Temperature variations at lower latitudes:* It has been remarked already that temperature variations at lower latitudes cannot be fitted well by equation (3.1). An example is shown for the latitude of 15 deg (Figure 3.8). Lower latitudes exhibit a more pronounced bimodality, or some other more complex behavior, depending on the altitude and latitude. Where bimodality is pronounced, a simple nonlinear version of the model could work, but this has not been tested extensively.

c. *Largest gradients in temperature variations:* For any given latitude, one can write from equation (3.1) that

$$\frac{dT}{dt} = \frac{\eta L - T}{\lambda}, \tag{3.6}$$

and compute dT/dt using the measured values of λ and η. The result is shown in Figure 3.9 for 50 deg latitude. The largest changes occur sometime in April and October, consistent with the known fact that the weather changes are most rapid in these months of Spring and Fall. If you are an outdoor jogger, you will need to change from long jumpers to shorts, and vice versa, sometime around these dates.

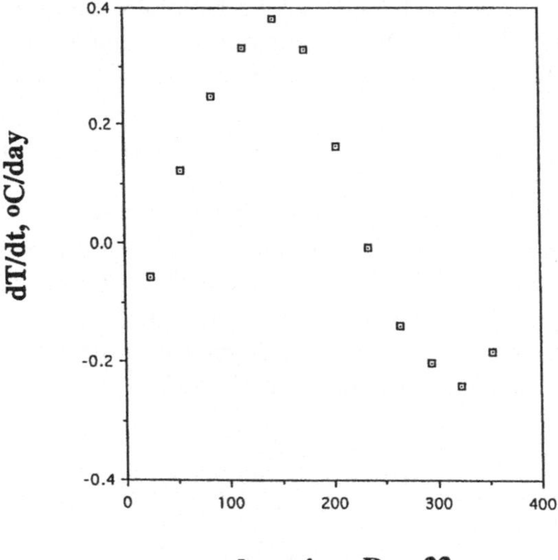

FIGURE 3.9. Temporal derivative of the temperature $T(t)$ through the year, as computed for 50 deg latitude using equation (3.6).

4 Discussion

A linear viscoelastic model describes several gross features of temperature variations in mid-latitudes. A prominent qualitative character of the model is the "memory" incorporated through the parameter λ. The best fit to the data shows that the parameter varies with latitude in a simple manner, as shown in Figure 3.5. (All these data correspond to a pressure altitude of 1000 mb. Slightly different definition of constant altitude could lead to slightly different numbers.) In the simple version, the present model averages temperature over all longitudes—and, at a given latitude, over many years—and is thus similar in spirit to general circulation models, though it is difficult to associate the two in detail.

To understand the parameter λ, let us make an energy balance for a thin slice of the globe centered around a given latitude ℓ. The radius of this slice, which is in the form of a disc, is $R\cos\ell$, R being the Earth's radius. The disc gets heated by the component of the solar radiation received normal to the surface, loses (gains) energy by convective heat loss to the neighboring discs as well as to the atmosphere, and stores (or releases) energy due to the imbalance between these two effects. This energy storage, associated with the cyclic heat-up and cool-down processes, occurs in the first few meters of the ground; as is well-known to divers and plumbers alike, one can identify a penetration depth below which the seasonal variations of temperatures are felt less conspicuously. The details of the energy loss (or gain) are not well understood, and the standard practice in the heat transfer literature is to represent this effect by means of a term $h(T - T^*)$, where one's ignorance is lumped into the heat transfer coefficient h; here, T^* is a reference temperature. The energy balance then takes the form

$$\frac{\rho c \Lambda}{h} \frac{d(T - T^*)}{dt} + (T - T^*) = \text{excess (deficit) energy supply}\,, \tag{4.1}$$

where ρ and c are the density and specific heat, respectively, of the soil in the upper few meters of the Earth's crust, and Λ is a characteristic length scale for the penetration of heat. It is clear that this characteristic length should be proportional to the radius of the disc under consideration. If we assume that the daily excess energy supply is proportional to L, then equation (4.1) coincides with equation (2.1) if

$$\lambda = \frac{\rho c \Lambda}{h}\,, \tag{4.2}$$

whence one may expect that

$$\frac{\lambda}{\cos\ell} = \text{constant}\,. \tag{4.3}$$

Table 4.1 shows the ratio is roughly constant with a value of about 64 days. With this information, and $A(\ell)$ given in the caption to Figure 3.2,

latitude, deg.	$\lambda/\cos\ell$
25	62.9
30	64.7
35	63.5
40	61.4
50	59.1
60	66
65	68.6

TABLE 4.1. The ratio $\lambda/\cos\ell$ at various latitudes. The ratio is a constant with a mean value of about 63.8 and a standard deviation of about 3.1.

equation (3.5) is an explicit formula for $\eta = \eta(\ell)$, whose graph resembles that shown in Figure 3.5.

Daily variations in temperature are caused mainly by ground absorption and release of heat. The thermal conductivity of the ground and its heat capacity are not very large, so (on a short term) it gives up at night what it receives during the day. Inserting reasonable values for ρ, c and h (or, equivalently, for the so-called Biot number, or for the soil diffusivity), one gets a plausible estimate for the penetration depth to be of the order of a meter. Clearly, more precise estimates of this depth will depend on the latitude, and on whether one is on land or in the ocean, and so forth.

Two further comments may be useful. First, our equation (2.1) is tantamount to assuming that the heating of the Earth by the Sun can be represented by the term $\eta L(t)$. In principle, this term can be computed exactly from the known information on the solar radiation arriving at the Earth. We have not attempted to do this. In effect, the "viscosity coefficient" η lumps these factors into an effective single number. Secondly, we have so far ignored temperature fluctuations about its mean value $\langle\theta\rangle$ at a given latitude; the model therefore has nothing to say about the latter. If one imagines that equation (2.1) without the derivative term holds for the mean temperature $\langle\theta\rangle$ as well, it is clear that one can define another constant η^* given by

$$\eta^* = \frac{\langle\theta\rangle}{\langle L^*\rangle}, \tag{4.4}$$

where the denominator $\langle L^*\rangle$, as already remarked, is approximately 12 hours at all latitudes. There is no reason to expect that this new coefficient η^* will be related to η. However, we find that the variation of η^* is roughly linear in ℓ and follows

$$\eta^* = 4.53\ell - 2.33. \tag{4.5}$$

5 Concluding remarks

A large number of non-equilibrium phenomena, involving dilute gases and a variety of rheological problems, have been represented successfully by relaxation (or memory) models. Various phenomena that occur in turbulent shear flows have also been modelled in this way. Two examples will suffice. In Narasimha and Prabhu [1972], a fully developed turbulent wake was distorted by a rapid pressure gradient, which was then released. The subsequent evolution of the flow, which is generally difficult to predict by conventional turbulence models, has been replicated well by a relaxation model. For a pipe flow with a periodically oscillating mass flow (Mao and Hanratty [1986]), the stress and rate of strain can be related quite well by a memory model. The present work has shown that the relaxation model performs similarly well for modelling seasonal variations of temperature. However, the ubiquitous success of memory models does not necessarily suggest that we understand the physics leading to relaxation effects. That understanding must come from the study of specific systems on hand.

Acknowledgments: Larry Sirovich is special for a number of reasons, one of which is that he takes delight in learning about new problems without being encumbered by their past foibles. We hope that the little problem discussed here will amuse him. KRS thanks the National Science Foundation, and DDJ thanks the US Army and DOE for supporting this work.

References

Crowley, T.J. and North, G.R. [1991], *Paleoclimatology*, chapter 1, Oxford University Press.

Joseph, D.D. [1990], *Fluid Dynamics of Viscoelastic Liquids*, Springer-Verlag, New York.

List, R.J. [1951], *Smithsonian Meteorological Tables*, Smithsonian Institute, Washington, D.C.

Mao, Z.-X. and Hanratty, T.J. [1986], "Studies of the wall shear-stress in a turbulent pulsating pipe-flow", *J. Fluid Mech.* **170**, 545.

Meeus, J. [1988], *Astronomical Formulae for Calculators*, Willmann–Bell, Inc.

Miller, A. and Thompson, J.C. [1979], *Elements of Meteorology*, Third Edition, Charles Merrill Publishing Company, Columbus.

Montenbruck, O. and T. Pfleger, [1994], *Astronomy on the Personal Computer*, Springer.

Narasimha, R. and Prabhu, A. [1972], "Equilibrium and relaxation in turbulent wakes", *J. Fluid Mech.* **54**, 1.

Oort, A.H. and Rasmuusen, E.M. [1971], *Atmospheric Circulation Statistics*, NOAA Professional Paper 5, US Dept of commerce, MD.

Smart, W.M. [1956], *Spherical Astronomy*, Cambridge University Press, p.151.

13

Simultaneously Band and Space Limited Functions in Two Dimensions, and Receptive Fields of Visual Neurons

Jonathan D. Victor
Bruce W. Knight

To Larry Sirovich, on the occasion of his 70th birthday.

ABSTRACT Functions of a single variable that are simultaneously band- and space-limited are useful for spectral estimation, and have also been proposed as reasonable models for the sensitivity profiles of receptive fields of neurons in primary visual cortex. Here we consider the two-dimensional extension of these ideas. Functions that are simultaneously space- and band-limited in circular regions form a natural set of families, parameterized by the "hardness" of the space- and band- limits. For a Gaussian ("soft") limit, these functions are the two-dimensional Hermite functions, with a modified Gaussian envelope. For abrupt space and spatial frequency limits, these functions are the two-dimensional analogue of the Slepian (prolate spheroidal) functions (Slepian and Pollack [1961]; Slepian [1964]). Between these limiting cases, these families of functions may be regarded as points along a 1-parameter continuum. These families and their associated operators have certain algebraic properties in common. The Hermite functions play a central role, for two reasons. They are good asymptotic approximations of the functions in the other families. Moreover, they can be decomposed both in polar coordinates and in Cartesian coordinates. This joint decomposition provides a way to construct profiles with circular symmetries from superposition of one-dimensional profiles. This result is approximately universal: it holds exactly in the "soft" (Gaussian) limit and in good approximation across the one-parameter continuum to the "hard" (Slepian) limit. These properties lead us to speculate that such two-dimensional profiles will play an important role in the understanding of visual processing in cortical areas beyond primary visual cortex. Comparison with published experimental results lends support to this conjecture.

Contents

1 Introduction

The understanding of the structure of neural computations, and how they are implemented in neural hardware, are major goals of neuroscience. The visual system often is used as a model for this purpose. For visual neurons, characterization of the spatial weighting of their various inputs is an important concrete step in this direction. For an idealized linear neuron, this spatial weighting—the sensitivity profile of the receptive field—fully describes the spatial integration performed by that neuron. Real neurons, including those of the retina and primary visual cortex, exhibit nonlinear combination of their inputs, but their receptive field profiles nevertheless provide a useful qualitative description of these neurons' response properties. For example, the circularly symmetric center/surround antagonism that characterizes typical retinal output neurons suggests a filtering process that removes overall luminance and other long-range correlations in the retinal image. The strongly oriented receptive field profiles encountered in primary visual cortex suggest extraction of one-dimensional features, oriented in corresponding directions.

The nervous system's design must represent a balance among multiple, often conflicting, demands. Efficiency (that is, representation of the spatiotemporal visual input with as few cells as possible, and with a firing rate that is, on average, as low as possible) may appear to be a main criterion for fitness. However, efficient schemes have hidden costs. These include the metabolic and morphological requirements of creating or decoding "efficient" representations (Laughlin, de Ruyter, and Anderson [1998]), the lack of robustness of "efficient" representations in the face of damage to the network, the length and complexity of connections that may be required to implement an efficient scheme, and the burden of specifying these connections in genetic material. Nevertheless, views that consider a biologically motivated notion of efficiency, along with general aspects of the statistics of natural visual images as factors in shaping the nervous system, can provide a successful account of receptive field properties, both in the retinal output (Atick and Redlich [1990]) and in primary visual cortex (Field [1994]). Regardless of whether efficiency plays a dominant role in shaping receptive fields in primary visual cortex, the empirical observation remains that to a good approximation, many receptive field profiles are well-described by a Gabor function (Daugman [1985]; Jones and Palmer [1987]; Marcelja [1980]), namely, a Gaussian multiplied by a sinusoid. Because the period of the sinusoid and the width of the Gaussian envelope of neurons encountered in primary visual cortex are comparable, these Gabor functions also can be well approximated by a Gaussian multiplied by a low-order oscillatory polynomial, such as a Hermite polynomial. Consequently, it may be that simultaneous confinement in space and in spatial frequency suffices to account for the shape of receptive field profiles in primary visual cortex, and that orientation plays no special role. Without recourse to detailed analyses of coding strategy and image statistics, one can argue that such receptive fields are good choices (Marcelja [1980]) to minimize wiring length and connectivity (by their confinement in space) and to analyze textures, features, and images at a particular spatial scale (by their confinement in spatial frequency).

It is not clear how to pursue the above line of inquiry beyond primary visual (striate) cortex. In extrastriate visual areas, neuronal properties become progressively less stereotyped, receptive fields are less well localized, and characterization methods based on standard systems-analysis procedures appear to be progressively less useful. We do not propose to solve this problem here. However, it is striking to note that within primary visual cortex, receptive field profiles appear suited for processing primarily along a single spatial dimension. On the other hand, many aspects of visual processing are essentially two-dimensional. These include extraction of low-level features such as T-junctions (Rubin [2001]), curvature (Wolfe, Yee, and Friedman-Hill [1992]), texture (Victor and Brodie [1978]) and shape (Wilkinson, Wilson, and Habak [1998]; Wilson and Wilkinson [1998]), as well as higher-level processes such as letter and face identification. Moreover, recordings from individual neurons beyond striate cortex reveal evidence of fundamentally two-dimensional processing (Gallant, Braun, and Van Essen [1993]; Gallant, Connor, Rakshit, Lewis, and Van Essen [1996]; Tanaka, Saito, Fukada, and Moriya [1991]). Gallant's work (Gallant, Braun, and Van Essen [1993]; Gallant, Connor, Rakshit, Lewis, and Van Essen [1996]) is particularly provocative in that it suggests that some of these neurons are tuned to various kinds of circular symmetry.

These considerations motivate exploration of the consequences of simultaneous space- and band-limitation in two dimensions. As elaborated in the Discussion, our notion of "confinement" is distinct from that of Daugman [1985], whose analysis, based on a quantum-mechanical notion of "uncertainty," led to the Gabor functions as playing an optimal role. Our notion of confinement has several parametric variations, and the family of functions that optimally achieve simultaneous confinement in space and spatial frequency depends on them. However, these families of functions also have algebraic properties and qualitative behavior that are independent of these details. Each family includes functions that are good models for receptive field profiles in primary visual cortex (that is, they resemble Gabor functions), but also functions that are intrinsically two-dimensional. One such family of functions is the Hermite functions. These functions serve as good asymptotic approximations for the other families. Their properties suggest how circularly symmetric receptive fields can be built out of simple combinations of the receptive fields encountered in primary visual cortex, V1.

2 Results

2.1 Definitions

Following the general notation of Slepian and Pollack [1961], we consider space-limiting operators D and band-limiting operators B. Both are linear operators on functions f on the plane. We define a general space-limiting operator D with shape parameter a and scale parameter d by

$$D_{a,d}f(x,y) = \exp\left[-\left(\frac{|x|^2+|y|^2}{d^2}\right)^{\frac{1}{2}a}\right] f(x,y)\,. \tag{2.1}$$

The corresponding band-limiting operators B are most readily defined in the frequency domain, by

$$B_{a,b}\,\tilde{f}(\omega_x,\omega_y) = \exp\left[-\left(\frac{|\omega_x|^2+|\omega_y|^2}{b^2}\right)^{\frac{1}{2}a}\right] \tilde{f}(\omega_x,\omega_y)\,, \tag{2.2}$$

where

$$\tilde{f}(\omega_x,\omega_y) = \frac{1}{2\pi}\iint f(x,y)\,\exp\left[-i(\omega_x x+\omega_y y)\right]dx\,dy\,, \tag{2.3}$$

whence

$$f(x,y) = \frac{1}{2\pi}\iint \tilde{f}(\omega_x,\omega_y)\exp\left[i(\omega_x x+\omega_y y)\right]d\omega_x\,d\omega_y\,. \tag{2.4}$$

Note that in the limit $a\to\infty$ that $D_{\infty,d}$ sets f to 0 outside of $|x|^2+|y|^2\le d^2$, and leaves f unchanged within the disk of radius d, and analogously for $B_{\infty,b}$. In equations (2.3) and (2.4), we use a symmetric form for the Fourier transform; this choice will prove convenient for a later analysis (see equation (2.108)).

2.2 Algebraic Properties

It follows immediately from the definition (equation 2.1) that $D_{a,t}D_{a,u} = D_{a,\nu}$, where $\frac{1}{t^a} + \frac{1}{u^a} = \frac{1}{\nu^a}$, and that $D_{\infty,d}$ is idempotent. This composition rule guarantees the existence of functional square roots,

$$(D_{a,t})^{\frac{1}{2}} = D_{a,\nu}, \text{where}\, \nu = 2^{\frac{1}{a}} t\,, \tag{2.5}$$

which will be important below. Corresponding relationships hold for B.

From here on, we will suppress subscripts whenever this does not lead to ambiguities. D and B are evidently self-adjoint operators. Since they do not commute, BD and DB are not self-adjoint, but combinations such as $D^m B^k D^m$ and $B^m D^k B^m$ are self-adjoint. Moreover, if ψ is an eigenvector of $D^u B D^\nu$ with eigenvalue λ, then $D^h \psi$ is an eigenvector of $D^{u+h} B D^{\nu-h}$, also with eigenvalue λ. This is because

$$D^{u+h} B D^{\nu-h} (D^h \psi) = D^h (D^u B D^\nu \psi) = D^h (\lambda \psi) = \lambda (D^h \psi). \tag{2.6}$$

In particular, the operators $D^u B D^v$ are isospectral for changes in u and ν which leave $u + \nu$ constant. These relations take a particularly simple form for D_∞ and B_∞, which are idempotent.

It now follows that BD and DB have only real eigenvalues (since they are isospectral with the self-adjoint operators $B^{\frac{1}{2}} D B^{\frac{1}{2}}$ and $D^{\frac{1}{2}} B D^{\frac{1}{2}}$). Moreover, their eigenvalues λ are necessarily between 0 and 1. $|\lambda| \leq 1$ follows from the observation that the operators D and B can only diminish the magnitude of a function, and $\lambda \geq 0$ follows from the inner-product calculation that for any function f,

$$\left(D^{\frac{1}{2}} B D^{\frac{1}{2}} f, f \right) = \left(B^{\frac{1}{2}} D^{\frac{1}{2}} f\,, B^{\frac{1}{2}} D^{\frac{1}{2}} f \right) \geq 0\,. \tag{2.7}$$

Consider now the problem of finding the function f that is most nearly preserved by successive application of space- and band-limits via D and B, respectively. One way of formulating this problem is as follows. Since application of D and B can only reduce $|f|^2 \equiv (f, f)$, we can formulate this problem as finding the function f that maximizes $|BDf|^2$ subject to $|f|^2 = 1$. We calculate

$$\max_{|f|^2=1} |BDf|^2 = \max_{(f,f)=1} (f, DB^2 Df)\,. \tag{2.8}$$

The Lagrange multiplier method, applied to the right hand side of equation (2.8), indicates that $|BDf|^2$ is extremized when f satisfies the eigenvalue equation

$$DB^2 Df = \lambda f\,. \tag{2.9}$$

By the remarks at the beginning of this section, the eigenfunctions of DB^2D can be obtained from the eigenfunctions of B^2D^2 by application of

D. The operator B^2D^2 is of the form BD, but with choices for the width parameters b and d reduced by a factor of $2^{1/a}$ (equation 2.5). Note that in the abrupt ($a = \infty$) case, the situation is particularly simple (Slepian and Pollack [1961]): this change in scale is trivial, and eigenfunctions of BD, operated on by D, are in turn eigenfunctions of DB.

Other than this change of scale and the application of D, the eigenfunction of BD with largest eigenvalue is the function f that is altered the least by successive application of space- and band-limits. Similarly, eigenfunctions of successively lower eigenvalue correspond to functions that, within subspaces orthogonal to eigenfunctions of higher eigenvalue, have this property.

The notion of a function that is simultaneously both space- and band-limited can be expressed in other ways, such as a function that maximizes $(DBDf, f)$ subject to $|f|^2 = 1$. By the above remarks, the solutions to the corresponding eigenvalue problems (equation 2.9) are readily obtained from the eigenfunctions of BD, and this relationship is particularly simple in the abrupt case. (Note that the one cannot merely ask for functions that maximize $(BDf, f) = 1$ subject to $|f|^2 = 1$. Since BD is not self-adjoint, this maximization does not lead to an eigenvalue problem for BD as did the extremum problem of equation (2.8).)

Thus, we direct our attention to finding the eigenfunctions of BD.

2.3 The Gaussian Case

We focus on the Gaussian ($a = 2$) case, for several reasons. First, the eigenfunctions of BD (and hence, DB) have a simple closed form. Second, the operators D and B generically have rotational symmetry, but in this case also separate in Cartesian coordinates. This leads to relationships between the eigenfunctions of the separated operators and eigenfunctions with rotational symmetry. Finally, the Gaussian case provides a reasonable approximate description of the general case, as anticipated from the slowly-varying linear operator theory of Sirovich and Knight (Knight and Sirovich [1982, 1986]; Sirovich and Knight [1981, 1982, 1985]).

We write $D_{gau,d} \equiv D_{2,d\sqrt{2}}$, an operator whose envelope is a product of one-dimensional Gaussians, each of standard deviation d. We use the analogous convention for $B_{gau,b} \equiv B_{2,b\sqrt{2}}$. Now it is convenient to define

$$D_{x,d}f(x, y) = \exp\left[-\frac{x^2}{2d^2}\right] f(x, y), \tag{2.10}$$

which space-limits only along the x-coordinate, and to analogously define $D_{y,d}$, $B_{x,b}$ and $B_{y,b}$. $D_{gau,d} = D_{x,d}D_{y,d}$, $B_{gau,d} = B_{x,b}B_{y,b}$, and the operators also have the commutation relations $D_{x,d}B_{y,b} = B_{y,b}D_{x,d}$ and $D_{y,d}B_{x,b} = B_{x,b}D_{y,d}$. Thus, there will be eigenfunctions $\psi(x, y)$ of

$B_{gau,b}D_{gau,d}$ with eigenvalue λ which have the form

$$\psi(x,y) = \psi_x(x)\psi_y(y)\,, \tag{2.11}$$

where the factors $\psi_x(x)$ and $\psi_y(y)$ are eigenfunctions of the one-dimensional operators $B_{x,b}D_{x,d}$ and $B_{y,b}D_{y,d}$ with eigenvalues λ_x and λ_y, and

$$\lambda = \lambda_x \lambda_y\,. \tag{2.12}$$

2.4 One Dimension

We now consider the one-dimensional operators $D_{gau,d} \equiv D_{2,d\sqrt{2}}$, $B_{gau,b} \equiv B_{2,b\sqrt{2}}$, and suppress the subscripts. As the Fourier transform of a Gaussian is another Gaussian, we have explicitly,

$$(BDf)(x) = \frac{b}{\sqrt{2\pi}} \int f(u) \exp\Bigl(-\frac{u^2}{2d^2}\Bigr) \exp\left(-\tfrac{1}{2}b^2(u-x)^2\right) du\,. \tag{2.13}$$

Following Knight and Sirovich [1982, 1986], we seek solutions of

$$BDf_n = \lambda_n f_n \tag{2.14}$$

in the form of a Hermite polynomial h_n scaled by a factor k, multiplied by a Gaussian of standard deviation α,

$$f_n(x) = h_n(kx) \exp\left(-\frac{x^2}{2\alpha^2}\right), \tag{2.15}$$

and anticipate eigenvalues of the form

$$\lambda_n = \eta^{n+\frac{1}{2}}\,. \tag{2.16}$$

It is convenient to use Hermite polynomials that are orthogonal with respect to a Gaussian of unit standard deviation, and whose highest coefficient is unity. This convention, more convenient for what follows, is different from the standard one (Abramowitz and Stegun [1964] equation 22.2.14) for the Hermite polynomials H_n; the relationship between these conventions is

$$h_n(x) = 2^{-n/2} H_n\left(\frac{x}{\sqrt{2}}\right). \tag{2.17}$$

Under our convention, the Hermite polynomials have the generating function

$$\sum_{n=0}^{\infty} \frac{z^n}{n!} h_n(x) = \exp\left(xz - \tfrac{1}{2}z^2\right). \tag{2.18}$$

The assumed form for f_n, equation (2.15), leads to a generating function for the right hand side of the eigenvalue equation (2.14):

$$\sum_{n=0}^{\infty} \frac{z^n}{n!} \lambda_n f_n(x) = \eta^{\frac{1}{2}} \exp\left(-\frac{x^2}{2\alpha^2} + k\eta x z - \tfrac{1}{2}\eta^2 z^2\right). \tag{2.19}$$

We can also write a generating function for the left-hand side of equation (2.14):

$$\sum_{n=0}^{\infty} \frac{z^n}{n!}(DBf_n)(x) = \sqrt{\frac{b^2}{b^2 + \frac{1}{d^2} + \frac{1}{\alpha^2}}} \exp\left[-\tfrac{1}{2}(b^2x^2 + z^2)\right]$$
$$\exp\left[\frac{1}{2}\frac{(b^2x + kz)^2}{b^2 + \frac{1}{d^2} + \frac{1}{\alpha^2}}\right]. \tag{2.20}$$

The two generating functions (equations (2.19) and (2.20)) have the same form. Equating the coefficients of $x^2, xz,$ and z^2 in the exponents, and also equating the two overall multiplicative factors, leads to constraints for the unknown quantities $\alpha, k,$ and η of equations (2.15) and (2.16). These constraints are satisfied by

$$k = b\sqrt{\frac{1-\eta^2}{\eta}}, \tag{2.21}$$

$$\alpha^2 = \frac{1 \pm \sqrt{1 + 4b^2d^2}}{2b^2}, \quad \text{and} \tag{2.22}$$

$$\eta = \frac{b^2}{b^2 + \frac{1}{d^2} + \frac{1}{\alpha^2}}. \tag{2.23}$$

Only the positive branch of equation (2.22) corresponds to eigenfunctions that approach zero for large x. The negative branch corresponds to imaginary α and to real eigenfunctions that diverge for large x. We will focus on the positive branch.

These equations can be placed in a more dimensionless form by taking $c = bd$, $\kappa = \frac{k}{b}$, and $\beta = \alpha b$. With these parameters,

$$\beta^2 = \frac{1 \pm \sqrt{1 + 4c^2}}{2}, \tag{2.24}$$

$$\eta = \frac{2c^2}{1 + 2c^2 + \sqrt{1 + 4c^2}} = \left(\frac{2c}{1 + \sqrt{1 + 4c^2}}\right)^2, \quad \text{and} \tag{2.25}$$

$$\kappa^2 = \frac{\sqrt{1 + 4c^2}}{c^2}. \tag{2.26}$$

After some algebra, the eigenfunctions (2.15) can be written

$$f_n\left(\frac{\chi}{k}\right) = h_n(\chi) \exp\left(-\tfrac{1}{4}\chi^2\right) \exp\left(\frac{\chi^2}{4\kappa^2c^2}\right)$$
$$= h_n(\chi) \exp\left(-\tfrac{1}{4}\chi^2\right) \exp\left(\frac{\chi^2}{4\sqrt{1 + 4c^2}}\right). \tag{2.27}$$

Equation (2.15) and the more explicit form (2.27) are thus the functions that extremize simultaneous space- and band-limited functions in the

Gaussian sense that we have defined. The combination

$$c = bd \tag{2.28}$$

is a product of the space limit and the bandwidth limit. As $c \to \infty$, the Gaussian envelope in equation (2.27) becomes progressively less prominent, and the eigenfunctions approach the unmodified Hermite functions $h_n(\chi) \exp\left(-\frac{1}{4}\chi^2\right)$.

2.5 Two Dimensions: Reorganization According to Rotational Symmetry

We now consider the two-dimensional Gaussian case. As a consequence of the Cartesian separation of equation (2.11), solutions of the two-dimensional eigenvalue problem equation (2.14) may be parameterized by integers n_x and n_y, with

$$f_{n_x,n_y}(x,y) = f_{n_x}(x)\, f_{n_y}(y)\,, \tag{2.29}$$

where the factors on the right hand side are given by equation (2.27). As a consequence of equations (2.12) and (2.16), the eigenvalue associated with f_{n_x,n_y} is

$$\lambda_{n_x,n_y} = \eta^{1+n_x+n_y}\,, \tag{2.30}$$

where η is given by equation (2.23) or (2.25).

Thus, eigenvalues are identical for eigenfunctions that share a common value of $n = n_x + n_y$. These $n+1$ eigenfunctions, namely $f_{0,n}$, $f_{1,n-1}$, $\ldots$, $f_{n-1,1}$, $f_{n,0}$, are readily reorganized into new linear combinations that exhibit polar symmetry, as one would expect from the polar symmetry of the operators B and D for the Gaussian ($a = 2$) case. To calculate these linear combinations explicitly, we combine generating functions with the umbral calculus of Rota and Taylor [1994]. The main steps are: (a) defining the umbral calculus, (b) writing a generating function for products of Hermite polynomials in Cartesian coordinates, (c) using the umbral calculus to reorganize this generating function in terms of polar coordinates, and (d) matching coefficients to arrive at the desired reorganization.

The umbral calculus is essentially an algebra of polynomials in several variables. Addition in this algebra is the usual addition. Multiplication in this algebra is a nonstandard operation that will be denoted $\otimes$. This operation is defined in terms of its action on products of Hermite polynomials (which form a basis), and then is extended to all polynomials via linearity. The linearity condition is equivalent to stating that $\otimes$ and addition obey the distributive law. For Hermite polynomials with identical formal arguments, we define

$$h_m(x) \otimes h_n(x) = h_{m+n}(x). \tag{2.31}$$

For Hermite polynomials with distinct arguments, $\otimes$ acts like ordinary multiplication:

$$h_m(x) \otimes h_n(y) = h_m(x)h_n(y) \tag{2.32}$$

We use exponential notation $p^{\otimes m} = p \otimes p \otimes \ldots \otimes p$ for iterated products of any polynomial p, and we also write $h = h_1$ so that $h^{\otimes n}(x) = h_n(x)$. For example, in this notation, the generating function (2.18) for Hermite polynomials takes the form

$$\sum_{n=0}^{\infty} \frac{z^n}{n!} h^{\otimes n}(x) = \exp\left(xz - \frac{1}{2}z^2\right). \tag{2.33}$$

Now consider a generating function $Q(z,t)$ defined by

$$Q(z,t) = \sum_{k=0}^{\infty}\sum_{l=0}^{\infty} \frac{z^k}{k!}\frac{t^l}{l!} q_{k,l}(x,y), \tag{2.34}$$

where

$$q_{k,l}(x,y) = [h(x) + ih(y)]^{\otimes k} \otimes [h(x) - ih(y)]^{\otimes l}. \tag{2.35}$$

With $k = a + r$ and $l = b + s$ and application of the binomial expansion to each term of equation (2.35) we find

$$Q(z,t) = \sum_{a,b,r,s=0}^{\infty} \frac{z^{a+r}}{a!\,r!}\frac{t^{b+s}}{b!\,s!} i^r(-i)^s\, h_{a+b}(x) \otimes h_{r+s}(y). \tag{2.36}$$

Each term is of the form of equation (2.32) at this step, so $\otimes$ becomes ordinary multiplication. Equation (2.36) now can be factored into

$$Q(z,t) = \left[\sum_{a,b=0}^{\infty} \frac{z^a t^b}{a!b!} h_{a+b}(x)\right] \cdot \left[\sum_{r,s=0}^{\infty} \frac{(iz)^r(-it)^s}{r!s!} h_{r+s}(y)\right]. \tag{2.37}$$

Application of the binomial expansion collapses each of these factors:

$$Q(z,t) = \left[\sum_{m=0}^{\infty} \frac{(z+t)^m}{m!} h_m(x)\right]\left[\sum_{m=0}^{\infty} \frac{(iz-it)^m}{m!} h_m(y)\right]. \tag{2.38}$$

It now follows from the generating function for h, equation (2.18), that

$$Q(z,t) = \exp\left[x(z+t) - \frac{(z+t)^2}{2}\right] \exp\left[y(iz-it) - \frac{(iz-it)^2}{2}\right], \tag{2.39}$$

or equivalently,

$$Q(z,t) = \exp[(x+iy)z + (x-iy)t - 2zt]. \tag{2.40}$$

With the usual polar substitutions $x = R\cos\theta$ and $y = R\sin\theta$, along with $\rho = \sqrt{zt}$ and $\sigma = \sqrt{\frac{z}{t}}$, equation (2.40) becomes

$$Q\left(\rho\sigma, \frac{\rho}{\sigma}\right) = \exp\left[R\rho\left(e^{i\theta}\sigma + \frac{1}{e^{i\theta}\sigma}\right) - 2\rho^2\right]. \tag{2.41}$$

We now form a Taylor series expansion of the right hand side:

$$\begin{aligned} Q\left(\rho\sigma, \frac{\rho}{\sigma}\right) &= \sum_{s=0}^{\infty}\frac{1}{s!}\left(R\rho\left(e^{i\theta}\sigma + \frac{1}{e^{i\theta}\sigma}\right) - 2\rho^2\right)^s \\ &= \sum_{s=0}^{\infty}\sum_{g=0}^{s}\frac{1}{g!(s-g)!}\left(R\rho\left(e^{i\theta}\sigma + \frac{1}{e^{i\theta}\sigma}\right)\right)^g\left(-2\rho^2\right)^{s-g} \\ &= \sum_{s=0}^{\infty}\sum_{g=0}^{s}\sum_{j=0}^{g}\frac{1}{(s-g)!j!(g-j)!}(R\rho)^g\left(-2\rho^2\right)^{s-g}\left(e^{i\theta}\sigma\right)^{2j-g}. \end{aligned} \tag{2.42}$$

From the middle line of equation (2.42), we see that any term that involves σ^μ must have $g \geq |\mu|$, and hence must be associated with $\rho^{2\nu+|\mu|}$ for some non-negative integer ν. We therefore collect terms that involve $\rho^{2\nu+|\mu|}\sigma^\mu$ in equation (2.42). These are the terms for which $j = \frac{1}{2}(|\mu| + g)$ and $s = \frac{1}{2}(|\mu| + g) + \nu$. Thus

$$\begin{aligned} Q\left(\rho\sigma, \frac{\rho}{\sigma}\right) &= \sum_{\mu=-\infty}^{\infty}\sum_{\nu=0}^{\infty}\rho^{2\nu+|\mu|}\sigma^\mu e^{i\mu\theta}R^{|\mu|} \\ &\quad\times\sum_g\frac{(-2)^{\frac{1}{2}(|\mu|-g)+v}R^{g-|\mu|}}{\left(\frac{1}{2}(|\mu|-g)+\nu\right)!\left(\frac{1}{2}(|\mu|+g)\right)!\left(\frac{1}{2}(g-|\mu|)\right)!} \end{aligned} \tag{2.43}$$

where the inner sum is over all values of g for which the arguments of the factorials are non-negative integers. With $p = \frac{1}{2}(g - |\mu|)$, we have

$$Q\left(\rho\sigma, \frac{\rho}{\sigma}\right) = \sum_{\mu=-\infty}^{\infty}\sum_{\nu=0}^{\infty}\rho^{2\nu+|\mu|}\sigma^\mu e^{i\mu\theta}R^{|\mu|}\sum_{p=0}^{\nu}\frac{(-2)^{\nu-p}R^{2p}}{(|\mu|+p)!p!(\nu-p)!}. \tag{2.44}$$

We now convert the expression in equation (2.34) for Q to polar form in another way. The definition of $\otimes$ leads to the identity

$$[h(x) + ih(y)] \otimes [h(x) - ih(y)] = h_2(x) + h_2(y). \tag{2.45}$$

This is a crucial step: the left-hand side is a product of Hermite polynomials in Cartesian coordinates, while the right-hand side depends only on the radius (as $h_2(u) = u^2 - 1$).

Repeated application of this identity to equation (2.35) yields

$$q_{k,l}(x, y) = [h_2(x) + h_2(y)]^{\otimes\min(k,l)} \otimes [h(x) \pm ih(y)]^{\otimes|k-l|}, \tag{2.46}$$

where the sign in the final term is chosen to match the sign of $k-l$.

Now consider the substitutions $\nu = \min(k,l)$ and $\mu = k-l$. As k and l each run from 0 to ∞, ν runs from 0 to ∞, and μ independently runs from $-\infty$ to ∞ (see Figure 2.1). Moreover, $k = \nu + \frac{\mu}{2} + \left|\frac{\mu}{2}\right|$ and $l = \nu - \frac{\mu}{2} + \left|\frac{\mu}{2}\right|$, so that (k,l) pairs of constant eigenvalue (constant $k+l$) correspond to constant values of $2v + |\mu|$. (This is the reason for the reorganization of terms between equations (2.42) and (2.43).) By use of the umbral identity (2.46), the expression (2.34) for $Q(z,t)$ can be transformed to

$$Q\left(\rho\sigma, \frac{\rho}{\sigma}\right) = \sum_{\mu=-\infty}^{\infty} \sum_{\nu=0}^{\infty} \frac{\rho^{2\nu+|\mu|}}{\left(\nu + \frac{\mu}{2} + \frac{|\mu|}{2}\right)!} \frac{\sigma^{\mu}}{\left(\nu - \frac{\mu}{2} + \frac{|\mu|}{2}\right)!} \times \left[h_2(x) + h_2(y)\right]^{\otimes\nu} \otimes \left[h(x) \pm ih(y)\right]^{\otimes\mu}, \quad (2.47)$$

where the sign in the final term is chosen to match the sign of μ.

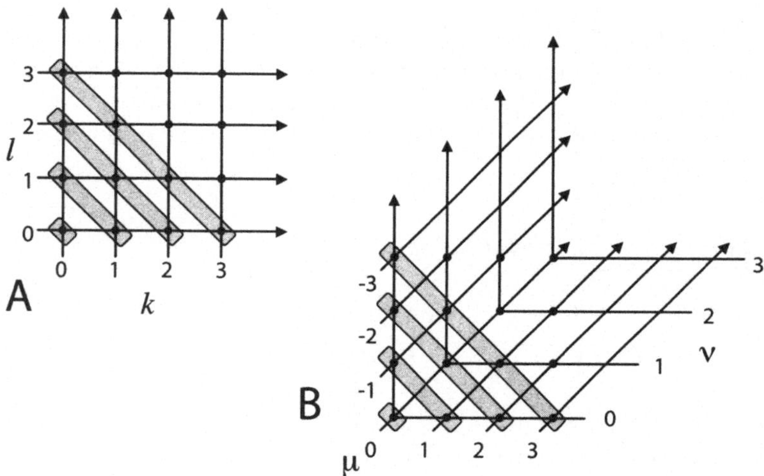

FIGURE 2.1. The change of indices from (k,l) (panel A) to (μ,ν) (panel B) via the substitutions $\nu = \min(k,l)$ and $\mu = k-l$. Coordinates that correspond to the eigenfunctions of equal eigenvalues are indicated by the gray enclosures.

We now equate the coefficient of $\rho^{2\nu+|\mu|}\sigma^{\mu}$ in equation (2.47) with the corresponding coefficient in equation (2.44). It suffices to consider $\mu \geq 0$. This yields

$$[h(x) + ih(y)]^{\otimes\mu} \otimes [h_2(x) + h_2(y)]^{\otimes\nu} = e^{i\mu\theta} R^{\mu} P_{\mu,\nu}(R^2), \quad (2.48)$$

where

$$P_{\mu,\nu}(r) = \sum_{p=0}^{\nu} (-2)^{\nu-p} \frac{(\mu+\nu)!\nu!}{(\mu+p)!p!(\nu-p)!} r^p. \quad (2.49)$$

These equations, along with equation (2.15) or (2.27) that convert the Hermite polynomials to the eigenfunctions of BD in one dimension, specify how the eigenfunctions in Cartesian separation that share a common eigenvalue can be reorganized into a polar separation. The right-hand side of equation (2.48) is separated in polar coordinates and has angular dependence $\exp(i\mu\theta)$. The radial dependence is R^{μ} times a polynomial $P_{\mu,\nu}$ of degree ν in R^2. Moreover, their relationship (see below) to the generalized Laguerre polynomials implies that there are ν nodes along each radius. The left-hand side of equation (2.48) is a sum of terms $h_{n_x}(x)\, h_{n_y}(y)$, as specified by the properties of $\otimes$. The indices n_x and n_y that would appear on the left- hand side in Cartesian form are all those that satisfy

$$n = n_x + n_y = 2\nu + |\mu|. \tag{2.50}$$

As an example of this reorganization, we take $\mu = 2$ and $\nu = 3$. Equation (2.48) becomes

$$[h(x) + ih(y)]^{\otimes 2} \otimes [h_2(x) + h_2(y)]^{\otimes 3} = e^{2i\theta} R^2 (R^6 - 30R^4 + 240R^2 - 480). \tag{2.51}$$

Reduction of the left-hand side via the definition of $\otimes$ leads to

$$\begin{aligned}
&[h(x) + ih(y)]^{\otimes 2} \otimes [h_2(x) + h_2(y)]^{\otimes 3} \\
&\quad = [h_2(x) + 2ih_1(x)h_1(y) - h_2(y)] \\
&\qquad \otimes [h_6(x) + 3h_4(x)h_2(y) + 3h_2(x)h_4(y) + h_6(y)] \\
&\quad = h_8(x) + 2ih_7(x)h_1(y) + 2h_6(x)h_2(y) + 6ih_5(x)h_3(y) \\
&\qquad + 6ih_3(x)h_5(y) - 2h_2(x)h_6(y) \\
&\qquad + 2ih_1(x)h_7(y) - h_8(y)\,.
\end{aligned} \tag{2.52}$$

Thus, the real part $h_8(x) + 2h_6(x)h_2(y) - 2h_2(x)h_6(y) - h_8(y)$ and the imaginary part $2h_7(x)h_1(y) + 6h_5(x)h_3(y) + 6h_3(x)h_5(y) + 2h_1(x)h_7(y)$ are the two polynomials associated with eigenfunctions of twofold axial symmetry ($\mu = 2$) and three radial nodes ($\nu = 3$). These eigenfunctions are illustrated in Figure (2.2) for $c = 4$. Figure (2.3) shows another example of this reorganization, with threefold axial symmetry ($\mu = 3$) and one radial node ($\nu = 1$), which emphasizes that the eigenfunctions in the polar separation may have symmetries manifested by none of the eigenfunctions in the Cartesian separation.

Properties of the polynomials $P_{\mu,v}$

The polynomials $P_{\mu,\nu}(R^2)$ that appear on the right-hand side of equation (2.48) are a doubly-indexed set with several interesting properties. Considered as functions on the plane, $e^{i\mu\theta} R^{\mu}\, P_{\mu,\nu}(R^2)$ form an orthogonal family with respect to a weight $\exp\left(-\frac{1}{2}R^2\right)$. This can be seen as follows. Two such functions that have different values of μ are orthogonal because of

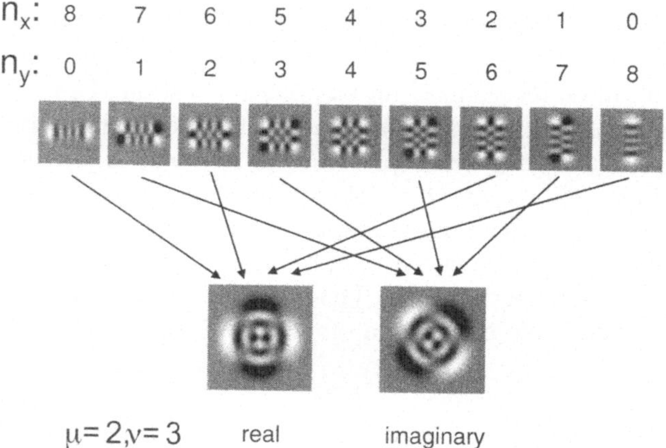

FIGURE 2.2. An example of the reorganization of eigenfunctions in the Cartesian separation into eigenfunctions in the polar separation. Top row: the nine eigenfunctions with $n_x + n_y = 8$. Bottom row: real and imaginary parts of polar separation with twofold axial symmetry ($\mu = 2$) and three radial nodes ($v = 3$), created from the Cartesian separation via equation (2.52). The space bandwidth product $c = 4$. The grayscale for each function is individually scaled.

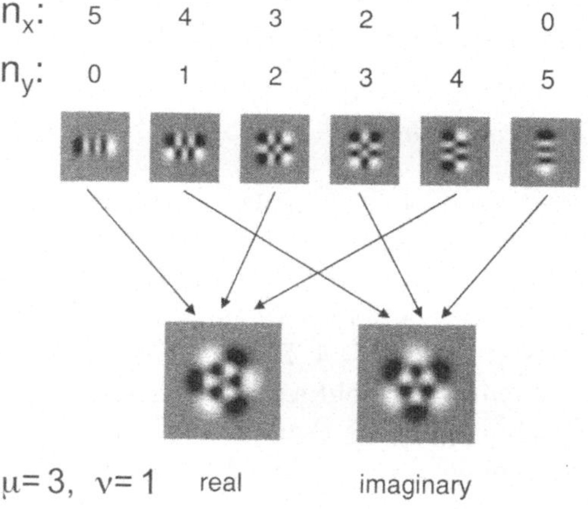

FIGURE 2.3. A second example of the reorganization of eigenfunctions in the Cartesian separation into eigenfunctions in the polar separation. Top row: the six eigenfunctions with $n_x + n_y = 5$. Bottom row: real and imaginary parts of polar separation with threefold axial symmetry ($\mu = 3$) and one radial node ($v = 1$). The space bandwidth product $c = 4$. The grayscale for each function is individually scaled.

their differing angular dependence. Two such functions that share a common value of μ but have different values of ν are orthogonal because their Cartesian decompositions are non-overlapping — since they have distinct eigenvalues (see Figure (2.1)).

Since the $e^{i\mu\theta} R^\mu P_{\mu,\nu}(R^2)$ are orthogonal on the plane with respect to $\exp\left(-\frac{1}{2}R^2\right)$, the polynomials $P_{\mu,\nu}(R^2)$ are orthogonal on the plane with respect to the weight $R^{2\mu}\exp\left(-\frac{1}{2}R^2\right)$. They thus can be considered to be generalized Hermite polynomials. Reduction by integration over circles shows that for fixed μ, these polynomials are also orthogonal with respect to a weight $R^{2\mu+1}\exp\left(-\frac{1}{2}R^2\right)$ on the half-line. With the transformation $\zeta = \frac{1}{2}R^2$, this weight becomes

$$R^{2\mu+1}\exp\left(-\tfrac{1}{2}R^2\right) dR = (2\zeta)^\mu \exp(-\zeta) d\zeta\,. \tag{2.53}$$

This demonstrates a relationship between the generalized Hermite polynomials, which have the weight on the left, and the familiar generalized Laguerre polynomials, which have the weight on the right. It extends the interrelationships expressed by equations 22.5.40 and 22.5.41 of Abramowitz and Stegun [1964].

We now find generating functions and normalization constants for these polynomials. We rewrite equation (2.41) as

$$Q\left(\rho\sigma\,,\frac{\rho}{\sigma}\right) = \exp\left(-2\rho^2\right) \exp\left[R\rho\left(e^{i\theta}\sigma + \frac{1}{e^{i\theta}\sigma}\right)\right] \tag{2.54}$$

and compare with the generating function for the ordinary Bessel functions $J_n(\chi)$ (equation 9.1.41 of Abramowitz and Stegun [1964]):

$$\sum_{n=-\infty}^{\infty} \tau^n J_n(\chi) = \exp\left[\tfrac{1}{2}\chi\left(\tau - \frac{1}{\tau}\right)\right]. \tag{2.55}$$

Taking $\tau = -ie^{i\theta}\sigma$ and $\chi = 2iR\rho$ leads to

$$Q\left(\rho\sigma\,,\tfrac{\rho}{\sigma}\right) = \exp\left(-2\rho^2\right) \sum_{n=\infty}^{\infty} \left(-ie^{i\theta}\sigma\right)^n J_n(2iR\rho)\,. \tag{2.56}$$

On the other hand, substitution of equation (2.48) into equation (2.47) gives a second expression for $Q\left(\rho\sigma, \frac{\rho}{\sigma}\right)$. Equating coefficients of $e^{i\mu\theta}$ now yields (for non-negative μ)

$$\sum_{\nu=0}^{\infty} \frac{\rho^{2\nu}}{(\nu+\mu)!\nu!} P_{\mu,\nu}(R^2) = \exp\left(-2\rho^2\right) \frac{J_\mu(2iR\rho)}{(iR\rho)^\mu}, \tag{2.57}$$

which is a generating function over ν for each series of polynomials $P_{\mu,\nu}$ with fixed ν.

To determine the normalization of the polynomials $P_{\mu,\nu}$, consider

$$\frac{1}{2\pi}\iint\sum_{0}^{\infty}\frac{z^k t^l q_{k,l}(x,y)}{k!\,l!}\frac{z'^{k'} t'^{l'} \overline{q}_{k',l'}(x,y)}{k'!\,l'!}\exp\left[-\tfrac{1}{2}(x^2+y^2)\right]dx\,dy$$
$$=\frac{1}{2\pi}\iint Q(z,t)\overline{Q}(z',t')\exp\left[-\tfrac{1}{2}(x^2+y^2)\right]dx\,dy \tag{2.58}$$

in which the sum on the left hand side is over all values of k, l, k', and l'. Via substitution of the expression (2.40) for the generating function Q and straightforward algebra, this is seen to be

$$\frac{1}{2\pi}\iint\sum_{0}^{\infty}\frac{z^k t^l q_{k,j}(x,y)}{k!\,l!}\frac{z'^{k'} t'^{l'} \overline{q}_{k',l'}(x,y)}{k'!\,l'!}\exp\left[-\tfrac{1}{2}(x^2+y^2)\right]dx\,dy$$
$$=\exp\left(2zz'+2tt'\right). \tag{2.59}$$

Consequently,

$$\frac{1}{2\pi}\iint\left|q_{k,l}(x,y)\right|^2\exp\left[-\tfrac{1}{2}(x^2+y^2)\right]dx\,dy=2^{k+l}k!\,l!\,, \tag{2.60}$$

and, as expected, cross-terms ($k\neq k'$ or $l\neq l'$) are zero. In polar form, recognizing (from equations (2.35) and (2.48)) that

$$q_{\mu+\nu,\nu}(x,y)=\exp\left(i\mu\theta\right)R^{\mu}P_{\mu,\nu}(R^2), \tag{2.61}$$

we find

$$\int_0^{\infty}\left|P_{\mu,\nu}(R^2)\right|^2R^{2\mu+1}\exp\left(-\tfrac{1}{2}R^2\right)dR=2^{\mu+2\nu}(\mu+\nu)!\,\nu!\,, \tag{2.62}$$

the normalization of the polynomials $P_{\mu,\nu}$.

2.6 The Non-Gaussian Case: One Dimension

We now return to the general band- and space-limiting operator B_aD_a, in which the profiles of the limiters are determined by the shape parameter a, as in equations (2.1) and (2.2). The eigenfunctions of the one-dimensional operator $B_\infty D_\infty$, the prolate spheroidal functions (Slepian functions, from Slepian and Pollack [1961]), closely resemble (Flammer [1957]; Xu, Haykin, and Racine [1999]) the eigenfunctions of the one-dimensional Gaussian operators B_2D_2, equation (2.27). A similar observation holds for the eigenfunctions of the corresponding two-dimensional operators (Slepian [1964]). This is rather remarkable, since the operator $B_\infty D_\infty$ limits abruptly in space and frequency, while the operator B_2D_2 applies smooth cutoffs. In this and the next sections, we provide a rationale for these similarities by drawing on the theory of Sirovich and Knight (Knight and Sirovich [1982,

1986]; Sirovich and Knight [1981, 1982, 1985]) of slowly varying linear operators. Our analysis applies not only to $B_\infty D_\infty$ but also to the operators $B_a D_a$ for intermediate exponents $a > 2$. This result can be viewed as an extension of the known asymptotic relationship between the Hermite functions and the prolate spheroidal functions (Flammer [1957]). We consider the one-dimensional case in some detail and then sketch how the arguments extend to two dimensions.

The theory of Sirovich and Knight yields asymptotic eigenvalues and eigenfunctions for integral kernels that have a slow dependence in a specific technical sense. It also delivers exact results for a broad class of kernels to which a quite general family of kernels are generically asymptotic over the principal part of their eigenspaces. It includes the WKB method for second-order differential equations as a special case. Reference Knight and Sirovich [1982] presents its application to problems related to the one here, and we now summarize the relevant part of that reference.

An integral kernel $K\{x, x'\}$ may be re-parameterized in terms of difference and mean variables

$$\nu = x - x' \quad \text{and} \quad q = \tfrac{1}{2}(x + x')\,. \tag{2.63}$$

In terms of these variables, the kernel is

$$K(\nu, q) = K\{q + \tfrac{1}{2}\nu\,, q - \tfrac{1}{2}\nu\}. \tag{2.64}$$

In the special case of a difference kernel $K\{x - x'\}$, the dependence on q in equation (2.64) is absent. The Wigner transform of $K\{x, x'\}$ is defined as

$$\begin{aligned} W(K\{x, x'\}) = \widetilde{K}(p, q) &= \int_{-\infty}^{\infty} e^{-ip\nu} K(\nu, q)\, d\nu \\ &= \int_{-\infty}^{\infty} e^{-ip\nu} K\left\{q + \tfrac{1}{2}\nu, q - \tfrac{1}{2}\nu\right\} d\nu\,. \end{aligned} \tag{2.65}$$

In the special case of a difference kernel $K\{x - x'\}$, equation (2.65) is simply a Fourier transform and the kernel's eigenfunctions e^{ipx} have eigenvalues $\widetilde{K}(p)$.

Note that if $K\{x, x'\}$ is symmetric in its arguments, then $\widetilde{K}(p, q)$ is real, and the implicit relation

$$\widetilde{K}(p, q) = \lambda\,, \tag{2.66}$$

for real λ, will yield a set of contour lines on the (p, q) plane. If these contour lines are *closed*, then we may pick a subset of them with specific enclosed areas:

$$\widetilde{K}(p, q) = \lambda_n\,, \quad \text{where } (p, q) \text{ encloses area } A(\lambda_n) = (2n + 1)\pi\,. \tag{2.67}$$

If $\lambda_{n+1} - \lambda_n$ is a stable small fraction of λ_n over a span of consecutive n's, then equation (2.67) gives a good estimate of the nth eigenvalue. (By experience, taking liberties with the criterion shows the estimate is robust.) Estimated eigenfunctions also emerge from the analysis. If we solve equation (2.67) for $p_n(q)$, then we find that locally at x, the nth eigenfunction vibrates with changing x at a frequency $2\pi p_n(x)$.

In one type of circumstance, the eigenvalues of equation (2.67) are *exact*: if the contours of equation (2.66) are concentric similar ellipses and if also, in (λ, p, q) 3-space the surface (2.66) is a paraboloid. In this case, the spacing between the consecutive eigenvalues is constant. In addition, in this case the exact eigenfunctions also may be specified in terms of Hermite functions.

This exact result leads to useful asymptotics in general situations that occur commonly. If a kernel transform $\widetilde{K}(p,q)$ is expanded about an extremum (p_0, q_0), then near that extremum the paraboloidal form is generic through second-order terms in $(p - p_0, q - q_0)$. In the examples below, $\widetilde{K}(p,q)$ is symmetric under reflection of either axis. Thus it is extremized at the origin, where Taylor expansion gives

$$\widetilde{K}(p,q) \approx \widetilde{K}(0,0) + \tfrac{1}{2}\widetilde{K}_{pp}(0,0)p^2 + \tfrac{1}{2}\widetilde{K}_{qq}(0,0)q^2 . \tag{2.68}$$

For the right-hand expression, equation (2.67) is exact and yields exact eigenfunctions. Over the set of areas $A(\lambda_n)$ in equation (2.67) for which $\widetilde{K}(p,q)$ is well approximated by equation (2.68), the exact eigenvalues of equation (2.68) will be good estimates of those sought, and similarly for the exact eigenfunctions of equation (2.68).

These observations imply that a smooth area-preserving transformation on the (p,q) plane must map $\widetilde{K}(p,q)$ to the Wigner transform of another kernel with the same asymptotic eigenvalues given by equation (2.67). As shown in Knight and Sirovich [1982], a more restrictive *unimodular affine* transformation on (p,q), which preserves straight lines as well as areas, yields a new kernel with *exactly* the same eigenvalues. The new kernel K' is related to the original kernel by a similarity transformation T:

$$K' = TKT^{-1} . \tag{2.69}$$

Here, T maps the orthonormal eigenfunctions of the original kernel to those of the new, and hence preserves inner products.

A unimodular affine transformation carries an ellipse to another ellipse with equal area. Given an ellipse, a unimodular affine transformation may be constructed which carries it to a circle centered at the origin. Construction of the corresponding similarity transformation T (equation (2.69)) is also straightforward and quite simple in the axis-symmetric case (equation (2.68)). The availability of the inverse transformation, from a centered circle to an arbitrary ellipse of equal area, reduces the eigenvalue problem for a kernel whose Wigner transform yields contour lines which are concentric similar ellipses, to the eigenvalue problem for a kernel whose Wigner transform contours are origin-centered concentric circles.

Thus (still following Knight and Sirovich [1982]), we examine the eigenvalue problem for a kernel whose Wigner transform is of the form

$$\widetilde{K}(p,q) = \widetilde{K}(p^2+q^2) = \widetilde{K}(J)\,. \tag{2.70}$$

As noted above, the paraboloidal special case

$$\widetilde{K}(p,q) = a + b(p^2+q^2) = a + bJ \tag{2.71}$$

satisfies the area rule (equation (2.67)) exactly, with

$$\lambda_n = a + b(2n+1)\,. \tag{2.72}$$

For other cases of equation (2.70), the area rule is *not exact*, and we may furnish a next-order error term

$$\lambda_n = \widetilde{K}(2n+1) + \tfrac{1}{2}\widetilde{K}_{JJ}(2n+1)\,. \tag{2.73}$$

Nonetheless for kernels with Wigner transforms of the form (2.70), the eigenvalue problem may still be solved *exactly*. This may be shown (Knight and Sirovich [1982]) by relating two classical generating function formulas. The orthonormal eigenfunctions that emerge from equation (2.70) are those encountered in the limiting case ($c \to \infty$) of equation (2.27), namely the normalized Hermite functions

$$u_n(x) = \frac{1}{\sqrt[4]{\pi}}\frac{1}{\sqrt{n!}}h_n(\sqrt{2}x)\exp\left(-\tfrac{1}{2}x^2\right). \tag{2.74}$$

In terms of these, the classical Mehler's formula (equation 22, section 10.3 in Erdelyi [1955]) is

$$\begin{aligned} G\{x,x'\} &= \frac{1}{\sqrt{\pi(1-z^2)}}\exp\left\{-\frac{\frac{1}{2}(z^2+1)(x^2+x'^2)-2zxx'}{1-z^2}\right\} \\ &= \frac{1}{\sqrt{\pi(1-z^2)}}\exp\left\{-\tfrac{1}{4}\Big(\frac{1+z}{1-z}(x-x')^2+\frac{1-z}{1+z}(x+x')^2\Big)\right\} \\ &= \sum_{n=0}^{\infty} z^n u_n(x)u_n(x')\,. \end{aligned} \tag{2.75}$$

The second line has been arranged in a form easier to compare to what's above and particularly with equation (2.63). Clearly $G\{x,x'\}$ is an integral kernel whose nth eigenfunction is $u_n(x)$ and nth eigenvalue is z^n, whence

$$Gu_n = z^n u_n\,. \tag{2.76}$$

Each term $u_n(x)u_n(x')$ on the right is a 1-dimensional projection kernel. Wigner transformation of Mehler's formula (2.75) (a straightforward "complete the squares" integral) yields

$$\widetilde{G}(p^2+q^2,z) = \frac{2}{1+z}\exp\left\{-\frac{1-z}{1+z}(p^2+q^2)\right\} = \sum_{n=0}^{\infty} z^n W\big(u_n(x)\,u_n(x')\big)\,. \tag{2.77}$$

We can expand the left-hand expression in powers of z, which shows that each projection kernel has a Wigner transform which is constant on concentric circular contour lines. The "concentric circular contours" property clearly is inherited by any weighted sum of functions of (p, q) which individually have that property. Thus a kernel of the form

$$K\{x,x'\} = \sum_{n=0}^{\infty} \lambda_n u_n(x)\,u_n(x') \tag{2.78}$$

will have a circular-contour Wigner transform as in equation (2.70). Is the converse true? Can any kernel which satisfies equation (2.70) be expressed in the form (2.78) (which solves the eigenvalue problem)? Compare equation (2.77) with the generating function for the orthonormal Laguerre functions $\mathrm{L}_n(x)$ (derived from the generating function for the standard Laguerre polynomials L_n from equation 22.9.15 of Abramowitz and Stegun [1964] with $\mathrm{L}_n(x) = (-1)^n e^{-\frac{1}{2}x} L_n(x)$):

$$\frac{1}{1+z}\exp\left\{-\tfrac{1}{2}\left(\frac{1-z}{1+z}\right)J\right\} = \sum_{n=0}^{\infty} z^n \mathrm{L}_n(J)\,. \tag{2.79}$$

We see that

$$W\big(u_n(x)\,u_n(x')\big) = 2\mathrm{L}_n(2J) \tag{2.80}$$

and these functions are a complete orthonormal set. Thus a projection integral applied to equation (2.70) evaluates the eigenvalue:

$$\lambda_n = \int_0^{\infty} \widetilde{K}(J)\cdot 2\mathrm{L}_n(2J)\,dJ\,. \tag{2.81}$$

Our converse holds because of the completeness of the Laguerre functions.

The Wigner transform of equation (2.78) is

$$\widetilde{K}(p^2+q^2) = \sum_{n=0}^{\infty} \lambda_n \cdot 2\mathrm{L}_n\big(2(p^2+q^2)\big)\,. \tag{2.82}$$

Each of the Laguerre functions here has a peaking form which gives a dominant contribution to the sum when $p^2 + q^2$ is near n, in qualitative agreement with the area rule (2.67). Below we will encounter kernels which are associated with the non-Gaussian space- and band-limited kernels and which share their eigenfunctions. The Wigner transforms of these kernels show a central regime of near-circular contours with rapid radial variation, and this feature will confirm their asymptotic agreement with the form of equation (2.70).

We will first apply the methodology of Sirovich and Knight to the space- and band-limited kernels themselves, which yields some insight but inconclusive results. We will then work with the more definitive associated kernels.

A first step in applying this theory is to focus on the self-adjoint operator $D^{\frac{1}{2}}BD^{\frac{1}{2}}$. We can write

$$D^{\frac{1}{2}}BD^{\frac{1}{2}}f(x) = \int K\{x,x'\}f(x')\,dx', \tag{2.83}$$

where

$$K\{x,x'\} = [D(x)]^{\frac{1}{2}}[D(x')]^{\frac{1}{2}}B(x-x'). \tag{2.84}$$

Here, $D(x)$ is the spatial profile which corresponds to the space-limiting operator D

$$D(x) = e^{-(|x|/d)^a}, \tag{2.85}$$

the one-dimensional analog of equation (2.1), and $B(x)$ is the Fourier transform of the analogous frequency-limiting profile of B, namely

$$B(x) = \frac{1}{\sqrt{2\pi}}\int_{-\infty}^{\infty}\exp\left[-\left(\frac{|\omega|}{b}\right)^a\right]d\omega, \tag{2.86}$$

with the convention following equation (2.4) for $a=\infty$. We make the substitutions $\nu = x - x'$ and $q = \frac{1}{2}(x+x')$, with the intent of considering K as varying slowly with q or rapidly with ν. This corresponds to the limit that the space-bandwidth product $c = bd$ is large. We next calculate the Wigner transform of K:

$$\begin{aligned}\widetilde{K}(p,q) &= \int_{-\infty}^{\infty} e^{-i\nu p}K(\nu,q)\,du \\ &= \frac{1}{\sqrt{2\pi}}\int_{-\infty}^{\infty}\left[D\left(\tfrac{1}{2}\nu+q\right)\right]^{\frac{1}{2}}\left[D\left(\tfrac{1}{2}\nu-q\right)\right]^{\frac{1}{2}}B(\nu)e^{-i\nu p}\,d\nu.\end{aligned} \tag{2.87}$$

For $a=2$, the Wigner transform is exactly a Gaussian,

$$\widetilde{K}(p,q) = \frac{bd}{\sqrt{1+b^2d^2}}\exp\left(-\frac{q^2}{d^2}-\frac{p^2d^2}{1+b^2d^2}\right). \tag{2.88}$$

(Here we have used $D = D_{2,d} \equiv D_{gau,d/\sqrt{2}}$ and similarly for B; the derivations from equations (2.13) to (2.27) used $D = D_{gau,d} \equiv D_{2,d\sqrt{2}}$.) The contour lines are concentric, similar ellipses around the origin. To follow the discussion above, under the unimodular transformation

$$q = \frac{d}{\sqrt[4]{1+(bd)^2}}\,\hat{q}, \quad p = \frac{\sqrt[4]{1+(bd)^2}}{d}\,\hat{p}, \tag{2.89}$$

equation (2.88) becomes

$$\widetilde{K}(\hat{p}^2+\hat{q}^2)=\frac{bd}{\sqrt{1+(bd)^2}}\exp\left(-\frac{1}{\sqrt{1+(bd)^2}}(\hat{p}^2+\hat{q}^2)\right). \tag{2.90}$$

This is just the general form of $\widetilde{G}$ in (2.77) above. In that expression, if we let

$$z=\frac{\sqrt{1+(bd)^2}-1}{\sqrt{1+(bd)^2}+1}, \tag{2.91}$$

we see that

$$\widetilde{K}(\hat{p}^2+\hat{q}^2)=\frac{bd}{1+\sqrt{1+(bd)^2}}\widetilde{G}\left(\hat{p}^2+\hat{q}^2,\frac{\sqrt{1+(bd)^2}-1}{\sqrt{1+(bd)^2}+1}\right). \tag{2.92}$$

Consequently, by equation (2.75), the eigenvalues are

$$\lambda_n=\frac{bd}{1+\sqrt{1+(bd)^2}}\left(\frac{\sqrt{1+(bd)^2}-1}{\sqrt{1+(bd)^2}+1}\right)^n. \tag{2.93}$$

The exact eigenfunctions likewise may be found from equation (2.75) and from the inverse transformation of the first member of equation (2.89).

If the space-bandwidth product bd is chosen to be large, we see from equation (2.93) that for early n, a succession of eigenvalues will lie near unity. Equation (2.90) similarly shows that for large bd, the Wigner transform will be near unity for an extended neighborhood around the origin, which extends to $\hat{p}^2+\hat{q}^2=bd$.

Figure 2.4 shows relief maps of the Wigner transform $\widetilde{K}$ for a range of shape parameters ($a=1,2,4,\infty$) and values of the space-bandwidth product $c=bd$ ($c=\frac{1}{4}\pi,\pi,4\pi$). The top row of each part of Figure 2.5 shows a top-down view. We see that the asymptotic result found analytically above for $a=2$ of an extended region at an altitude near unity is already manifest at the modest value of $c=\pi$ and is more pronounced for $a>2$. In fact, for the Slepian case of $a=\infty$, the known eigenvalue spectrum has early values near unity and a sudden plunge to near zero at a critical n which depends on the space-bandwidth parameter c. By applying the area rule to the area of the plateau near unity in this case, we can get a good estimate of the critical n where the plunge occurs. However, the very flatness of the plateau reflects the non-generic feature of numerous almost-degenerate eigenvalues, and this confounds attempts to deduce the features of the eigenfunctions from the features of the contour lines. In the $a=\infty$ case this problem is particularly severe: the Wigner transform essentially involves the band-limited Fourier inversion integral of a function which suddenly jumps to zero. The consequent inevitable Gibbs phenomenon, which simply reflects the location of this jump, is the most prominent altitude feature on the otherwise almost flat plateau.

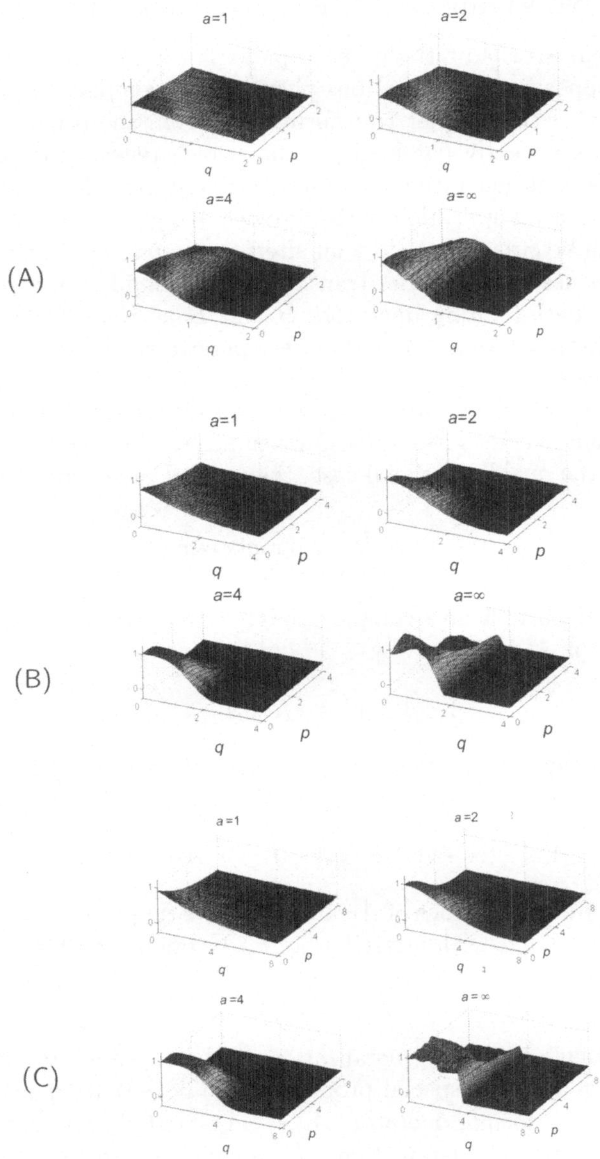

FIGURE 2.4. Wigner transforms $\widetilde{K}(p,q)$ of the operator $D^{\frac{1}{2}}BD^{\frac{1}{2}}$ for four values of the shape parameter a. The scale parameters b and d are given by $b = d = \sqrt{c}$, where $c = \frac{1}{4}\pi$ (panel A), $c = \pi$ (panel B), and $c = 4\pi$ (panel C). Only one quadrant is shown, since the transforms have even symmetry in both arguments. In each plot, the color scale runs from blue (at minimum amplitude) to deep red (at maximum amplitude).

2.7 The Non-Gaussian Case: Another Viewpoint

The above analysis is admittedly incomplete. For large values of the exponent a, the appearance of the Gibbs phenomenon complicates the asymptotic behavior of the Wigner transform away from the origin. Moreover, the presence of a nearly flat high plateau, which reflects the presence of numerous nearly degenerate eigenvalues (a non-generic feature), disrupts the straightforward application of the Sirovich–Knight methodology. These difficulties can be circumvented by an alternative approach. This approach makes explicit use of the Fourier transform relationship between the band-limiting and space-limiting operators B and D as well as the fact that taking the functional square root of either operator is equivalent to making a change in scale.

An examination of the band- and space-limiting kernel in terms of its underlying components gives some further insight into its mathematical structure. In the one-dimensional case, the natural inner product is

$$(f, g) = \int_{-\infty}^{\infty} f^*(x) g(x)\, dx\,, \tag{2.94}$$

where the asterisk indicates complex conjugation. For an operator A, its adjoint operator $A^\dagger$ is defined by

$$(A^\dagger f, g) = (f, Ag)\,. \tag{2.95}$$

The adjoint of the adjoint operator is the original operator. An operator A may be resolved as

$$A = \tfrac{1}{2}(A + A^\dagger) + \tfrac{1}{2}(A - A^\dagger) = A_R + i\, A_I\,. \tag{2.96}$$

Substitution of the definitions of A_R and A_I above (in place of A) into equation (2.95) shows that both are self-adjoint. An operator which commutes with its adjoint

$$AA^\dagger = A^\dagger A \tag{2.97}$$

is called "normal"; this is a generalization of "self-adjoint" and leads similarly to several valuable special properties. We note that equation (2.96) implies that for a normal operator, A, $A^\dagger$, A_R, and A_I all commute and have a common set of eigenfunctions. As A_R and A_I are self-adjoint, the eigenfunctions may be chosen to be orthonormal. If an eigenfunction assigns the eigenvalue $\lambda^{(R)}$ to A_R and $\lambda^{(I)}$ to A_I, then evidently the action of A upon that eigenfunction yields the eigenvalue

$$\lambda = \lambda^{(R)} + i\lambda^{(I)}\,. \tag{2.98}$$

An operator U which respects inner products

$$(Uf, Ug) = (f, g)\,. \tag{2.99}$$

is "unitary." If U and V are both unitary, clearly the concatenation UV inherits this property. We may regard the left expression of equation (2.99) as a particular case of the right expression of equation (2.95), whence

$$(U^\dagger U f, g) = (f, g)\,, \tag{2.100}$$

so

$$U^\dagger = U^{-1} \tag{2.101}$$

for a unitary operator. Since U^{-1} and U commute, a unitary operator is normal and has orthonormal eigenfunctions. In equation (2.99) we may choose both f and g to be eigenfunctions of U, and observe that the eigenvalues of U lie on the unit circle. (The above material is reviewed in many places, for example Halmos [1942]).

Now let us consider three particular operators:

(i) The parity operator P defined by

$$Pf(x) = f(-x)\,. \tag{2.102}$$

Clearly, $P^2 = 1$, whence

$$P^{-1} = P\,. \tag{2.103}$$

If (Pf, Pg) is made explicit by use of equations (2.102) and (2.94), substitution of the integration variable

$$x = -x' \tag{2.104}$$

confirms that P is unitary.

(ii) The scaling operator S_γ defined by

$$S_\gamma f(x) = \frac{1}{\sqrt{\gamma}}\, f\left(\frac{x}{\gamma}\right)\,. \tag{2.105}$$

We note

$$S_\gamma^{-1} = S_{1/\gamma}\,. \tag{2.106}$$

If $(S_\gamma f, S_\gamma g)$ is made explicit by use of equations (2.105) and (2.94), substitution of the integration variable

$$x = \gamma x' \tag{2.107}$$

shows that S_γ is unitary.

(iii) The Fourier transform operator F defined by

$$(Ff)(x) = \frac{1}{\sqrt{2\pi}} \int e^{-ixx'}\, f(x')\, dx'\,. \tag{2.108}$$

Evidently the action of the operator P on equation (2.108) exchanges $(-i)$ for $(+i)$ and yields the inverse Fourier operator. Thus,

$$PF = F^{-1} \quad \text{whence} \quad PF^2 = 1 \quad \text{and } F^2 = P\,. \tag{2.109}$$

Furthermore,

$$F^4 = 1\,, \tag{2.110}$$

so F has four eigenvalues which are the fourth-roots of unity: $i, -1, -i, 1$.

By equation (2.108), the familiar bilinear Parseval relation between a pair of functions and their Fourier transforms may be written

$$(Ff, Fg) = (f, g)\,, \tag{2.111}$$

which is an example of equation (2.99), so that F is unitary.

Direct calculation verifies that the three operators defined above have simple commutation relationships:

$$FP = PF\,, \quad PS_\gamma = S_\gamma P\,, \quad S_\gamma F = FS_{1/\gamma}\,. \tag{2.112}$$

The first two pairs simply commute, and so have a common set of eigenfunctions. The combination $F_\gamma \equiv S_\gamma F$ defined by

$$(F_\gamma f)(x) = \frac{1}{\sqrt{2\pi\gamma}} \int e^{-ixx'/\gamma} f(x')dx'\,, \tag{2.113}$$

the "scaled Fourier transform operator", will be used below to demonstrate that a set of results holds with even broader generality than is immediately apparent.

Two further properties of the Fourier operator F will prove important below. It is fairly well known (see for example Vilenkin [1968] p. 565, sec. 4, equation (1)) that the orthonormal Hermite functions $u_n(x)$ which we introduced in equation (2.74) are eigenfunctions of the Fourier operator, which order the eigenvalues we found at equation (2.110) by

$$Fu_n = (-i)^n u_n\,. \tag{2.114}$$

This is a special case of equation (2.76) for

$$z = i\,. \tag{2.115}$$

This substitution in equation (2.75) reduces G to the definition of F given in equation (2.108). The second important property of F is the expression for its Wigner transform. Direct evaluation is straightforward, or, substitution of equation (2.115) into $\widetilde{G}$ (equation 2.77) gives

$$\widetilde{F}(p^2 + q^2) = (1+i)e^{-i(p^2+q^2)}\,. \tag{2.116}$$

Thus, $\widetilde{F}$ has a constant amplitude on the (p, q) plane, and a phase which is constant on circles and accelerates as a quadratic with increasing radius.

We next use the machinery developed above to further elucidate the structure of the operator $K\{x, x'\}$ in equation (2.83). The kernel given in equation (2.83) corresponds to the sequence of operators

$$K = D^{\frac{1}{2}} B D^{\frac{1}{2}}\,, \tag{2.117}$$

where D corresponds to simple function multiplication and B to convolution. Until further notice let us specialize by choosing the same limitation function for both space and frequency:

$$b = d = \sqrt{c} \tag{2.118}$$

for equations (2.83) and (2.84). Then, in our present notation,

$$B = F^{-1}DF \tag{2.119}$$

and so

$$K = D^{\frac{1}{2}}F^{-1}DFD^{\frac{1}{2}}\,. \tag{2.120}$$

It is convenient to adopt the notation

$$\hat{D}(x) = \big[D(x)\big]^{\frac{1}{2}} \tag{2.121}$$

and to define the convolution operator

$$\hat{B} = F^{-1}\hat{D}F\,. \tag{2.122}$$

Then B in equation (2.119) may be expressed as the iterated convolution

$$\begin{aligned} B &= \big(\hat{B}\big)^2 = \big(F^{-1}\hat{D}F\big)\big(F^{-1}\hat{D}F\big) \\ &= F^{-1}\hat{D}^2F = F^{-1}DF\,. \end{aligned} \tag{2.123}$$

Thus, equation (2.117) becomes

$$K = \hat{D}\hat{B}^2\hat{D} = \hat{D}F^{-1}\hat{D}^2F\hat{D}\,. \tag{2.124}$$

Now let us note that $\hat{D}(x)$ is an even function of x, so its multiplicative action commutes with the action of the parity operator P, defined in equation (2.102):

$$\hat{D}P = P\hat{D}\,. \tag{2.125}$$

If we insert $F^{-1} = PF$ (equation (2.109)) in equation (2.124) we find

$$\begin{aligned} K &= \hat{D}PF\hat{D}^2F\hat{D} = P\hat{D}F\hat{D}\hat{D}F\hat{D} \\ &= P\big(\hat{D}F\hat{D}\big)^2. \end{aligned} \tag{2.126}$$

Because P commutes with both $\hat{D}$ and F, from equation (2.126) we observe that the eigenfunctions of K are the same as those of its associated operator

$$Z = \hat{D}F\hat{D}. \tag{2.127}$$

As $\hat{D}$ is self-adjoint, and as F is unitary, we may now show that Z is a normal operator:

$$Z^\dagger Z = \big(\hat{D}F^{-1}\hat{D}\big)\big(\hat{D}F\hat{D}\big) = \big(\hat{D}PF\hat{D}\big)\big(\hat{D}F\hat{D}\big)$$

$$= (\hat{D}F\hat{D})(\hat{D}PF\hat{D}) = ZZ^{\dagger} \tag{2.128}$$

and so we expect that the operator $Z = \hat{D}F\hat{D}$ will endow the eigenfunctions of K with *complex* eigenvalues. The possible separation of $\hat{D}F\hat{D}$ into a combination of two commuting self-adjoint operators (equations (2.96), (2.97)) corresponds to the possible separation of F into Fourier cosine and sine transforms,

$$F = F_R + iF_I\,. \tag{2.129}$$

The Fourier cosine component F_R is a self-adjoint operator and hence has a Wigner transform that is real. It matches F on the subspace spanned by the even-order eigenvectors and annihilates the subspace spanned by the odd-order eigenvectors. Similarly, the Fourier sine component F_I is a self-adjoint operator and has a Wigner transform that is real, and iF_I matches F on the subspace spanned by the odd-order eigenvectors.

Corresponding statements hold for the integral kernel Z. Represented as an integral kernel, $Z = \hat{D}F\hat{D}$ takes the form

$$Z\{x,x'\} = \frac{1}{\sqrt{2\pi}}\,\hat{D}(x)\,e^{-ixx'}\,\hat{D}(x')\,. \tag{2.130}$$

Noting the steps in equation (2.96), we may write

$$\begin{aligned} Z\{x,x'\} &= \frac{1}{2\pi}\left(\hat{D}(x)\cos(xx')\hat{D}(x') - i\hat{D}(x)\sin(xx')\hat{D}(x')\right) \\ &= Z_R\{x,x'\} + iZ_I\{x,x'\} \end{aligned} \tag{2.131}$$

where both Z_R and Z_I are manifestly symmetric kernels (and, as noted above, will thus have Wigner transforms which are real).

Much structural information about the operator K can be extracted from equation (2.130). From equations (2.121) and (2.85), we have that $\hat{D}(x)$ is of the form

$$\hat{D}(x) = e^{-\frac{1}{2}\left(\frac{|x|}{d}\right)^{a}} = e^{-\Gamma(x)}, \tag{2.132}$$

where $\Gamma(x)$ is even, zero at $x = 0$, and monotone upward to infinity. We note that as x increases, $\hat{D}(x)$ is near unity until the value of $\frac{|x|}{d}$ achieves a fair fraction of unity. If d is large, there will be a fair range of values x, x' over which $Z\{x,x'\}$ is reasonably close to $F\{x,x'\}$. As the first several eigenfunctions of F (2.74) are quite well confined to the neighborhood of the origin by their quadratic exponential factor, there is room to suspect that they might well-approximate the near-the-origin eigenfunctions of Z (which are those of K as well). We note that this was indeed the case for the "soft" Gaussian-based limiter kernel whose eigenfunctions and eigenvalues were derived exactly above. In the "hard" limit of Slepian this is likewise true: the Slepian eigenfunctions satisfy the second-order "prolate spheroidal" ordinary differential equation, which for large bandwidth becomes asymptotic to the "parabolic cylinder" ordinary differential equation whose eigenfunctions are the Hermite functions. This is elaborated by Flammer [1957] (and has been exploited by Xu, Haykin, and Racine [1999] for the reduction of electroencephalographic data). We now pursue the conjecture for those frequency- and space-limiting kernels that lie between the "soft" and "hard" limits.

We have seen that the eigenvalue spectrum of the kernel K must lie between 1 and 0. We further noted that K commutes with the parity operator P (equations (2.112),(2.126)). Thus the eigenfunctions of K have even or odd symmetry, and assign to P the eigenvalues ± 1 respectively. In both the Slepian and Gaussian cases these eigenfunctions, unsurprisingly, alternate in parity with descending eigenvalues of K. When we factor P from K (equation (2.126)) the remaining operator Z^2 thus must have real eigenvalues which are positive or negative according to the eigenfunction's parity. Consequently the eigenvalue equation

$$Z\varphi_n = \zeta_n \varphi_n \tag{2.133}$$

must have eigenvalues which are positive or negative real for even parity, and are pure imaginary for odd parity. If we consider a sequence of space- and bandwidth-limiting operators K for which the bandwidth goes to infinity, we have

$$\hat{D} \to 1 \quad \text{and} \quad Z \to F\,. \tag{2.134}$$

This establishes that in the same limit the eigenvalues of Z go to $\{\pm 1, \pm i\}$ though it does not yet establish the choice of ± 1 on the even eigenfunctions nor $\pm i$ on the odd ones; and does not establish the eigenfunctions (2.74), because linear combinations with a common eigenvalue have not been ruled out. However, the exact solution for K in the "soft" (Gaussian) case does yield the eigenfunctions of equation (2.74) in the limit and consequently the eigenvalue sequence of equation (2.114).

The Wigner transform of the kernel Z is

$$\begin{aligned} \widetilde{Z}(p,q) &= \int_{-\infty}^{\infty} \hat{D}\big(q+\tfrac{1}{2}\nu\big)\,\hat{D}\big(q-\tfrac{1}{2}\nu\big)\,\frac{e^{-i\left(q^2-\frac{1}{4}\nu^2+p\nu\right)}}{\sqrt{2\pi}}\,d\nu \\ &= e^{-i(p^2+q^2)}\int_{-\infty}^{\infty} \hat{D}\big(q+\tfrac{1}{2}\nu\big)\hat{D}\big(q-\tfrac{1}{2}\nu\big)\frac{e^{i\left(\frac{1}{2}\nu-p\right)^2}}{\sqrt{2\pi}}\,d\nu \\ &= e^{-i(p^2+q^2)}\int_{-\infty}^{\infty} \hat{D}\big(q+p+\tfrac{1}{2}\nu'\big)\hat{D}\big(q-p-\tfrac{1}{2}\nu'\big)\frac{e^{i\left(\frac{1}{2}\nu'\right)^2}}{\sqrt{2\pi}}\,d\nu'\,, \end{aligned} \tag{2.135}$$

where on the second line the square has been completed in the exponent, and on the third line $\nu = \nu' + 2p$ has shifted the integration origin; these are the two critical steps in the closed evaluation of $\widetilde{F}$ from F. In the last form of equation (2.135), the oscillations of the exponential term in the integrand accelerate as ν' advances, and eventually their signed contributions to the integral cancel strongly. If $\hat{D}(x)$ is slow enough (from a large enough choice of the cutoff d), then for fixed (p,q), this will happen before either $\hat{D}$ factor in equation (2.135) departs enough from unity to lend a substantial (p,q)-dependence to the integral. More specifically, for large enough d, equation (2.132) may be approximated by

$$\hat{D}(x) \approx 1 - \tfrac{1}{2}\left(\frac{|x|}{d}\right)^a, \tag{2.136}$$

whose substitution in equation (2.135) yields

$$\widetilde{Z}(p,q) \approx \widetilde{F}(p^2+q^2) - \frac{1}{d^a\sqrt{2\pi}}\int_{-\infty}^{\infty} e^{is^2}\left\{|s+p+q|^a + |s+p-q|^a\right\} ds \tag{2.137}$$

(where we have let $\nu' = 2s$). As the integral on the right is a function of only p, q, and a, once any choice of these has been made, a large enough choice of the cutoff d will make $\widetilde{Z}(p,q) \to \widetilde{F}(p^2+q^2)$. For what follows, we recall that $d = \sqrt{c}$ relates our cutoff to the space-bandwidth product.

The second and third rows of Figure 2.5 examine the limits of validity of this asymptotic approximation. Again for the modest value of $c = \pi$, and for $a \geq 2$, an amplitude plateau appears surrounding the origin and has an altitude near $\sqrt{2}$, consistent with equation (2.116). The circularity of phase contours near the origin, for $a = 4$, persists from orange through dark blue, corresponding to a half cycle. By equation (2.114), this half cycle signals dominant contributions from the first two eigenfunctions. These confirming features remain stable as we proceed to $c = 4\pi$. (Note that it is the radial scale that has been changed in Figure 2.5 to accommodate the appearance of further significant features.) However, for $a = 4$ (compare the amplitude picture with the previous) the area of the plateau has increased by a factor of 4, indicating that the number of eigenfunctions which make major contributions to this region's concordance with equation (2.116) has increased from 2 to 8. The corresponding phase picture shows that in this region both radial frequency and phase agree with equation (2.116) as well. The $a = \infty$ (Slepian) case also now shows a region of clear concurrence. Though a little shaky (when compared to $a = 4$), it extends through at least $2\frac{1}{4}$ cycles, or through the first 9 eigenfunctions. As mentioned above, as $c = bd$ increases, the early Slepian functions are *known* to approach the Hermite functions (Flammer [1957]). The figure's better concordance for an intermediate value of the shape parameter a (which supports the discussion of the asymptotics of equation (2.137)) is strong evidence that across the range of values of a, the eigenfunctions are asymptotic to the Hermite functions.

Return to the general self-adjoint operator $(D_d)^{\frac{1}{2}} B_b (D_d)^{\frac{1}{2}} = \hat{D}_d B_d \hat{D}_d$, where D_d and B_b are defined by the one-dimensional analogs of equations (2.1) and (2.2), respectively. To show that the above considerations apply even if $b \neq d$, we reconsider the scaled Fourier transform operator F_γ, as

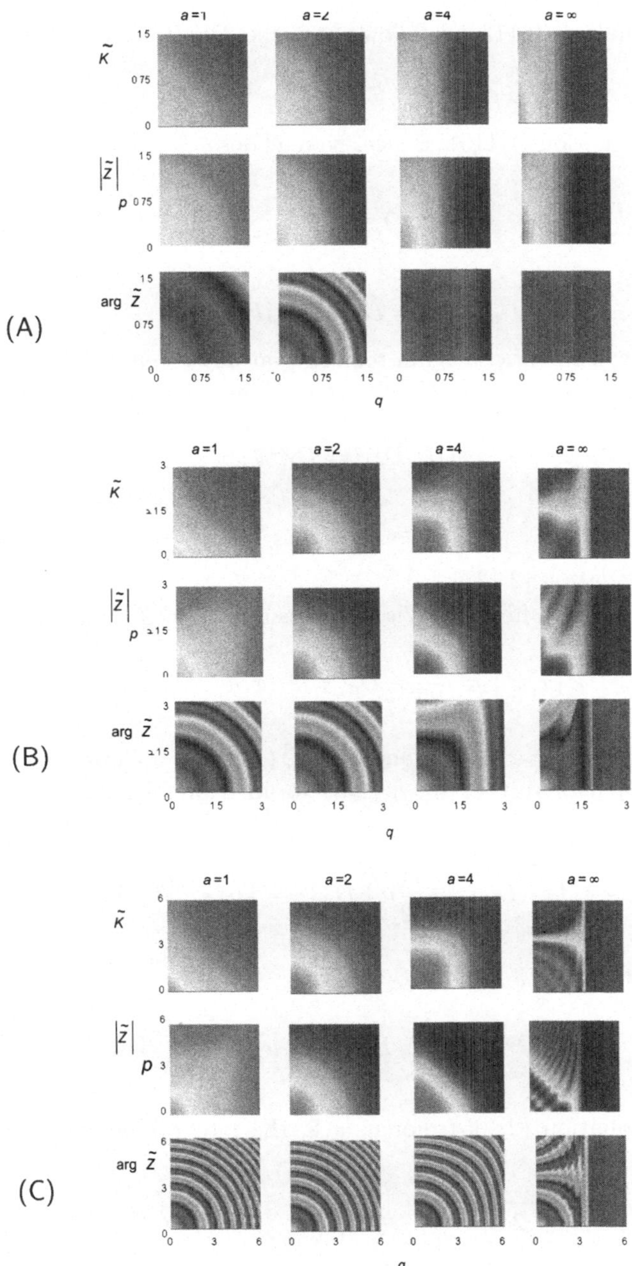

FIGURE 2.5. Comparison of the Wigner transforms $\widetilde{K}(p,q)$ and $\widetilde{Z}(p,q)$ for the values of the shape parameter a and scale parameters b and d of Figure 2.4. In each panel, first row: $\widetilde{K}(p,q)$; second row: amplitude of $\widetilde{Z}(p,q)$; third row: phase of $\widetilde{Z}(p,q)$. A common color scale for amplitude, running from blue to deep red, is used for the first two rows. Color scale for third row: red corresponds positive real, yellow to positive imaginary, green to negative real, blue to negative imaginary.

defined by equation (2.113). It follows from the definitions of B and D that

$$B_b = PF_\gamma D_{b\gamma} F_\gamma = (F_\gamma)^{-1} D_{b\gamma} F_\gamma \,, \tag{2.138}$$

generalizing equation (2.119). Hence, if we choose

$$\gamma = \frac{d}{b} \,, \tag{2.139}$$

we find

$$\hat{D}_d B_d \hat{D}_d = \hat{D}_d P F_\gamma D_d F_\gamma \hat{D}_d \,. \tag{2.140}$$

Symmetry considerations again require that any eigenfunction ψ of $\hat{D}_d B_b \hat{D}_d$ is either even- or odd-symmetric. Thus, if ψ is an eigenfunction of $\hat{D}_d B_b \hat{D}_d$

$$\hat{D}_d B_b \hat{D}_d \psi = P Z^2 \psi, \tag{2.141}$$

where we have defined

$$Z = \hat{D}_d F_\gamma \hat{D}_d \,, \tag{2.142}$$

generalizing equation (2.127).

We now proceed to find the Wigner transform $\widetilde{Z}$ of Z, where

$$Z\{x, x'\} = \frac{1}{\sqrt{2\pi\gamma}} \hat{D}_d(x) \hat{D}_d(x') e^{-ixx'/\gamma} \,. \tag{2.143}$$

The substitutions $\nu = x - x'$ and $q = \frac{1}{2}(x + x')$, followed by Fourier transformation with respect to ν, leads to an expression for the Wigner transform of Z,

$$\widetilde{Z}(p, q) = \frac{1}{\sqrt{2\pi\gamma}} \int_{-\infty}^{\infty} \hat{D}_d \left(q + \frac{1}{2}\nu \right) \hat{D}_d \left(q - \tfrac{1}{2}\nu \right) e^{-i(q^2 - \frac{1}{4}\nu^2)/\gamma} e^{-i\nu p} \, d\nu \,, \tag{2.144}$$

which is rearranged to

$$\widetilde{Z}(p, q) = \frac{1}{\sqrt{2\pi\gamma}} e^{-i(q^2/\gamma + p^2\gamma)} \int_{-\infty}^{\infty} \hat{D}_d \left(q + \tfrac{1}{2}\nu \right) \hat{D}_d \left(q - \tfrac{1}{2}\nu \right) e^{i(\nu - 2p\gamma)^2/4\gamma} \, d\nu \,, \tag{2.145}$$

generalizing equation (2.135). For $a = 2$, the integral can be performed exactly:

$$\widetilde{Z}(p, q) = \sqrt{\frac{2}{\frac{1}{bd} - i}} \exp\left(-\frac{q^2}{d^2}(1 + ibd) - \frac{p^2 d^2}{1 - ibd} \right) . \tag{2.146}$$

In the limit of $a \to \infty$, the second derivatives of the integral (2.145) at the origin behave similarly to those of $\widetilde{K}$: $\frac{\partial^2 \widetilde{Z}}{\partial p^2}$ is analytic, $\frac{\partial^2 \widetilde{Z}}{\partial p \partial q}$ is zero, and $\frac{\partial^2 \widetilde{Z}}{\partial q^2}$ is undefined. Thus, a Gaussian approximation for $\widetilde{Z}$ is no more justified than for $\widetilde{K}$. However, as in the special case of $b = d$ considered above, the integral (2.145) can be approximated directly for the general $a > 2$, if, within the vicinity of $\nu = 2p\gamma$, the functions D_d are slowly varying. This condition translates to

$$|q \pm p\gamma| \ll 2^{1/a}\, d\,. \tag{2.147}$$

(The factor $2^{1/a}$ originates from the square root of the operator D_d.) This also takes the more symmetric form

$$|\frac{q}{\sqrt{\gamma}} \pm p\sqrt{\gamma}| \ll 2^{1/a}\sqrt{bd}\,. \tag{2.148}$$

The above condition is satisfied if both $|q| \ll d$ and $|p| \ll b$, that is, within the band- and space-limits specified by D and B, respectively. In this regime, the dominant contribution to the integral (2.145) can be obtained by replacing D with its peak value, 1. This leads to

$$\widetilde{Z}(p,q) \approx (1+i)e^{-i(q^2/\gamma + p^2\gamma)}\,, \tag{2.149}$$

generalizing the approximation of equation (2.135) by equation (2.116).

The approximate Wigner transform, equation (2.149), has contour lines that are ellipses parallel to the (q,p) coordinate axes. Moreover (similar to equation (2.89) above), these contour lines can be made circular by the symplectic transformation $q' = \frac{q}{\sqrt{\gamma}}, p' = p\sqrt{\gamma}$. Z is not Hermitian, but it is normal — that is, it commutes with its adjoint. As we have seen above, this suffices for the analysis of Knight and Sirovich [1986] to be applicable. In particular, we can conclude that within the regime specified by equation (2.148), the eigenfunctions of Z, and hence of $\hat{D}_d B_d \hat{D}_d$,are approximated by the Hermite functions.

Thus, we have generalized the asymptotic relationship of Flammer [1957] between the Hermite functions and the Slepian functions (which extremize space and band limits in the sense of $a = \infty$) to functions which extremize space and band limits for any choice of the shape parameter a in $(0, \infty)$. As the next section shows, certain features of our analysis apply even more generally, and in particular do not require specification of the shape for the space or band-limiter.

A Perturbation Analysis

For the particular case of $a = 2$ (the Gaussian case), we have an exact solution for the eigenvalues and eigenfunctions of Z. We have seen that the spectrum separates naturally into four subsets which respectively are given by $\{1, -i, -1, i\}$, each multiplied by a set of positive numbers less than unity and descending to zero. Since $K = PZ^2$ (equations (2.126) and (2.127)), this characterization of the eigenfunctions and eigenvalues of Z determines the behavior of the eigenfunctions and eigenvalues of K.

Here we use a perturbation argument to show that this behavior is generic. In contrast to the previous arguments, this aspect of the analysis is rather general, and does not depend on the precise form of the space-limiting function D or the band-limiting function B, provided that they are even-symmetric and Fourier transforms of each other.

Under these conditions, it suffices to consider the integral kernel of equation (2.130), $Z = \hat{D}F\hat{D}$, which can be recast in the more general form

$$Z\{x, x'\} = \frac{1}{\sqrt{2\pi}} e^{-\Gamma(x,t)} e^{-ixx'} e^{-\Gamma(x',t)}, \tag{2.150}$$

where t is a parameter that smoothly specifies the shape of the space-limiting function D (or $\hat{D}$), and Γ is real. In particular, we can specify that $t = 0$ corresponds to the Gaussian ("soft") case, and $t = 1$ corresponds to the Slepian ("hard") case, but for the present argument, the shapes specified by intermediate values of t need not be specified. Several choices of the parameterization by t are of interest, which includes allowing t to control a or d, the parameters that already appear in the definition (2.85) of D. Our goal is to determine how the eigenvalues ζ_n of equation (2.133) depend on t.

Since the eigenvectors φ_n may be chosen to be orthonormal, it follows from the eigenvalue equation (2.133) that

$$\zeta_n = (\varphi_n\,, Z\varphi_n)\,. \tag{2.151}$$

Consequently,

$$\frac{d\zeta_n}{dt} = -\big(\varphi_n\,, (\Gamma_t Z + Z\Gamma_t)\varphi_n\big) = -2\zeta_n(\varphi_n\,, \Gamma_t\varphi_n)\,, \tag{2.152}$$

where $\Gamma_t \equiv \frac{\partial}{\partial t}\Gamma(x,t)$. (We have used $(\varphi_n\,, \frac{\partial}{\partial t}\varphi_n) = 0$ which follows from orthonormality.) The differential equation (2.152) is readily integrated to obtain an expression for the eigenvalues ζ_n:

$$\zeta_n(t) = \zeta_n(0)e^{-2\int_0^t c_n(t')\,dt'}, \tag{2.153}$$

where

$$c_n(t) = \big(\varphi_n(t)\,, \Gamma_t\varphi_n(t)\big)\,. \tag{2.154}$$

Thus, as the kernel Z of equations (2.130) and (2.150) varies parametrically with t, its eigenvalues vary according to equation (2.153).

The integral in equation (2.153) is real. This means that the eigenvalues $\zeta_n(t)$ necessarily have the same phase as $\zeta_n(0)$. By choosing $\hat{D} = 1$ at $t = 0$, (so $Z = F$, the infinite space-bandwidth limit) we can thus infer from equation (2.114) that the phase of $\zeta_n(t)$ is equal to $-\frac{1}{2}n\pi$ for the general Z. Finally, we note that if the parametric dependence of γ on t is monotone increasing (corresponding, for example, to space-limiting functions $\hat{D}$ that are successively narrower), then equation (2.153) implies that as t increases, each of the eigenvalues $\zeta_n(t)$ shrinks monotonically towards zero along a cardinal axis.

2.8 The Non-Gaussian Case: Two Dimensions

The above analysis of the operator Z can be applied directly to the two-dimensional operators $D^{\frac{1}{2}}BD^{\frac{1}{2}}$. This implies a corresponding asymptotic relationship between the eigenfunctions for the general operators $D^{\frac{1}{2}}BD^{\frac{1}{2}}$ (or BD) and those for the Gaussian case.

In two dimensions, the operators B, D, and their products do not separate in Cartesian coordinates for $a \neq 2$. However, they do have exact rotational symmetry (including $a \neq 2$). This symmetry, and the separation into polar coordinates that it implies, means that one can find a complete set of eigenfunctions, all of which have an angular dependence of the form $e^{i\mu\theta}$ for some integer μ. Conversely, for each integer μ, we anticipate a discrete set of eigenfunctions $\psi_{\mu,\nu}(R,\theta) = e^{i\mu\theta}\zeta_{\mu,\nu}(R)$, where ν counts the number of zero-crossings in the radial dependence $\zeta_{\mu,\nu}(R)$. In the Gaussian case, the corresponding eigenvalue $\lambda_{\mu,\nu}$ depends only on $2\nu + |\mu|$. This degeneracy is a consequence of the dual separation into polar and Cartesian coordinates, equation (2.50).

In sum, the above analysis shows that for the general shape parameter a, the two-dimensional functions are also parameterized by a non-negative integer ν that counts the number of zero-crossings along a radius, and a second integer μ (which may be negative) that describes the angular dependence. In view of the asymptotic relationship of the general one-dimensional functions and the Hermite functions, we anticipate that the eigenvalue associated with ν and μ asymptotically depends only on $2\nu + |\mu|$.

2.9 V4 Receptive Fields

One of our motivations for studying functions that are simultaneously space- and band-limited is to gain some insight into neural processing of visual images beyond primary visual cortex (V1), and especially in V4. The extrastriate area V4 appears to be an important area for the analysis of shape Merigan [1996]. The behavior of neurons in V4 differs dramatically from those of its inputs, V1 and V2. Neurons in V1 and V2 usually can be characterized as having a single preferred orientation, and respond well to a grating of an appropriately chosen spatial frequency, especially if the grating is limited in spatial extent. Known differences between V1 and V2 neurons are relatively subtle (Levitt, Kiper, and Movshon [1994]), and consist mainly of differences in overall receptive field size, contrast sensitivity, and the prominence of responses to illusory contours (Grosof, Shapley, and Hawken [1993]; von der Heydt, Peterhans, and Baumgartner [1984]). In contrast, V4 neurons do not have a clear orientation preference, are often difficult to stimulate with gratings, and typically require complex stimuli for strong responses (Kobatake and Tanaka [1994]).

Consequently, systematic study of the spatial structure of the receptive fields of V4 neurons has been difficult. One of the few successful examples is the provocative study by Gallant, Braun, and Van Essen [1993] and Gallant, Connor, Rakshit, Lewis, and Van Essen [1996]. V4 neurons were stimulated with patches of standard ("Cartesian") gratings, and also patches of "polar" gratings and patches of "hyperbolic" gratings (Figure 2.6). The polar gratings include target- and pinwheel-like stimuli; the hyperbolic gratings consist of alternating bands of light and dark that form rectangular hyperbolae with shared asymptotes. The main finding was that many V4 neurons were well-stimulated by particular examples of non-Cartesian stimuli. Although some neurons responded substantially better to members of one class of gratings than to the other two, most neurons had broad tunings across these classes (that is, they responded almost as well to some stimuli of the non-preferred class as to the best stimulus). Moreover, the typical neuron responded relatively poorly to most stimuli, even within its preferred class.

To determine whether this kind of behavior is expected from neurons that filter the image in a manner that is both space and band-limited, we performed a crude simulation. Receptive field profiles were constructed from the two-dimensional eigenfunctions for BD (Gaussian case), for values of the circular symmetry index $\mu \in \{0, 1, 2, 4\}$ and for a range of radial nodes $\nu \in \{0, 1, 2, 3\}$. We chose the space-bandwidth product $c = bd = 4$. The resulting profiles are shown in Figure 2.7. Raw "responses" were calculated simply by evaluating the inner product of each stimulus in Figure 2.6 with the receptive field profile. (The stimulus was considered to have a real part, as specified by Gallant, Braun, and Van Essen [1993]; Gallant, Connor, Rakshit, Lewis, and Van Essen [1996], and a corresponding imaginary part, obtained by replacing cosines by sines. Thus, this highly schematized model essentially posits a quadrature pair operation that removes the effect of the spatial "phase" of the Cartesian or non-Cartesian grating.) For each receptive field profile, the raw responses were then normalized by the largest quadrature pair response encountered to any of the stimuli.

A histogram of the normalized responses obtained for each receptive field profile is shown in Figure 2.8. Several features are immediately apparent. As in the data of Gallant, Braun, and Van Essen [1993]; Gallant, Connor, Rakshit, Lewis, and Van Essen [1996], most receptive fields responded poorly to most stimuli – as evidenced by the large peaks in most of the histograms at a response size of 0. The receptive fields corresponding to $\mu = 1$ responded almost exclusively to Cartesian gratings. This is a consequence of the symmetry of these receptive fields. Among the other profiles, many ($\mu = 4$) had optimal responses to polar gratings, and one ($\mu = 0, \nu = 1$) had an optimal response to a hyperbolic grating. The profiles corresponding to $\mu = 2$ tended to have large responses to some Cartesian and some polar gratings, while the profiles corresponding to $\mu = 4$ tended to have largest responses to polar gratings, with next-largest responses to hyperbolic gratings.

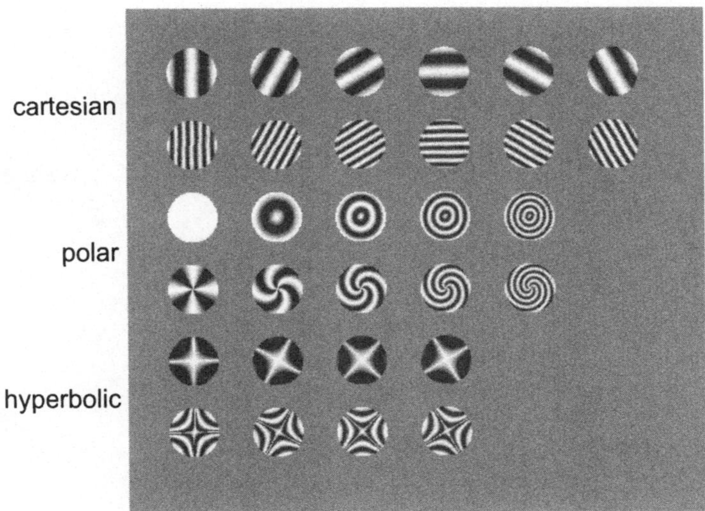

FIGURE 2.6. Some of the stimuli used by Gallant, Braun, and Van Essen [1993]; Gallant, Connor, Rakshit, Lewis, and Van Essen [1996]. The full stimulus set included three other spatial frequencies for each stimulus class, and also counter-clockwise variants of the "polar" stimuli — for a total of 30 Cartesian gratings, 45 polar gratings, and 20 hyperbolic gratings.

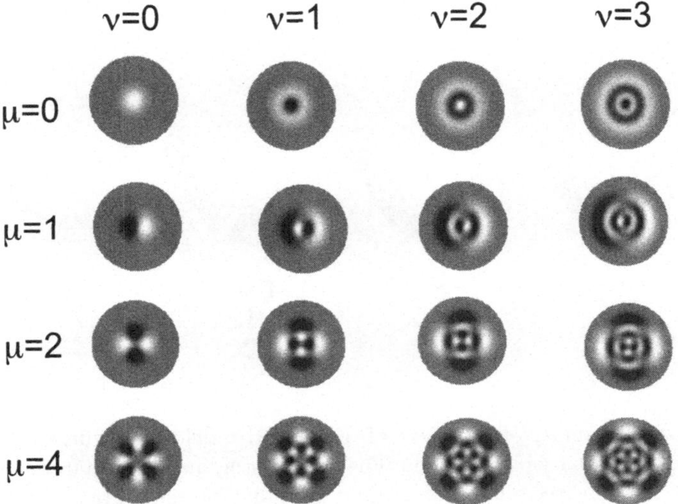

FIGURE 2.7. Receptive field sensitivity profiles (real parts) used for the simulation. All profiles had a space bandwidth product $c = 4$. The size of the disk corresponds to the region of space that covered the test stimuli, taken from Gallant, Braun, and Van Essen [1993]; Gallant, Connor, Rakshit, Lewis, and Van Essen [1996] (see Figure 2.6).

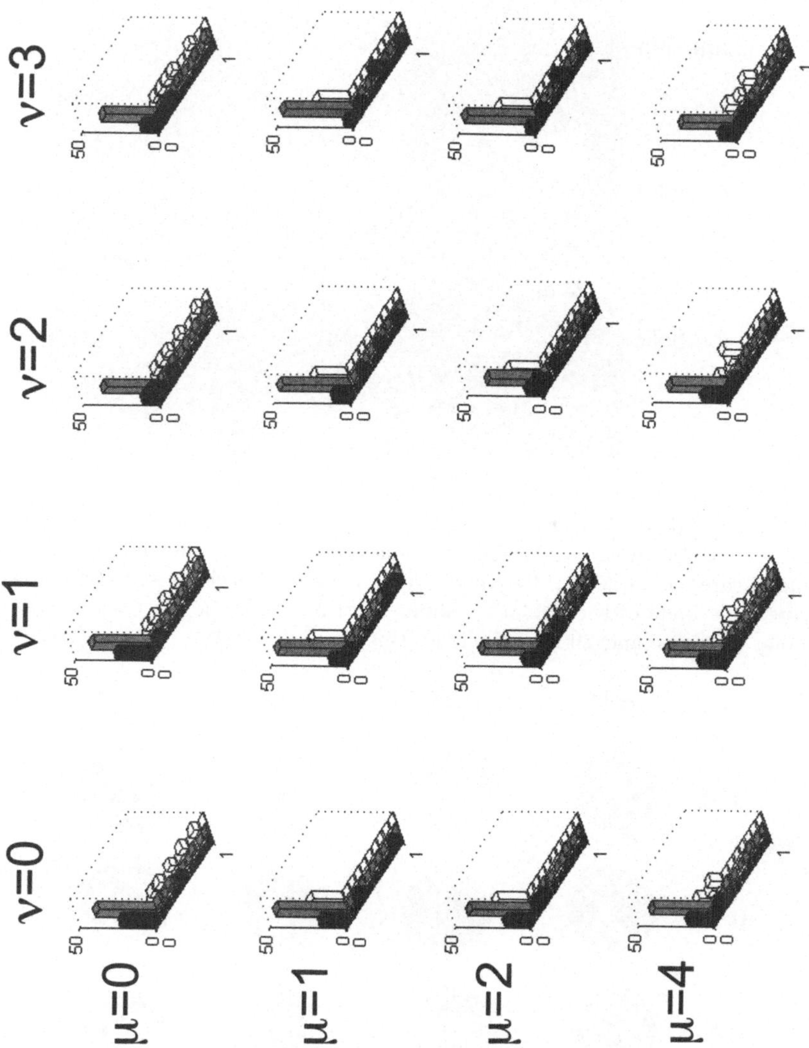

FIGURE 2.8. Summary of responses of the receptive fields of Figure 2.7 to Cartesian, polar, and hyperbolic gratings. The three histograms in each frame represent responses to the three classes of grating stimuli: Cartesian gratings (30 total), dark; polar gratings (45 total), gray; hyperbolic gratings (20 total), light. The horizontal axis represents response size, normalized by the largest quadrature pair response across all three classes. The height of each histogram bar represents the number of stimuli that elicited a response of each (normalized) size.

3 Discussion

This paper describes two-dimensional profiles that are simultaneously limited in spatial extent and in bandwidth. The main motivation is to generalize ideas that have benefited the study of visual processing in primary visual cortex, V1. As outlined above, V1 neurons, to a first approximation, are often thought of as one-dimensional spatial analyzers. They generally have a well-defined orientation preference, and typically they can be characterized by their responses to windowed gratings. However, there are indications that visual processing by individual neurons, even in V1, is not restricted to a single orientation (Purpura, Victor, and Katz [1994]). Moreover, certain aspects of early visual processing (whose physiologic locus is uncertain) make explicit use of two-dimensional information, such as extraction of T-junctions (Rubin [2001]), curvature (Wolfe, Yee, and Friedman-Hill [1992]), texture (Victor and Brodie [1978]) and shape (Wilkinson, Wilson, and Habak [1998]). Finally, neurons beyond V1 and V2 are generally not well-stimulated by standard gratings or other simple stimuli, which makes quantitative study of such neurons a challenge (Gallant, Braun, and Van Essen [1993]; Gallant, Connor, Rakshit, Lewis, and Van Essen [1996]; Kobatake and Tanaka [1994]; Tanaka, Saito, Fukada, and Moriya [1991]).

Our notion of simultaneous confinement in space and spatial frequency is distinct from the notion in Daugman [1985] of minimizing uncertainty in two important respects. First, (see equations (2.8) and (2.9) and surrounding material), our notion seeks profiles that are least altered by application of linear operators that truncate influences that are non-local in space (D) or spatial frequency (B). In contrast, Daugman's (1985) notion directly adopts Gabor's (1946) idea of seeking profiles that minimize joint spread in space and spatial frequency, as quantified by a Heisenberg uncertainty product. Secondly, our approach considers the profiles themselves as the subject for extremization. Though the profiles naturally group into real and imaginary pairs (lower portions of Figure 2.2 and Figure 2.3), they are intrinsically real, and their positive and negative lobes are intrinsic to their extremal properties. Daugman's approach focuses on the magnitude-squared of the profiles (Stork and Wilson [1990]); the positive and negative lobes of the derived profiles only arise because of phase factors $e^{i\omega x}$ that are irrelevant to the extremal properties.

Interestingly, applying the Gabor/Daugman approach to real-valued functions (Gabor [1946]; Stork and Wilson [1990]) identifies one-dimensional Hermite functions as playing an extremal role, but this role proves to be that of maximizing uncertainty within the space of polynomials multiplied by Gaussians, rather than minimizing it (Klein and Beutter [1992]).

The families of functions presented here have attributes that recommend them as a natural extension of windowed gratings. Since each family forms a complete orthogonal set, its members can serve as a basis for characterization of the quasilinear aspects of a neuron's response, either as individual stimuli presented in discrete trials or as members of a rapidly-presented stimulus sequence (Ringach, Sapiro, and Shapley [1997]). More importantly, these functions can serve as a common stimulus set to determine how spatial processing evolves from V1 to V4. From the point of view of V1, these families of functions contain elements (those with one index 0 in Cartesian separation) that are very similar to Gabor patches (Swanson, Wilson, and Giese [1984]). From the point of view of V4, the families (in polar separation) contain elements that resemble the non-Cartesian gratings used by Gallant and co-workers. With these stimuli, hypotheses for the organization and relationship of V1 and V4 receptive fields can readily be tested. For example, if indeed V1 neurons are one-dimensional analyzers, there should be little response to functions which, in the Cartesian separation $\psi_{n_x}(x)\psi_{n_y}(y)$, have both indices n_x and n_y nonzero. On the other hand, if V1 neurons only *appear* to be one-dimensional analyzers because they have a limited space-bandwidth product, then neurons that respond to an intrinsically two-dimensional function such as $\psi_1(x)\psi_1(y)$ should be about as common as neurons that respond to a one-dimensional multilobed function, such as $\psi_2(x)\psi_0(y)$, of equal eigenvalue. The hypothesis that processing in V4 is intrinsically two-dimensional, rather than just characterized by an increased space-bandwidth product, makes the converse prediction.

As noted above, the "Gabor" profiles of typical neurons in V1 have only one or two major lobes, and equally well might be replaced by functions such as $\psi_1(x)\psi_0(y)$ and $\psi_2(x)\psi_0(y)$, with an appropriate choice of axes. Wilson [1999], drawing on anatomical work by Schoups, Tootell, and Orban [1995], has suggested that V4 neurons obtain their properties by pooling of V1 responses across multiple orientations, and that circularly symmetric V4 receptive fields might result from uniform pooling of one-dimensional profiles in V1. Appropriately weighted combinations of profiles at orthogonal orientations also can lead to receptive field sensitivity profiles with n-fold rotational symmetry, for arbitrary n, as seen from the relationship between the Cartesian and polar separations described above. Neurons with preferences for n-fold rotational symmetry have been encountered in extrastriate cortices (Tanaka, Saito, Fukada, and Moriya [1991]). It is interesting (and also highly speculative) to note that the relationship between the polar separation and the Cartesian separation provides an economical way to construct receptive fields of arbitrary rotational symmetry.

Our crude analysis (Figure 2.8) of simulated neurons based on the eigenfunctions in polar separation indicates that these receptive field profiles indeed can be tuned for Cartesian and non-Cartesian gratings, as Gallant, Braun, and Van Essen [1993]; Gallant, Connor, Rakshit, Lewis, and Van Essen [1996] found for real V4 neurons. Given the highly schematized and simple nature of our simulation, a detailed correspondence with physiology would be unexpected. Indeed, only certain aspects of Gallant et al.'s results are accounted for by this scheme. Gallant et al. found tuning to hyperbolic gratings that was more prominent than our simulations suggest. They also found that responses were relatively insensitive to translations of the stimulus. This suggests that the very simple scheme we used here (inner product followed by a quadrature-pair calculation) needs to be augmented by additional stages of processing, perhaps in the form of receptive field subunits. More detailed modeling in this direction requires additional physiologic data, and is beyond the scope of this work. However, it is hoped that the considerations presented here will serve as a basis for a sound theoretically-motivated experimental approach to understanding visual processing beyond the earliest cortical stages.

4 Appendix

Normalization of the Eigenfunctions in the Gaussian Case. The normalization of the eigenfunctions (2.29) requires integrals similar to equation (2.58), but with respect to Gaussians of variance other than unity. The generating-function strategy used for the evaluation of the integrals of equation (2.58) readily extends to this situation. First, consider

$$\begin{aligned}\frac{1}{2\pi V}\iint\sum_{0}^{\infty}&\frac{z^k t^l\, q_{k,l}(x,y)}{k!\,l!}\,\frac{z'^{k'}t'^{l'}\,\overline{q}_{k',l'}(x,y)}{k'!\,l'!}\exp\left[-\frac{1}{2V}(x^2+y^2)\right]dx\,dy\\ &=\frac{1}{2\pi V}\iint Q(z,t)\overline{Q}(x',t')\exp\left[-\frac{1}{2V}(x^2+y^2)\right]dx\,dy \qquad (4.1)\end{aligned}$$

in which the sum on the left-hand side is over all values of k, l, k' and l'. Proceeding in the same manner as with equation (2.58), we find

$$\begin{aligned}\frac{1}{2\pi V}\iint\sum_{0}^{\infty}&\frac{z^k t^l q_{k,l}(x,y)}{k!\,l!}\,\frac{z'^{k'}t'^{l'}\overline{q}_{k',l'}(x,y)}{k'!\,l'!}\exp\left[\frac{1}{2V}(x^2+y^2)\right]dx\,dy\\ &=\exp\left[2V(zz'+tt')\right]\exp\left[2(V-1)(zt+z't').\right] \qquad (4.2)\end{aligned}$$

Because of the polar symmetry, terms for which $k-l\neq k'-l'$ are necessarily 0. Extracting and equating terms corresponding to $z^{\mu+\nu}t^{\nu}(z')^{\mu+\omega}(t')^{\omega}$ leads to

$$\begin{aligned}\frac{1}{2\pi V}\iint &q_{\mu+\nu,\nu}(x,y)\,\overline{q}_{\mu+\omega,\omega}(x,y)\exp\left[-\frac{1}{2V}(x^2+y^2)\right]dx\,dy\\ &=(\mu+\nu)!\nu!(\mu+\omega)!\omega!\sum_{\alpha,\beta,\gamma,\delta}\frac{2^{\alpha+\beta+\gamma+\delta}(V-1)^{\alpha+\beta}V^{\gamma+\delta}}{\alpha!\,\beta!\,\gamma!\,\delta!}, \qquad (4.3)\end{aligned}$$

where the sum is over all indices α, β, γ and δ satisfying

$$\begin{aligned}\alpha+\gamma&=\mu+\nu\\ \alpha+\delta&=\nu\end{aligned}$$

$$\begin{aligned}\beta+\gamma&=\mu+\omega\\ \beta+\delta&=\omega\,.\end{aligned}\tag{4.4}$$

Equation (4.3) leads to

$$\frac{1}{2\pi V}\iint q_{\mu+\nu,\nu}(x,y)\,\overline{q}_{\mu+\omega,\omega}(x,y)\,\exp\Big[-\frac{1}{2V}(x^2+y^2)\Big]\,dx\,dy$$
$$=2^{\mu+\nu+\omega}(\mu+\nu)!\,\nu!\,(\mu+\omega)!\,\omega!\sum_{s}^{\min(\nu,\omega)}\frac{(V-1)^{\nu+\omega-2s}\,V^{\mu+2s}}{(\nu-s)!\,(\omega-s)!\,s!\,(\mu+s)!}\,.\tag{4.5}$$

The general result (4.5) can also be recast as

$$\frac{1}{V}\int_0^\infty P_{\mu,\nu}(R^2)\,P_{\mu,\omega}(R^2)\,R^{2\mu+1}\,\exp\Big(-\frac{R^2}{2V}\Big)\,dR$$
$$=2^{\mu+\nu+\omega}(\mu+\nu)!\,\nu!\,(\mu+\omega)!\,\omega!\sum_{s}^{\min(\nu,\omega)}\frac{(V-1)^{\nu+\omega-2s}V^{\mu+2s}}{(\nu-s)!\,(\omega-s)!\,s!\,(\mu+s)!}\,.\tag{4.6}$$

To see the relationship to the case $V=1$ of equation (2.58), note that the only nonzero contributions to the sum in equation (4.3) are for $\alpha=\beta=0$ which forces $\gamma=\mu+\nu=\mu+\omega$ and $\delta=\nu=\omega$. In equations (4.5) and (4.6), this corresponds to the sole term $s=\nu=\omega$, which establishes the reduction to equation (2.60) with $\kappa=\mu+\nu$ and $l=\nu$ and to equation (2.62).

Modest Non-Orthogonality After Shifts. The extent to which shifting a function spoils the orthogonality can be captured by considering the Hermite-function limit (large c) of the one-dimensional eigenfunctions. These eigenfunctions (see equations (2.21) and (2.27)) are given by

$$\varphi_n(x)=h_n(kx)\,\exp\big(-\tfrac{1}{4}k^2x^2\big)\,,\tag{4.7}$$

from which it follows that

$$\sum_{n=0}^{\infty}\frac{z^n}{n!}\varphi_n(x)=\exp\big(-\tfrac{1}{4}k^2x^2+kxz-\tfrac{1}{2}z^2\big)\,.\tag{4.8}$$

To determine the non-orthogonality of the mth and nth eigenfunctions after positional shifts s, define

$$a_{m,n}(x)=\frac{k}{\sqrt{2\pi}}\int_{-\infty}^{\infty}\varphi_m(x-s)\,\varphi_n(x+s)\,dx\,.\tag{4.9}$$

Consider a generating function defined by

$$A(s,y,z)=\sum_{m=0}^{\infty}\sum_{n=0}^{\infty}\frac{y^m}{m!}\,\frac{z^n}{n!}\,a_{m,n}(s)\,.\tag{4.10}$$

From the generating function (4.8) for φ, it follows that

$$A(s,y,z) = \frac{k}{\sqrt{2\pi}} \int_{-\infty}^{\infty} \exp\left(-\tfrac{1}{4}k^2(x-s)^2 + k(x-s)y - \tfrac{1}{2}y^2\right) \exp\left(-\tfrac{1}{4}k^2(x+s)^2 + k(x+s)z - \tfrac{1}{2}z^2\right) dx\,, \tag{4.11}$$

and hence,

$$\begin{aligned} A(s,y,z) &= \frac{k}{\sqrt{2\pi}} \exp\left(-\tfrac{1}{2}k^2s^2 + ks(-y+z)\right) \\ &\quad \int_{-\infty}^{\infty} \exp\left(-\tfrac{1}{2}k^2x^2 + kx(y+z) - \tfrac{1}{2}y^2 + z^2\right) dx \\ &= \exp\left(-\tfrac{1}{2}k^2s^2 + ks(-y+z) + yz\right). \end{aligned} \tag{4.12}$$

Equating coefficients of terms $y^m z^n$ in equations (4.10) and (4.12) yields:

$$a_{m,n}(s) = \sum_{j=0}^{\min(m,n)} (-1)^{m-j} j! \binom{m}{j} \binom{n}{j} (ks)^{m+n-2j} \exp\left(-\tfrac{1}{2}k^2s^2\right). \tag{4.13}$$

Note that when $s = 0$, the only nonzero term on the right-hand side of equation (4.13) has $m + n - 2j = 0$, which can only occur if $m = n$ (thus, recovering orthogonality when there is no shift). For nonzero s, the values of $a_{m,n}(s)$ are controlled by the Gaussian envelope.

Acknowledgments: This work was supported by NIH NEI EY9314 and DARPA MDA972-01-1-0028. The authors thank Partha Mitra and Larry Sirovich for several helpful discussions, Stan Klein and Hugh Wilson for insightful comments on the manuscript, and Adam Kahn for programming assistance.

References

Abramowitz, M. and I. A. Stegun [1964], *Handbook of Mathematical Functions.* National Bureau of Standards. Reprinted by Dover, New York, 1970.

Atick, J. and A. Redlich [1990], Towards a theory of early visual processing, *Neural Comput.* **2**, 308–320.

Daugman, J. G. [1985], Uncertainty relation for resolution in space, spatial frequency, and orientation optimized by two-dimensional visual cortical filters, *J Opt Soc Am [A].* **2**, 1160–9.

Erdelyi, A. [1955], *Higher Transcendental Functions.* Bateman Manuscript Project, Vol. II. New York: McGraw-Hill.

Field, D. J. (1994), What is the goal of sensory coding? *Neural Comput.* **6**: 559–601.

Flammer, C. [1957], *Spheroidal Wave Functions.* Stanford: Stanford University Press.

Gabor, D. [1946], Theory of communication, *J. Inst. Elect. Eng.* **93**, 429–457.

Gallant, J. L., J. Braun, and D. C. Van Essen [1993], Selectivity for polar, hyperbolic, and Cartesian gratings in macaque visual cortex, *Science*, **259**, 100–3.

Gallant, J. L., C. E. Connor, S. Rakshit, J. W. Lewis, and D. C. Van Essen [1996], Neural responses to polar, hyperbolic, and Cartesian gratings in area V4 of the macaque monkey, *J Neurophysiol.* **76**: 2718–39.

Grosof, D. H., R. M. Shapley, and M. J. Hawken [1993], Macaque V1 neurons can signal 'illusory' contours, *Nature.* **365**, 550–2.

Halmos, P. [1942], *Finite Dimensional Vector Spaces.* Princeton: Princeton University Press.

Jones, J. P. and L. A. Palmer [1987], An evaluation of the two-dimensional Gabor filter model of simple receptive fields in cat striate cortex, *J Neurophysiol.* **58**, 1233–58.

Klein, S. A. and B. Beutter [1992], Minimizing and maximizing the joint space-spatial frequency uncertainty of Gabor-like functions: comment, *J. Opt. Soc. Am. A* **9**, 337–40.

Knight, B. W. and L. Sirovich [1982], The Wigner transform and some exact properties of linear operators, *SIAM J. Appl. Math.* **42**, 378–389.

Knight, B. W. and L. Sirovich [1986], The eigenfunction problem in higher dimensions: Exact results, *Proc. Natl. Acad. Sci. USA* **83**, 527–530.

Kobatake, E. and K. Tanaka, [1994], Neuronal selectivities to complex object features in the ventral visual pathway of the macaque cerebral cortex, *J Neurophysiol.* **71**, 856–67.

Laughlin, S. B., R. R. de Ruyter van Steveninck, and J. C. Anderson [1998], The metabolic cost of neural information, *Nat Neurosci.* **1**, 36–41.

Levitt, J. B., D. C. Kiper, and J. A. Movshon[1994], Receptive fields and functional architecture of macaque V2, *J Neurophysiol.* **71**, 2517–42.

Marcelja, S. [1980], Mathematical description of the responses of simple cortical cells, *J Opt Soc Am.* **70**, 1297–1300.

Merigan, W. H. [1996], Basic visual capacities and shape discrimination after lesions of extrastriate area V4 in macaques, *Vis Neurosci.* **13**, 51–60.

Percival, D. and A. Walden [1993], *Spectral Analysis for Physical Applications: Multitaper and Conventional Univariate Techniques.* Cambridge: Cambridge University Press.

Purpura, K. P., J. D. Victor, and E. Katz [1994], Striate cortex extracts higher-order spatial correlations from visual textures, *Proc Natl Acad Sci U S A.* **91**, 8482–6.

Ringach, D. L., G. Sapiro, and R. Shapley [1997], A subspace reverse-correlation technique for the study of visual neurons, *Vision Res.* **37**, 2455–2464.

Rota, G.-C. and B. D. Taylor [1994] The classical umbral calculus, *SIAM J. Math. Anal.* **25**, 694–711.

Rubin, N. [2001], The role of junctions in surface completion and contour matching, *Perception* **30**, 339–366

Schoups, A. A., R. B. Tootell, W. Vanduffel, and G. A. Orban [1995], Use of the double-label deoxyglucose approach to map the orientation columnar system beyond area V1 and V2 in the macaque, and its plasticity, *Soc. Neurosci. Abstr.* **21**, 18.

Sirovich, L. and B. W. Knight [1981], On the eigentheory of operators which exhibit a slow variation, *Quart. Appl. Math.* **38**, 469–488.

Sirovich, L. and B. W. Knight [1982], Contributions to the eigenvalue problem for slowly varying operators, *SIAM J. Appl. Math.* **42**, 356–377.

Sirovich, L. and B. W. Knight [1985], The eigenfunction problem in higher dimensions: Asymptotic Theory, *Proc. Natl. Acad. Sci. USA.* **82**, 8275–8278.

Slepian, D. and H. O. Pollack [1961], Prolate spheroidal wave functions, Fourier analysis and uncertainty I. *Bell Syst. Tech. J.* **40**, 43–64.

Slepian, D. [1964], Prolate spheroidal wave functions, Fourier analysis and uncertainty IV: Extensions to many dimensions; generalized prolate spheroidal functions, *Bell Syst. Tech. J.* **43**, 3009–3057.

Stork, D. G. and H. R. Wilson [1990], Do Gabor functions provide appropriate descriptions of visual cortical receptive fields?, *J. Opt. Soc. Am. A* **7**, 1362–73.

Swanson, W. H., H. R. Wilson, and S. C. Giese [1984], Contrast matching data predicted from contrast increment thresholds, *Vision Res.* **24**, 63–75.

Tanaka, K., H. Saito, Y. Fukada, and M. Moriya [1991], Coding visual images of objects in the inferotemporal cortex of the macaque monkey, *J Neurophysiol.* **66**, 170–89.

Victor, J. D. and S. Brodie [1978], Discriminable textures with identical Buffon needle statistics, *Biological Cybernetics* **31**, 231–234.

Vilenkin, N. J. [1968], *Special Functions and the Theory of Group Representations.* American Mathematical Society.

von der Heydt, R., E. Peterhans, and G. Baumgartner [1984], Illusory contours and cortical neuron responses, *Science* **224**, 1260–2.

Wilkinson, F., H. R. Wilson, and C. Habak [1998], Detection and recognition of radial frequency patterns, *Vision Res.* **38**, 3555–68.

Wilson, H. R. [1999] Non-Fourier cortical processes in texture, form, and motion perception. In *Cerebral Cortex, vol. 13*, pages 445–477. ed: Ulinski et al. Kluwer Academic, New York.

Wilson, H. R. and F. Wilkinson [1998], Detection of global structure in Glass patterns: implications for form vision, *Vision Res.* **38**, 2933–47.

Wolfe, J. M., A. Yee, and S. R. Friedman-Hill, [1992], Curvature is a basic feature for visual search tasks, *Perception* **21**, 465–80.

Xu, Y., S. Haykin, and R. J. Racine [1999], Multiple window time-frequency distribution and coherence of EEG using Slepian sequences and Hermite functions, *IEEE Trans. Biomed. Engrg.* **46**, 861–866.

14

Pseudochaos

G. M. Zaslavsky
M. Edelman

To Larry Sirovich, on the occasion of his 70th birthday.

ABSTRACT A family of the billiard-type systems with zero Lyapunov exponent is considered as an example of dynamics which is between the regular one and chaotic mixing. This type of dynamics is called "pseudochaos". We demonstrate how the fractional kinetic equation can be introduced for the pseudochaos and how the main critical exponents of the fractional kinetics can be evaluated from the dynamics. Problems related to pseudochaos are discussed: Poincaré recurrences, continued fractions, log-periodicity, rhombic billiards, and others. Pseudochaotic dynamics and fractional kinetics can be applied to streamlines or magnetic field lines behavior.

Contents

1 Introduction

There are many examples that show transition to turbulence through spatio-temporal chaos. The case of 2-dimensional Kolmogorov flow (Platt, Sirovich, and Fitzmaurice [1991]) displays a complicated alternation of different laminar and chaotic regimes following changes of a control parameter, which precede the dynamics known as the turbulent one. While a common opinion focuses on an exceptional role of coherent structures presented even in turbulent flows, the structures are still elusive and we still do not have a universal method for their depiction. Interesting ideas of the diagnostic of coherent structures can be found in recent publications (Sirovich [1989]; Weiss, Provenzale, and McWilliams [1998]; Haller [2000]). An indirect analysis of flows can be developed from the trajectories of passive particles (tracers). This approach is known also as the Lagrangian one versus the Eulerian one.

While the study of tracer dynamics is not sufficient to provide full information about the corresponding flow, it can reveal an important information about the flow transitions from one regime to another. An important example of the tracer dynamics appears for the so-called ABC (Arnold-Beltrami-Childress) flow (Arnold [1978]) for which the tracer dynamics coincides with streamline trajectories, is Hamiltonian (Zaslavsky, Sagdeev, Usikov, and Chernikov [1991]), and consists of regular and chaotic domains in phase space for all nonzero values of the flow parameters (Zaslavsky, Sagdeev, Usikov, and Chernikov [1991]; Henon [1966]). The way to derive Hamiltonian equations for tracers is based on the condition of incompressibility ($div\ \mathbf{v} = 0$). The same idea can be extended to more general compressible helical flows (Morgulis, Yudovich, and Zaslavsky [1995]). The area of chaotic streamlines depends on the symmetry of flow, flow parameters, type of the perturbation, etc. (Agullo, Verga, and Zaslavsky [1997]; Benkadda, Gabbai, and Zaslavsky [1997]; Beyer and Benkadda [2001]; Zaslavsky, Sagdeev, Usikov, and Chernikov [1991]).

Similar to the tracer dynamics problem is the behavior of magnetic field lines for a given vector-field $\mathbf{B}(\mathbf{r})$ (Rosenbluth, Sagdeev, Taylor, and Zaslavsky [1966]; Zaslavsky, Sagdeev, Usikov, and Chernikov [1991]). There is one-to-one correspondence between the chaos of magnetic field lines and the chaos of Lagrangian particles in incompressible flow. Magnetic chaos was actively studied with respect to the fusion program. This analogy has been used to study a new problem: randomness without the exponential sensitivity to the perturbation of initial conditions (Zaslavsky and Edelman [2001]). Field-lines of $\mathbf{v}$ or $\mathbf{B}$ can wind invariant surfaces of a topological genus more than one. In this case equations for the field lines are not integrable even if the Lyapunov exponent $\sigma_L = 0$ (Kozlov [1996]). We call trajectories to be *pseudochaotic* if $\sigma_L = 0$ but the equations that define the trajectories are not integrable.

It was shown for some cases of pseudochaos (Zaslavsky and Edelman [2001]) that the ensemble of trajectories can be considered in a statistical manner and can be described by a kind of kinetic equation with fractional derivatives. Kinetics of trajectories reflects self-similarity of dynamics and superdiffusive equations for the moments of a distribution function along coordinates, i.e. superdiffusive transport. This article continues the work of

Zaslavsky and Edelman [2001] replacing the problem of field-lines behavior by a billiard type problem. There is no one-to-one correspondence between the field-line trajectories problem and the billiard one. Nevertheless, the study of particle dynamics in the billiard-type systems can provide a realistic insight on different possibilities of the dynamics as it has happened with the Sinai billiard.

We consider different billiard-type models with zero Lyapunov exponent. These models can be applied to such physical problems as ray dynamics in a media with complex (fractal) boundary (Sapoval, Gobron, and Margolina [1991]; Hebert, Sapoval, and Russ [1999]), light propagation in the optical wave-guides (Wilkinson, Fromhold, Taylor, and Micolich [2001a,b]), sound propagation in nonuniform media (Zaslavsky and Abdullaev [1997]), magnetic field lines behavior in toroidal plasmas (Zaslavsky and Edelman [2001]), tracer dynamics in reconnected vortices (Zaslavsky and Edelman [2001]), etc. Similar problems proved to be interesting in problems of quantum chaos (Richens and Berry [1981]; Wiersig [2000]; Artuso, Guarneri, and Rebuzzini [2000]; Artuso, Casati, and Guarneri [1997]) and general problems of anomalous transport (Artuso, Guarneri, and Rebuzzini [2000]; Artuso, Casati, and Guarneri [1997]; Zwanzig [1983]; Zaslavsky and Edelman [2001]). There are also important works on the polygonal billiards (Galperin and Zemlyakov [1990]; Katok [1980]; Katok and Hasselblatt [1995]; Gutkin [1996]) and on the interval exchange transformation (IET) (Katok [1980]; Zorich [1997]) which considers complexity, mixing, and transport. In addition to the given references, let us mention that: there exists a claim of existence of positive Lyapunov exponent (Vega, Uzer, and Ford [1993]), polynomial asymptotics for the rate of mixing for a piecewise affine interval map were obtained in Isola [1999], additional material on the distribution properties of continued fractions can be found in Mayer [1990], and multifractal features of continued fractions were proved in Pollicott and Weiss [1999]. The latter result is consistent with our numerical demonstration of scaling properties of continued fractions.

In this paper we briefly review our previous results on the kinetics and transport properties of the trajectories in a family of polygonal billiards and the corresponding Lorentz-type gases (Zaslavsky and Edelman [2001]), and present new results on the transport exponents and more complicated billiard-type models. It is worthwhile to mention here, that while there is a way of evaluating the transport exponents for IET (Zorich [1997]), these results cannot be applied immediately to the transport exponent of trajectories, and a special analysis of trajectories and their ensemble is necessary.

There are three separate problems related to the pseudochaotic kinetics, that will be briefly discussed here: transport in the rhombic billiards, log-periodic properties of continued fractions, and the origin of specific exponents for the Poincaré recurrences.

2 Filamented and Merged Surfaces, Billiards, and Weak Mixing

We will speak about the particle dynamics having in mind a tracer or a field line as a particle trajectory. Examples of the filamented and merged surfaces are given in Figures 2.1(a) and (b) respectively. The case (a) appears in the toroidal plasma confinement (Zaslavsky and Edelman [2001]; Morozov and Soloviev [1966]) while the case (b) can appear in both plasma and fluids. Under some conditions, the problem of the geodesics along surfaces can be reduced to the problem of trajectories in elastic billiards (Galperin and Zemlyakov [1990]; Katok and Hasselblatt [1995]). Some examples of billiards are given in Figure 2.2. Billiards must have rational angles to be reducible to the equivalent dynamics along surfaces. Each of the billiards in Figure 2.2 can be periodically continued in two or one directions (Figures 2.3 and 2.4) creating a "generalized Lorentz gas" (GLG). As an example, let us point an evident connection between geodesics in Figure 2.1(b), and trajectories in Figures 2.2(a) and 2.3(a).

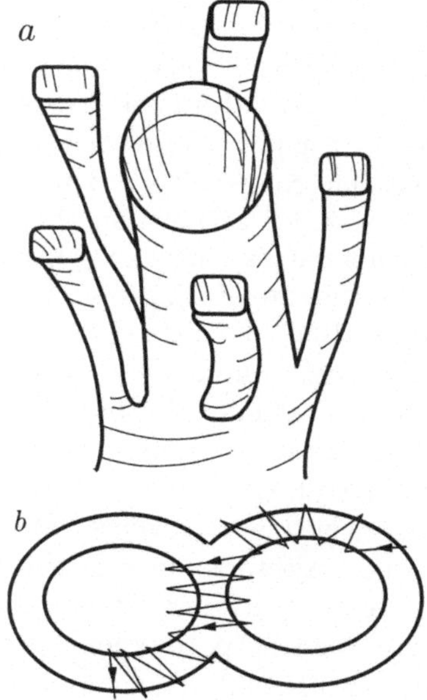

FIGURE 2.1. Examples of filamented (a) and merged (b) surfaces.

Billiard trajectories can be studied by introducing a map of some interval into itself. For the case in Figure 2.2(a) one can consider the map of the slit into itself, which belongs to the type of the interval exchange transformation: (IET) (Galperin and Zemlyakov [1990]; Zorich [1997]). There are some common features between all four types of problems mentioned above: geodesics along complex surfaces, billiards, GLG, and IET. All of them correspond to the dynamics with zero Lyapunov exponent, i.e. two initially close trajectories diverge not faster than a power of time. All of them describe the so-called pseudo-integrable (but not integrable) situation (Katok [1980]; Katok and Hasselblatt [1995]). All of them correspond to the dynamics with weak-mixing (Katok [1980]; Katok and Hasselblatt [1995]) (more accurate formulation can be found in Gutkin [1996]).

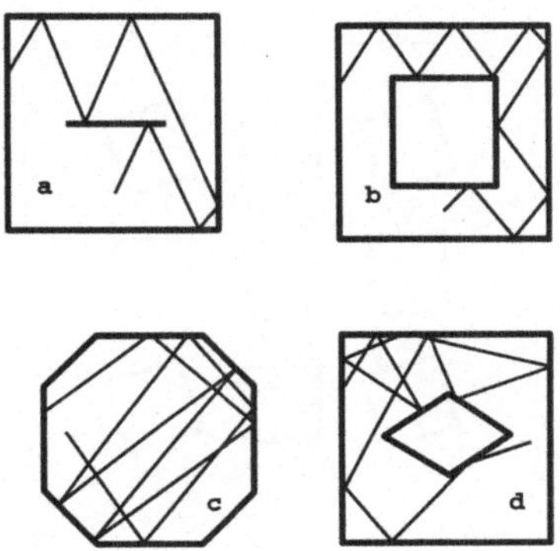

FIGURE 2.2. Examples of billiards of different types.

Let us recall (Cornfeld, Fomin, and Sinai [1982]) that for almost all square integrable functions $G_1(x)$ and $G_2(x)$ of the coordinate x in phase space, the weak mixing means that

$$\lim_{n\to\infty} \frac{1}{n} \sum_{k=0}^{n-1} \left[\langle G_1(\hat{T}^k x) G_2(x)\rangle - \langle G_1\rangle\langle G_2(x)\rangle\right]^2 = 0 \tag{2.1}$$

while the mixing, or strong mixing, means

$$\lim_{n\to\infty} \left[\langle G_1(\hat{T}^n x) G_2(x)\rangle - \langle G_1\rangle\langle G_2(x)\rangle\right] = 0 \tag{2.2}$$

where $\hat{T}$ is a time-shift operator and $\langle \ldots \rangle$ means the averaging over a corresponding measure. Weak mixing can be accompanied by arbitrary large and long-lasting bursts, i.e. fluctuations with large deviations from zero of the correlation function

$$R_n = \langle \hat{T}^n x \cdot x\rangle - \langle x\rangle^2 . \tag{2.3}$$

These fluctuations exist simultaneously with the property that time average of R_n^2 is zero. A sample of a trajectory for the billiard in Figure 2.2(a) and the corresponding GLG in Figure 2.3(a) is given in Figure 2.5 indicating long-lasting almost periodic pieces of the trajectory.

There is no one-to-one correspondence between four different types of the above-mentioned systems: geodesics on a compact surface, billiard, GLG,

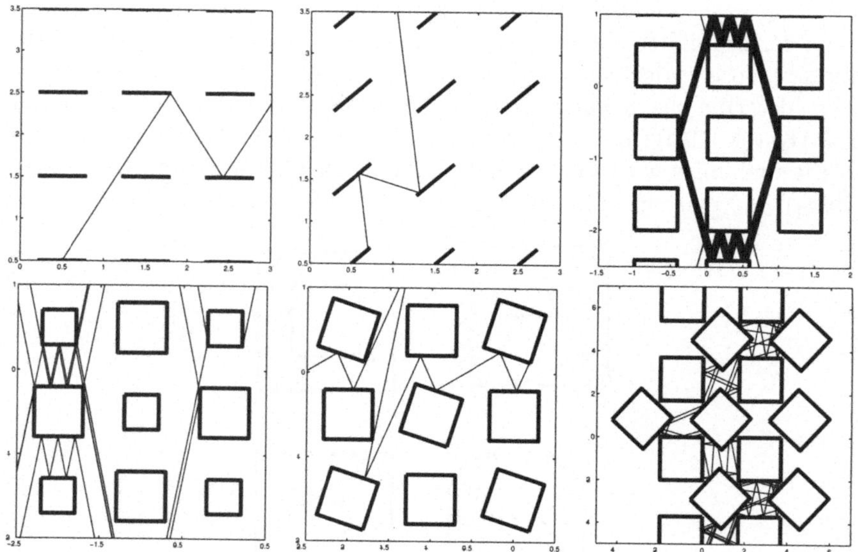

FIGURE 2.3. Periodically continued billiards form a "generalized Lorentz gas" (GLG).

and IET. For example, properties of the IET obtained in Zorich [1997] are not sufficient to describe kinetics of trajectories in the corresponding billiard since the equation for the mapping time is not included in IET. This explains a major difficulty in studying the billiards and GLG. Let us mention that IET dynamics and polygonal billiard dynamics is ergodic and weak mixing (Katok [1980]; Gutkin [1996]), but there is only a conjecture that the complexity of the billiard trajectories is at most polynomial (Gutkin [1996]; Zaslavsky and Edelman [2001]; Afraimovich and Zaslavsky [2002]). The complexity of trajectories is considered in the sense of (Gutkin [1996]), i.e. as a number of different "histories" of a specified length of sequences that are coding the trajectories. Due to this, we will call *pseudochaotic* the dynamics of trajectories in billiards and GLG, when their Lyapunov exponent is zero and their complexity is, at most, algebraic. As it will be shown, such billiards can reveal fairly good statistical properties.

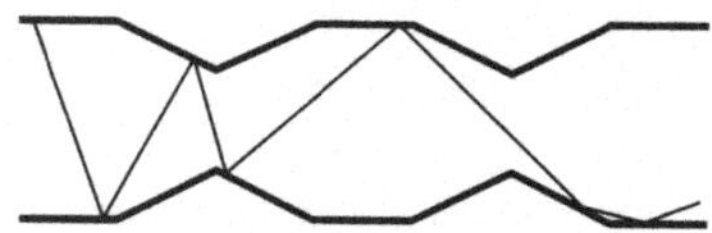

FIGURE 2.4. Ray propagation billiard model with a piecewise periodic nonuniformity.

Let us provide a few examples for the billiard with a slit in Figure

2.2(a) and the corresponding GLG in Figure 2.3(a) (Zaslavsky and Edelman [2001]). Consider an element ℓ of the slit in Figure 2.2(a) and introduce a normalized distribution of the Poincaré recurrences to ℓ as

$$P_{\text{rec}}(t) = \lim_{\ell \to 0} \frac{1}{\ell} P_{\text{rec}}(\ell, t) \tag{2.4}$$

where $P_{\text{rec}}(\ell, t)$ is a probability density to return a trajectory at time $t \in (t, t + \Delta t)$ to the element ℓ, does not matter from which side of ℓ. Due to the above-mentioned conjecture, we can expect that

$$P_{\text{rec}}(t) \sim 1/t^{\gamma}, \qquad (t \to \infty) \tag{2.5}$$

with a recurrence exponent γ that satisfies the condition

$$2 < \gamma \tag{2.6}$$

due to the Kac lemma (Cornfeld, Fomin, and Sinai [1982]):

$$\tau_{\text{rec}} \equiv \int_0^{\infty} dt\ t\ P(t) < \infty\ , \tag{2.7}$$

i.e. the mean recurrence time is finite for the bounded area preserving dynamics.

To study transport properties in billiards, consider a bundle of trajectories, launched with initially close angles, as an ensemble (Zaslavsky and Edelman [2001]). For the same case in Figure 2.3(a) we can consider moments $\langle y^{2m} \rangle$ with an integer m. It was shown in Zaslavsky and Edelman [2001] that

$$\langle y^{2m} \rangle \sim t^{\mu(m)} \tag{2.8}$$

with

$$\mu(m) \approx m\mu(1), \qquad \mu(1) \approx 1.7, \qquad \gamma \approx 2.7, \tag{2.9}$$

and with the connection

$$\gamma \approx \mu(1) + 1 \tag{2.10}$$

A similar value of $\mu(1)$ was obtained in Artuso, Guarneri, and Rebuzzini [2000]. Some deviations of $\mu(m)$ from the linear law (2.9) will not be discussed here (see more in Zaslavsky and Edelman [2001]). The values (2.10) are almost independent on the length of the slit and the launching angle. This universality will be commented later as well as deviations from (2.8) – (2.10). The results (2.8) — (2.10) can be conjectured to the billiards in Figure 2.2(b) and to the GLG in Figures 2.3(b) and (c), since trajectories do not change initial tangent of the angles during the scattering. Formula (2.8) shows anomalous transport ($\mu(1) \neq 1$) of the superdiffusion type and an approximate self-similarity since $\mu(m) \sim m\mu(1)$. All these properties and deviations from them will be discussed in the forthcoming sections.

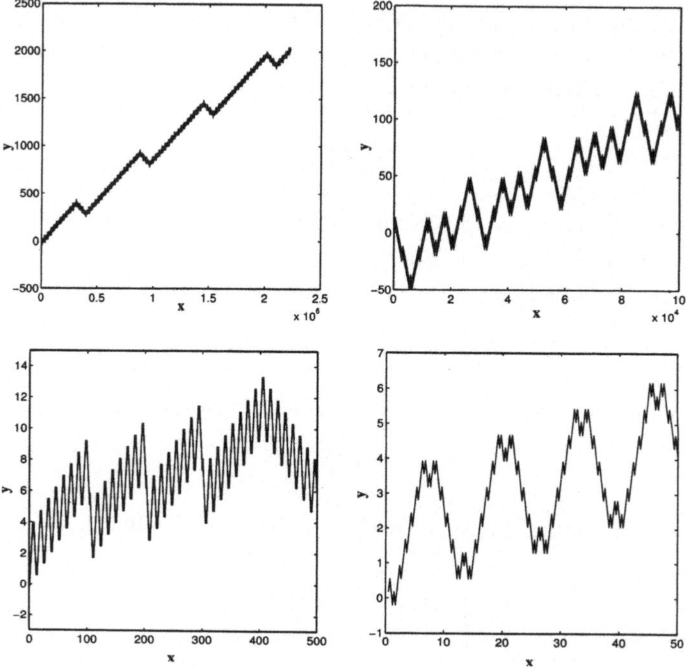

FIGURE 2.5. Samples of a trajectory with different space (time) scales for the billiard in Figure 2.2(a).

3 More Billiards

In this section we would like to present more different types of billiards and their statistical properties. These results can be considered as a kind of "experimental material."

Let us start from the two equal-squares-lattices in Figures 2.3(e,f) with infinite and finite horizons, respectively. Some squares are rotated by an angle of irrational tangent with respect to the others. Due to that, almost all trajectories rotate ergodically in the coordinate space in contrary to the billiards in Figures 2.3(a-d). Samples of trajectories are given in Figures 2.1 and 2.2. There are two important features common for both figures. First, the trajectories have arbitrary long flights. We call "a flight" any long part of the trajectory that, after small scale averaging, is almost regular. Such pieces of trajectories correspond to an intermittent dynamics and they are responsible for the distribution functions with tails and for fractional kinetics (see a review on the connection between flights and tails of the distribution function in Montroll and Shlesinger [1984], and, particularly, for dynamical systems in Zaslavsky [1994a,b]; Saichev and Zaslavsky [1997]). The trajectories have similarity in small and in large time and space scales. The self-similarity is less evident in Figures 3.1, 3.2 than in

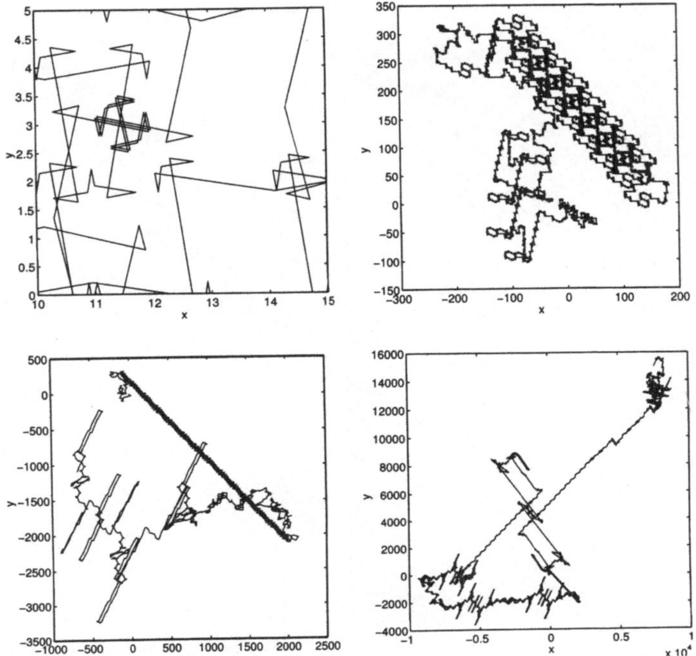

FIGURE 3.1. A trajectory for the billiard in Figure 2.3(e) with an infinite horizon. Its different zooms show flights that last about 10^4.

Figure 2.5. More delicate description of the trajectories can be obtained from simulations.

In Figure 3.2 we present two types of statistical properties of trajectories for the finite horizon billiard. The moments of the coordinate displacements

$$\langle x^{2m} \rangle \approx t^{\mu_x(m)}, \qquad \langle y^{2m} \rangle \approx t^{\mu_y(m)} \tag{3.1}$$

with

$$\mu_x(m) = \mu_y(m) \equiv \mu(m) \approx m\mu(1) \tag{3.2}$$

and

$$\mu(1) \approx 1.5 \ . \tag{3.3}$$

Considering one cell of the lattice (similar to the billiards in Figure 2.1, one can obtain a distribution function of Poincaré recurrences (see (2.5)). The simulation gives

$$\gamma \approx 2.4 \tag{3.4}$$

in a good agreement with the relation (2.10) within the accuracy of computations. Similar to (3.3), values of $\mu(1) \approx 1.5$ were obtained for the billiard with the infinite horizon (Figure 2.3(e)) with the recurrence exponent $\gamma \approx 2.5$.

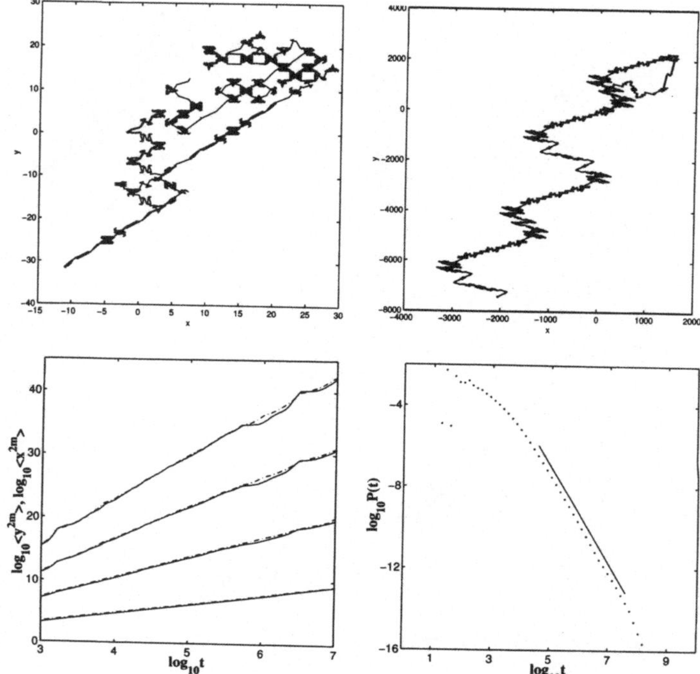

FIGURE 3.2. A trajectory with two, small and large, time intervals (two top plates), and the moments $\langle x^{2m} \rangle$ (regular lines), $\langle y^{2m} \rangle$ (dash-point lines), and distribution of the recurrences $P(t)$. The data for the two bottom plates are obtained after averaging over 8,192 trajectories and time $3.4 \cdot 10^8$ for each. Four curves of the moments vs. time correspond to $m = 1, 2, 3, 4$ from the bottom.

4 Continued Fractions and Scalings

It is convenient to consider first a theory for the billiard with a slit in Figures 2.2(a) or 2.3(a) since the kinetics in the billiard is one-dimensional along y and velocity v_x along x is constant. This billiard attracted attention fairly long ago (see Zwanzig [1983] and following publications by Hannay and McCraw [1990]; Richens and Berry [1981]; Wiersig [2000]; Artuso, Guarneri, and Rebuzzini [2000]; Artuso, Casati, and Guarneri [1997]). In our work Zaslavsky and Edelman [2001] we use the properties of the continued fractions (Khinchin [1964]; Levy [1937]) to obtain the recurrences distribution and kinetic properties of trajectories. In this section we will repeat and extend the results of Zaslavsky and Edelman [2001].

Let us recall the samples of the trajectory in Figure 2.5. Trajectories with an angle ϑ to x will be called rational/irrational if $\tan\vartheta$ is rational/irrational. We will consider only irrational ensembles, i.e. bunches of trajectories with irrational $\tan\vartheta_j$ and $\vartheta_j \in (\vartheta - \Delta\vartheta/2, \vartheta + \Delta\vartheta/2)$ within an interval $\Delta\vartheta$.

Let

$$\tan\vartheta = a_0 + \xi \ , \tag{4.1}$$

where a_0 is the integer part of $\tan\vartheta$, and $\xi \in (0,1)$ can be written as a continued fraction

$$\xi = 1/(a_1 + 1/a_2 + \ldots)) \equiv [a_1, a_2, \ldots] \tag{4.2}$$

For irrational ξ the sequence $[a_1, a_2, \ldots]$ is infinite and its finite approximate is

$$\xi_n \equiv [a_1, \ldots, a_n] = p_n/q_n \tag{4.3}$$

with coprime p_n, q_n. In the following, we use three important results on continued fractions (Levy [1937]):

(a) Estimate for the accuracy of the n-th convergent

$$|\xi - p_n/q_n| \leq 1/q_n q_{n+1} \ , \tag{4.4}$$

(b) Asymptotic property of the sequence $\{a_j\}$ (in measure)

$$\lim_{n\to\infty}(a_1 \ldots a_n)^{1/n} = \prod_{k=1}^{\infty}\left(1 + \frac{1}{k^2+2k}\right)^{\ln k/\ln 2} = 2.63\ldots \tag{4.5}$$

(c) Asymptotic property of the sequence $\{q_j\}$ (in measure)

$$\lim_{n\to\infty}\left(\frac{1}{n}\ln q_n\right) = \pi^2/12\ln 2 = 1.186\ldots \tag{4.6}$$

The formulas (4.5) and (4.6) indicate a remarkable scaling structure of the continued fractions

$$\begin{aligned} a_1 \ldots a_n &\sim \lambda_a^n g_a(n) \\ q_n &\sim \lambda_q^n g_q(n) \end{aligned} \tag{4.7}$$

with scaling parameters λ_a, λ_q and subexponential functions $g_a(n), g_q(n)$.

For an arbitrary irrational trajectory with an angle ϑ, consider its n-th approximate ξ_n and a corresponding rational trajectory with an angle $\vartheta^{(n)}$ such that $\tan\vartheta^{(n)} = a_0 + \xi_n$. This trajectory is periodic with the period (Galperin and Zemlyakov [1990])

$$T_n = \text{const} \cdot q_n \tag{4.8}$$

with the const = 1 or 2. The approximate trajectory is close to the original one for a time much larger than T_n due to the fast convergence (4.4). That is just the property that one can see from the samples in Figure 2.5. The property (4.8) was used in Zaslavsky and Edelman [2001] to get the scaling for different quasi-periods of any irrational trajectory

$$T_n \sim \lambda_T^n g_T(n) \tag{4.9}$$

which is similar to (4.7) with the time-scaling parameter λ_T

$$\ln \lambda_T = \ln \lambda_q = \pi^2/12 \ln 2 = 1.186\ldots \tag{4.10}$$

Simulation for an arbitrary chosen $\vartheta = 4.153087\ldots$ gives $\ln \lambda_T = 1.17 \pm 0.08$ in an excellent agreement with (4.10).

Expression (4.10) defines hierarchical structures of different flights in time. Corresponding hierarchical structures should exist for the lengths ℓ_n of the flights along y. Nevertheless, there exists an ambiguity of the flight lengths ℓ_n since trajectories can have strong coordinate oscillations for the same time duration of flights. To find a corresponding scaling parameter λ_y, we assume that

$$\ln \lambda_y = \overline{\ln \lambda_{\text{den}}} \tag{4.11}$$

where λ_{den} means scaling parameter of different possible denominators of the rational convergent of ξ, obtained at the same hierarchical level along a trajectory, and the bar means averaging over all such possibilities. For example, at the hierarchical level n, ξ_n has q_n as the minimal denominator and $a_1 \ldots a_n$ as the maximal denominator since

$$(a_1 \ldots a_n)^{1/n} \geq (q_n)^{1/n}\ , (n \to \infty) \tag{4.12}$$

what follows from the recurrent formula $q_n = a_n q_{n-1} + q_{n-2}$ and $q_0 = 1$, $q_{n-1} = 0$. Applying (4.7) we obtain

$$\min \lambda_{\text{den}} = \lambda_T\ , \qquad \max \lambda_{\text{den}} = \lambda_a \tag{4.13}$$

The simplest estimate is

$$\ln \lambda_y = \overline{\ln \lambda_{\text{den}}} \approx \frac{1}{2}(\ln \lambda_T + \ln \lambda_a) \approx 1.07\ldots \tag{4.14}$$

The obtained information can be applied to construct the kinetic evolution of an ensemble of trajectories.

All the numbers (4.10), (4.14) should be the same for the billiards with a slit (Figures 2.2(a), 2.3(a)) and with equal squares (Figures 2.2(b), 2.3(c)).

5 Fractional Kinetics in Billiards

The fractional space-time kinetic equation was introduced in Zaslavsky [1992] and detailed for dynamical systems in Zaslavsky [1994a,b]; Saichev and Zaslavsky [1997] in order to describe self-similar non-Gaussian processes with a strong intermittency. For a billiard of the type in Figures 2.2(a,b) and 2.3(a-c), the equation has the form

$$\frac{\partial^\beta f(y,t)}{\partial t^\beta} = \mathcal{D}\frac{\partial^\alpha f(y,t)}{\partial |y|^\alpha} \tag{5.1}$$

where $f(y,t)$ is a distribution function, and fractional derivatives can be considered using their Fourier transform (F.T.):

$$\text{F.T.}\ \frac{\partial^\beta}{\partial t^\beta} = (i\omega)^\beta\ , \qquad \text{F.T.}\ \frac{\partial^\alpha}{\partial |y|^\alpha} = (i|k|)^\alpha, \tag{5.2}$$

Usually, we are interested in asymptotics $t \to \infty$, $|y| \to \infty$. Since the moments

$$\langle |y|^{2m} \rangle = \int y^{2m} f(y,t)\, dy \tag{5.3}$$

diverge for $2m > \alpha$ (Zaslavsky [1994a,b]; Saichev and Zaslavsky [1997]), it is convenient to consider truncated distribution $f(y,t)$, ($y < y_{\max}$) and moments with $m \le m_{max}$ and $m_{max} > 2$, all of which are finite and present a self-similar evolution (see for example Figure 3.2 which shows $m = 1,2,3,4$). With this comment we can get from (5.1)

$$\langle |y|^\alpha \rangle = \text{const.}\ t^\beta \tag{5.4}$$

and expect that

$$\langle |y|^{2m} \rangle \sim t^{\mu(m)} \tag{5.5}$$

with

$$\mu(m) \sim m\mu(1) \tag{5.6}$$

(compare to (3.2)).

In fact, the situation is more complicated (Zaslavsky and Edelman [2000]). The self-similarity of $f(y,t)$ means that equation (5.1) is invariant under the transformation

$$t \to \lambda_T t\ , \qquad y \to \lambda_y y \tag{5.7}$$

with appropriate values of the scaling parameters λ_T, λ_y. Then from (5.1), (5.7) we obtain the fixed-point condition

$$\lim_{n\to\infty} (\lambda_y^\alpha / \lambda_T^\beta)^n = 1 \tag{5.8}$$

or

$$\beta \ln \lambda_T = \alpha \ln \lambda_y + 2\pi i k \tag{5.9}$$

with an integer k. Then the solution for (5.1) can be written in the form (Zaslavsky and Edelman [2001, 2000])

$$\langle |y|^\alpha \rangle = \sum_{k=-\infty}^{\infty} C_k t^{\beta_k} \tag{5.10}$$

$$\beta_k = \alpha\mu(1)/2 + 2\pi i k / \ln \lambda_T$$

with

$$\mu = 2 \ln \lambda_y / \ln \lambda_T , \tag{5.11}$$

and some expansion coefficients C_k. The expression (5.10) can be transformed into

$$\langle |y|^\alpha \rangle^{\frac{2}{\alpha}} = t^\mu \sum_{k=0}^{\infty} \mathcal{D}_k \cos \left(2\pi k \frac{\ln t}{\ln \lambda_T} + \psi_k \right) \tag{5.12}$$

with new coefficients $\mathcal{D}_k$ and phases ψ_k. This expression shows the so-called log-periodicity with a period

$$T_{\log} = \ln \lambda_T \tag{5.13}$$

and with corresponding terms $\mathcal{D}_{k \neq 0}$ that typically are small. For the self-similar behavior of the moments of $f(y,t)$ we also expect

$$\langle y^{2m} \rangle \sim t^{\mu(m)} \sum_{k=0}^{\infty} \bar{\mathcal{D}}_k \cos \left(2\pi k \frac{\ln t}{\ln \lambda_T} + \bar{\psi}_k \right) \tag{5.14}$$

with coefficients $\bar{\mathcal{D}}_k$ and phases $\bar{\psi}_k$, and

$$\mu(m) \sim m\mu(1) = m\mu = 2m \ln \lambda_y / \ln \lambda_T . \tag{5.15}$$

Expressions (5.13), (5.14) are just the ones that should be tested by simulations. As it was mentioned in the previous section, the results should be the same for the billiards in Figures 2.2(a), 2.3(a) and 2.2(b), 2.3(c). The results of simulations are presented in Figure 5.1. In Figure 5.1(a) we observe power law decay of the distribution of Poincaré recurrences with $\gamma \approx 2.7$. There are also some oscillations which make deviations from this value of γ. Figure 5.1(b) shows moments for $m = 1, 2, 3, 4$ and the self-similarity with $\mu = \mu(1) \approx 1.7$. There are small deviations from the linear dependence of $\mu(m)$ on m for $m = 2, 3$ and 4, $\mu(2) = 3.5$; $\mu(3) = 5.5$; $\mu(4) = 7.4$.

Let us consider the formula (5.15) and use for λ_T and λ_y the values predicted by the continued fraction theory (4.10) and (4.14). It gives $\mu = \mu(1) = 1.8$ which is in a good agreement with the results of simulation in Figure 5.1(b). The values of

$$\gamma \approx 2.7 , \qquad \mu \approx 1.7 \tag{5.16}$$

are also in agreement with (2.10). It is a more delicate issue to evaluate the log-periodicity. There are small oscillations in the second moments $\langle x^2 \rangle, \langle y^2 \rangle$ time dependence. In Figure 5.1(c) we show their amplitude

$$\Delta(x^2) = \langle x^2 \rangle - \text{const.}\, t^\mu . \tag{5.17}$$

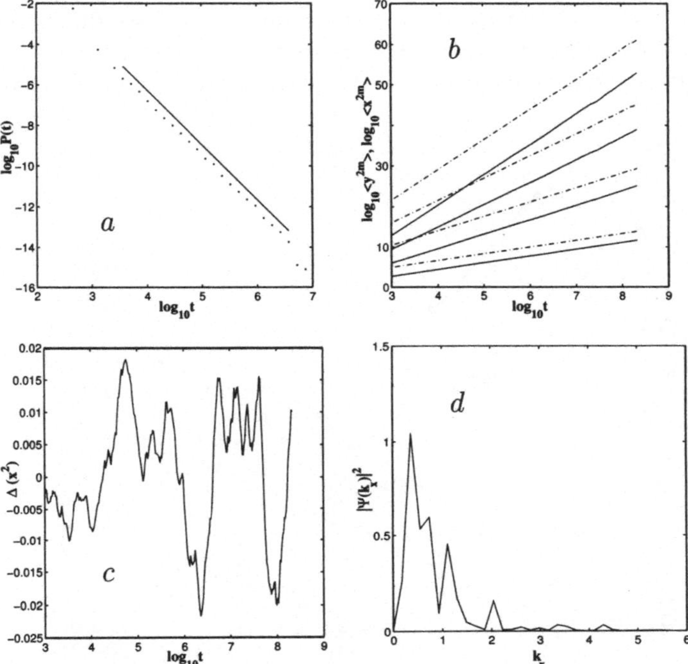

FIGURE 5.1. Statistical properties of the square billiards: (a) density distribution function of the recurrences $P(t)$; (b) moments of x (full lines) and y (point-dash lines) for $m = 1, 2, 3, 4$ starting from the bottom; (c) fluctuations of the second moment of x vs. time; and (d) their Fourier spectrum. The data are obtained for 4,048 trajectories during the time 10^8 for each.

These oscillations do not look absolutely random due to the presence of structures. To find it, consider the Fourier transform of (5.17), i.e.

$$\Psi(k_x) = \int dt\ e^{2\pi i k_x \log_{10} t} \Delta(x^2) \tag{5.18}$$

Figure 5.1(d) shows the almost continuous spectrum with some regular peaks. The characteristic width of the regular part of the spectrum is $\Delta k_x \sim 2$. From (5.13) and (4.10), the width of the spectrum in the decimal logarithm basis should be $1/\log_{10} \lambda_T \approx 2$ in the excellent agreement with the data of simulations. Similar agreement was obtained in Zaslavsky and Edelman [2001] for the billiard of Figure 2.2(a), 2.3(a) type.

6 Rhombic Billiards

This type of billiard is shown in Figure 2.2(c,d). Its one-dimensional periodic continuation (Figure 2.4) can be interesting for different applications such as wave/ray propagation in nonuniform media. Two-dimensional periodic rhombic scatterers with an external field and small values of the Lyapunov exponents were considered in Lepri, Rondoni, and Benettin [2000].

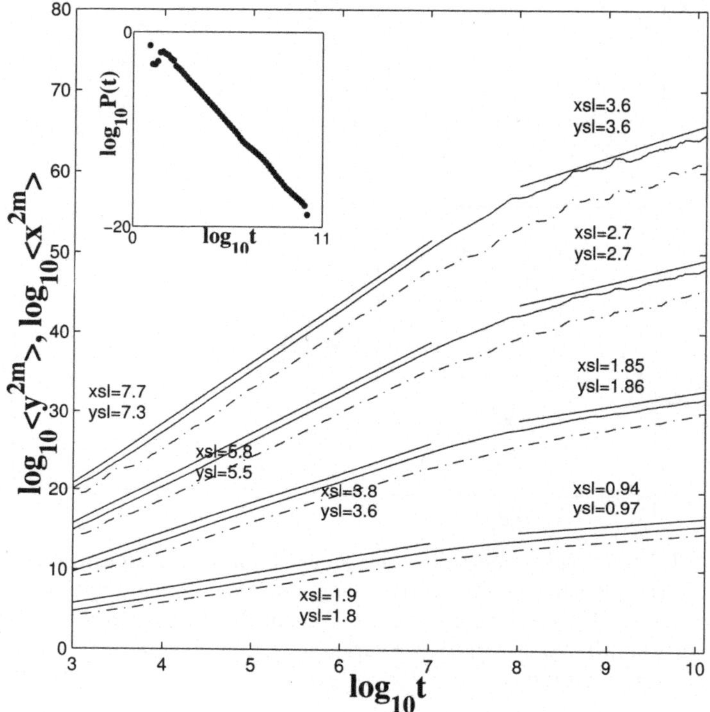

FIGURE 6.1. Moments and distribution density $P(t)$ of the recurrences for the rhombic billiard. The data are obtained from 512 trajectories during the time 10^{10} for each. The curves correspond $m = 1, 2, 3, 4$.

The most interesting case is the irrational billiard, i.e. the case with an irrational ratio of the rhombus diagonals. The rhombic billiard inherits properties of the triangular billiards considered in Artuso, Guarneri, and Rebuzzini [2000]; Artuso, Casati, and Guarneri [1997]; Casati and Prosen [1999]. In Figure 6.1 we presented the recurrences distribution $P(t)$ for which the slope recurrence exponent is $\gamma = 2 \pm 0.1$, and the coordinates moments $\langle x^{2m} \rangle, \langle y^{2m} \rangle$ versus time. The latter shows two exponents: during the time interval $0 < t < 10^7$, $\mu(1) \approx 1.8 \div 1.9$, and after that time $10^7 < t < 10^{10}$, $\mu(1) \approx 0.95 \pm 0.05$. We cannot guarantee that this exponent is not an intermediate one, but it satisfies the condition (2.10) and both values of $gamma$ and $\mu(1)$ for large t are close to the values obtained in Casati and Prosen [1999] for the triangular billiard. These values will be discussed in Section 8.

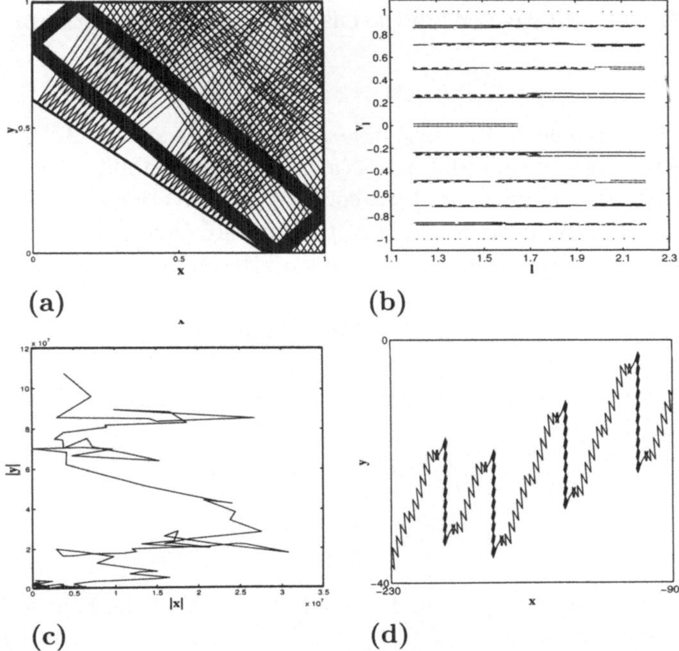

FIGURE 6.2. Rhombic billiard with the x-diagonal $=$ 1.6 and y-diagonal $= 1.2290472$. (a) A sample trajectory in a square shows slow evolution; (b) phase plane ℓ, v_ℓ (ℓ is the upper side of the square in (a)) shows slow evolution in the angle (i.e. along v_ℓ); (c) a sample trajectory in large scale; and (d) its zoom.

In Figure 6.2 we present samples of trajectories which show trappings and very slow escape from the trapping domains. This feature of the trajectories can be especially well observed from Figure 6.2(b) of the Poincaré section: each point corresponds to a trajectory crossing of the interval $\ell:\ y = 1$, $x \in (0, 1)$ in Figure 6.2(a). There is the extremely slow process of mixing along the velocity v_ℓ and the filling of the phase space is performed mainly along x for some special values of v_ℓ. We will speculate on the recurrences estimate in Section 8, applying for this type of dynamics.

7 Back to the Continued Fractions

Here we present a comment to the continued fractions properties. Since trajectories in the billiard with a slit or a square bear scaling properties of the continued fractions from one side, and the log-periodicity from another side, one can expect that the statistical features of continued fractions should have similar type of the log-periodic oscillations.

Consider a large number (ensemble) of irrational numbers $\{\xi^{(\nu)} < 1\}$.

For each representative $\xi^{(\nu)}$ of the ensemble consider its n-th approximant, i.e.

$$\xi_n^{(\nu)} = [a_1^{(\nu)}, a_2^{(\nu)}, \dots, a_n^{(\nu)}] = p_n^{(\nu)}/q_n^{(\nu)} . \tag{7.1}$$

The new ensemble is the ensemble of the denominators $\{q_n^{(\nu)}\}$. We can introduce a distribution function $\Phi(q;n)$, i.e. a probability density to have value q for the denominator of the convergent of irrational numbers at n-th step. The variable n plays a role of the discrete time. $\Phi(q;n)$ should be a universal function for which we can introduce moments

$$\langle q_n^{2m} \rangle = \int q^{2m} \Phi(q;n) dq \tag{7.2}$$

The asymptotic property (4.7) suggests that

$$\frac{1}{2m} \ln \langle q_n^{2m} \rangle = nc(1 + d_n) \tag{7.3}$$

with

$$c = \ln \lambda_q \tag{7.4}$$

and d_n depending slowly on n. The conjecture is that the Fourier spectrum of d_n consists of two parts: quasi-periodic and continuous, similar to what we have for the billiards.

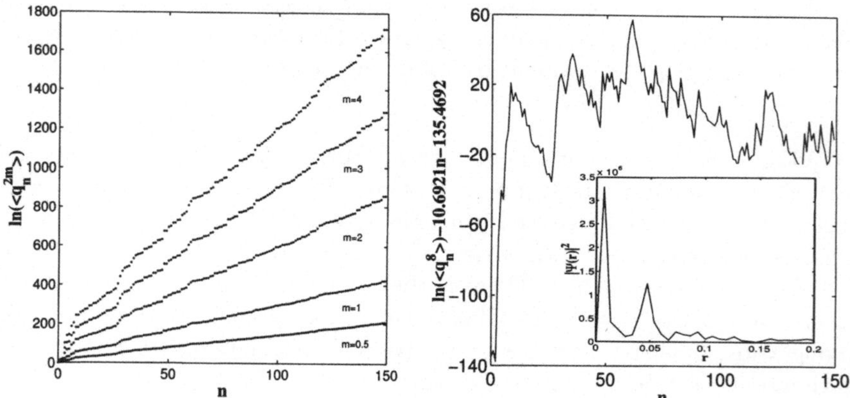

FIGURE 7.1. Statistical features of the continued fractions: moments of the denominators $\langle q_n^{2m} \rangle$ vs. number n of the approximation level and fluctuations of the 8-th moment with the spectral function.

To verify the conjecture, we considered 10^3 different irrational numbers $\xi^{(\nu)} < 1$ ($\nu = 1, \dots, 10^3$) and their approximants with the quadruple precision. The behavior of moments $\langle q_n^{2m} \rangle$ is presented in Figure 7.1 up to $n = 150$. The plot shows oscillations which are stronger for higher moments. The slope corresponds to the value (7.4) with an accuracy up to the level of oscillations. The value of oscillations with respect to the straight line is shown in Figure 7.1(a), and their Fourier transform is in Figure 7.1(b). The figures show finite, quickly decaying peaks of modulation of the continuous type spectrum. The continuous spectrum property has been proved in Ibragimov and Linnik [1971] as the asymptotic one. Log-periodicity means that the spectrum is multifractal, what is consistent with the general result of Pollicott and Weiss [1999] and, in addition, specifies the type of multifractality.

8 Estimates for the Distribution of Recurrences

In this section we provide rough estimates for the distribution of recurrences $P(t)$ or, more accurately, for possible values of the recurrence exponent γ when $P(t)$ behaves algebraically. As it was mentioned in (2.6) the value of γ should exceed 2 for the bounded Hamiltonian dynamics that will be considered below.

A qualitative way to obtain $P(t)$ is to consider an enveloping phase volume $\bar{\Gamma}(t)$ that can be obtained from the initial phase volume Γ_0 by some smoothing operation over a small ball of a size ϵ along a trajectory. This operation is known as the coarse-graining. While $\Gamma_0 = const$ (area-preserving dynamics), $\bar{\Gamma} > \Gamma_0$ as it was known since Gibbs. There are different possibilities:

$$\begin{aligned}
\bar{\Gamma}_1(t) &= \text{const.} \cdot t^{1+\delta_1} \\
\bar{\Gamma}_2(t) &= \text{const.} \cdot t^{3/2 \pm \delta_2} \\
\bar{\Gamma}_3(t) &= \text{const.} \cdot t^{2 \pm \delta_3} \\
\bar{\Gamma}_4(t) &= \text{const.} \cdot t^{3 \pm \delta_4}
\end{aligned} \tag{8.1}$$

where δ_j are considered as corrections to four different main values. The case of $\bar{\Gamma}_1(t)$ corresponds to almost one-dimensional coarse-grained dynamics. It seems that this case happens in the rhombic and triangular billiards and Figures 6.2(b,c) confirm this statement. The same case can also appear in one-dimensional dynamics (see Young [1999] and some rigorous results therein).

The case of $\bar{\Gamma}_2(t)$ corresponds to the linear evolution along one coordinate and the diffusional type evolution along the other coordinate along which the phase volume growth is proportional to $t^{1/2}$. This is one of the most typical case for many systems with intermittent chaotic dynamics (Zaslavsky and Edelman [2000]). The third case of $\bar{\Gamma}_3(t)$ corresponds to the almost linear phase space coverage along one coordinate and then the similar coverage along another coordinate. The alternations of the coordinates can be randomly distributed but the result provides almost t^2-growth of $\bar{\Gamma}_3(t)$. It seems that it is the case of the original Sinai billiard when $\delta_3 = 0$ (Chernov and Young [2000]). The case of $\bar{\Gamma}_4(t)$ can appear in special situations of 1 1/2 degrees of freedom or in some higher dimension system (Rakhlin [2001]).

From (8.1) we obtain the return probability during the time interval $(0, t)$ as $\Gamma_0/\bar{\Gamma}_j(t)$ and finally for the probability density of the return at $t \in (t, t+dt)$:

$$P(t) \approx \left| \frac{d}{dt} \frac{\Gamma_0}{\bar{\Gamma}_j(t)} \right| \sim \text{const.}/t^{\gamma_j} \tag{8.2}$$

with the corresponding values of γ_j

$$\gamma_j \approx \begin{cases} 2+\delta_1 \\ 5/2 \pm \delta_2 \\ 3 \pm \delta_3 \\ 4 \pm \delta_4 \end{cases} \tag{8.3}$$

where δ_j represents the corrections to the main exponents. We have to recall that for some cases exact values of the exponents have no meaning since the log-periodic oscillations of the distributions $P(t)$.

9 Conclusions

The randomness with zero Lyapunov exponent, that we call pseudochaos, can appear in numerous applications. Its analysis provides a link between the structure of streamlines of flows, magnetic field lines, billiards, continued fractions, and fractional kinetics. It seems that the dynamical processes that can be described by the fractional type kinetic equation reveal a polynomial complexity i.e. a polynomial separation of close trajectories (Afraimovich and Zaslavsky [2002]), while the chaotic hyperbolic dynamics has the exponential growth of the coarse-grained phase volume and the corresponding exponential separation.

Acknowledgments: We appreciate V. Afraimovich and L.-S. Young for numerous and useful discussions, B. Sapoval and J. Wiersig for correspondence and information about their publications.

This work was supported by the U.S. Navy Grants No. N00014-96-1-0055, N00014-97-1-0426, and the U.S. Department of Energy Grant No. DE-FG02-92ER54184. The simulation was supported in part by the National Science Foundation (NSF) cooperative agreement No. ACI-9619020 through computing resources provided by the National Partnership for Advanced Computational Infrastructure at the San Diego Supercomputer Center, and by NERSC.

References

Afraimovich, V. and G. M. Zaslavsky [2002], Complexity of trajectories in chaotic dynamics, Preprint, 40 pages.

Agullo, O., A. D. Verga, and G. M. Zaslavsky [1997], Chaotic advection and transport in helical Beltrami flows: A Hamiltonian system with anomalous diffusion, *Phys. Rev. E* **55**, 5587-5596.

Arnold, V. I. [1978], *Mathematical Methods in Classical Mechanics.* Springer, New York.

Artuso, R., J. Casati, and I. Guarneri [1997], Numerical study on ergodic properties of triangular billiards, *Phys. Rev. E* **55**, 6384-6390.

Artuso, R., I. Guarneri, and L. Rebuzzini [2000], Spectral properties and anomalous transport in a polygonal billiard, *Chaos* **10**, 189-194.

Beyer, P. and S. Benkadda [2001], Advection of passive particles in the Kolmogorov flow, *Chaos* **11**, 774-779.

Benkadda, S. S., P. Gabbai, and G. M. Zaslavsky [1997], Passive particle dynamics in a flow exhibiting transition to turbulence, *Phys. Plasmas* **4**, 2864–2870.

Casati, G. and T. Prosen [1999], Mixing property of triangular billiards, *Phys. Rev. Lett.* **83**, 4729-4732.

Chernov N. and L.-S. Young [2000], *Decay of correlations in Lorentz gases and hard balls*, preprint, 32 pages.

Cornfeld, I. P., S. V. Fomin, and Ya. G. Sinai [1982], *Ergodic Theory.* Springer, New York.

Galperin, G. A. and A. N. Zemlyakov [1990], *Mathematical Billiards.* Nauka, Moscow (in Russian).

Gutkin, E. [1986], Billiards in polygons, *Physica D* **19**, 311–333.

Gutkin, E. [1996], Billiards in polygons: Survey of recent results, *J. Stat. Phys.* **83**, 7–26.

Haller, G. [2000], Finding finite-time invariant manifolds in two-dimensional velocity fields, *Chaos* **10**, 99-108.

Hannay, J. H. and R. J. McCraw [1990], Barrier billiards — a simple pseudo-integrable system, *J. Phys. A* **23**, 887-899.

Hebert, B., B. Sapoval, and S. Russ [1999], Experimental study of a fractal acoustical cavity, *J. Acoust. Soc. Am.* **105**, 1567–1574.

Henon, M. [1966], Sur la topologie des lignes de courant dans un cas particulier, *Compt. Rend. Acad. Sci. A Math.* **262**, 312–314.

Ibragimov, I. H. and Yu. V. Linnik [1971], *Independent and Stationary Sequences of Variables.* Wolters-Noordhoff Publ., Groningen.

Isola, S. [1999], Renewal sequences and intermittency, *J. Stat. Phys.* **97**, 263–280.

Katok, A. [1980], Interval exchange transformations and some special flows are not mixing, *Isr. J. Math.* **35**, 301–310.

Katok, A. [1987], The growth-rate for the number of singular and periodic orbits for a polygonal billiard, *Commun. Math. Phys.* **111**, 151–160.

Katok, A. and B. Hasselblatt [1995], *Introduction to the Modern Theory of Dynamical Systems.* Cambridge Univ. Press, Cambridge.

Khinchin, A. Ya. [1964], *Continued Fractions.* University of Chicago Press, Chicago.

Kozlov, V. V. [1996], *Symmetries, Topology, and Resonances in Hamiltonian Mechanics.* Springer, Berlin.

Lepri, S., L. Rondoni, and G. Benettin [2000], The Gallavotti-Cohen fluctuation theorem for a nonchaotic model, *J. Stat. Phys.* **99**, 857–872.

Levy, P. [1937], *Théorie de l'Addition des Variables Aletoires.* Gauthier-Villiers, Paris.

Mayer, D. [1990], On the thermodynamic formalism for the Gauss map, *Commun. Math. Phys.* **130**, 311-333.

Montroll, E. W. and M. F. Shlesinger [1984], On the Wonderful World of Random Walks. In *Studies in Statistical Mechanics*, eds. J. Lebowitz and E. Montroll, **11**, pages 1-121, North-Holland, Amsterdam.

Morgulis, A. I., V. I. Yudovich, and G. M. Zaslavsky [1995], Compressible Helical Flows, *Commun. Pure and Appl. Math.* **XLVIII**, 571–582.

Morozov, A. I. and L. S. Soloviev [1966], The structure of magnetic fields, in *Reviews of Plasma Physics*, ed. M.A. Leontovich. Consultants Bureau, New York, vol. 2, p. 1–101.

Platt, N, L. Sirovich, and N. Fitzmaurice [1991], An investigation of chaotic Kolmogorov flows, *Phys. Fluids A* **3**, 681–696.

Pollicott, M. and M. Weiss [1999], Multifractal analysis of Lyapunov exponent for continued fraction and Manneville-Pomeau transformations and applications to Diophantine approximation, *Commun. Math. Phys.* **207**, 145–171.

Rakhlin, D. [2001], Enhanced diffusion in smoothly modulated superlattices *Phys. Rev. E* **63**, 011112.

Richens, P. J. and M. V. Berry [1981], Pseudointegrable systems in classical and quantum-mechanics, *Physica D* **2**, 495–512.

Rosenbluth, M., R. Z. Sagdeev, J. B. Taylor, and G. M. Zaslavsky [1966], Destruction of magnetic surfaces by magnetic field irregularities, *Nucl. Fusion* **6**, 297–300.

Saichev, A. I. and G. M. Zaslavsky [1997], Fractional kinetic equations: solutions and applications, *Chaos* **7**, 753–764.

Sapoval B., Th. Gobron, and A. Margolina [1991], Vibrations of fractal drums, *Phys. Rev. Lett.* **67**, 2974–2977.

Sirovich, L. [1989], Chaotic dynamics of coherent structures, *Physica D* **37**, 126–145.

Vega, J. L., T. Uzer, and J. Ford [1993], Chaotic billiards with neutral boundaries, *Phys. Rev. E* **48**, 3414–3420.

Weiss, J. B., A. Provenzale, and J. C. McWilliams [1998], Lagrangian dynamics in high-dimensional point-vortex systems, *Phys. Fluids* **10**, 1929–1941.

Wiersig, J., [2000], Singular continuous spectra in a pseudointegrable billiard, *Phys. Rev. E* **62**, R21–R24.

Wilkinson, P. B., T. M. Fromhold, R. P. Taylor, and A. P. Micolich [2001a], Electromagnetic wave chaos in gradient refractive index optical cavities, *Phys. Rev. Lett.* **86**, 5466–5469.

Wilkinson, P. B., T. M. Fromhold, R. P. Taylor, and A. P. Micolich [2001b], Effects of geometrical ray chaos on the electromagnetic eigenmodes of a gradient index optical cavity, *Phys. Rev. E.* **64**, 026203.

Young, L.-S. [1999], Recurrence times and rates of mixing, *Israel J. Math.* **110**, 153–188.

Zaslavsky, G. M. [1992], Anomalous transport and fractal kinetics. In *Topological Aspects of the Dynamics in Fluids and Plasmas*, eds. H. K. Moffatt et al., pages 481–500, Kluwer, Dordrecht.

Zaslavsky, G. M. [1994a], Fractional kinetic-equation for hamiltonian chaos, *Physica D* **76**, 110–122.

Zaslavsky, G. M. [1994b], Renormalization group theory of anomalous transport in systems with Hamiltonian chaos, *Chaos* **4**, 25–33.

Zaslavsky, G. M. and S. S. Abdullaev [1997], Chaotic transmission of waves and "cooling" of signals, *Chaos* **7**, 182–186.

Zaslavsky, G. M. and M. Edelman [1997], Maxwell's demon as a dynamical model, *Phys. Rev. E* **56**, 5310–5320.

Zaslavsky, G. M. and M. Edelman [2000], Hierarchical structures in the phase space and fractional kinetics: I. Classical systems, *Chaos* **10**, 135–146.

Zaslavsky, G. M. and M. Edelman [2001], Weak mixing and anomalous kinetics along filamented surfaces, *Chaos* **11**, 295–305.

Zaslavsky, G. M., R. Z. Sagdeev, D. A. Usikov, and A. A. Chernikov [1991], *Weak Chaos and Quasi-Regular Patterns*. Cambridge Univ. Press, Cambridge.

Zorich, A. [1997], Deviation for interval exchange transformations, *Ergodic Theory and Dynamic Systems* **17**, 1477–1499.

Zwanzig, R. [1983], From classical dynamics to continuous-time random-walks, *J. Stat. Phys.* **30**, 255–262.